MW01346420

The Relay Testing Handbook

Principles and Practice

The Relay Testing Handbook

Principles and Practice

Chris Werstiuk

Professional Engineer
Journeyman Power System Electrician
Electrical Technologist

Valence Electrical Training Services

7450 W. 52nd Ave, M330
Arvada, CO 80002
www.ValenceOnline.com

First printing in 2012.

The Relay Testing Handbook: Principles and Practice

ISBN: 978-1-934348-20-8

Library of Congress Control Number: 2012934627

Published By: **Valence Electrical Training Services LLC**

7450 W. 52nd Ave., M330, Arvada, CO, 80002, U.S.A. | (303) 219-0797 | **info@valenceonline.com**

Distributed by: **www.valenceonline.com**

Edited by: **One-on-One Book Production, West Hills, CA**

Cover Art: **© James Steidl. Image from BigStockPhoto.com**

Interior Design and Layout: **Adina Cucicov, Flamingo Designs**

ATTENTION CORPORATIONS, UNIVERSITIES, COLLEGES, and PROFESSIONAL ORGANIZATIONS: Quantity discounts are available on bulk purchases of this book. Special books or book excerpts can also be created to fit specific needs.

Printed in the United States of America.   c

Author's Note

This book has grown from a 45-minute paper presentation at the 2001 InterNational Electrical Testing Association (NETA) conference into a decade-long project. What started as a simple paper about protective relay logic for microprocessor based relays has blossomed into a comprehensive training manual covering all aspects of relay testing. I am grateful for the countless hours my colleagues spent during the peer review process, and for their invaluable contributions to this endeavor.

Traditional protective relay books are written by engineers as a resource for engineers to use when modeling the electrical system or creating relay settings, and they often have very little practical use for the test technician in the field. *The Relay Testing Handbook* is a practical resource written by a relay tester for relay testers; it is a comprehensive series of practical instructional manuals that provides the knowledge necessary to test most modern protective relays. The complete handbook combines basic electrical fundamentals, detailed descriptions of protective elements, and generic test plans with examples of real-world applications, enabling you to confidently handle nearly any relay testing situation. Practical examples include a wide variety of relay manufacturers and models to demonstrate that you can apply the same basic fundamentals to most relay testing scenarios.

The Relay Testing Handbook is a nine-part series that covers virtually every aspect of relay testing. Eight books of the series have been compiled into this volume that explain the underlying principles of relay testing, including electrical theory, relay testing philosophies, digital logic and test plan creation and implementation. After the foundation is laid, you will find practical step-by-step procedures for testing the most common protection applications for: voltage, overcurrent, differential, and line distance relays.

Thank you for supporting this major undertaking. I hope you find this and all other installments of *The Relay Testing Handbook* series to be a useful resource. This project is ongoing and we are constantly seeking to make improvements. Our publishing model allows us to quickly correct errors or omissions and implement suggestions. Please contact us at info@valenceonline.com to report a problem. If we implement your suggestion, we'll send you an updated copy and a prize. You can also go to www.valenceonline.com/updates to see what's changed since the *The Relay Testing Handbook* was released in 2012.

Acknowledgments

This book would not be possible without support from these fine people

Philip B. Baker
Electrical Technician

Eric Cameron, B.E.Sc.
Ainsworth Power Services
www.ainsworth.com

Bob Davis, CET PSE
Northern Alberta Institute of Technology
GET IN GO FAR
www.nait.ca

David Snyder
Hydropower Test and Evaluation
Bonneville Lock and Dam
www.nwp.usace.army.mil/op/b/

John Hodson
Electrical Power Systems Consultant
Calgary, Alberta
Do it right the first time
www.leadingedgesales.ca, www.psams.ca,
www.powerecosystems.com, www.arcteq.fi,
www.magnaiv.com

David Magnan
Project Manager
PCA Valence Engineering Technologies Ltd.
www.pcavalence.com

Mose Ramieh III
Level IV NETA Technician
Level III NICET Electrical Testing Technician
Vice President of Power & Generation Testing, Inc.
Electrical Guru and all around nice guy.

Les Warner, C.E.T.
PCA Valence Engineering Technologies Ltd.
www.pcavalence.com

Lina Dennison

Ken Gibbs, C.E.T.
PCA Valence Engineering Technologies Ltd.
www.pcavalence.com

Roger Grylls, CET
Magna IV Engineering
Superior Client Service. Practical Solutions
www.magnaiv.com

Jamie MacLean
Electrical Engineer
TransAlta Utilities

Table of Contents

Table of Figures

Chapter 1

Electrical Fundamentals

Every relay technician should thoroughly understand the basics of electricity and power systems so they can apply this knowledge to their relay testing tasks. This first chapter is dedicated to these topics.

1. The Three-Phase Electrical System

Many people have difficulty understanding or visualizing a three-phase electrical system. Understanding the relationships between phases is not necessary for simple current, voltage, and/or frequency relay testing. However, it is imperative when testing more complex protective elements. The following explanations use shortcuts to demonstrate the principles described and may not be technically correct. We want you to focus on the information necessary to understand the three-phase electrical system and encourage you to seek out other material to better understand the details for any given subject.

A) Generation

Almost all electrical systems are supplied by generators of one sort or another. Although generators can operate using gas, diesel, steam, water, wind, etc.; actual generator construction is basically the same despite the input fuel. A generator rotor is inserted within the center of stator poles that are installed on the outside edge of a circular generator. The rotor is a large electromagnet rotated within the stator poles via a prime mover which can be turned by any of the fuels described previously. The generator voltage is determined by the number of coils wrapped around each iron core within the stator poles and the strength of the rotor's magnetic field. As the rotor rotates within the stator poles, the magnetic interaction between the rotor and stator creates a voltage. The rotor/stator magnetic interaction, and subsequent voltage, varies as the rotor's alignment to the poles change. Figure 1-1 shows a simplified generator sine wave. A real AC generator would have additional stator poles and rotor magnets. Notice how the horizontal scale of the sine wave graph corresponds to the rotor position inside the stator. When the magnetic north interacts with the stator pole, the induced voltage is positive and the magnetic south creates negative voltage. In our example a cycle is equivalent to one full rotation of the rotor.

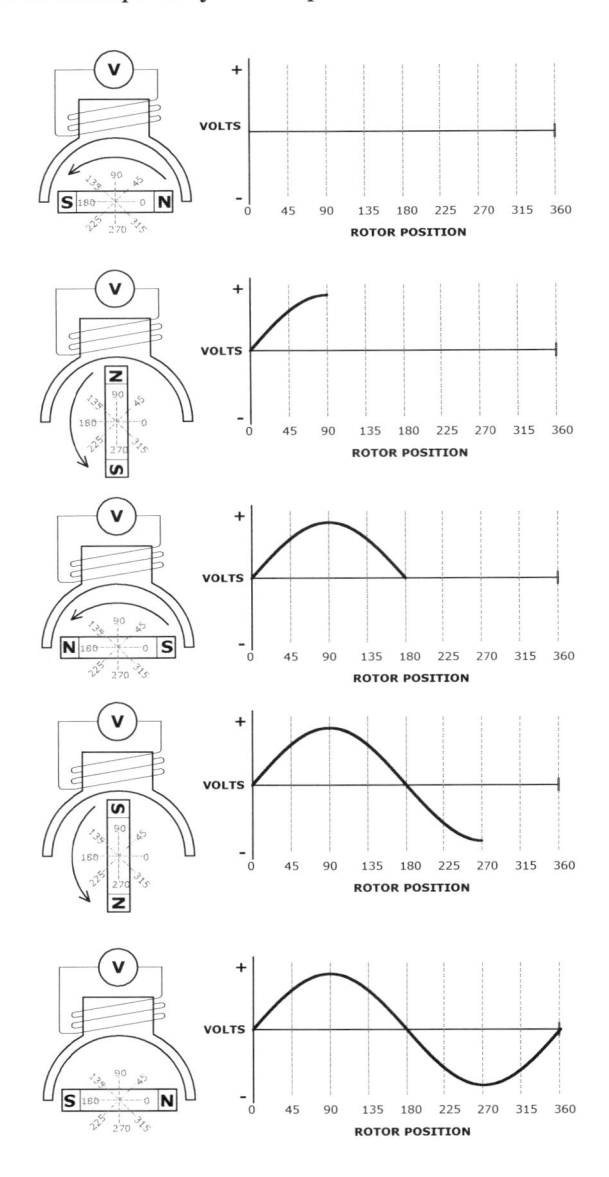

Figure 1-1: Simple Single-Phase Generator

Most electrical systems use three-phase voltages which are created by adding additional poles around the rotor with overlapping edges to ensure smooth transitions. The three poles in this simplified example (real generators have many more poles and rotor magnetic fields) are arranged 120° apart and create the sine wave shown in Figure 1-2. The rotor interacts with more than one pole simultaneously. Pay attention to the north *and* south poles. Notice how the horizontal axis is marked in degrees, but the axis units could also be time when the rotor speed is used for reference instead of rotor position. Most applications in North America rotate at 60 cycles per second or 60 hertz as shown in Figure 1-2.

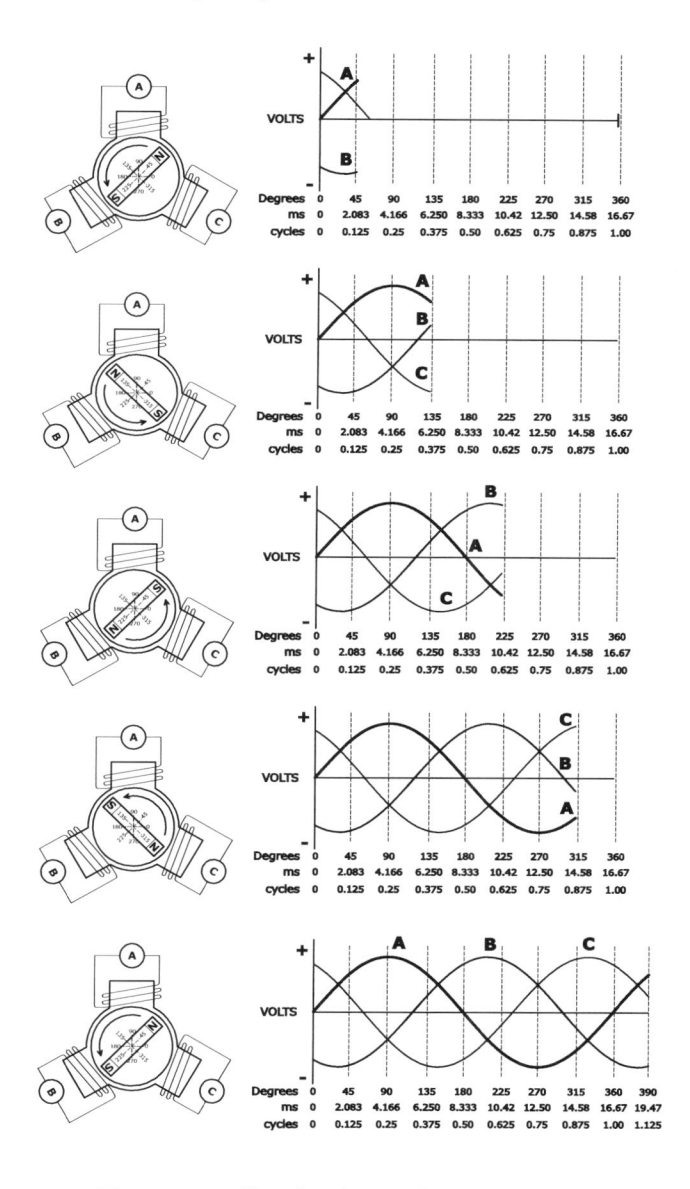

Figure 1-2: Simple Three-Phase Generator

Three-phase systems can be either clockwise (A-B-C) or counterclockwise (A-C-B) rotation. The phase rotation of the system can be determined by the rotor direction of rotation, or the position of the windings. To correctly determine phase rotation, imagine that the sine waves are constantly moving from right to left and pick a vertical line as a reference. Notice which order the positive peaks pass through your reference. The "A" wave will cross first followed by "B ", and then "C" which is A-B-C rotation. We always write A-B-C and A-C-B as a standard even though B-C-A and C-A-B are technically possible in an A-B-C system: and C-B-A and B-A-C are technically correct for an A-C-B system. Use the examples in Figure 1-3 to help understand phase rotation.

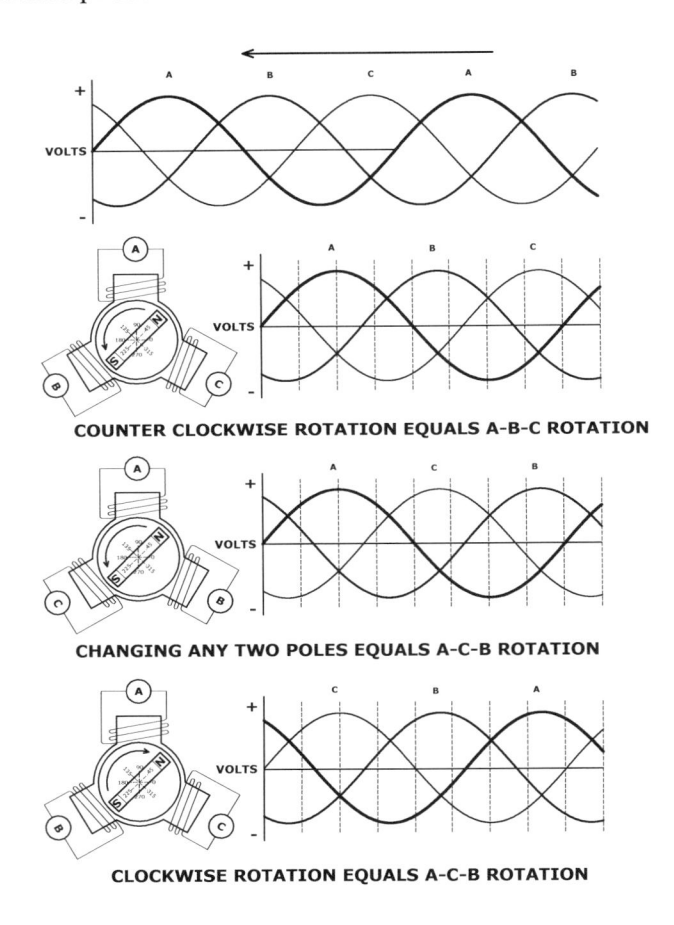

COUNTER CLOCKWISE ROTATION EQUALS A-B-C ROTATION

CHANGING ANY TWO POLES EQUALS A-C-B ROTATION

CLOCKWISE ROTATION EQUALS A-C-B ROTATION

Figure 1-3: Phase Sequence Examples

B) Frequency

Frequency is generically defined as the rate at which something occurs or is repeated over a particular period of time or in a given sample. The electrical frequency is defined as the number of cycles that occur in 1 second. In North America, the typical system frequency is 60 Hz or 60 cycles per second. In Europe, the typical frequency is 50 Hz or 50 cycles per second. There are more cycles per second in North America which means that each cycle is generated in 16.666ms. In Europe, each cycle is generated in 20ms.

The difference between these standards is shown in Figure 1-4 below. Notice that while the time reference remain the same regardless of the 50 or 60Hz waveform, the cycle and degrees reference are relative to the generated cycle.

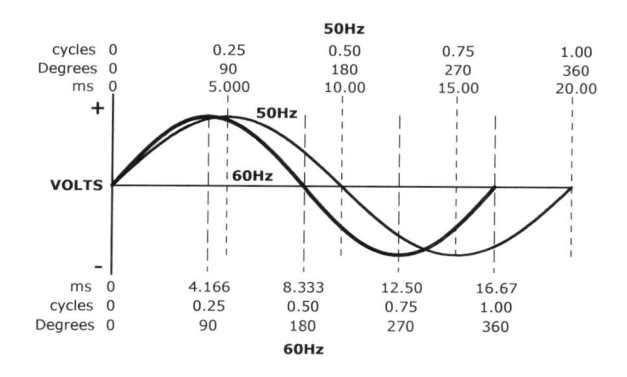

Figure 1-4: Phasor Diagram Example

The formulas to convert cycles-to-seconds and seconds-to-cycles are:

$$Frequency = \frac{cycles}{seconds}$$

$$seconds = \frac{cycles}{Frequency}$$

$$Frequency = \frac{cycles}{seconds}$$

$$cycles = Frequency \times seconds$$

$$seconds = \frac{1\ cycle}{60\ Hz}$$

$$cycles = 60 \times 0.01\overline{66}$$

$$seconds = 0.01\overline{66}$$

$$cycles = 1$$

C) Phasor Diagrams

While sine-wave drawings display the actual output of a generator, they are hard to draw and take up a lot of space. They also become more difficult to read as more information is added. We need an easier way to display the relationships between phases as well as the relationships between voltages and currents. Phasor diagrams were designed to quickly understand the relationships in an electrical system. A properly constructed phasor diagram displays all of the pertinent information we need to visually analyze an electrical system and can also be called a vector diagram. Each phasor (vector) on a phasor diagram is composed of the following:

- A line drawn to scale representing the rms voltage or current of the positive-half-cycle of the waveform. The line length is usually referred to as the magnitude.
- Phasors are typically drawn starting at the center of the diagram (the origin). The phasor's angle is determined by the location of the peak voltage or current in the positive-half-cycle of the waveform.
- An arrow is drawn on the tip of the phasor. There is no official standard for phasor arrows but whatever standard that is chosen should be consistent. Typically a voltage phasor has an open arrow and a current phasor has a closed arrow.
- The phasor is labeled with the electrical parameter (e.g. Volts = "V," "E") it represents, and the connection (e.g. A-N phase = "AN," "an"). The first letter of the connection represents the arrow tip and the second letter represents the origin. In Figure 1-5 the "n" designations of Van, Vbn, and Vcn are all connected together at the origin of the phasor diagram.

See Figure 1-5 to help understand how phasor diagrams relate to the electrical system. Those of you familiar with phasor diagrams may be quick to point out the phasor diagram does not appear to be correct at first glance, but pay attention to the angle reference in each drawing.

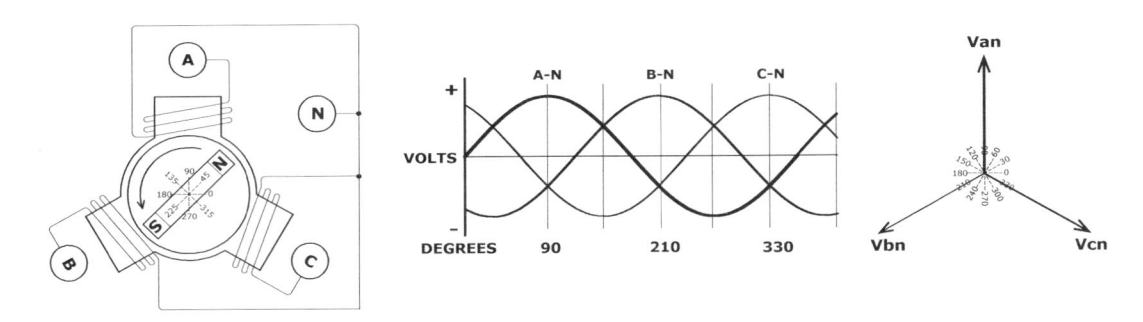

Figure 1-5: Electrical Diagram Comparison

The phasor diagram looks static on the page but actually represents values rotating in the counter-clockwise direction. You can visualize phasor diagrams by imagining a tire turning at 60 revolutions per second. If a strobe light shines onto the tire 60 times per second, the tire will appear to be standing still. If we mark three points on the tire 120 degrees apart and start it spinning again with the strobe light, the three marks are the points on a phasor diagram. The marks appear to be standing still because of the strobe light but they are actually turning in the direction of phase rotation.

Phasor diagrams need a reference and the A-phase voltage is usually used as the reference which is typically the Van phasor. We can modify Figure 1-5 to use Van as the reference by making the Van-rms-maximum-of-positive-half-cycle 0° on all diagrams and plotting the other phasors relative to Van as shown on Figure 1-6. Phasor diagram purists, pay attention to the angles again.

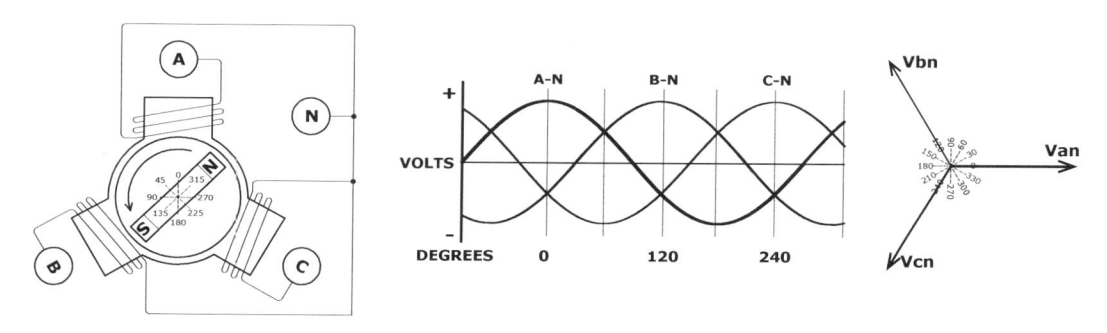

Figure 1-6: Electrical Diagram Comparison with Van Reference

Phasor diagrams need to be consistent in order to be effective so we will modify the previous examples to create a more traditional phasor diagram. Remember that the generator examples have been simplified to make generator operation easier to understand and we will only use the sine-wave and phasor diagrams in the future. Phasor diagrams typically rotate with counter-clockwise direction. Figure 1-7 demonstrates a more realistic phasor diagram with Van as the reference at 0°, counter-clockwise rotation, A-B-C rotation, and a lagging angle reference (Where the phasor angles are displayed 0-360° clockwise).

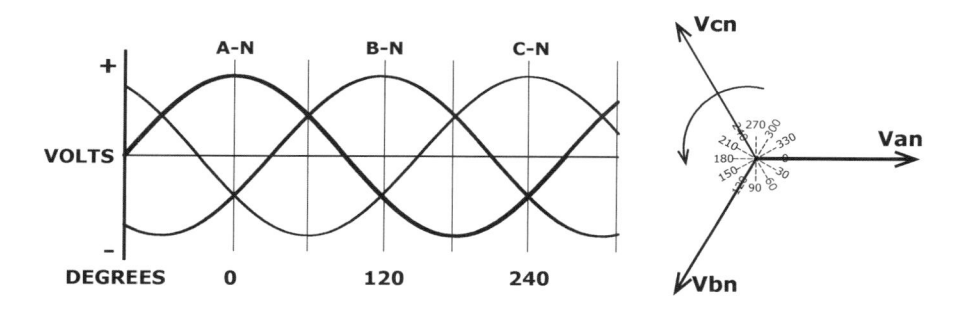

Figure 1-7: Phasor Diagram 1

The big differences between Figure 1-7 and the previous examples are the reference angles inside the diagram and the location on the Vbn and Vcn phasors. Remember that you can determine the phase rotation on a sine-wave drawing by selecting a reference and pretending that the waves are flowing towards the left. The phase rotation for this example is A-B-C. You can tell the phase rotation on a phasor diagram using the same technique. Use 0° as the reference and rotate the phasors in a counter-clockwise direction. The Van phasor crosses first followed by the Vbn phasor and finally the Vcn phasor. Both diagrams indicate A-B-C rotation.

The angles on the Phasor Diagram are relative and depend on your background. The angles shown in Figure 1-7 make a lot of sense if you decide that your reference is 0° and all angles are considered to be "lagging" the reference. GE/Multilin relays use this system and a typical, three-phase electrical system is defined as Van@0°, Vbn@120°, and Vcn@240° as shown in Figure 1-7.

Many organizations and engineers define the exact same electrical system as Van@0°, Vbn@240°, and Vcn@120° as shown in Figure 1-8. Notice that the only difference between Figures 1-7 and 1-8 is the angle references. The actual phasors have not changed position because counter-clockwise rotation is the standard for phasor diagrams unless defined otherwise.

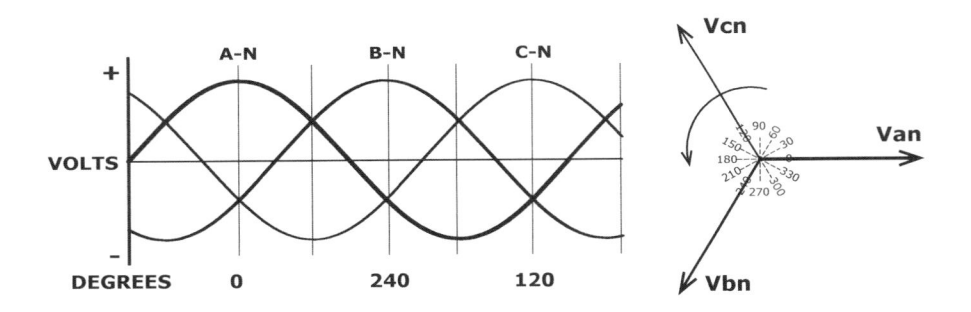

Figure 1-8: Phasor Diagram 2

Because having two standards for defining was not confusing enough, we can define the exact same electrical system as Van@0°, Vbn@-120°, and Vcn@120° as shown in Figure 1-9. This is the standard that was used by my instructors and will be the standard used for the majority of this book. Remember that there is no right or wrong standard as long as they are applied consistently and you are aware what standard is in-use.

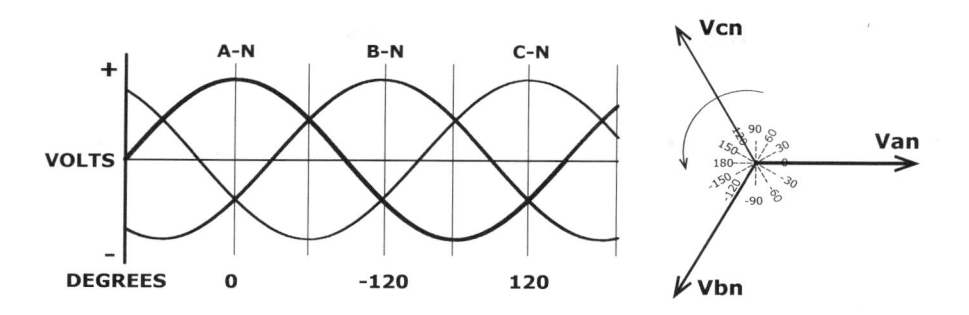

Figure 1-9: Phasor Diagram 3

Once again phase rotation can be changed from A-B-C to A-C-B by changing the rotation or changing any two phases. A-B-C systems rotate counter clockwise and can be verified on the phasor diagram by using zero degrees as a reference and imagining the phasors rotating counter-clockwise. When rotating counter clockwise, A-N phasor crosses first followed by B-N and finally C-N. The phasors are always rotating. Any phasor diagram represents one moment in time. Review Figure 1-10 to see the differences between A-B-C and A-C-B systems.

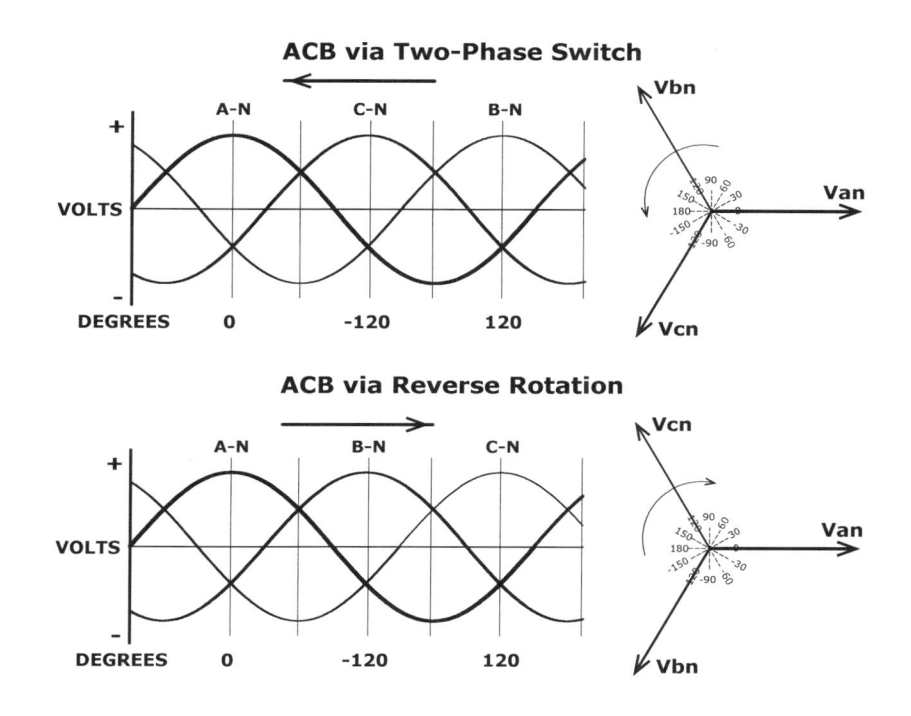

Figure 1-10: A-B-C / A-C-B Phasor Diagrams

Up until now, we have exclusively focused on the voltage waveforms and phasors that represent the potential energy across two points as measured by a voltmeter. Current waveforms represent the current flow through a device as measured by an ammeter in series with a load. The waveforms and phasor diagrams are drawn in the same manner as the voltage waveforms and phasors but use Amperes (Amps / A) as their scale. The current phasors are drawn with closed arrows to differentiate them from voltage phasors. Review Figure 1-11 see current and voltage values plotted simultaneously.

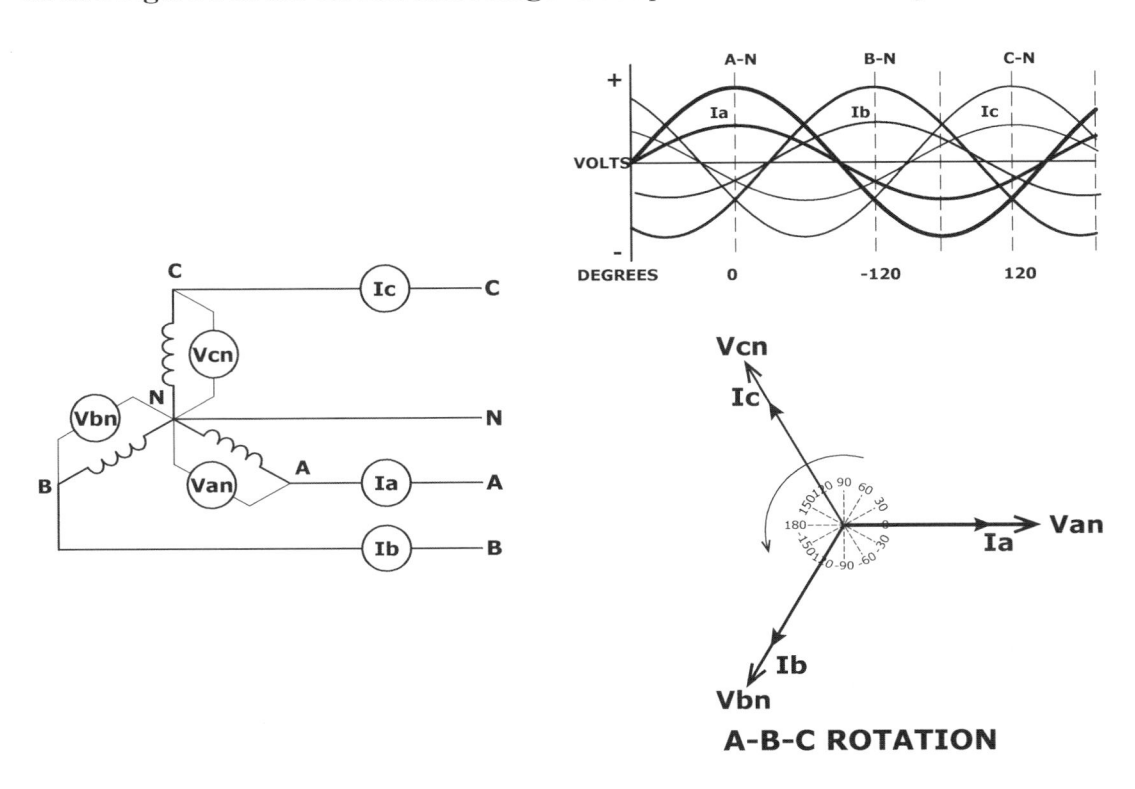

Figure 1-11: Current and Voltage Phasors

I like to place zero degrees at the 3 o'clock position, but it can easily be placed at the 12 o'clock position if it helps you visualize the diagram. It is simply a matter of personal preference.

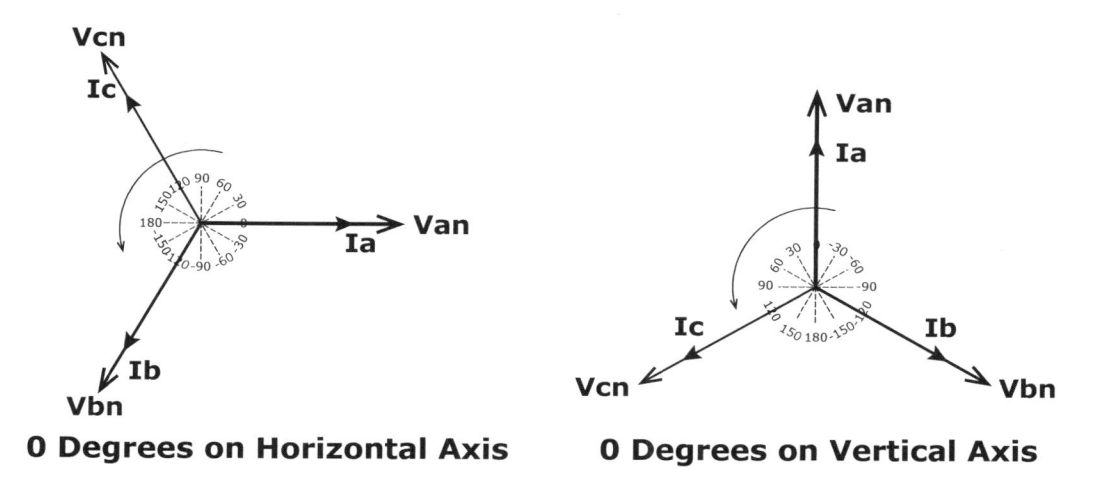

Figure 1-12: Vector Diagrams on Different Axis

i) Adding or Subtracting Phasors

There will be times when phasors need to be added together to help understand the properties of the electrical system such as Delta-Wye conversions in the next section or Sequence Components later in the chapter.

Phasor addition is probably the most difficult electrical calculation and there are three methods that you can employ:

a. Use a calculator – Some calculators will allow you to add 1@0° and 1@60° using the interface. This functionality is rare and is sometimes found on scientific calculators that can add complex numbers using polar quantities. You can also download the "Dot Point Learning Systems Science Calculator Free Version" at **http://www.dplsystems.com.au/downloads.htm** and use the "Vector calculator" tool. The answer is 1.732@30°.

b. Convert the Phasor (Polar) to Rectangular – Adding and subtracting phasors in polar format (magnitude and angle) is very difficult. But if we can convert the phasors into rectangular format (real and imaginary numbers or horizontal and vertical magnitudes) first, the addition or subtraction is quite easy.

Those who thought that trigonometry would never be useful in the real world may be surprised because we can apply the Pythagorean Theorem to any phasor using the following formulas:

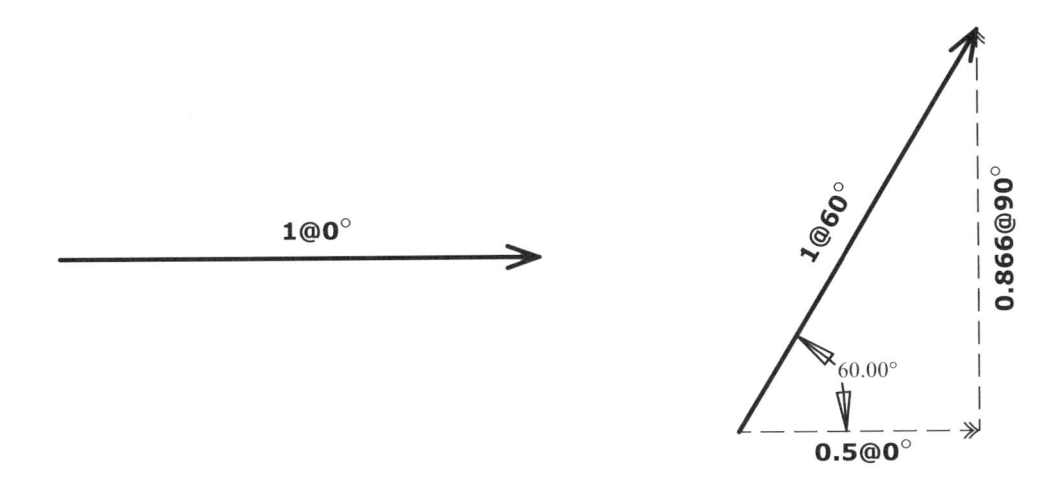

$$\text{Real} = \text{Magnitude} \times \cos(\text{Angle})$$

$$\text{Real} = 1 \times \cos(0°)$$

$$\text{Real} = 1 \times 1$$

$$\text{Real} = 1$$

$$\text{Imaginary} = \text{Magnitude} \times \sin(\text{Angle})$$

$$\text{Imaginary} = 1 \times \sin(0°)$$

$$\text{Imaginary} = 1 \times 0$$

$$\text{Imaginary} = 0$$

$$\text{Rectangular} = \text{Real} + "j" \text{Imaginary} = x + jy$$

$$\text{Rectangular} = 1 + j0$$

$$\text{Real} = \text{Magnitude} \times \cos(\text{Angle})$$

$$\text{Real} = 1 \times \cos(60°)$$

$$\text{Real} = 1 \times 0.5$$

$$\text{Real} = 0.5$$

$$\text{Imaginary} = \text{Magnitude} \times \sin(\text{Angle})$$

$$\text{Imaginary} = 1 \times \sin(60°)$$

$$\text{Imaginary} = 1 \times 0.866$$

$$\text{Imaginary} = 0.866$$

$$\text{Rectangular} = \text{Real} + "j" \text{Imaginary} = x + jy$$

$$\text{Rectangular} = 0.5 + j0.866$$

Now we can add the real (x-axis) values together, then the imaginary numbers (y-axis or "j" in formulas), and then use the Pythagorean Theorem to convert the value back to a phasor:

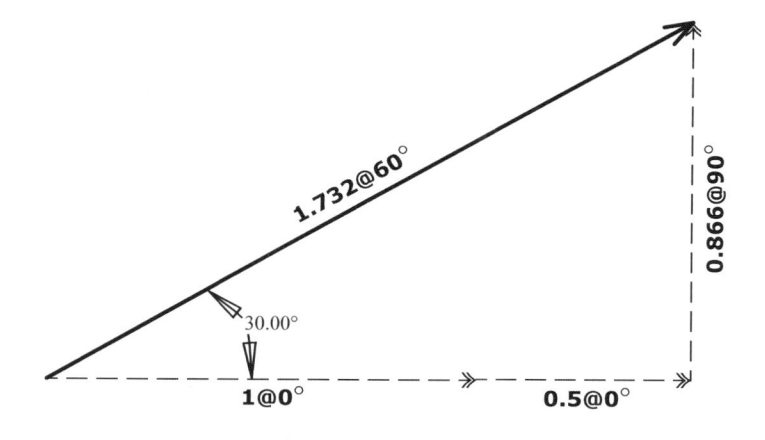

$$\text{Real(Sum)} = \text{Real}(1@0°) + \text{Real}(1@60°)$$

$$\text{Real(Sum)} = 1 + 0.5$$

$$\text{Real(Sum)} = 1.5$$

$$\text{Imaginary(Sum)} = " j " \text{Imaginary}(1@0°) + " j " \text{Imaginary}(1@60°)$$

$$\text{Imaginary(Sum)} = j0 + j0.866$$

$$\text{Imaginary(Sum)} = j0.866$$

$$\text{Sum} = 1.5 + j0.866$$

$$\text{Magnitude(Sum)} = \sqrt{\left(\text{Real(Sum)}^2 + \text{Imaginary(Sum)}^2\right)}$$

$$\text{Magnitude(Sum)} = \sqrt{\left(1.5^2 + 0.866^2\right)}$$

$$\text{Magnitude(Sum)} = \sqrt{\left(2.25 + 0.749956\right)}$$

$$\text{Magnitude(Sum)} = \sqrt{2.999956}$$

$$\text{Magnitude(Sum)} = 1.732$$

$$\text{Angle(Sum)} = \arctan\left(\frac{\text{Imaginary(Sum)}}{\text{Real(Sum)}}\right)$$

$$\text{Angle(Sum)} = \arctan\left(\frac{0.866}{1.5}\right)$$

$$\text{Angle(Sum)} = \arctan\left(0.577\right)$$

$$\text{Angle(Sum)} = 30°$$

c. Add the Vectors Graphically – If you draw the vectors to scale with a CAD program or a ruler and protractor, you can draw one vector on the tip on the other and measure the distance and angle from the first vector's source as shown in the following section.

D) Three-Phase Connections

There are two main connections when using three-phase systems, Delta and Wye (Star). Until now, we have been dealing with Wye connections where all three phases have a common point. With this connection, the line and phase currents are the same, but the phase and line voltages are different as shown in Figure 1-13:

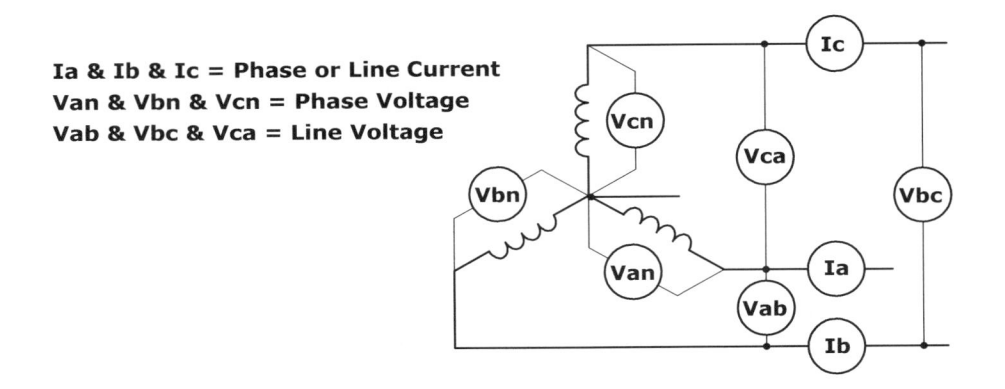

Ia & Ib & Ic = Phase or Line Current
Van & Vbn & Vcn = Phase Voltage
Vab & Vbc & Vca = Line Voltage

Figure 1-13: Wye Connection

In order to draw phasors of the three new line voltages (Vab, Vbc, Vca), we need to follow their connections through the Wye connection. Vca is the vector sum of Vcn and Vna using the vector conventions we discussed earlier. Vcn is still drawn at 120° because it has not changed. Vna is drawn at 180° because it is the opposite of Van at 0 degrees. When performing vector addition, the first vector is drawn (Vna) because "a" is the reference of a Vca phasor. The second vector (Vcn) is drawn at the correct angle starting on the tip of the first vector. The sum of the two vectors is Vcn + Vna or Vcnna. We can remove the "nn" and the final phasor designation is Vca.

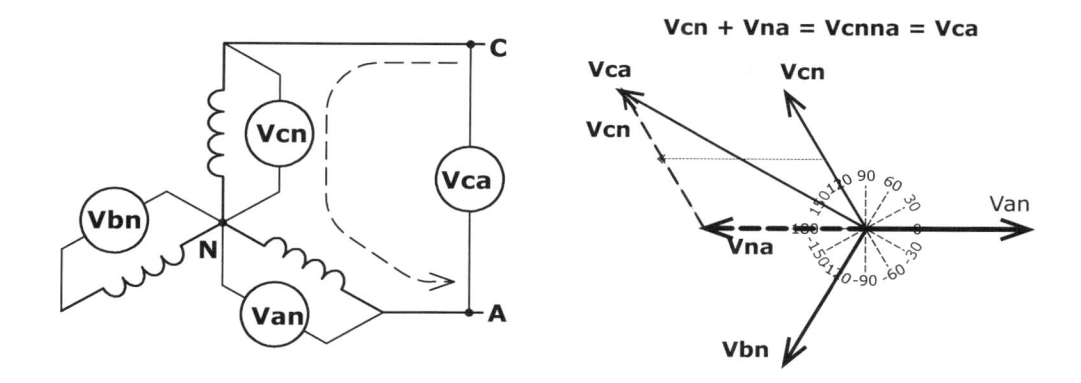

Figure 1-14: Phase to Line Conversion

Experience has shown that when two vectors with the same magnitude are added when they are 60 degrees apart, the sum will be √3 larger at the midpoint between the two vectors. If the vector diagram was drawn to scale, we could measure the resultant vector. (Something to remember when you forget your fancy calculator.) Therefore, Vca is √3Vcn@150°.

$$Vna = 180°, Vcn = 120°$$

$$Vna - Vcn = 180° - 120° = 60°$$

$$Vca = Vcn + \frac{60°}{2}$$

$$Vca = 120° + 30°$$

$$Vca = 150°$$

Based on experience, any conversion between Delta and Wye balanced systems (identical magnitudes 120 degrees apart) will increase or decrease by √3 and be displaced by 30 degrees.

The vector diagram for a balanced Wye connected three-phase system would look like Figure 1-15 where:

WYE TO DELTA		DELTA TO WYE	
DELTA	**WYE**	**DELTA**	**WYE**
Van @ 0°	Vab = √3Van @ 30° (Van+30°)	Vab @ 30°	$Van = \dfrac{Vab}{\sqrt{3}}$ @ 0° (Vab-30°)
Vbn @ -120°	Vbc = √3Vbn @ -90° (Vbn+30°)	Vbc @ -90°	$Vbn = \dfrac{Vbc}{\sqrt{3}}$ @ –120° (Vbc-30°)
Vcn @ 120°	Vca = √3Vcn @ 150° (Vcn+30°)	Vca @ 150°	$Vcn = \dfrac{Vca}{\sqrt{3}}$ @ 120° (Vca-30°)

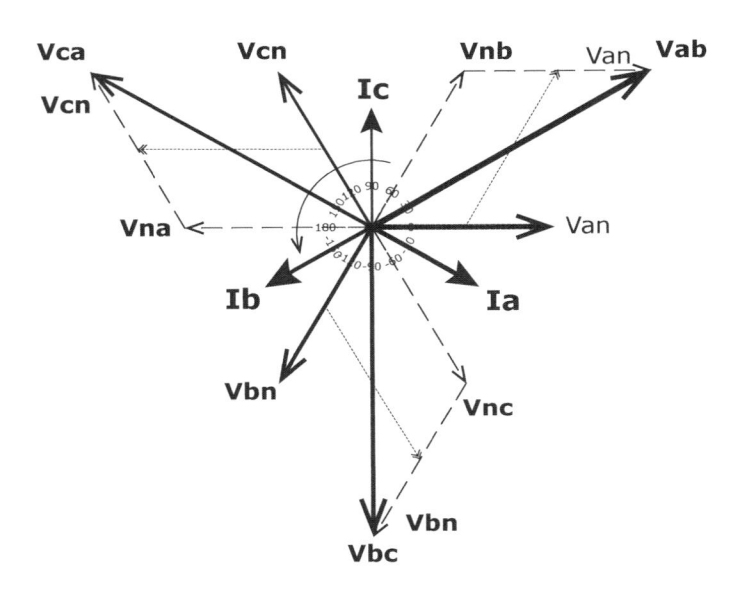

Figure 1-15: Delta/Wye Connected Vector Diagram

A delta connected system is shown in Figure 1-16. Using this connection, the line and phase voltages are the same, but there is a difference between the phase and line currents.

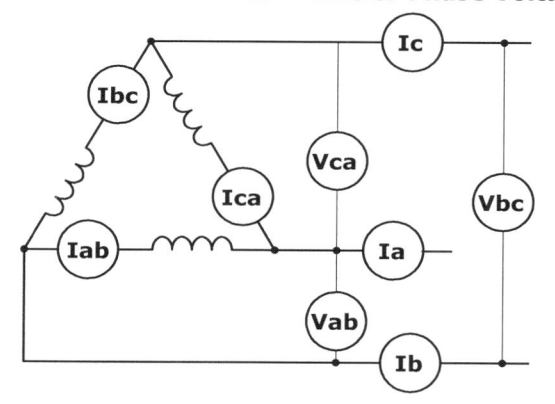

Figure 1-16: Delta Connection

Notice that regardless of the three-phase system the line voltage and line current connections remain the same. Phase current is rarely measured, and we won't waste any more time on phase current in delta systems. The delta connection vector diagram is identical to the Wye connected diagram without the neutral connections as shown in Figure 1-17.

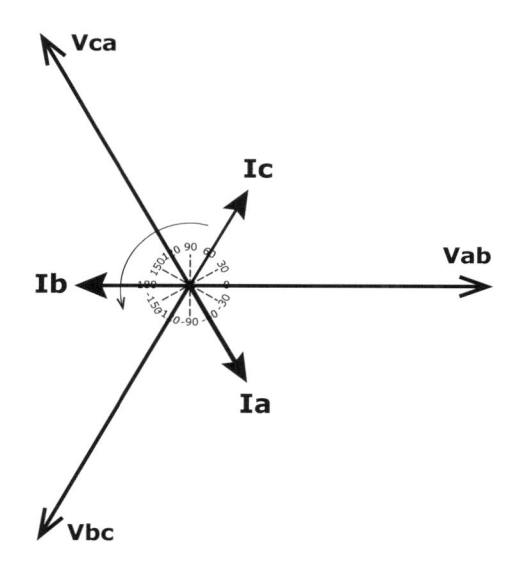

Figure 1-17: Delta Connection Vector Diagram for Balanced System with 30° Lag

E) Watts

Generators are turned by prime movers that are controlled by a governor. The governor operates like the cruise control in your car and tries to maintain the same speed (usually 60Hz). If the load increases or decreases, the generator will momentarily slow down or speed up before the governor can respond to compensate for the load change. The generator's speed controls the real power output of the generator that is measured in Watts when connected to the grid. Watts are a measure of the real power consumed by a load like actual torque in a motor, heat from a heater, or light and heat from an incandescent light bulb. Watts do all of the work in an electrical system.

A system that is generating 100% Watts are displayed on sine-wave and phasor diagrams with the phase voltage and phase current at the same phase angle as displayed in Figure 1-18.

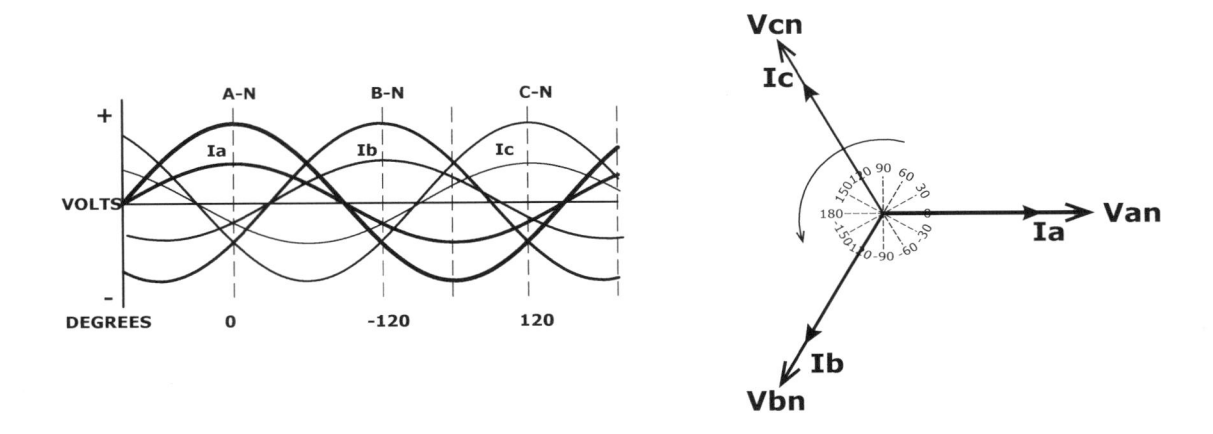

Figure 1-18: Watts - Current and Voltage Are In-Phase

F) Inductive VARS

In the early years of electricity, a major debate was raised between the widespread use of AC or DC in electrical systems. Ultimately AC won the war and is the standard throughout most of the world because of its ability to transform lower voltages into higher voltages. This was an important distinction because the higher voltages require less current to transmit the same amount of power, which drastically reduced transmission costs. This transformation ability uses induction to create a magnetic field when conductors are coiled around a magnetic core.

Inductive VARs represent the energy created by the magnetic fields and is created by the interaction of electrical coils and magnetic cores in equipment such as AC motors, transformers, and generators.

Ohm's Law in DC systems explains the relationship of Volts, Amps, and Resistance in a circuit and it is updated in AC systems to include Reactance. Inductive reactance is a component of the Inductive VARs and will resist any *change in* current as opposed to resistance which resists the entire flow of current. The Inductive Reactance causes the current to lag the voltage due to the opposing force in inductive systems. Most systems are inductive due to electric motors and current usually lags the voltage in most systems.

Figure 1-19 simplifies the AC waveform by only showing one phase to help you visualize the lag between voltage and current. The other two phases will change accordingly in a balanced system. Remember that the axis remains in place while the waveform moves from right to left. The voltage peak will cross the y-axis before the current peak and, therefore, the current lags the voltage (or the voltage leads the current). In the phasor diagram, the vectors rotate counter-clockwise and the voltage will cross the 0 degree axis before the current phasor.

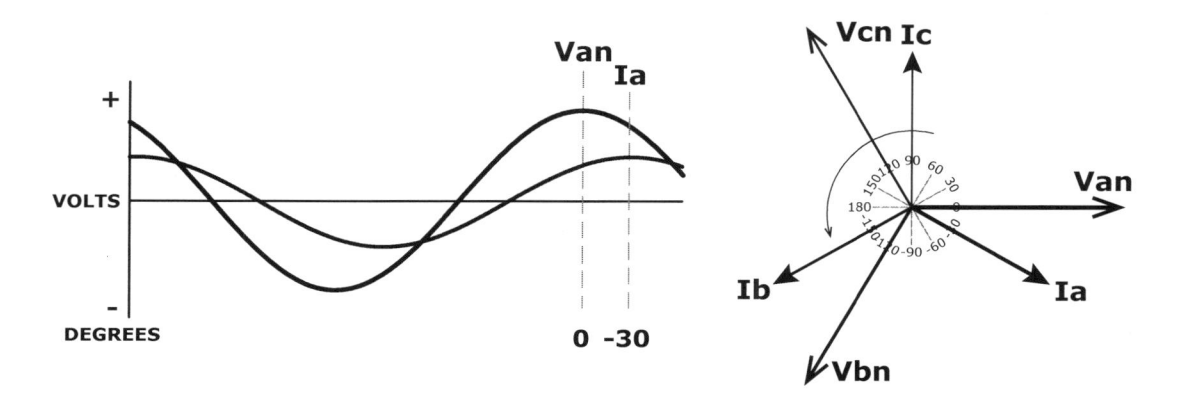

A-B-C ROTATION WITH 30 DEGREE LAG

Figure 1-19: Inductive VARs - Current Lags the Voltage by 30°

G) Capacitive VARS

Capacitive VARs represent the reactive power stored in capacitors, over-excited generators, cables or power lines, or other sources that cause the current to lead the voltage. While inductive VARs resist any change in current, capacitive VARs give the voltage an extra boost. Inductive and capacitive VARs operate in opposition and can cancel each other out. (See "The Power Triangle" in the next section for details.) In a system where the capacitive VARs are greater than the inductive VARs, the current will lead the voltage.

While the generator governor controls the power output of the generator as described previously, the generator exciter controls the generator output voltage. If the generator is not connected to the grid, the exciter can raise or lower the rotor current that, in turn, raises or lowers the generator voltage in direct proportion. However, when the generator is connected to a grid the voltage is relatively fixed by the grid and raising and lowering the rotor current causes the generator to input or export VARs. Generators supply most of the VARs required by inductive machines like transformers or motors.

Figure 1-20 simplifies the AC waveform by showing only one phase to help you visualize how the current leads the voltage. The other two phases will change accordingly in a balanced system. Remember, the axis remains in place while the waveform moves from right to left. The current peak will cross the y-axis before the voltage peak and, therefore, the current leads the voltage. In the phasor diagram, the vectors rotate counter-clockwise and the current will cross the 0 degree axis before the voltage phasor.

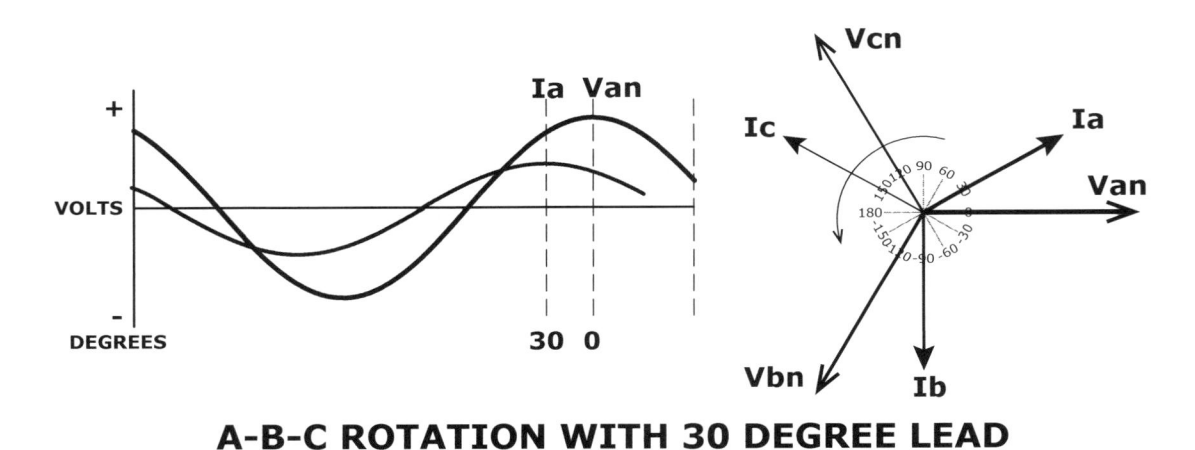

A-B-C ROTATION WITH 30 DEGREE LEAD

Figure 1-20: Capacitive VARs - Current Leads the Voltage

H) The Power Triangle

In an AC electrical system, all power can be broken down into three separate components. VA is the apparent power supplied and is calculated by multiplying the voltage and current. The real energy produced by an electrical system is measured in Watts and is the amount of current in phase with the voltage and multiplied by the voltage. Reactive VARs is the combination of capacitive and inductive VARs produced by capacitors and inductive machines and are +/-90° out of phase with the voltage respectively. Watts and VARs are combined to create VA. It can be hard to grasp how these electrical measurements interact, but the power triangle helps put it into perspective. The standard formulas for three-phase power are shown in the following figure:

$$VA = (Van \times Ia) + (Vbn \times Ib) + (Vcn \times Ic)$$

$$WATTS = \left[Van \times Ia \times \cos(Van° - Ia°)\right] + \left[Vbn \times Ib \times \cos(Vbn° - Ib°)\right] + \left[Vcn \times Ic \times \cos(Vcn° - Ic°)\right]$$

$$VARS = \left[Van \times Ia \times \sin(Van° - Ia°)\right] + \left[Vbn \times Ib \times \sin(Vbn° - Ib°)\right] + \left[Vcn \times Ic \times \sin(Vcn° - Ic°)\right]$$

Figure 1-21: Three-Phase Power Formulas

All of the formulas in Figure 1-22 can be used to calculate three-phase power assuming that all three-phase currents and voltages are balanced. While no system is perfectly balanced, these equations will put you in the ballpark. Make sure you use the correct phasors when determining relationships. "ANGLE" in the VAR and Watt formulas are the difference between the phase-to-neutral voltage and the line current (ANGLE = V_{L-N} angle° – I_{L-N} angle°) phase-to-phase voltages can be used to determine ANGLE in balanced systems only by assuming that the V_{L-L} leads V_{L-N} by 30 degrees (ANGLE = V_{L-L} angle° - 30° - I_{L-N} angle°).

$$VA = \sqrt{3} \times LineVolts \times LineAmps$$

$$VARS = \sqrt{3} \times LineVolts \times LineAmps \times \sin(ANGLE)$$

$$WATTS = \sqrt{3} \times LineVolts \times LineAmps \times \cos(ANGLE)$$

Figure 1-22: Three-Phase Power Formulas Assuming Balanced Conditions

In a perfect electrical system where only Watts exist, the power triangle is a straight line as no VARs are produced. See Figure 1-23 to see a power triangle where only pure power is produced.

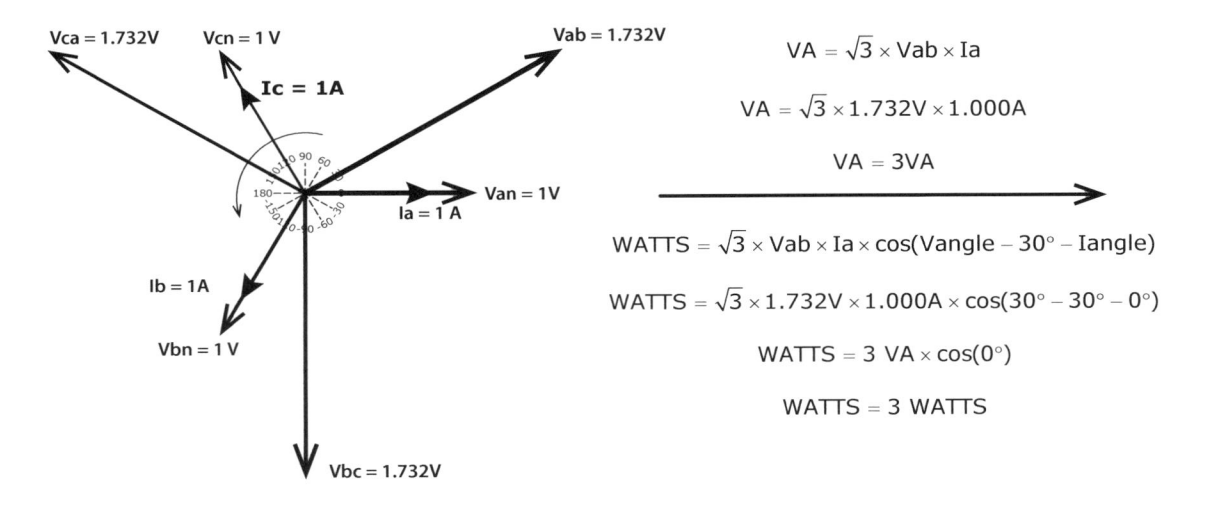

$$VA = \sqrt{3} \times Vab \times Ia$$
$$VA = \sqrt{3} \times 1.732V \times 1.000A$$
$$VA = 3VA$$

$$WATTS = \sqrt{3} \times Vab \times Ia \times \cos(Vangle - 30° - Iangle)$$
$$WATTS = \sqrt{3} \times 1.732V \times 1.000A \times \cos(30° - 30° - 0°)$$
$$WATTS = 3\ VA \times \cos(0°)$$
$$WATTS = 3\ WATTS$$

Figure 1-23: Three-Phase Power Triangle - Watts Only

When the electrical system supplies a purely inductive source, the current will lag the voltage by 90 degrees, and the power triangle is a vertical line as shown in Figure 1-24:

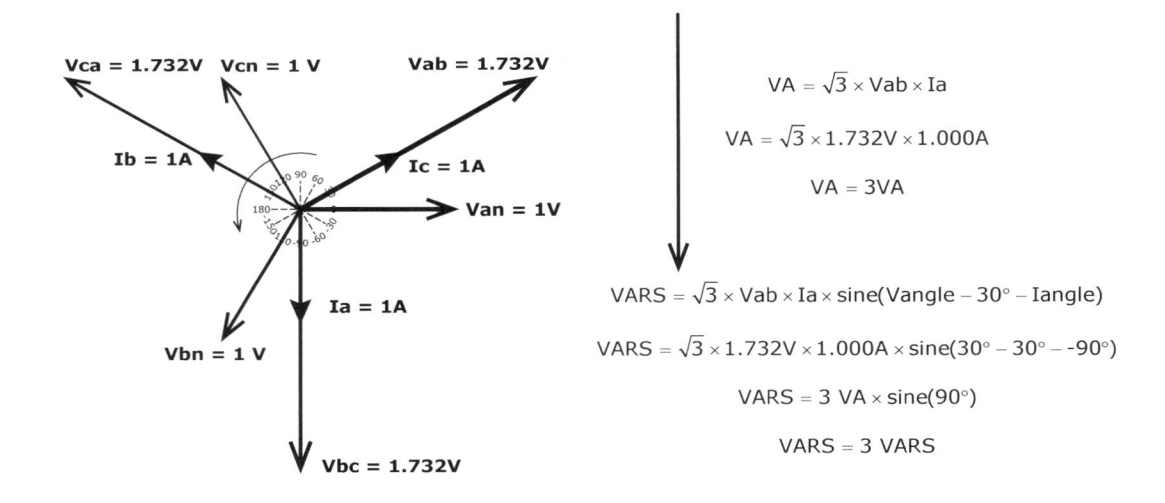

$$VA = \sqrt{3} \times Vab \times Ia$$
$$VA = \sqrt{3} \times 1.732V \times 1.000A$$
$$VA = 3VA$$

$$VARS = \sqrt{3} \times Vab \times Ia \times \text{sine}(Vangle - 30° - Iangle)$$
$$VARS = \sqrt{3} \times 1.732V \times 1.000A \times \text{sine}(30° - 30° - -90°)$$
$$VARS = 3\ VA \times \text{sine}(90°)$$
$$VARS = 3\ VARS$$

Figure 1-24: Power Triangle - Only Inductive VARs

When the electrical system supplies a purely capacitive source, the current will lead the voltage by 90 degrees, and the power triangle is a vertical line as shown in Figure 1-25:

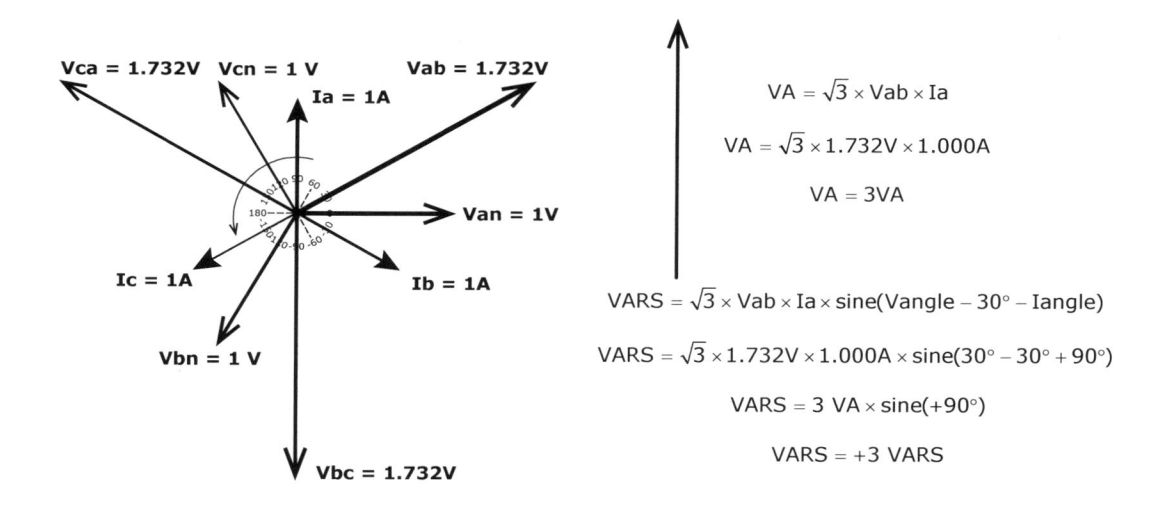

$$VA = \sqrt{3} \times Vab \times Ia$$

$$VA = \sqrt{3} \times 1.732V \times 1.000A$$

$$VA = 3VA$$

$$VARS = \sqrt{3} \times Vab \times Ia \times sine(Vangle - 30° - Iangle)$$

$$VARS = \sqrt{3} \times 1.732V \times 1.000A \times sine(30° - 30° + 90°)$$

$$VARS = 3\ VA \times sine(+90°)$$

$$VARS = +3\ VARS$$

Figure 1-25: Power Triangle - Only Capacitive VARs

Our next example is closer to real life with the current lagging the voltage by 30 degrees. Most power systems have all three parts of the power triangle with Watts, VARs, and VA as shown in Figure 1-26.

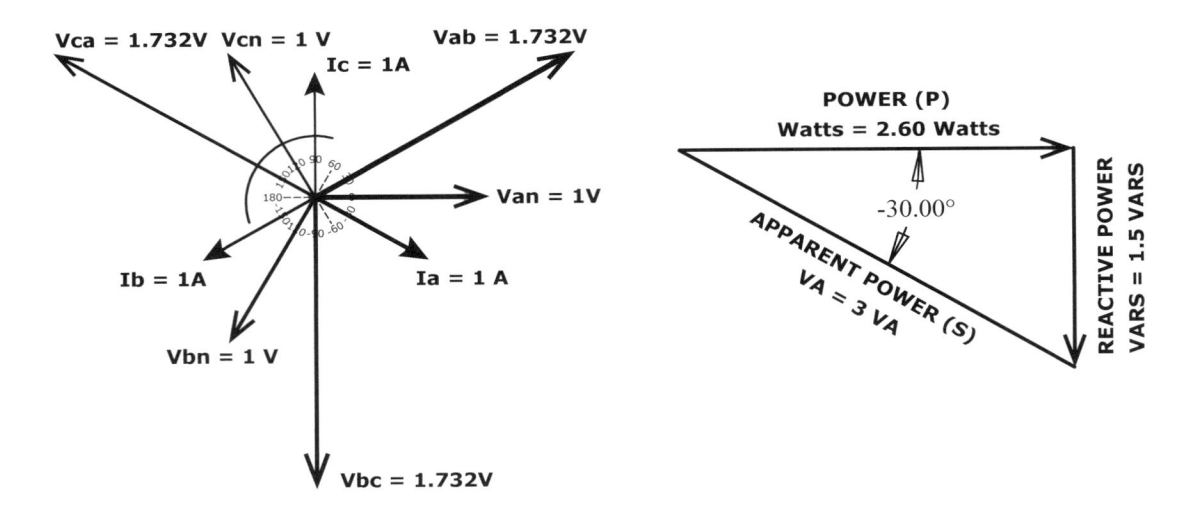

Figure 1-26: Power Triangle

$$\text{WATTS} = \sqrt{3} \times \text{Vab} \times \text{Ia} \times \cos(\text{Vangle} - 30° - \text{Iangle})$$

$$\text{WATTS} = \sqrt{3} \times 1.732\text{V} \times 1.0\text{A} \times \cos(30° - 30° - \text{-}30°)$$

$$\text{WATTS} = 3 \times \cos(30°)$$

$$\text{WATTS} = 3 \times 0.866$$

$$\text{WATTS} = 2.60 \text{ WATTS}$$

$$\text{VA} = \sqrt{3} \times \text{Vab} \times \text{Ia}$$

$$\text{VA} = \sqrt{3} \times 1.732\text{V} \times 1.0\text{A}$$

$$\text{VA} = 3\text{VA}$$

$$\text{VARS} = \sqrt{3} \times \text{Vab} \times \text{Ia} \times \text{sine}(\text{Vangle} - 30° - \text{Iangle})$$

$$\text{VARS} = \sqrt{3} \times 1.732\text{V} \times 1.0\text{A} \times \text{sine}(30° - 30° - \text{-}30°)$$

$$\text{VARS} = 3 \times \text{sine}(30°)$$

$$\text{VARS} = 3 \times 0.5$$

Figure 1-27: Example Three-Phase Power/VARS/VA Formula Calculations

You can apply the Pythagorean Theorem in nearly every aspect of phasor diagrams and other electrical properties as shown in the following example:

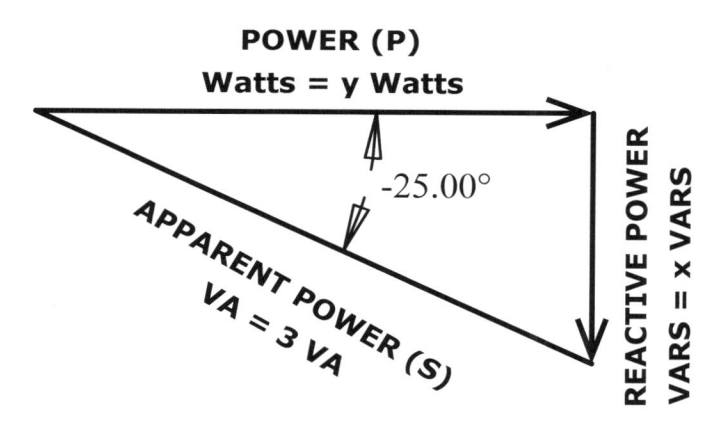

$$WATTS = \sqrt{3} \times Vab \times Ia \times \cos(Vangle - 30° - Iangle)$$

$$WATTS = VA \times \cos(30° - 30° - \text{-}25°)$$

$$WATTS = 3 \times \cos(25°)$$

$$WATTS = 3 \times 0.906$$

$$WATTS = 2.72 \ WATTS$$

$$\cos\theta = \frac{y}{VA}$$

$$\cos(\text{-}25) = \frac{y}{3VA}$$

$$y = \cos(\text{-}25) \times 3VA = 0.906 \times 3VA$$

$$y = 2.72 \ WATTS$$

$$VARS = \sqrt{3} \times Vab \times Ia \times \text{sine}(Vangle - 30° - Iangle)$$

$$VARS = VA \times \text{sine}(30° - 30° - \text{-}25°)$$

$$VARS = 3 \times \text{sine}(25°)$$

$$VARS = 3 \times 0.423$$

$$VARS = 1.268 \ VARS$$

$$\text{sine}\theta = \frac{x}{VA}$$

$$\text{sine}(\text{-}25) = \frac{x}{3VA}$$

$$x = \text{sine}(\text{-}25) \times 3VA = 0.423 \times 3VA$$

$$x = 1.268 \ VARS$$

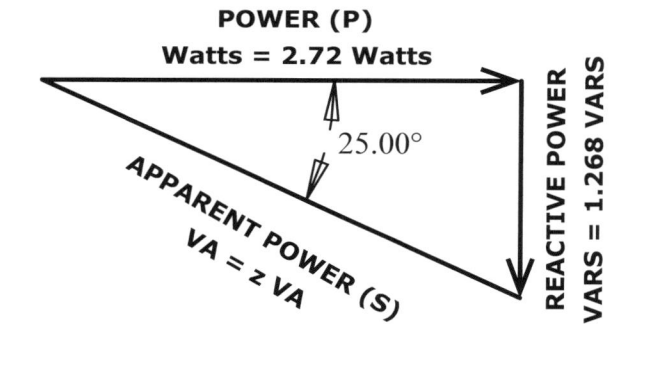

$$VA = \sqrt{(WATTS^2 + VARS^2)}$$

$$VA = \sqrt{(2.72^2 + 1.268^2)}$$

$$VA = \sqrt{(7.3984 + 1.6078)}$$

$$VA = \sqrt{(9.00)} = 3 \ VA$$

I) Power Factor

An electrical system's power factor represents the angle between the Watts and VA of the electrical system. It can also be the angle between voltage and current in any phase when calculated on a phase-by-phase basis. Power factor is expressed as the Cosine of the angle, which can be immediately converted into efficiency by multiplying the power factor by 100.

The meaning of a positive and negative power factor is another one of those situations where the correct answer depends on the location and background of the measuring device. The following statements are usually true in North America (unless you are in a generator plant) and it is always a good idea to represent the power factor as importing or exporting VARS if you are communicating with someone who might use a different standard.

When the current leads the voltage or the system has capacitive VARs, the power factor is considered leading, negative, or exporting. When the system is primarily inductive and the current lags the voltage, the power factor is positive, lagging, or importing. Review the following examples of power factor calculations:

i) Unity Power Factor Calculation (Voltage and Current in Phase)

$$PowerFactor = \cos(Angle) \qquad Efficiency = PowerFactor \times 100$$

$$PowerFactor = \cos(0) \qquad Efficiency = 1.00 \times 100$$

$$PowerFactor = 1.00 \qquad Efficiency = 100\%$$

(All of the energy supplied is real power or Watts)

ii) Inductive System (Current Lags Voltage by 30°)

$$\text{PowerFactor} = \cos(\text{Angle})$$

$$\text{PowerFactor} = \cos(-30)$$

$$\text{PowerFactor} = 0.867 \text{ lag or} + 0.867$$

$$\text{Efficiency} = \text{PowerFactor} \times 100$$

$$\text{Efficiency} = 0.867 \times 100$$

$$\text{Efficiency} = 86.7\%$$

(86.7% of the supplied energy is real power or Watts)

iii) Capacitive System (Current Leads Voltage by 30°)

$$\text{PowerFactor} = \cos(\text{Angle})$$

$$\text{PowerFactor} = \cos(30)$$

$$\text{PowerFactor} = 0.867 \text{ lead or} - 0.867$$

$$\text{Efficiency} = \text{PowerFactor} \times 100$$

$$\text{Efficiency} = 0.867 \times 100$$

$$\text{Efficiency} = 86.7\%$$

(86.7% of the supplied energy is real power or Watts)

2. Transformers

As we discussed previously, the ability to transform smaller voltages to higher voltages is the primary reason why we use AC electricity today. Using higher voltages to transfer electricity over long distances saves money, because less current is required at higher voltages to transfer the same amount of energy. Less current means smaller conductors which cost less than large conductors and weigh less so the support structures are also smaller and cheaper. Online savings are realized through reduced line losses as less current flows through the conductors. Transformers are nearly perfect (ideal) machines that require very little energy to create the transformation.

The simplest transformers are constructed around a core made of magnetic material. Two different sets of copper or aluminum coils are wound around the core, and the number of turns determines the rated voltages of the transformer. The set of coils with the most turns is called the "High" side because it will have a higher voltage than the "Low" side. Some people automatically refer to the high side as the transformer primary even though the primary of a transformer is defined by normal load flow. A step-up transformer steps a lower voltage to a higher voltage and the low side is the transformer primary. The opposite is true with a step down transformer.

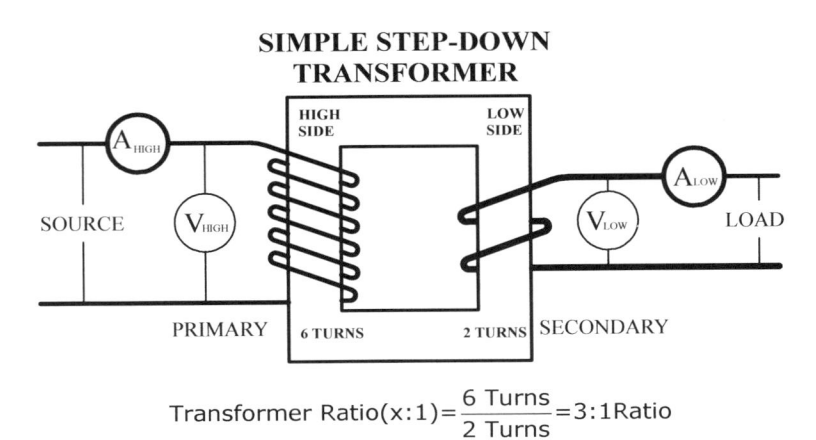

$$\text{Transformer Ratio}(x:1) = \frac{6 \text{ Turns}}{2 \text{ Turns}} = 3:1 \text{ Ratio}$$

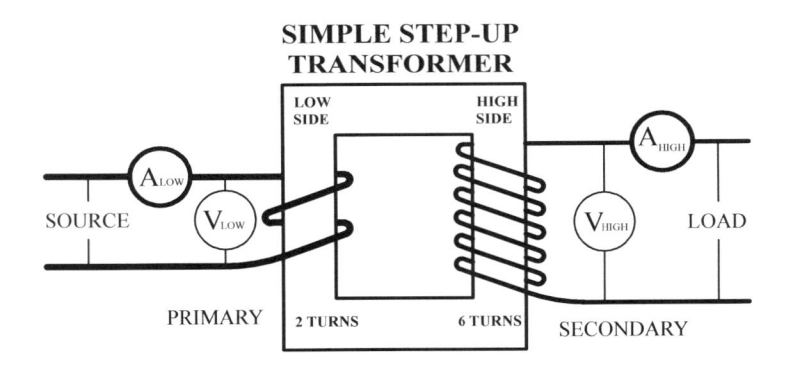

Figure 1-28: Simple Transformer Construction

There are two main principals that apply to all ideal transformers:

- The transformer ratio is absolute
- Power-in equals Power-out

The transformer ratio is pre-defined when the transformer is built and is based on the transformer turns ratio. Transformer ratios are defined on the transformer nameplate by high and low side voltages. For example, a 3:1 transformer ratio could be listed on the transformer nameplate as 345,000V (345kV): 115,000V (115kV). Transformer Ratio $(x:1) = \dfrac{345,000V}{115,000V} = 3:1$

$$\text{Transformer Ratio } (x:1) = \frac{\text{High Side Voltage}}{\text{Low Side Voltage}} \quad \textbf{OR}$$

(handwritten: $\frac{69}{12}$ $\frac{5.75}{1}$)

$$\text{High Side Voltage} = \text{Transformer Ratio } (x:1) \times \text{Low Side Voltage} \quad \textbf{OR}$$

$$\text{Low Side Voltage} = \frac{\text{High Side Voltage}}{\text{Transformer Ratio } (x:1)}$$

Figure 1-29: Transformer Voltage Ratio Calculation

Three-phase power is calculated by multiplying the system voltage, current, and power factor.

$$\text{3 Phase Power (watts)} = \sqrt{3} \times V_{LINE} \times I_{LINE} \times \text{Power Factor}$$

If a transformer has two different voltage levels and the power-in equals the power-out, the current must also be transformed. Transformer current is inversely proportional to the voltage. Therefore, the high-voltage side of the transformer has less current flowing through its winding and more current will flow through the low-voltage windings. If the high-side current of our example transformer was 35A, the low-side would be rated for 105A.

$$\left(\text{Low Side Current} = 35A \times 3(x:1) = 105A \qquad \right):$$

$$\text{Transformer Ratio}(x:1) = \frac{\text{Low Side Current}}{\text{High Side Current}} \quad \textbf{OR}$$

$$\text{High Side Current} = \frac{\text{Low Side Current}}{\text{Transformer Ratio}(x:1)} \quad \textbf{OR}$$

$$\text{Low Side Current} = \text{High Side Current} \times \text{Transformer Ratio}(x:1)$$

Figure 1-30: Transformer Current Ratio Calculation

Transformers can be connected wye, delta or with a combination of wye and delta windings. Remember that typical delta systems only have ratings for Line (line-to-line) voltages and currents. Wye systems have Phase (phase-to-neutral) and Line voltage and line current ratings. The transformer nameplate will display the line and phase (if applicable) voltages for each winding along with the apparent (VA) power rating of the transformer. The transformer ratio will always use the Line-Line voltages.

There are often several VA (E.g. 25/35/45/50 MVA) ratings listed and each VA rating corresponds to a different cooling stage. The smallest VA rating applies when the transformer is air cooled only. Each higher rating applies as another cooling method is applied. The following items might be true if four VA ratings are listed on a transformer's nameplate:

- Smallest VA = Air Cooled
- Next largest VA = Stage 1 Fans
- Next Largest VA = Stage 1 and Stage 2 Fans
- Largest VA = Stage 1 and Stage 2 Fans and Re-circulating Pump

There are two schools of thought to determine which rating is used. A conservative approach uses the minimum VA rating because this is the transformer rating if all cooling methods failed for whatever reason. The more common approach uses the largest VA rating because this is the actual transformer rating under normal conditions.

The most important information for transformer related relay testing includes:

- Rated voltage for all windings,
- Rated current for all windings,
- Winding connection (Delta or Wye),
- And VA.

The rated current isn't always shown on a transformer's nameplate but it is easily calculated using the following formula. Review the following figures for examples of the most typical transformer applications and calculations that apply to three-phase balanced systems. Always remember to use only line voltages in calculations and the following unit modifiers apply.

V, A, VA	= Listed Value x 1 (e.g. 2A = 2A)
kV, kA, kVA	= Listed Value x 1,000 (e.g. 5kV = 5,000 V)
MVA, MW, MVARS	= Listed Value x 1,000,000 (e.g. 10MVA = 10,000,000 VA)

$$VA = \sqrt{3} \times V_{LINE} \times I_{LINE} \quad OR \quad I_{LINE} = \frac{VA}{\sqrt{3} \times V_{LINE}} \quad OR \quad V_{LINE} = \frac{VA}{\sqrt{3} \times I_{LINE}}$$

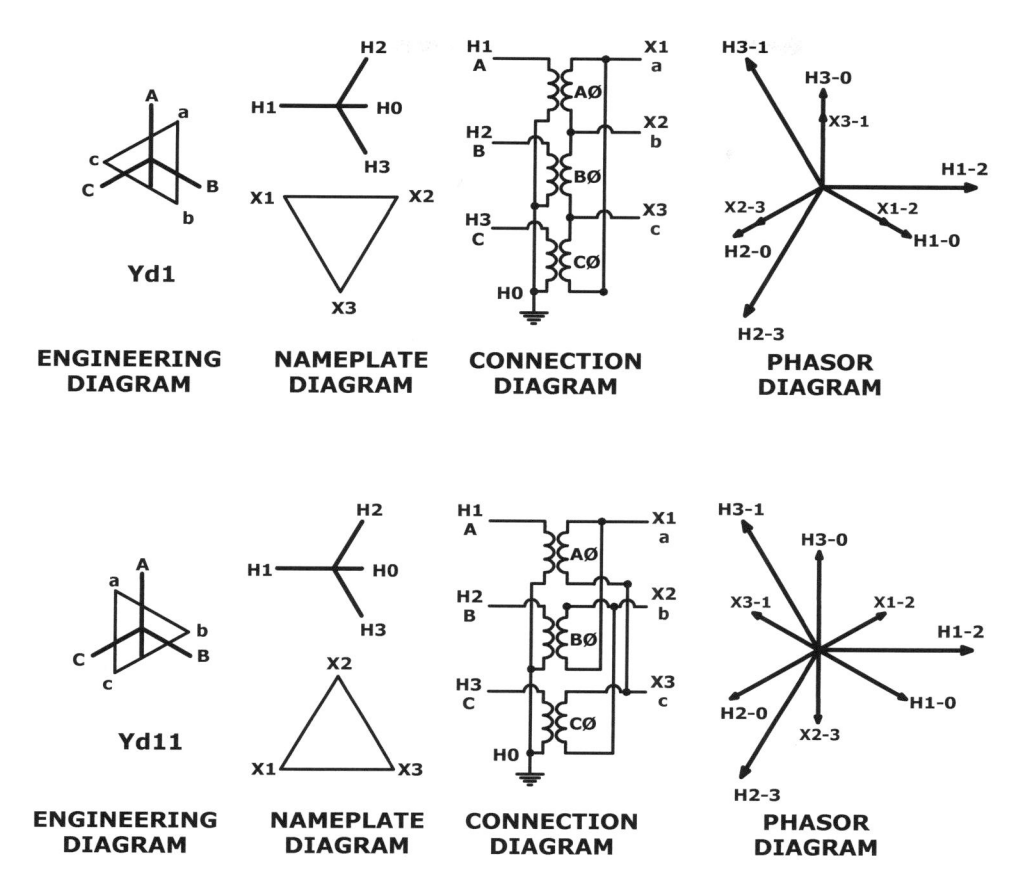

Figure 1-31: Wye-Delta Transformer Connections

VA = 1 MVA

Voltage: 4160/2400 Y : 600V

$$I_{LINE(High)} = \frac{VA}{\sqrt{3} \times V_{LINE}} = \frac{1,000kVA}{\sqrt{3} \times 4.16kV} = 138.79A$$

$$I_{LINE(Low)} = \frac{VA}{\sqrt{3} \times V_{LINE}} = \frac{1,000,000VA}{\sqrt{3} \times 600V} = 962.28A$$

$$\text{Transformer Ratio (x:1)} = \frac{\text{High Side Voltage}}{\text{Low Side Voltage}} = \frac{4160V}{600V} = 6.9333 : 1$$

$$\text{Transformer Ratio (x:1)} = \frac{\text{Low Side Current}}{\text{High Side Current}} = \frac{962.28A}{138.79A} = 6.9333 : 1$$

Figure 1-32: Wye-Delta Transformer Calculations

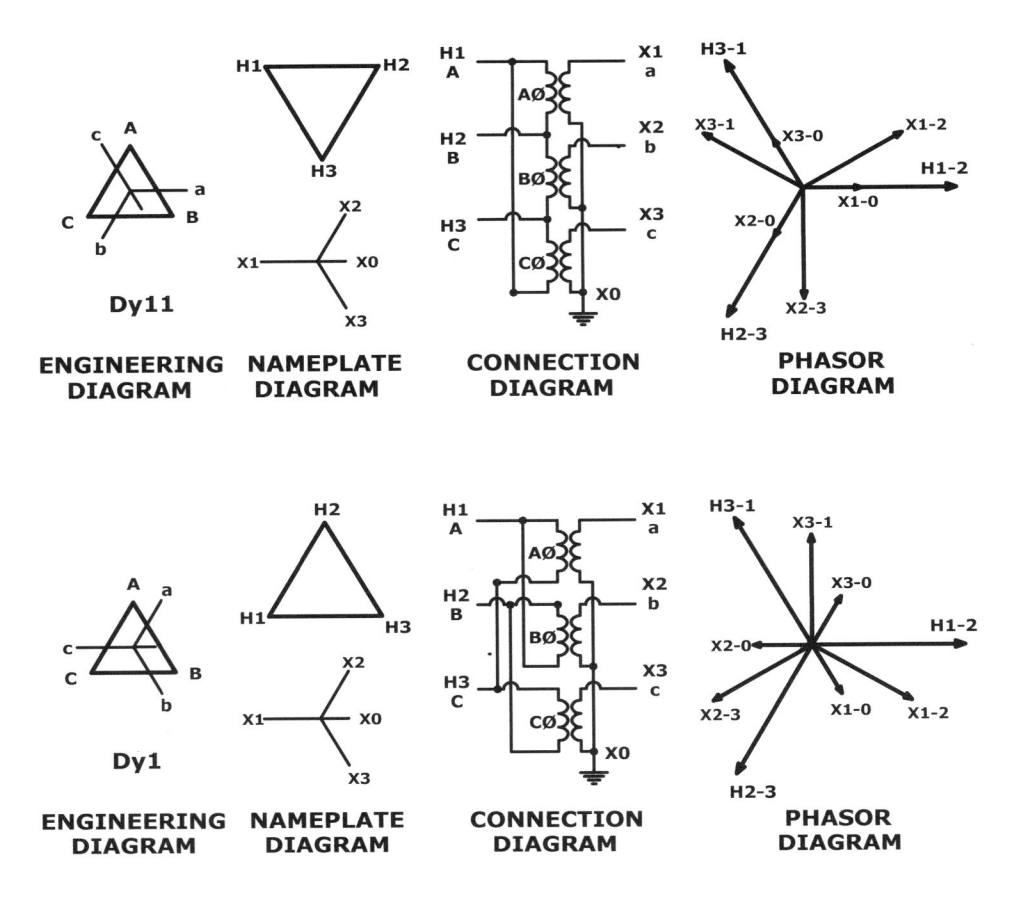

Figure 1-33: Delta-Wye Transformer Connections

VA = 1 MVA

Voltage: 25kV : 480V/247 V

$$I_{LINE(High)} = \frac{VA}{\sqrt{3} \times V_{LINE}} = \frac{1,000kVA}{\sqrt{3} \times 25kV} = 23.09A$$

$$I_{LINE(Low)} = \frac{VA}{\sqrt{3} \times V_{LINE}} = \frac{1,000kVA}{\sqrt{3} \times 480V} = 1.203kA$$

$$\text{Transformer Ratio (x:1)} = \frac{\text{High Side Voltage}}{\text{Low Side Voltage}} = \frac{25,000V}{480V} = 52.08:1$$

$$\text{Transformer Ratio(x:1)} = \frac{\text{Low Side Current}}{\text{High Side Current}} = \frac{1,203A}{23.09A} = 52.08:1$$

Figure 1-34: Delta-Wye Transformer Calculations

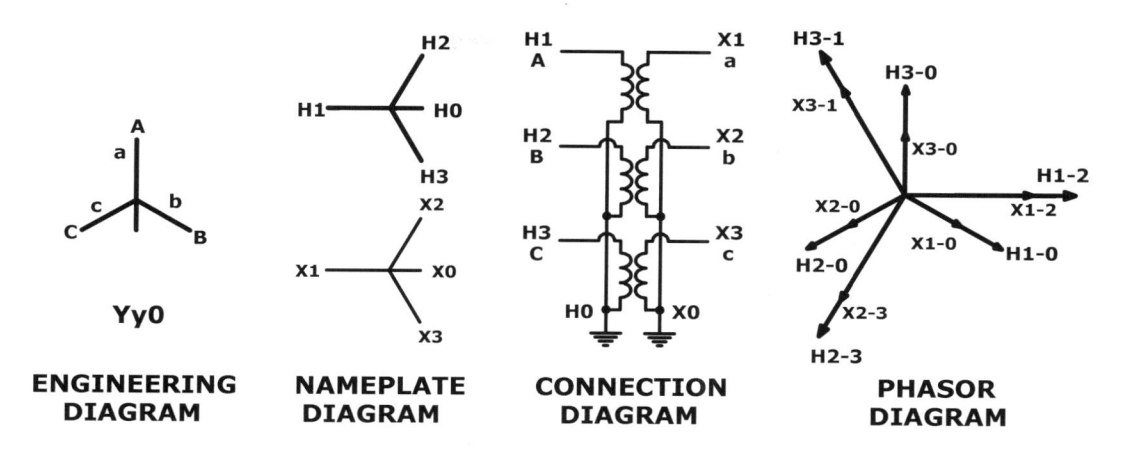

Figure 1-35: Wye-Wye Transformer Connections

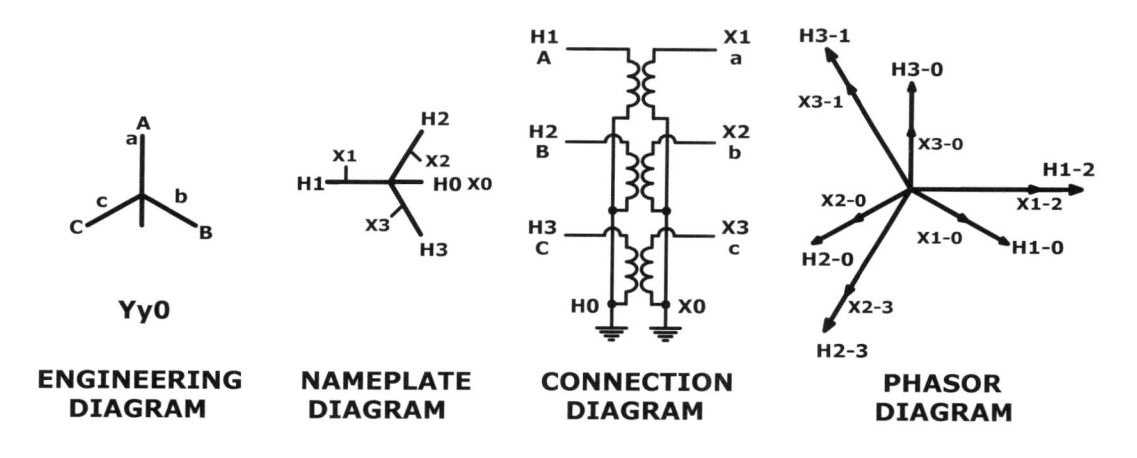

Figure 1-36: Auto-Transformer Connections

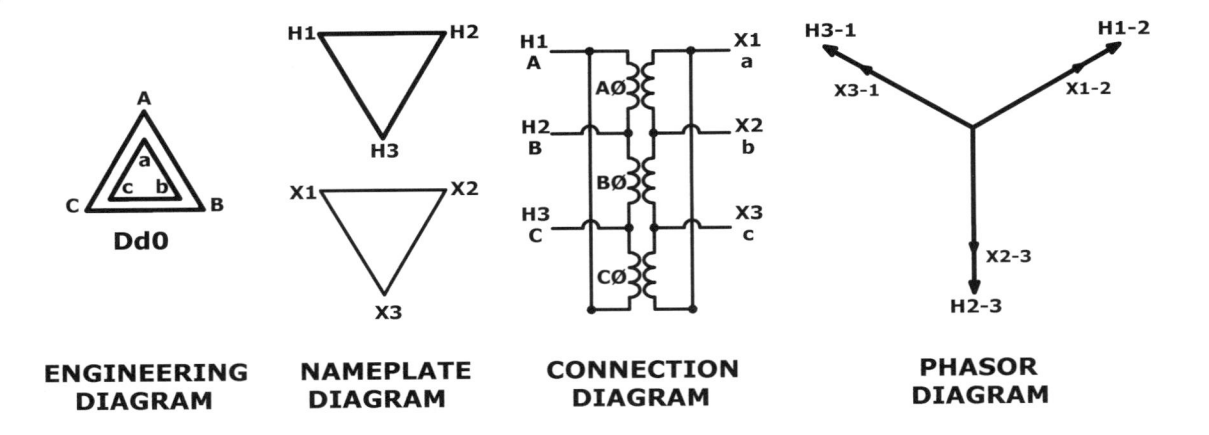

ENGINEERING DIAGRAM **NAMEPLATE DIAGRAM** **CONNECTION DIAGRAM** **PHASOR DIAGRAM**

Figure 1-37: Delta-Delta Transformer Connections

VA = 30 MVA

Voltage: 115kV : 13.8 V

$$I_{LINE(High)} = \frac{VA}{\sqrt{3} \times V_{LINE}} = \frac{30,000kVA}{\sqrt{3} \times 115kV} = 150.62A$$

$$I_{LINE(Low)} = \frac{VA}{\sqrt{3} \times V_{LINE}} = \frac{30MVA}{\sqrt{3} \times 13.8kV} = 1.255kA$$

$$\text{Transformer Ratio (x:1)} = \frac{\text{High Side Voltage}}{\text{Low Side Voltage}} = \frac{115kV}{13.8kV} = 8.33 : 1$$

$$\text{Transformer Ratio(x:1)} = \frac{\text{Low Side Current}}{\text{High Side Current}} = \frac{1,255.15A}{150.62A} = 8.333 : 1$$

Figure 1-38: Wye-Wye and Delta-Delta Transformer Calculations

3. Instrument Transformers

Instrument transformers are specially designed to transform high system voltages and currents into smaller voltages and currents for remote metering and protection. Instrument transformers can be extremely accurate to allow the relay to protect the electrical system and still be isolated from dangerously high currents and voltages. The following sections detail the different types of instrument transformers and typical connections.

A) Current Transformers (CTs)

As the name suggests, current transformers transform high currents to a smaller value. CT construction can be similar to the transformers explained previously but are specially designed for current transformations. There are three important values on the CT nameplate:

i) Ratio

The CT ratio determines the ratio between the primary and secondary current. The CT primary (High Side) should be rated at least 125% higher than the full load rating of the downstream or transmission equipment to maintain accuracy throughout the full range of operation. The CT secondary (Low Side) nominal rating is typically 5A in North America and 1A elsewhere. If secondary has a different rating…stop and make sure the correct information is presented to you and/or the correct equipment is installed.

CT ratios are typically displayed as a primary and secondary nominal current. (2000:5 A) The CT ratio is determined by dividing the primary value by the secondary rating.

Once you know the CT ratio, convert between primary and secondary currents by plugging the appropriate number into the formulas in Figure 1-39.

$$CT\ Ratio = \frac{CT\ Primary}{CT\ Secondary} \quad OR \quad CT\ Secondary = \frac{CT\ Primary}{CT\ Ratio} \quad OR$$

$$CT\ Primary = CT\ Secondary \times CT\ Ratio$$

$$\frac{CT\ Secondary}{CT\ Primary} = \frac{CT\ Rated\ Secondary}{CT\ Rated\ Primary}$$

$$CT\ Primary = \frac{CT\ Rated\ Primary \times CT\ Secondary}{CT\ Rated\ Secondary}$$

Figure 1-39: CT Ratio Calculations

ii) Polarity

CT polarity is extremely important for protective devices that require a specific phase angle or direction of current in order to operate. However, it is also important to know how the CT is physically installed. If a CT has been installed in the reverse direction, it is easily fixed with a quick change of the secondary wiring even if the installed polarity marks do not match the specifications.

The CTs on electrical drawings should have a polarity mark or labeling standard to relate primary current direction to secondary current direction. Figure 1-40 displays typical polarity marks. Many people get confused when the polarity markings are not where they expect. The best way to determine current flow is to repeat the following mantra "current in equals current out." Trace the direction of normal primary current flow and note which side of the CT it enters. The current will flow out the secondary terminal corresponding with the primary terminal regardless of polarity marks. If the current flows into H2, it will flow out X2. The last CT circuit in the example shows this.

Arrows indicate direction of current flow

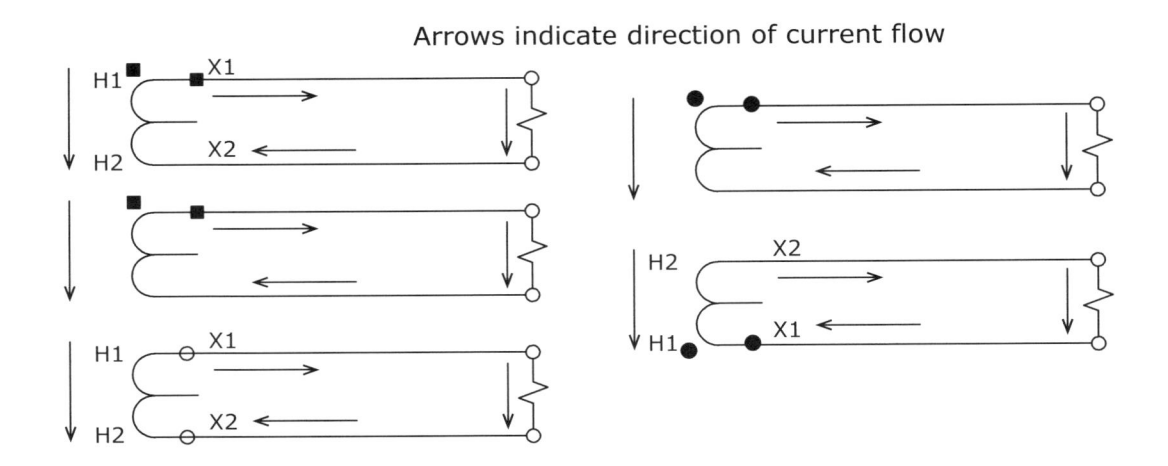

Figure 1-40: CT Polarity Connections

iii) Burden / Accuracy Class

There are metering class, protection class, or dual classified CTs. Metering class CTs are designed for maximum accuracy within the typical range of a CT secondary (0-5A). Protection class CTs are accurate within a wide range of current (up to 20x its rating). Obviously, protection CTs should be used for relay applications to ensure operation over a wide range of fault currents.

The CT burden is the AC resistance of the CT's secondary circuit. The CT burden must be less than the CTs rating or the CT may saturate and fail to operate at higher currents. In the past, current coils of electromechanical relays had relatively high resistance values and the CT burden was a very hot topic. Today's relays have very low burden coils and CT burden is not the priority that it once was. The CT burden is calculated from nameplate data as shown in Figure 1-41.

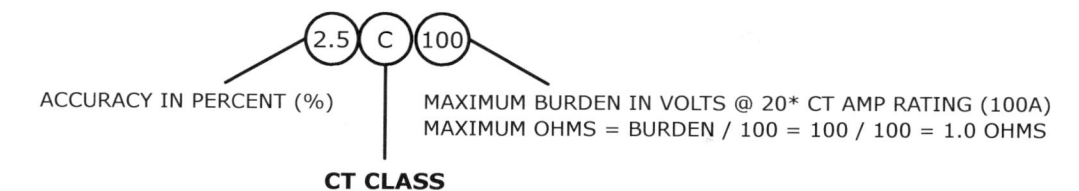

ACCURACY IN PERCENT (%)   MAXIMUM BURDEN IN VOLTS @ 20* CT AMP RATING (100A)
MAXIMUM OHMS = BURDEN / 100 = 100 / 100 = 1.0 OHMS

CT CLASS
"C" = DONUT CT WITH CALCULATED % ACCURACY
"T" = % ACCURACY TESTED IN FACTORY
"H" = ACCURACY APPLIES UP TO 20x CT NOMINAL AMPS
"L" = ACCURACY CAN VARY UP TO 4x RATING DEPENDING ON LOAD

Figure 1-41: CT Accuracy Class

B) Potential or Voltage Transformers (PTs or VTs)

PT construction is very similar to the transformers described in the "Transformers" section of this chapter and are designed to transform high system voltages into smaller voltages. The smaller voltages are measured by the relay to accurately calculate the primary voltages. There are two important aspects of PTs when applied to protective relaying.

i) PT Ratio

The PT ratio is pre-defined when it is built and is based on the transformer turns ratio. The ratio is defined on the transformer nameplate by high and low side voltages. For example, a 35:1 PT ratio could be listed on the transformer nameplate as 4.2kV:120V.

$$\left(\text{PT Ratio} \, (x:1) = \frac{4,200 \, \text{V}}{120 \, \text{V}} = 35:1 \right)$$

$$\text{PT Ratio} \, (x:1) = \frac{\text{High Side Voltage}}{\text{Low Side Voltage}} \quad \text{OR}$$

$$\text{High Side Voltage} = \text{Transformer Ratio} \, (x:1) \times \text{Low Side Voltage} \quad \text{OR}$$

$$\text{Low Side Voltage} = \frac{\text{High Side Voltage}}{\text{PT Ratio} \, (x:1)}$$

Figure 1-42: PT Voltage Ratio Calculation

ii) PT Connection

There are two most common PT combinations on three-phase, three-wire systems as shown in Figure 1-43. The vector diagrams are identical between PT configurations and the Open Delta is more common because it only requires two PTs, thereby dropping the cost by one-third.

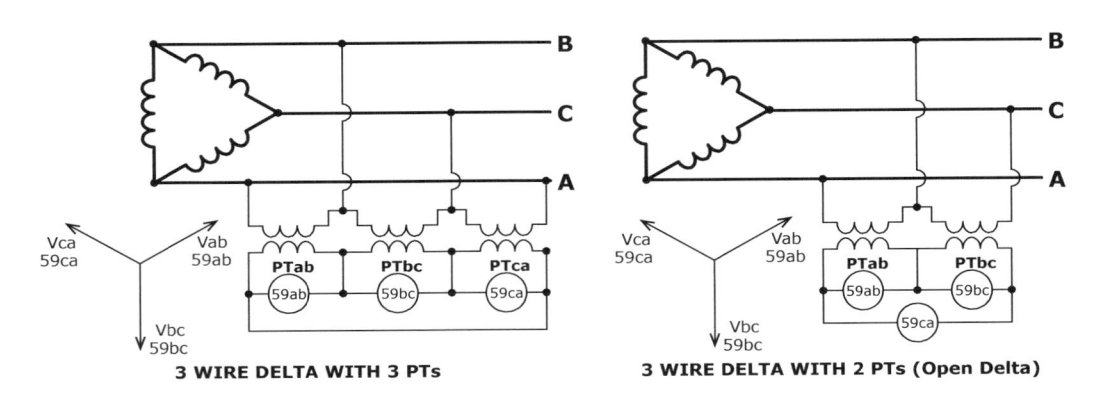

3 WIRE DELTA WITH 3 PTs   3 WIRE DELTA WITH 2 PTs (Open Delta)

Figure 1-43: 3-Wire Delta PT Configurations

The two most common PT configurations for four-wire, Wye systems are shown in Figure 1-44. The line-to-line voltage magnitudes and angles between the two configurations are identical and Open-Delta configurations commonly used if line-to-neutral voltages are not required by the relay or application.

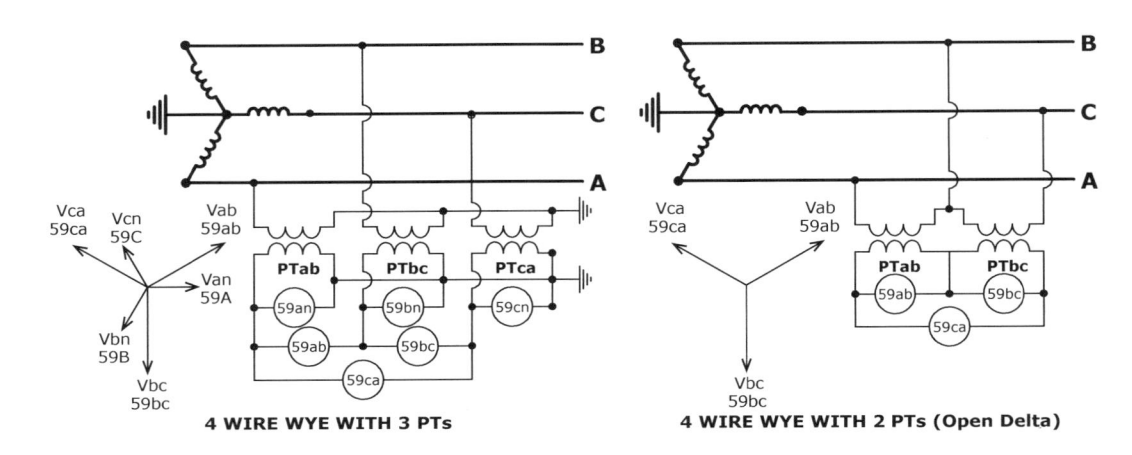

4 WIRE WYE WITH 3 PTs   4 WIRE WYE WITH 2 PTs (Open Delta)

Figure 1-44: 4-Wire Wye PT Configurations

Confusing line-to-ground and line-to-line setting values is the most common mistake related to voltages and relays. Review the manufacturer's specifications to make sure the correct voltage levels are expected and set. Use the chart in Figure 1-45 for the most common PT secondary voltages. Calculate nominal PT secondary voltages by dividing the rated primary voltage by the PT ratio. Never mix Line-to-Line and Line-to-Ground voltages when performing PT ratio calculations. For example, medium voltage PT ratios are typically rated using line voltages (eg. 4200V / 120V = PT ratio 35:1) but substation PTs are typically rated using phase-to-neutral voltage ratios and secondary nominal voltages (E.g. 1200:1 PT ratio @ 67V = 80,400V / 67V Line-to-Neutral or 138,000V / 115V Line-to-Line)

	CONNECTIONS	LINE-GROUND NOMINAL	LINE-LINE NOMINAL
3-Wire Delta 2 PTs	V_{AB}, V_{BC}, V_{CA} V_{AB}, V_{CB}	N/A	120V or 115V
4-Wire Wye 3 PTs	V_{AB}, V_{BC}, V_{CA} V_{AN}, V_{BN}, V_{CN}	69.28V or 66.40V or 67.0V	120V or 115V or 116.04V
4-Wire Wye 2 PTs	V_{AB}, V_{BC}, V_{CA}	N/A	120V or 115V

Figure 1-45: Nominal PT Voltages

4. Fault Types

The electrical system can be affected by many problems, but protective relays are primarily designed to protect the electrical system and/or equipment from fault situations. The three-phase electrical system typically has three phases that are connected together by high-impedance loads. A low impedance connection between phases and/or ground can cause extremely high fault currents, arc flashes, heat, and explosive forces, and catastrophic damage to equipment and personnel. A low impedance connection between phases and/or ground could be caused by a tree falling onto a transmission line, foreign material falling across a live bus, very humid conditions with insufficient clearance between phases, or failing insulation.

There are an unlimited number of scenarios that can cause an electrical fault which are typically categorized into four fault types. It is very common for an electrical fault to start in one category and cascade into another if the problem is not cleared in time. The most common fault types are:

A) Three-Phase Faults

Three-phase (3Ø or 3P) Faults occur when all three phases are connected together with low impedance. We typically create balanced 3Ø Faults when creating simulations which are the equivalent of leaving safety grounds on a system and then energizing that system.

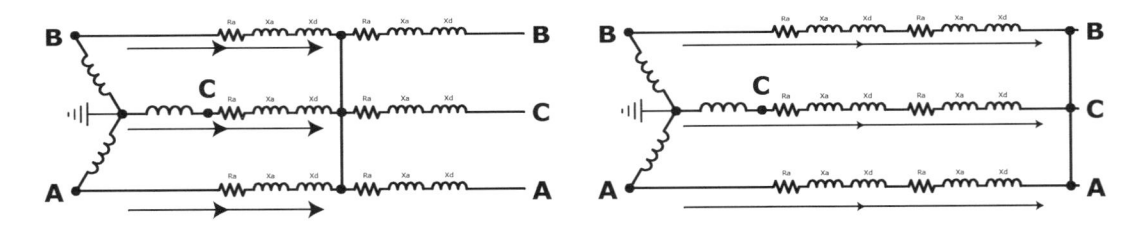

Figure 1-46: Three-Phase Fault 3-Line Drawing

A three-phase fault causes all three currents to increase simultaneously with equal magnitudes. The magnitude of *fault current* will depend on the impedance and location of the fault, as well as the strength of the electrical system. The safety grounds connected to the grid in our example will cause much higher fault currents than the same safety grounds connected to an isolated generator. Similarly, a fault closer to the source (Figure 1-46 – Left Side) will produce more fault current than a fault at the end of the line (Figure 1-46 – Right Side) because there will be more impedance between the source and the fault in the second situation. The current will lag the voltage by some value usually determined by the voltage class of the system.

All three voltages will decrease simultaneously with equal magnitudes. The magnitude of *fault voltage* will depend on the impedance and location of the fault, as well as the strength of the electrical system. The voltage drop will not be as significant if the safety grounds are connected to the grid compared to the huge voltage drop that will be observed in an isolated system which could cause the voltage to drop to zero. Similarly, a fault closer to the source will create a larger voltage drop than a fault at the end of the line because there will be more impedance between the source and the fault.

The following phasor diagrams demonstrate the difference between a "normal" system and the same system with a three-phase fault. Notice that all phasors are still 120° apart, the voltages stayed at the same at the same phase angles, and the current magnitudes increased with a greater lag from the voltages.

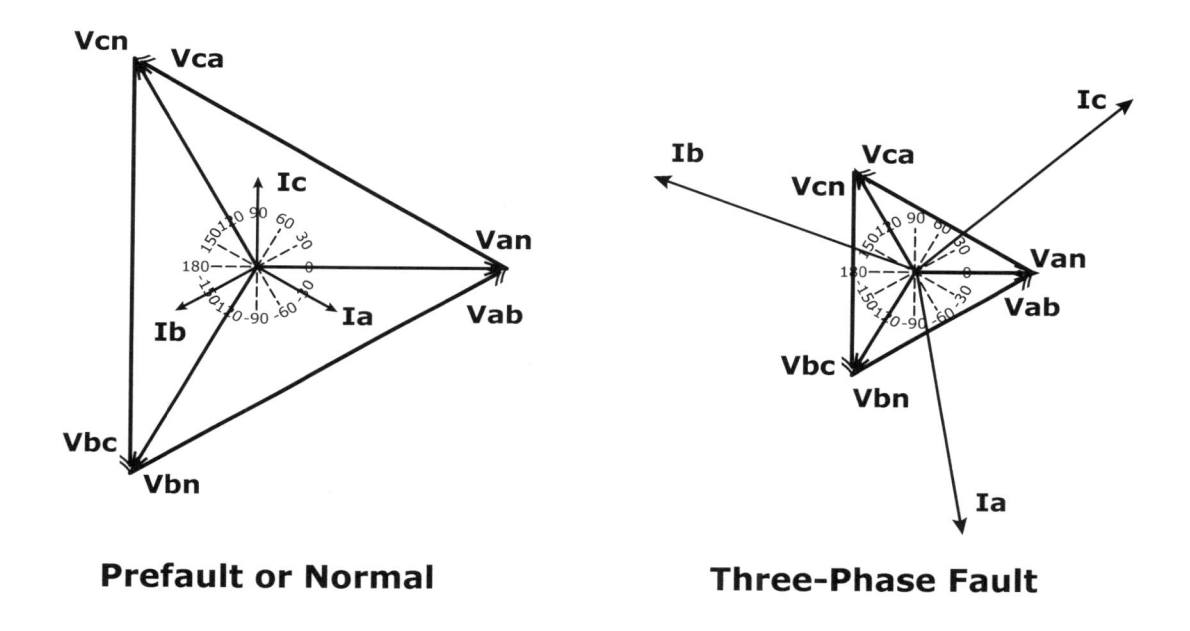

Prefault or Normal   **Three-Phase Fault**

Figure 1-47: Three-Phase Fault Phasor Diagram

It is always a good idea to simulate conditions as close to a true fault when performing tests and a 3Ø fault simulation should have the following characteristics:

- **Prefault**: nominal: 3Ø voltage magnitudes with 120° between phases in the normal phase rotation.
- **Fault Voltage**: Reduced, but identical, voltage magnitudes on all three phases with no change in phase angles from the prefault condition.
- **Fault Current**: Larger than nominal current on all three phases with identical magnitudes and 120° between phases with the normal phase rotation. The current should lag the voltage by 60 to 90°.

B) Phase-to-Phase Faults

Phase-to-phase (P-P or Ø-Ø) faults occur when two phases are connected together with low impedance. A Ø-Ø fault can occur when a bird flies between two transmission conductors and its wing-tips touch both conductors simultaneously. Phase-to-phase faults are described by the phases that are affected by the low-impedance connection. The B-C fault in Figure 1-48 is a Ø-Ø fault that connects B-phase and C-phase.

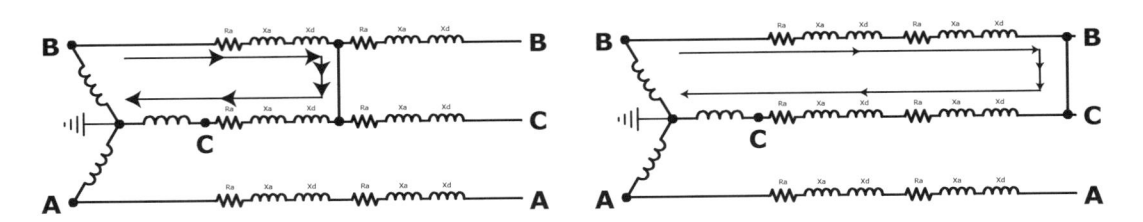

Figure 1-48: Phase-Phase Fault 3-Line Drawing

The magnitude of *fault current* will depend on the impedance and location of the fault, as well as the strength of the electrical system. A fault closer to the source will produce more fault current than a fault at the end of the line because there will be more impedance between the source and the fault in the second situation. The *fault current* will lag the *fault voltage* by some value usually determined by the voltage class of the system.

If you follow the flow of current in Figure 1-48, you should notice that the current flows from the B-phase source into the fault, and then returns to the source via C-phase. Basic electrical theory states that the current flowing in a circuit must be equal, so the B-phase and C-phase currents must have the same magnitudes. However, relays monitor current leaving the source so the relay will see this fault as two equal currents with opposite polarity. Therefore, when we simulate a P-P fault, the currents injected into a relay must have the same magnitudes and be 180° apart from each other. The current flowing through the actual fault will be equal to 2x the injected currents.

Injectected Current = Ib = Ic

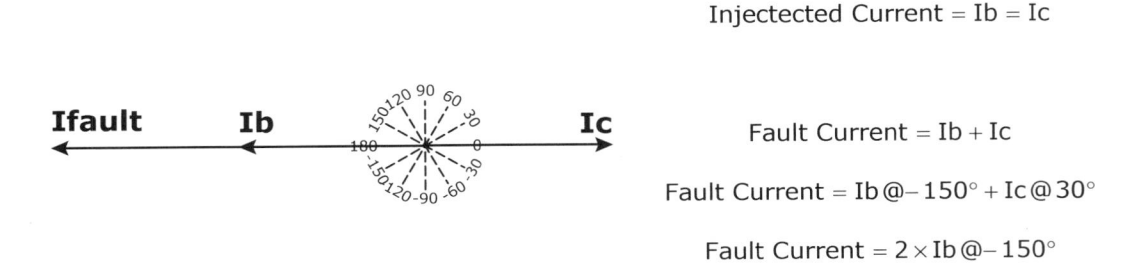

Fault Current = Ib + Ic

Fault Current = Ib @ $-150°$ + Ic @ $30°$

Fault Current = $2 \times$ Ib @ $-150°$

Figure 1-49: Fault Current vs. Injected Current

The effect of a P-P fault on the faulted voltages is even more complex. The faulted voltages will have equal magnitudes because the impedance between the source and the fault on each faulted phase should be equal. A fault closer to the source will create a larger voltage drop than a fault at the end of the line because there will be more impedance between the source and the fault.

The faulted voltage angles are also affected because the ratio of reactance and resistance in the circuit changes when the fault is introduced. There is a lot of information to consider when creating the correct voltage magnitudes and angles and we discuss the calculations in detail in *Chapter 15: Line Distance (21) Element Testing*. For our purposes in this chapter, it is important that you be able to recognize a Ø-Ø fault as the following phasor diagrams demonstrate. Notice that the faulted voltages have collapsed and come together to change the voltage triangle from an equiangular/equilateral triangle (three equal magnitudes and angles) in prefault to an acute, isosceles triangle (two equal magnitudes, two equal angles, and all angles are less than 90°).

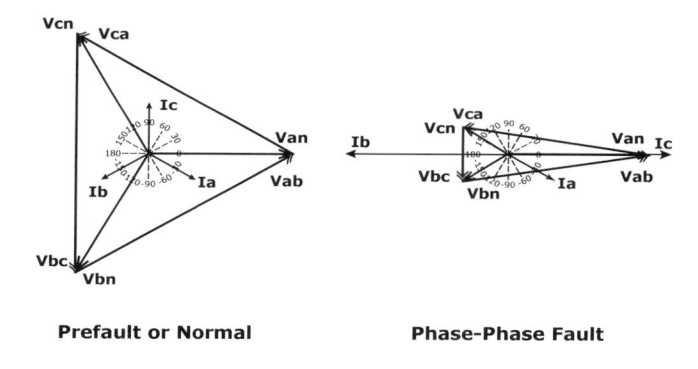

Prefault or Normal   Phase-Phase Fault

Figure 1-50: Phase-Phase Fault

It is always a good idea to simulate conditions as close to a true fault when performing tests and a Ø-Ø fault simulation should have the following characteristics:

1. Prefault: nominal, 3Ø voltage magnitudes with 120° between phases in the normal phase rotation.

2. Fault Voltage: Reduced, but identical, voltage magnitudes on both phases affected by the fault. The phase angle between the faulted phases should be less than 120°.

3. The unaffected phase voltage should be identical to the prefault condition.

4. Fault Current: Increased, but identical, current magnitudes on both phases affected by the fault. The phase angle between the faulted phases should be 180°.

5. The unaffected phase current does not change between prefault and fault.

C) Phase-to-Ground Faults

Phase-to-ground (P-G, P-N, Ø-G or Ø-N), or single-phase faults occur when any one phase is connected to the ground or neutral conductor with low impedance. A Ø-N fault can occur when a tree falls on a transmission line. Phase-to-ground faults are described by the phase that is affected by the low-impedance connection. The A-N fault in Figure 1-51 is a Ø-N fault that connects A-phase to the neutral.

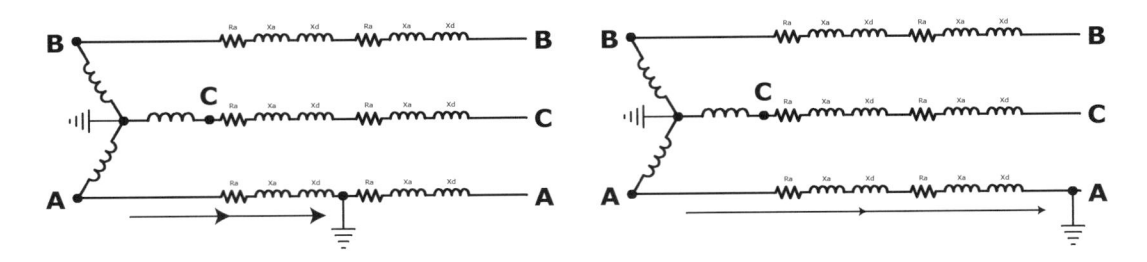

Figure 1-51: Phase-Ground Fault 3-Line Drawing

The magnitude of *fault current* will depend on the impedance and location of the fault, as well as the strength of the electrical system. A fault closer to the source will produce more fault current than a fault at the end of the line because there will be more impedance between the source and the fault in the second situation. The *fault current* will lag the *fault voltage* by some value usually determined by the voltage class of the system.

The fault only affects A-phase voltages and currents, so the B-phase and C-phase currents are nearly identical to the prefault values. The A-phase current will increase and lag the voltage at some value between 60 and 90°.

The B-phase and C-phase voltages will also be unaffected by the fault and they will have the same magnitude and phase angle during the fault as existed in prefault. The A-phase voltage magnitude will decrease, but the phase angle will not change between prefault and fault.

A Ø-N fault example is demonstrated in Figure 1-52. Notice that the faulted voltage has collapsed to change the voltage triangle from an equiangular/equilateral triangle (three equal magnitudes and angles) in prefault to an isosceles triangle.

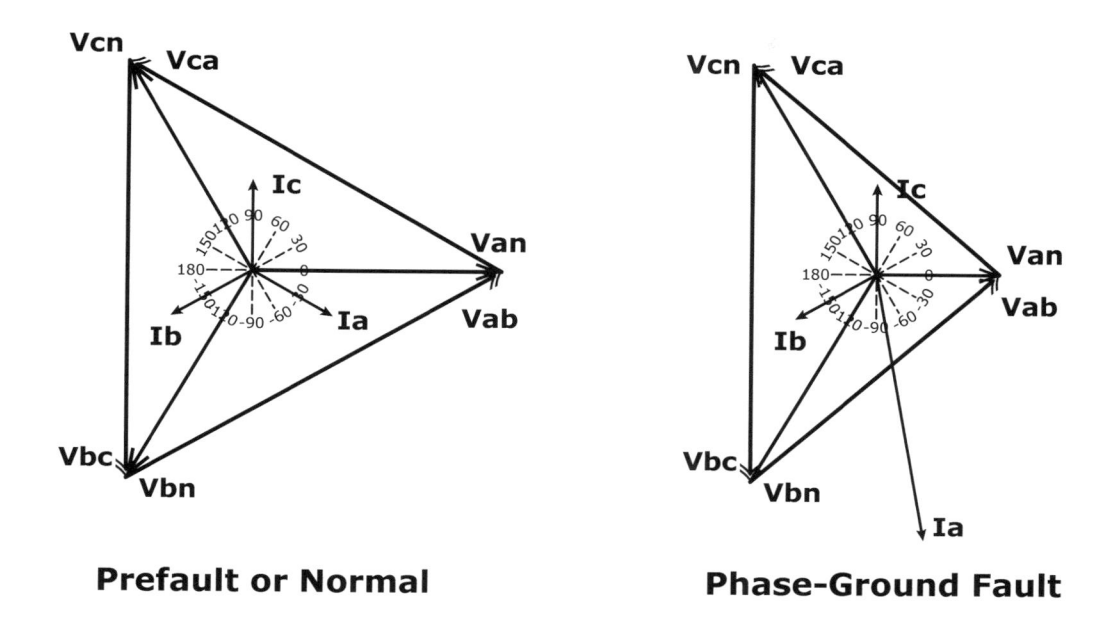

Figure 1-52: Phase-Ground Fault

It is always a good idea to simulate conditions as close to a true fault when performing tests and a Ø-N fault simulation should have the following characteristics:

1. Prefault: nominal, 3Ø voltage magnitudes with 120° between phases in the normal phase rotation.

2. Fault Voltage: Reduced voltage the phase affected by the fault. The voltage angle is identical to prefault.

3. The unaffected phase voltages should be identical to the prefault condition.

4. Fault Current: Increased magnitude on the affected phase. The phase angle should lag the phase voltage by a value between 60 and 90°.

5. The unaffected phase currents do not change between prefault and fault.

D) Phase-to-Phase-to-Ground Faults

Phase-to-phase-to-ground (P-P-G, P-P-N, Ø-Ø-G or Ø-Ø-N), or two-phase-to-ground faults occur when any two phases are connected to the ground or neutral conductor with low impedance. A Ø-N fault can occur when a tree falls on two phases of a transmission line. Two phase-to-ground faults are described by the phases that are affected by the low-impedance connection. The C-A-G fault in Figure 1-53 is a Ø-Ø-N fault that connects C and A-phases to Ground.

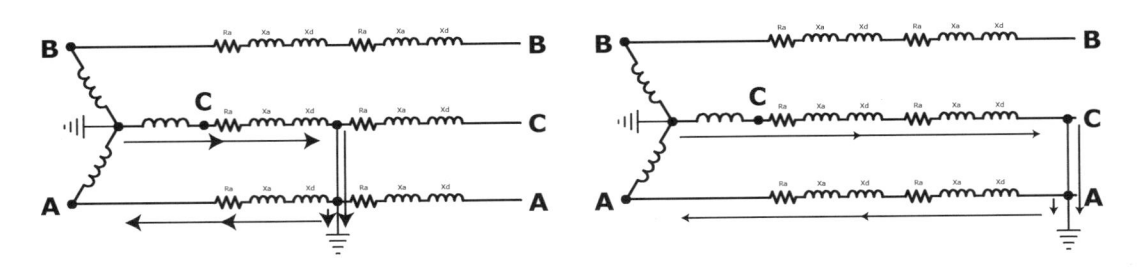

Figure 1-53: Phase-Phase-Ground Fault 3-Line Drawing

P-P-G ground are very complex faults that are very difficult to reproduce without modeling software to calculate all of the fault quantities involved which is beyond the scope of this book. However, you should be able to see the characteristics of a P-P-G fault as shown in Figure 1-54.

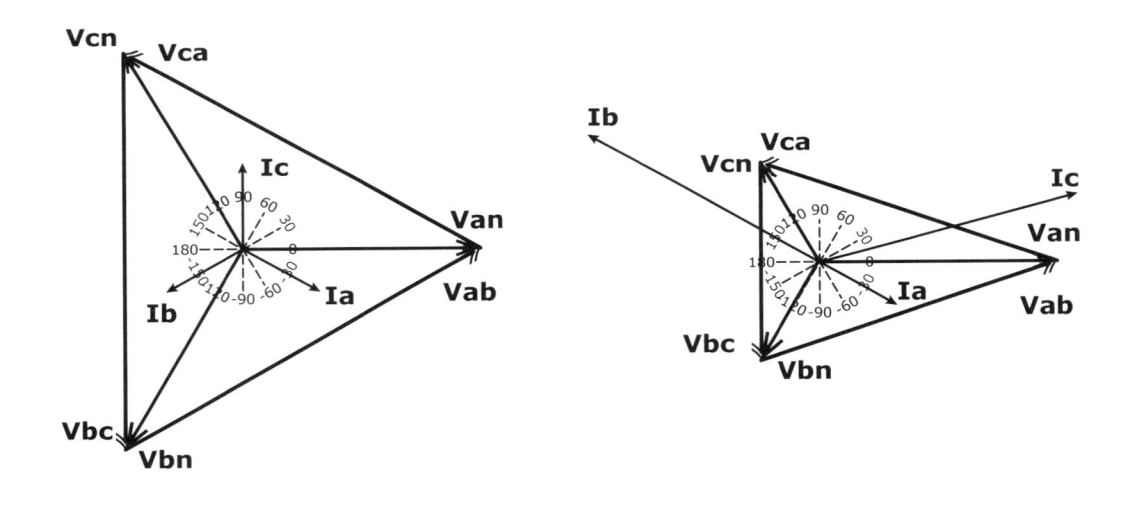

Prefault or Normal   **Phase-Phase-Ground Fault**

Figure 1-54: Phase-Phase-Ground Fault

A Ø-Ø-N fault simulation has the following characteristics:

1. Prefault: nominal, 3Ø voltage magnitudes with 120° between phases in the normal phase rotation.

2. Fault Voltage: Reduced, but identical, voltage magnitudes on both phases affected by the fault. The faulted phase voltage angles are nearly identical to prefault angles.

3. The unaffected phase voltage should be identical to the prefault condition.

4. Fault Current: The effected phase currents magnitudes will increase. The phase current angles lag the phase voltages but appear to be random and not the orderly 180° apart as found in a P-P fault.

5. The unaffected phase current does not change between prefault and fault.

5. Grounding

There are many ways to ground a system or detect ground faults. A quick review of the various grounding methods follow.

A) Solid Ground

Systems that are solidly grounded have their star (common) point grounded through a conductor. This method can allow extremely high fault currents and the ground conductor must be correctly sized for the calculated fault current. Ground fault detection for these systems is normally performed by placing a CT inline or around the ground conductor to measure any current flowing through it as shown in Figure 1-55:

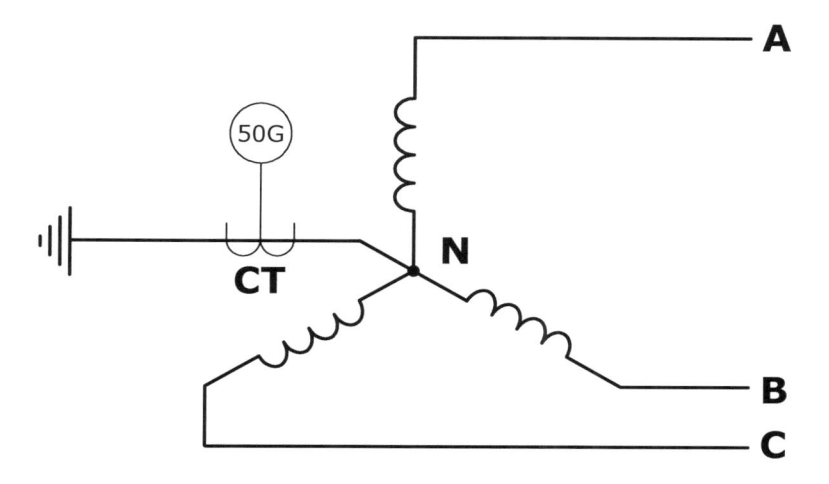

Figure 1-55: Solidly Grounded System

B) Resistor Ground

Resistive ground systems are used to control and limit the amount of possible ground current. This grounding method has a Neutral Grounding Resistor (NGR) connected between the star point and ground. The resistor limits the ground fault current depending on the line-to-ground system voltage and the NGR's resistance. Some NGRs can be rated for a maximum amount of time and any protective device must trip before this maximum time is reached or the NGR may be damaged. NGRs can also be rated for continuous duty and allow ground fault current to flow indefinitely.

Ground faults are normally detected in this system using two different methods. The first method is identical to the solid grounded system by placing a CT in series with the NGR to monitor ground current. The relay should be set to pickup at a value below the NGR rating, because that is the maximum allowable ground fault current.

The second method for ground fault detection monitors the voltage drop across the NGR that is created when current flows through the resistor. On low voltage systems, the voltage can be monitored directly. A Potential Transformer (PT or VT) is required on medium voltage systems to reduce the measured voltage to safe levels.

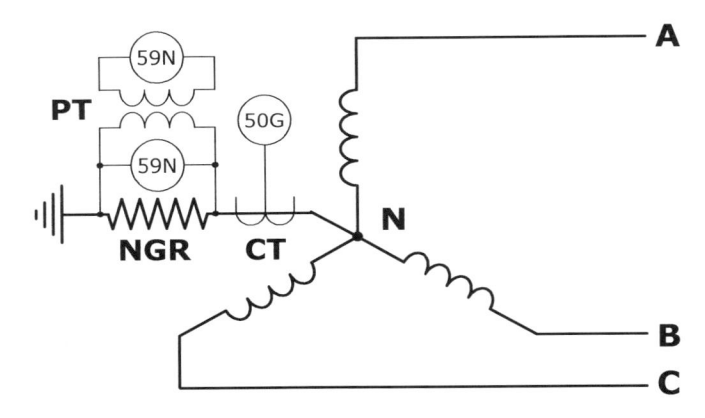

Figure 1-56: Resistive Grounded System

$$MaximumGroundFaultAmps = \frac{LinetoNeutralVoltage}{NGRResistance}$$

Figure 1-57: NGR Maximum Ground Fault Current Formula

$$MaximumGroundFaultVolts = \frac{LinetoNeutralVoltage \times PTSecondaryVolts}{PTPrimaryVolts}$$

Figure 1-58: NGR Maximum Ground Fault Voltage Formula

$$GroundFaultVolts = \frac{GroundFaultAmps \times NGRResistance \times PTSecondaryVolts}{PTPrimaryVolts}$$

Figure 1-59: NGR Ground Fault Voltage Formula

C) Impedance Grounded Systems

Higher voltage NGRs can be very expensive and impedance grounded systems use the principles of the resistive grounded system at a reduced cost. A transformer primary winding is connected in series between the neutral and ground with a resistor installed across the secondary winding. Thanks to the miracle of transformers, the small resistance on the secondary is an effectively larger resistance on the primary winding in direct relation to the transformer ratio. Ground fault detection is performed using a CT in series with the neutral grounding transformer or by monitoring the voltage on the secondary of the neutral grounding transformer.

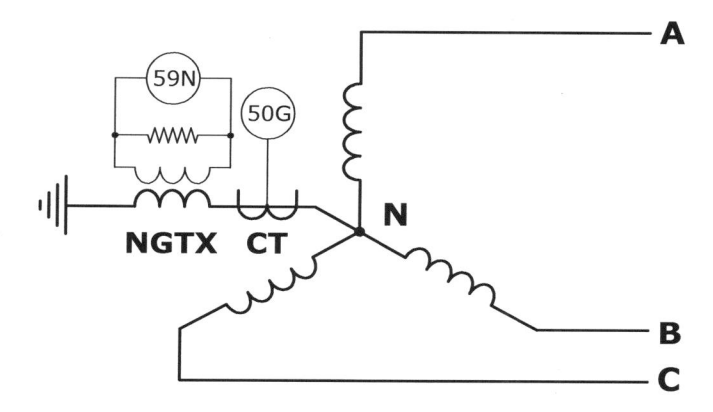

Figure 1-60: Neutral Grounding Transformer

$$\text{Neutral Grounding Impedance} = \frac{\text{Secondary Resistance} \times \text{NGTX Primary Volts}}{\text{NGTX Secondary Volts}}$$

Figure 1-61: NGTX Effective Resistance Formula

$$\text{Ground Fault Volts} = \frac{\text{Ground Fault Amps} \times \text{PT Primary Volts}}{\text{PT Secondary Volts}} \times \text{NGTX Resistance}$$

Figure 1-62: NGTX Ground Fault Voltage Formula

D) Ungrounded Systems

Ungrounded systems have no intentional connection to ground and can also be called 3-wire systems. These systems are unique because the first low-impedance connection to ground will not cause any problems to the electrical system and will simply create a phase-to-ground reference in the system. The other two phases now have a line-to-ground reference that is $\sqrt{3}$ larger than a similarly configured, traditional 4-wire system with an intentional ground connection. A second low-impedance connection to ground in the 3-wire system can be more dramatic because the line-to-ground voltage in this situation is larger than which typically causes larger fault currents.

Ground detection systems are implemented to detect the first line-to-ground connection in a 3-wire system so that it can be removed before a second line-to-ground connection occurs. There are several methods for detecting a ground in ungrounded systems and the most common method for medium voltage systems is the broken delta or open-corner delta connection. The primaries of three potential transformers (PTs) are connected in Wye configuration and the secondaries are connected in Delta with one open connection. A resistor and protective relay are placed across the open connection to monitor any unbalance condition that will be created by a grounded phase. If all three phase/line voltages are balanced, we can see that the voltage across the open delta is zero using vector addition.

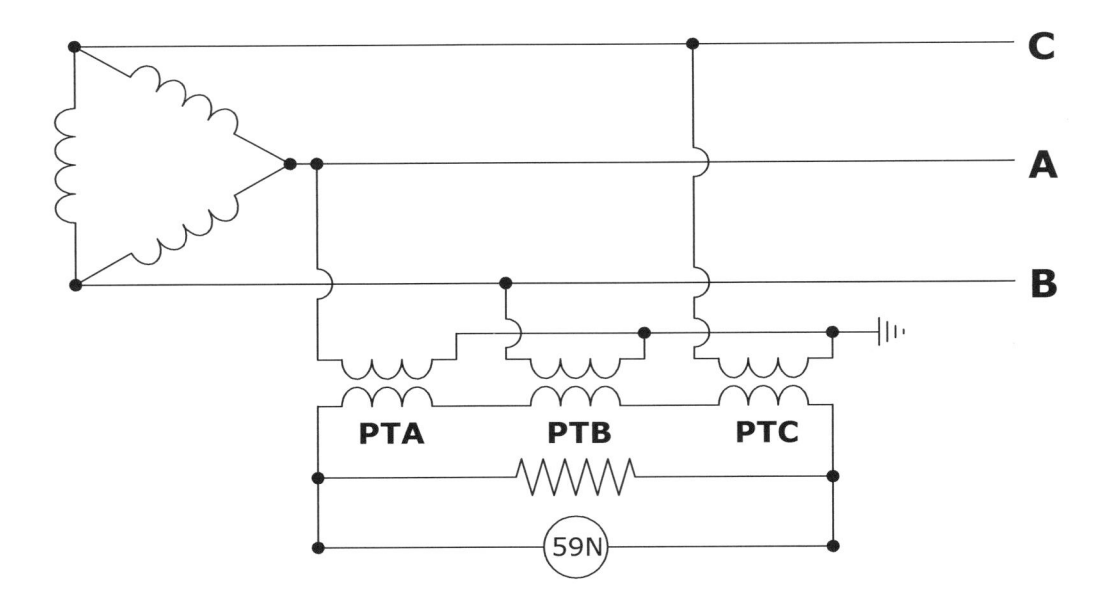

Figure 1-63: Open Corner Delta Ground Detector

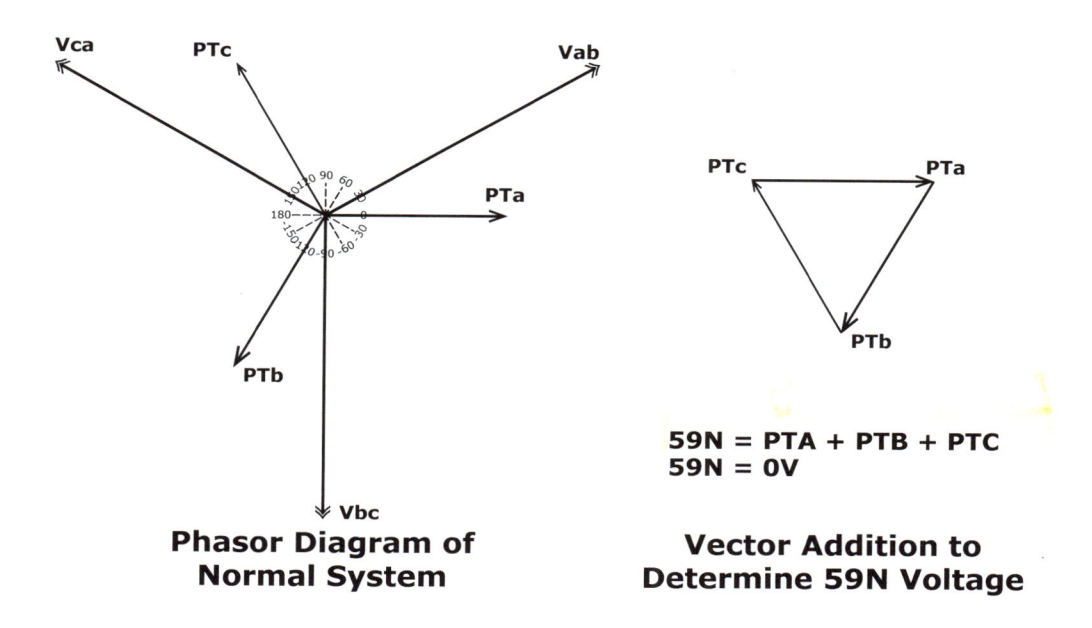

59N = PTA + PTB + PTC
59N = 0V

**Phasor Diagram of
Normal System**

**Vector Addition to
Determine 59N Voltage**

Figure 1-64: Open Corner Delta Normal Phasors

Now if one phase is grounded, the voltage will drop to zero and the vector addition shows us that the open corner will equal the missing phasor. Any unbalance voltage will be reflected in the open-corner delta voltage.

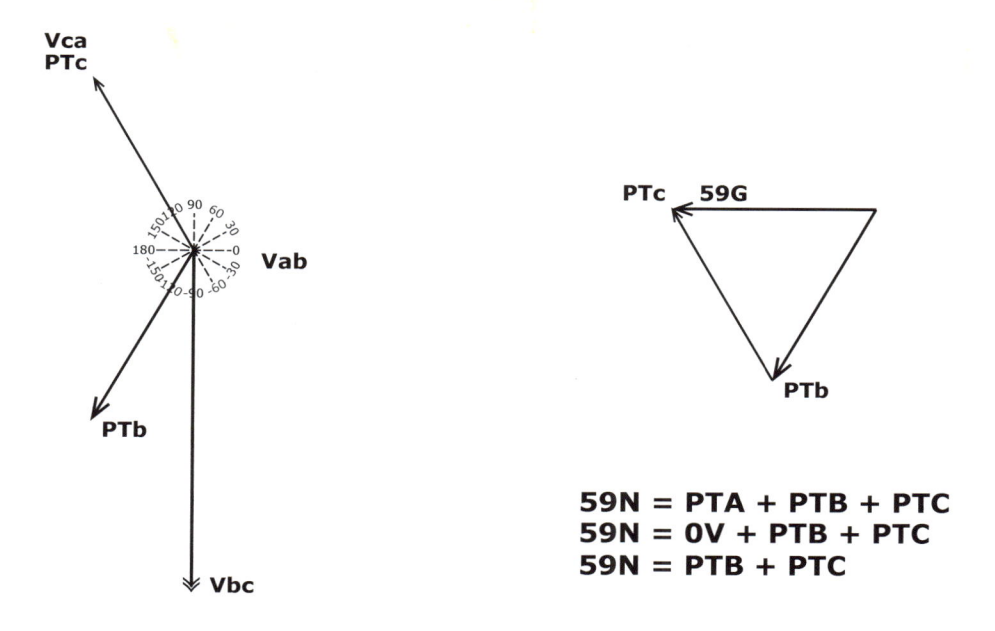

59N = PTA + PTB + PTC
59N = 0V + PTB + PTC
59N = PTB + PTC

Figure 1-65: Open Corner Delta with One Phase Grounded

6. Sequence Components

The vector addition examples covered in this book are very simple when compared to the variables that can occur during a real fault. Trying to use simple vector addition to determine the cause of a fault or even understand and predict fault characteristics can become a nightmare, especially when you realize that most of the electrical principles we use today were developed before computers or even calculators. Sequence components were introduced in 1918 by Dr. C. L. Fortescue to help simplify the analysis of unbalanced systems to better understand fault characteristics so that protective relays could better protect the electrical system.

Sequence components are a mathematical model of an electrical system to help understand electrical networks during faults. These theories were designed to interpret faults, so it makes sense that more advanced relays use sequence components to detect faults and many do. Sequence components cannot be measured directly. They must be calculated using measurements of individual phase magnitudes and angles.

A) Positive Sequence

A positive sequence system is a completely balanced three-phase electrical system with identical magnitudes between all three phases that are exactly 120° apart with the correct system rotation. This is an ideal and never happens in real life. All electrical systems have positive, negative, and zero sequence networks for currents and voltages. Positive sequence values are designated with the subscript 1 (E.g. V1, I1, Z1) and can be calculated using the formulas in Figure 1-66 where "a" rotates the value immediately following it by 120° and "a2" rotates the value immediately following it by 240°.

$$V_1 = \frac{V_a + aV_b + a^2V_c}{3} \qquad I_1 = \frac{I_a + aI_b + a^2I_c}{3} \qquad Z_1 = \frac{Z_a + aZ_b + a^2Z_c}{3}$$

Figure 1-66: Positive Sequence Formulas

Use the examples in Figure 1-67 to better understand how Positive Sequence is calculated for a perfectly balanced system. Va = 120V @ 0°, Vb = 120V @ -120°, and Vc = 120V @ 120°.

$$V_1 = \frac{V_a + aV_b + a^2V_c}{3}$$

$$V_1 = \frac{120V \text{ @ } 0° + (1\text{@}120° \times 120V \text{ @ } -120°) + (1\text{@}240° \times 120V \text{ @ } 120°)}{3}$$

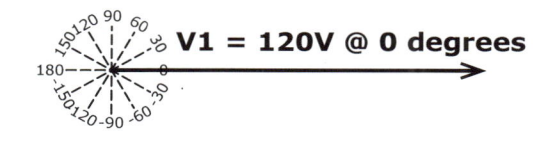

$$V_1 = \frac{120V \text{ @ } 0° + 120V \text{ @ } 0° + 120V \text{ @ } 0°}{3}$$

$$V_1 = \frac{360V}{3} = 120V$$

Figure 1-67: Positive Sequence Calculation

B) Negative Sequence

Negative sequence components exist if there is an unbalance in the three-phase system. Unbalances can be caused by different loading between phases, eddy currents inside a generator or transformer, or mismatched poles. The mathematical models of negative sequence components operate with opposite phase rotation that exist simultaneously with positive and zero sequence components. (A-B-C system: negative sequence rotation is A-C-B) Negative sequence components are not desired because they can cause overheating in almost all connected equipment and are not productive. Negative Sequence values are designated with the subscript 2 (E.g. V_2, I_2, Z_2) and can be calculated using the formulas in Figure 1-68 where "a" rotates the following value by 120° and "a_2" rotates the following value by 240°.

$$V_2 = \frac{V_a + a^2 V_b + a V_c}{3} \qquad I_2 = \frac{I_a + a^2 I_b + a I_c}{3} \qquad Z_2 = \frac{Z_a + a^2 Z_b + a Z}{3}$$

Figure 1-68: Negative Sequence Formulas

In a perfect system, the negative sequence voltage equals zero as shown in Figure 1-69

$$V_2 = \frac{V_a + a^2 V_b + a V_c}{3}$$

$$V_1 = \frac{120V @ 0° + (1 @ 240° \times 120V @ -120°) + (1 @ 120° \times 120V @ 120°)}{3}$$

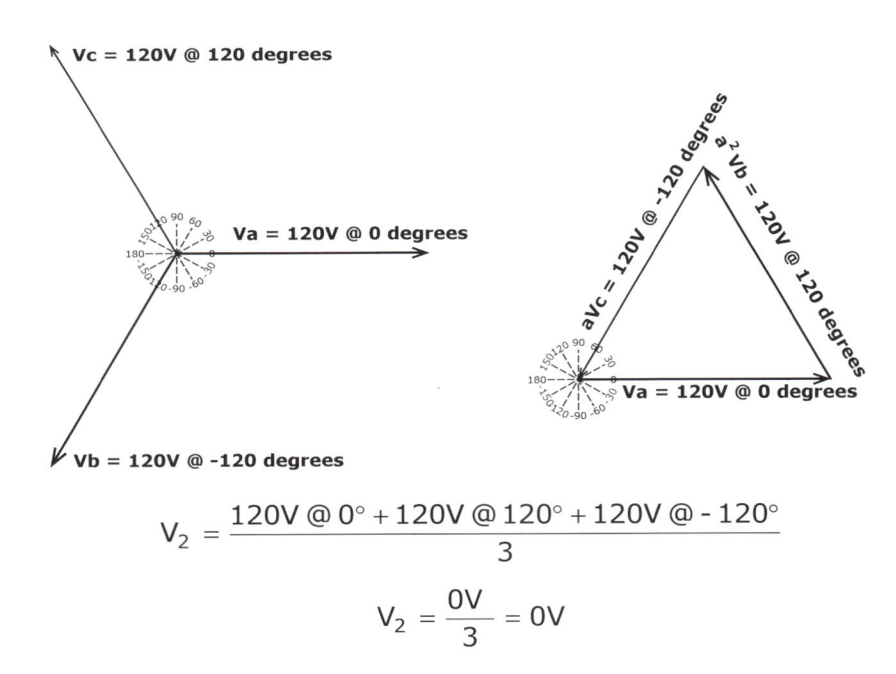

$$V_2 = \frac{120V @ 0° + 120V @ 120° + 120V @ -120°}{3}$$

$$V_2 = \frac{0V}{3} = 0V$$

Figure 1-69: Negative Sequence Calculation 1

If the system was 100% negative sequence (A-C-B instead of A-B-C), use the examples in Figure 1-70.

$$V_2 = \frac{V_a + a^2 V_b + a V_c}{3}$$

$$V_2 = \frac{120V @ 0° + (1 @ 240° \times 120V @ 120°) + (1 @ 120° \times 120V @ -120°)}{3}$$

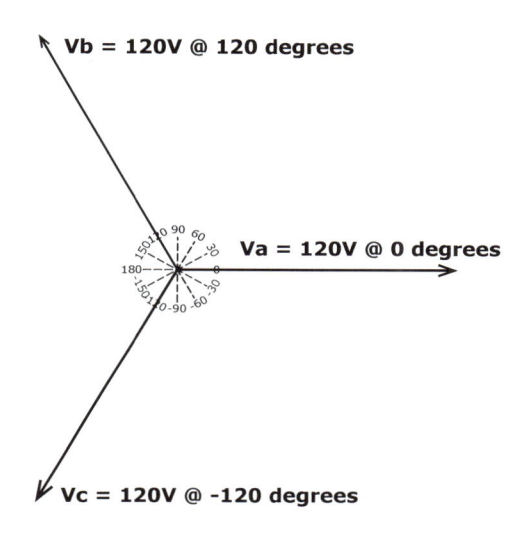

$$V_2 = \frac{120V @ 0° + 120V @ 0° + 120V @ 0°}{3}$$

$$V_1 = \frac{360V}{3} = 120V$$

Figure 1-70: Negative Sequence Calculation 2

C) Zero Sequence

Zero sequence components exist whenever current flows between the electrical system and ground. Zero sequence components are always in phase regardless of the source phase. The zero sequence currents in each phase add together in the neutral or ground connection. There is an important distinction between ground-fault current and zero-sequence current. Ground-fault current can be directly measured using an ammeter or other measuring device. Zero-sequence current is a calculated value and is often 1/3 the ground fault current. The formula for zero-sequence components is:

$$V_0 = \frac{V_a + V_b + V_c}{3} \qquad I_0 = \frac{I_a + I_b + I_c}{3} \qquad Z_0 = \frac{Z_a + Z_b + \bar{z}}{3}$$

Figure 1-71: Zero Sequence Formulas

In a perfect system, the zero-sequence voltage equals zero as shown in Figure 1-72..

$$V_0 = \frac{V_a + V_b + V_c}{3}$$

$$V_0 = \frac{120V @ 0° + 120V @ -120° + 120V @ 120°)}{3}$$

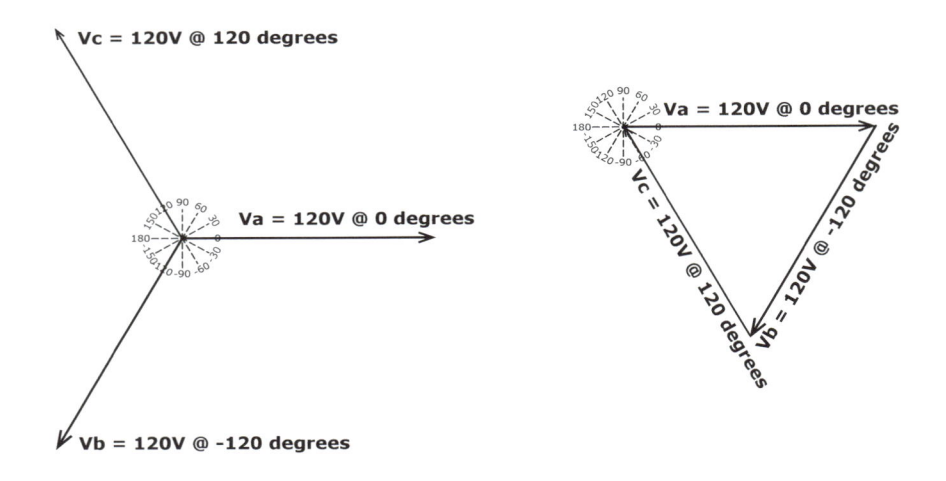

$$V_0 = \frac{120V@0° + 120V@-120° + 120V@120°}{3}$$

$$V_0 = \frac{0V}{3} = 0V$$

Figure 1-72: Zero Sequence Calculation 1

If one phase was solidly grounded (V_{an} = 0V), the calculation would be as follows.

$$V_0 = \frac{V_a + V_b + V_c}{3}$$

$$V_0 = \frac{0V @ 0° + 120V @ -120° + 120V @ 120°)}{3}$$

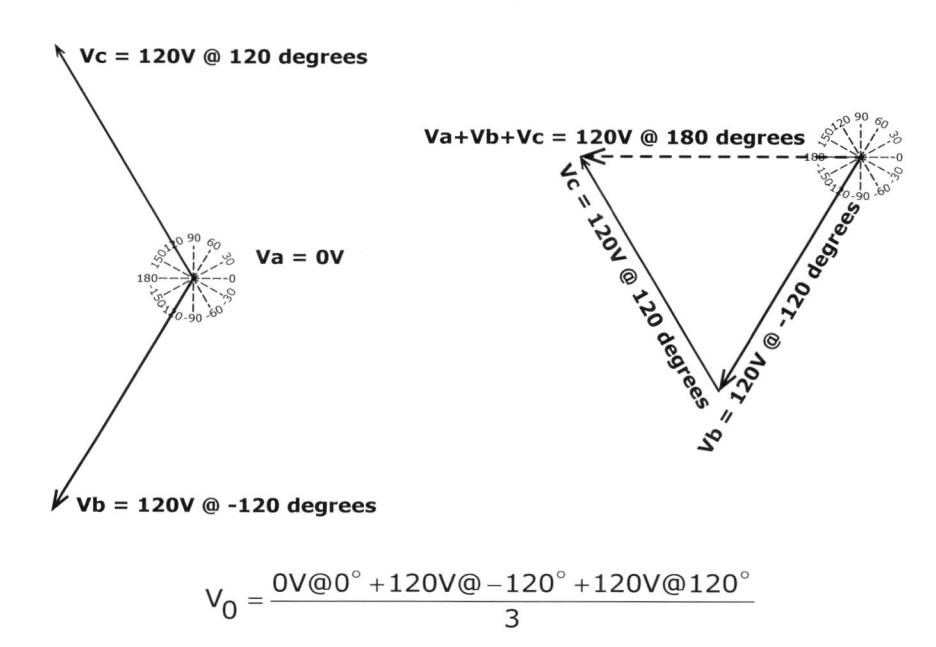

$$V_0 = \frac{0V@0° + 120V@-120° + 120V@120°}{3}$$

$$V_0 = \frac{120V}{3} = 40V$$

Figure 1-73: Zero Sequence Calculation 2

7. Fault Types and Sequence Components

The primary purpose for sequence components is to analyze unbalanced systems, so let's look at some examples and see what sequence components are generated by different fault types.

A) Phase-to-Ground Faults

Phase-to-ground faults are characterized by an increased current and decreased voltage on the faulted phase. The other two phases not associated with the fault stay close to their prefault values and, to make the examples simple, will remain at prefault values. We'll use the following values for our calculations which simulate a 30V fault voltage, and a 10A fault current lagging angle by 65°.

	PREFAULT	FAULT
Va	69.28V @ 0°	30V @ 0°
Vb	69.28V @ -120°	69.28V @ -120°
Vc	69.28V @ 120°	69.28V @ 120°
Ia	2A @ 30°	10A @ -65°
Ib	2A @ -150°	2A @ -150°
Ic	2A @ 90°	2A @ 90°

We do not need to perform any calculations on the prefault values because they are the definition of a balanced 3Ø system which means that the voltage and current are 100% positive sequence with no negative or zero sequence components. The negative and zero sequence components can only exist in an unbalanced system.

The following calculations demonstrate all of the information we've presented so far and can get quite intense. See if you can follow the calculations, but remember that there are many sequence calculation programs available. It is more important that you understand the process rather than perform every single calculation. We'll start with the positive sequence fault voltage calculation:

$$V1 = \frac{V_a + aV_b + a^2V_c}{3}$$

$$V1 = \frac{30V @ 0° + a(69.28V @ -120°) + a^2(69.28V @ 120°)}{3}$$

$$V1 = \frac{30V @ 0° + 69.28V @ (-120° + 120°) + 69.28V @ (120° + 240°)}{3}$$

$$V1 = \frac{30V @ 0° + 69.28V @ 0° + 69.28V @ 0°}{3}$$

$$V1 = \frac{168.56V @ 0°}{3}$$

$$V1 = 56.19V @ 0°$$

The negative sequence fault voltage calculation is next:

$$V2 = \frac{V_a + a^2 V_b + a V_c}{3}$$

$$V2 = \frac{30V@0° + a^2\left(69.28V@{-}120°\right) + a\left(69.28V@120°\right)}{3}$$

$$V2 = \frac{30V@0° + 69.28V@\left({-}120° + 240°\right) + 69.28V@\left(120° + 120°\right)}{3}$$

$$V2 = \frac{30V@0° + 69.28V@120° + 69.28V@240°}{3}$$

$$V2 = \frac{\left(Va\times\cos\theta + jVa\times\sin e\theta\right) + \left(Vb\times\cos\theta + jVb\times\sin e\theta\right) + \left(Vc\times\cos\theta + jVc\times\sin e\theta\right)}{3}$$

$$V2 = \frac{\left(30\times\cos(0) + j30\times\sin e(0)\right) + \left(69.28\times\cos(120) + j69.28\times\sin e(120)\right) + \left(69.28\times\cos(240) + j69.28\times\sin e(240)\right)}{3}$$

$$V2 = \frac{\left(30 + j0\right) + \left({-}34.64 + j60.00\right) + \left({-}34.64 + j{-}60\right)}{3}$$

$$V2 = \frac{\left(30 + {-}34.64 + {-}34.64\right) + \left(j0 + j60.00 + j{-}60\right)}{3}$$

$$V2 = \frac{\left({-}39.28\right) + \left(j0\right)}{3}$$

$$V2 = \frac{39.28V@180°}{3}$$

$$V2 = 13.09V@180°$$

The zero sequence fault voltage calculation is next:

$$V0 = \frac{V_a + V_b + V_c}{3}$$

$$V0 = \frac{30V@0° + \left(69.28V@{-}120°\right) + \left(69.28V@120°\right)}{3}$$

$$V0 = \frac{\left(Va\times\cos\theta + jVa\times\sin e\theta\right) + \left(Vb\times\cos\theta + jVb\times\sin e\theta\right) + \left(Vc\times\cos\theta + jVc\times\sin e\theta\right)}{3}$$

$$V0 = \frac{\left(30\times\cos(0) + j30\times\sin e(0)\right) + \left(69.28\times\cos({-}120) + j69.28\times\sin e({-}120)\right) + \left(69.28\times\cos(120) + j69.28\times\sin e(120)\right)}{3}$$

$$V0 = \frac{\left(30 + j0\right) + \left({-}34.64 + j{-}60.00\right) + \left({-}34.64 + j60\right)}{3}$$

$$V0 = \frac{\left(30 + {-}34.64 + {-}34.64\right) + \left(j0 + j{-}60.00 + j60\right)}{3}$$

$$V0 = \frac{\left({-}39.28\right) + \left(j0\right)}{3}$$

$$V0 = \frac{39.28V@180°}{3}$$

$$V0 = 13.09V@180°$$

The positive sequence fault current calculation is next:

$$I1 = \frac{I_a + aI_b + a^2I_c}{3}$$

$$I1 = \frac{10A\,@-65° + a\left(2A@\text{-}150°\right) + a^2\left(2A@90°\right)}{3}$$

$$I1 = \frac{10A\,@-65° + 2A@\left(\text{-}150° + 120°\right) + 2A@\left(90° + 240°\right)}{3}$$

$$I1 = \frac{10A\,@-65° + 2A@\text{-}30° + 2A@330°}{3}$$

$$I1 = \frac{\left(Ia\times\cos\theta + jIa\times\sin e\theta\right) + \left(Ib\times\cos\theta + jIb\times\sin e\theta\right) + \left(Ic\times\cos\theta + jIc\times\sin e\theta\right)}{3}$$

$$I2 = \frac{\left(10\times\cos\left(-65\right) + j10\times\sin e\left(-65\right)\right) + \left(2\times\cos\left(-30\right) + j2\times\sin e\left(-30\right)\right) + \left(2\times\cos\left(330\right) + j2\times\sin e\left(330\right)\right)}{3}$$

$$I1 = \frac{\left(4.23 + j - 9.06\right) + \left(1.732 + j - 1\right) + \left(1.732 + j - 1\right)}{3}$$

$$I1 = \frac{\left(4.23 + 1.732 + 1.732\right) + \left(j - 9.06 + j - 1 + j - 1\right)}{3}$$

$$I1 = \frac{\left(7.69\right) + \left(j - 11.06\right)}{3}$$

$$I1 = \frac{\left(\sqrt{7.69^2 + -11.06^2}\right)@\arctan\left(\dfrac{-11.06}{7.69}\right)}{3}$$

$$I1 = \frac{13.473A\,@-55.20°}{3}$$

$$I1 = 4.491\,@-55.20°$$

The negative sequence fault current calculation is next:

$$I2 = \frac{I_a + a^2 I_b + aI_c}{3}$$

$$I2 = \frac{10A@-65° + a(2A@\text{-}150°) + a^2(2A@90°)}{3}$$

$$I2 = \frac{10A@-65° + 2A@(-150° + 240°) + 2A@(90° + 120°)}{3}$$

$$I2 = \frac{10A@-65° + 2A@90° + 2A@210°}{3}$$

$$I2 = \frac{(Ia \times \cos\theta + jIa \times \sin e\theta) + (Ib \times \cos\theta + jIb \times \sin e\theta) + (Ic \times \cos\theta + jIc \times \sin e\theta)}{3}$$

$$I2 = \frac{(10 \times \cos(-65) + j10 \times \sin e(-65)) + (2 \times \cos(90) + j2 \times \sin e(90)) + (2 \times \cos(210) + j2 \times \sin e(210))}{3}$$

$$I2 = \frac{(4.23 + j-9.06) + (0 + j2) + (-1.732 + j-1)}{3}$$

$$I2 = \frac{(4.23 + 0 + -1.732) + (j-9.06 + j2 + j-1)}{3}$$

$$I1 = \frac{(2.494) + (j-8.06)}{3}$$

$$I2 = \frac{\left(\sqrt{2.494^2 + -8.06^2}\right)@\arctan\left(\frac{-8.06}{2.494}\right)}{3}$$

$$I2 = \frac{8.440A@-72.81°}{3}$$

$$I2 = 2.813A@-72.81°$$

The zero sequence fault current calculation is next:

$$I0 = \frac{I_a + I_b + I_c}{3}$$

$$I0 = \frac{10A@-65° + (2A@-150°) + (2A@90°)}{3}$$

$$I0 = \frac{(Ia \times \cos\theta + jIa \times \sin e\theta) + (Ib \times \cos\theta + jIb \times \sin e\theta) + (Ic \times \cos\theta + jIc \times \sin e\theta)}{3}$$

$$I0 = \frac{(10 \times \cos(-65) + j10 \times \sin e(-65)) + (2 \times \cos(-150) + j2 \times \sin e(-150)) + (2 \times \cos(90) + j2 \times \sin e(90))}{3}$$

$$I0 = \frac{(4.226 + j - 9.06) + (-1.732 + j - 1.00) + (0.000 + j2.00)}{3}$$

$$I0 = \frac{(4.226 + -1.732 + 0.000) + (j - 9.06 + j - 1.00 + j2.00)}{3}$$

$$I0 = \frac{(2.494) + (j - 8.06)}{3}$$

$$I0 = \frac{\left(\sqrt{2.494^2 + -8.06^2}\right)@\arctan\left(\frac{-8.06}{2.494}\right)}{3}$$

$$I0 = \frac{8.440A@-72.81°}{3}$$

$$I0 = 2.813A@-72.81°$$

A summary of the calculations includes:

SEQUENCE	VOLTAGE	CURRENT
POSITIVE	56.19V @ 0°	4.491A @ -55.20°
NEGATIVE	13.09V @ 180°	2.813A @ -72.81°
ZERO	13.09V @ 180°	2.813A @ -72.81°

A large part of fault analysis is pattern recognition and we can make the following conclusions regarding a Phase-to-neutral fault:

- All three sequence components are present during the fault.
- The negative sequence and zero sequence values are almost identical.

B) Phase-to-Phase Faults

Phase-to-phase faults are characterized by an increased current and decreased voltage on the faulted phases. The other phase not associated with the fault will stay close to prefault values and, to make the examples simple, will remain at prefault values. We'll use the following values for our calculations which simulate a 30V fault voltage, and a 10A fault current lagging angle by 65°.

	PREFAULT	FAULT
Va	69.28V @ 0°	37.75V @ -36.59°
Vb	69.28V @ -120°	37.75V @ -83.41°
Vc	69.28V @ 120°	69.28V @ 120°
Ia	2A @ 30°	10A @ -35°
Ib	2A @ -150°	10A @ -145°
Ic	2A @ 90°	2A @ 90°

The following table was created with a sequence component calculator:

SEQUENCE	VOLTAGE	CURRENT
POSITIVE	43.30V @ 0°	3.4541A @ -58.64°
NEGATIVE	-12.99V @ -22.50°	2.372A @ -14.278°
ZERO	0.00V	0.667A @ 90.00°

A large part of fault analysis is pattern recognition and we can make the following conclusions regarding a phase-to-phase fault:

- Only Positive and Negative Sequence voltages are present during the fault.
- There are significant Positive and Negative Sequence currents and a small amount of Zero Sequence currents.

C) Three-Phase Faults

Three-Phase faults are characterized by an increased current and decreased voltage on three phases. We'll use the following values for our calculations which simulate a 30V fault voltage, and a 20A fault current lagging angle by 65°.

	PREFAULT	FAULT
Va	69.28V @ 0°	30.00V @ 0°
Vb	69.28V @ -120°	30.00V @ -120°
Vc	69.28V @ 120°	30.00V @ 120°
Ia	2A @ 30°	10A @ -65°
Ib	2A @ -150°	10A @ 175°
Ic	2A @ 90°	10A @ 55°

The following table was created with a sequence component calculator:

SEQUENCE	VOLTAGE	CURRENT
POSITIVE	30.00V @ 0°	6.331A @ -61.54°
NEGATIVE	0.00V	0.00A
ZERO	0.00V	0.00A

A large part of fault analysis is pattern recognition and we can make the following conclusions regarding a phase-to-phase fault:

- Only Positive Sequence voltages and currents exist

D) Fault Types and Sequence Components Summary

We can summarize the results in this section with the following table.

SEQUENCE	PHASE TO NEUTRAL	PHASE-TO-PHASE	THREE-PHASE
POSITIVE	Yes	Yes	Yes
NEGATIVE	Yes	Yes	No
ZERO	Yes	No	No

We can use this information to quickly analyze a fault and, more importantly, a relay design engineer can use these patterns to create more reliable relays that can respond to changes in sequence components rather than looking at specific voltages and currents as previous generations of relays were forced to do. Unfortunately for the relay tester, relays that use sequence components in their calculations often make traditional test procedures obsolete and requires the relay tester to create more complex test plans that compensate for the relay's advanced features.

Chapter 2

Introduction to Protective Relays

Protective relays are the police force of any electrical system. They constantly look for electrical faults or abnormal conditions and stand ready to quickly isolate problem areas from the rest of the system before too much damage or instability occurs. As with all other parts of society, protective relaying has evolved over time and has fully embraced the digital age. Modern digital relays are accurate, reliable, sophisticated, powerful, and full of features you'll probably never use. Unfortunately, the extra benefits of digital relays also increase complexity, and digital relay testing can be daunting to even the most experienced relay technician. The average relay technician's necessary skills have shifted from an of understanding magnetic and mechanical systems, to an understanding digital communications and/or logic systems.

1. What are Protective Relays?

Protective Relays monitor electrical systems via current and/or voltage inputs to detect and isolate electrical faults before catastrophic damage can occur. If the relay detects a fault or abnormal system condition, it initiates a trip signal to isolate the fault or will signal an alarm to warn operators. There should be minimal disruption to non-faulted sections of the electrical system when a relay operates. Relay schemes are designed to:

A) Protect Equipment from Damage

Most electrical equipment can withstand short duration faults or overload currents. As the severity of the fault or overload increases, the potential damage increases. Therefore, many protective relays incorporate time curves to vary the time delay between fault detection and operation in relation to the severity of the fault. Different types of equipment have different damage withstand characteristics and several unique characteristic curves can be applied.

Equipment or electrical systems can also be adversely affected by other electrical parameters such as voltage, frequency, or direction of current; and protective relays can be applied to protect systems or equipment from these abnormal conditions as well.

B) Be Selective

As society becomes more modern, its dependence on electricity increases and sudden losses of power can have wide repercussions from a mere nuisance (start looking for the candles) to life and death situations (hospitals). Whether it is the utility's loss of revenue or the loss of production at a plant, financial losses are the most common repercussion of electrical system interruptions. Protective relays should only operate when absolutely necessary, and there should be minimum disruption to the rest of the electrical system. The following characteristics help relays meet these criteria:

i) Zones of Protection

Electrical systems can be divided into five primary categories:

- Generating Plants (Generators and/or generators with step-up transformers)
- Transformers
- Buses
- Transmission or Distribution (Transmission lines or Feeder Cables)
- End Use Equipment (Motors, MCCs, etc.)

Each of these categories has its own unique characteristic and requires specialized protection. Combinations of protective elements are applied in zones surrounding the equipment within an electrical system as shown in Figure 2-1. Zones of protection ensure that the equipment in that zone receives optimum protection, and the protective devices do not interfere with any other part of the electrical system. A transmission-line fault should not affect bus protection and vice-versa for all different zones. Today's protective relays incorporate multiple protective elements, and a single relay will protect an entire zone.

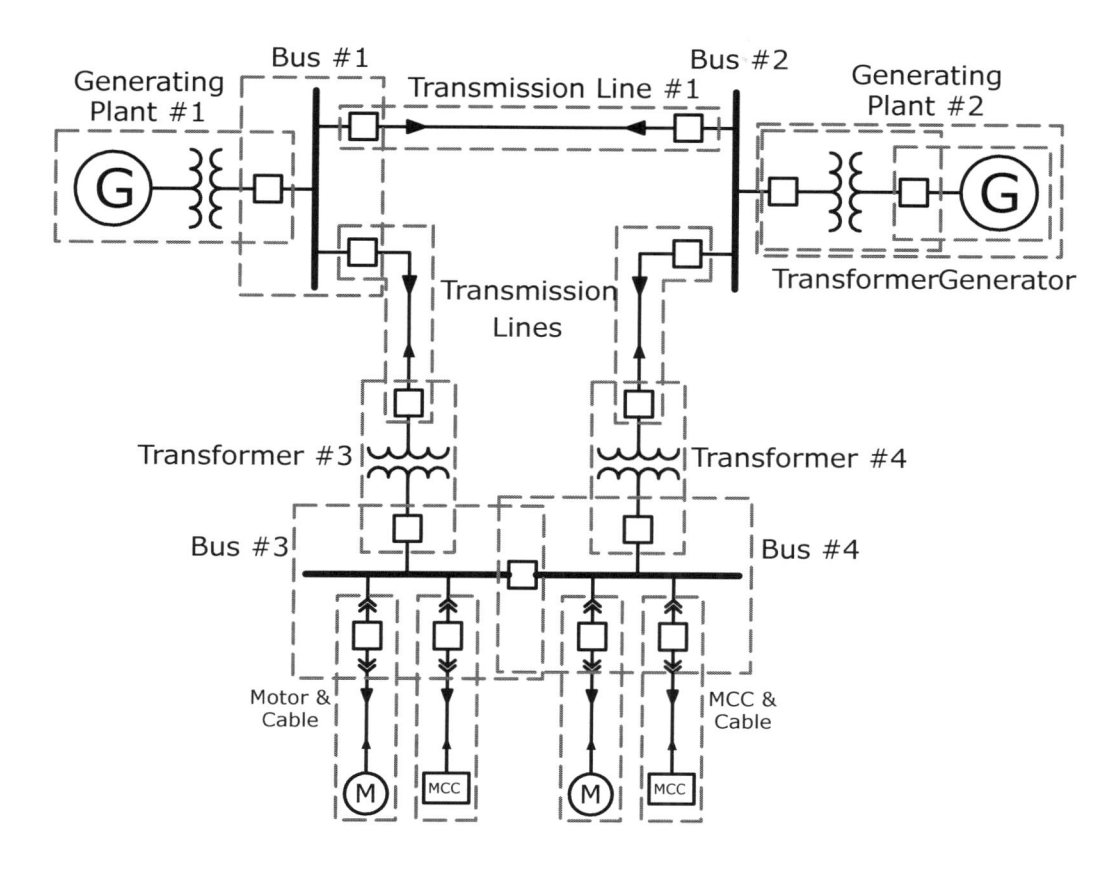

Figure 2-1: Zones of Protection

ii) Zone Overlap

A quick review of the previous "Zones of Protection" diagram reveals that every circuit breaker is associated with at least one zone of protection. A CT input between the circuit breaker and each associated zone's protective relay is required for the protective schemes to operate. The relays can use the same CT input if they are connected in series.

Overlapping zones are important to ensure a fault is cleared, even if the primary protective system fails, to limit the damage and prevent system instability. A fault on Bus #1 will open all of the breakers associated with Bus #1 and reduce impact to the rest of the system. It would be difficult to predict what would happen on Bus #2 because there is no Bus protection enabled, and any number of breakers could operate depending on their settings.

Overlapping is achieved by carefully selecting the circuit breaker's CT location for the zone. A close up of the Tie Breaker between Bus #3 and Bus #4 demonstrates how overlapping is achieved. By using the CT on the opposite side of the Tie Breaker, the bus protection is sure to operate if a fault occurs within the Tie Breaker. A Tie Breaker flashover fault could exist for an extended period and cause exponentially more damage if the CT inputs were reversed as shown in Figures 2-2 and 2-3.

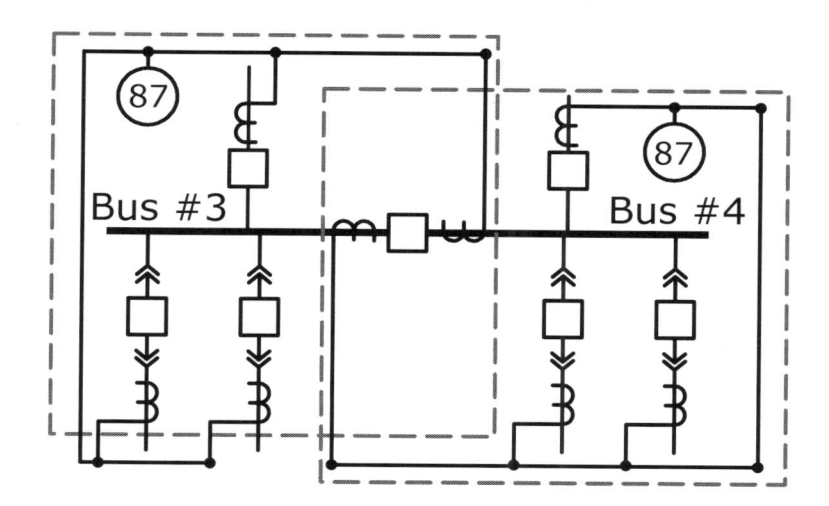

Figure 2-2: Overlapping Zones of Protection

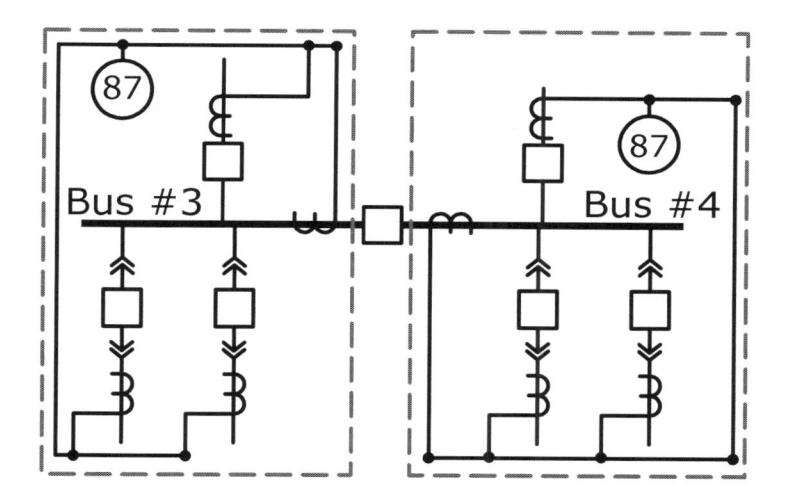

Figure 2-3: Non-Overlapping Zone of Protection Example

iii) Protective Relay Elements

In the past, protective relays were designed to operate if a certain type of electrical fault occurred. An individual relay would monitor a phase or line value and operate if a fault was detected. The most common protective relay elements are listed below along with their IEEE reference numbers. The IEEE numbers were created to provide a quick and easy reference to the relay function on electrical drawings.

- Impedance Protection (21) – The relay monitors system impedance and will trip if it falls within a predetermined characteristic.
- Undervoltage Protection (27) – If the voltage drops below a predetermined voltage, the relay will operate.
- Reverse Power Protection (32) – The relay monitors the direction of power flow and will operate if power flows in the wrong direction.
- Loss of Field Protection (40) – The relay monitors generator output impedance and will trip if it falls within a predetermined loss of field characteristic.
- Negative Sequence Overcurrent (46) – The relay calculates the negative sequence current and trips if it exceeds a preset value.
- Overload Protection (49) – The relay models the thermal characteristics of the protected equipment and measures the input current. The relay will trip if excessive current over time exceeds the thermal capacity of the relay.

- Instantaneous Overcurrent (50) – The relay measures current and trips if it exceeds a preset value with no intentional time delay.
- Time Overcurrent (51) – The relay measures current and trips if it exceeds a preset value for a period of time. The time delay can be proportional to measured current.
- Voltage Controlled/Restrained Time Overcurrent (51V) – The relay measures voltage and current. The relay will trip if the current exceeds a preset value for a period of time. The preset value of current varies in relation to the measured voltage.
- Overvoltage Protection (59) – If the voltage rises above a predetermined setpoint, the relay will operate.
- Loss of Potential (60) – The relay monitors system voltages and operates if it detects that a PT fuse has opened.
- Directional Overcurrent (67) – The relay monitors current direction and will operate if the current flows in the wrong direction above a setpoint.
- Frequency Protection (81) – The relay monitors system frequency and will operate if an abnormal frequency is detected.
- Differential Protection (87) – The relay monitors current flowing in and out of a device and will operate if there is a difference between input and output current that indicates a fault inside the equipment itself.

C) Coordinate with Other Protective Devices

Relays or other protective devices are usually installed at every isolating device in an electrical system to provide optimum protection and versatility. In a perfect world, every device would isolate a fault with no disruption to the unaffected parts of the system.

Figure 2-4 displays a perfectly coordinated system where a fault downstream of PCB7 was isolated from the rest of the system by Relay R7 that opened PCB7. If PCB7 did not operate because of malfunction or incorrect settings, PCB3 should open after an additional time delay. However, there are repercussions if PCB3 operated to clear a fault downstream of PCB7:

- Any equipment downstream of PCB6 would also be offline even though there is no relation between the fault and de-energized equipment.
- Additional damage to equipment is likely because the fault duration will be longer.
- Restoration of power will likely be delayed trying to locate the cause of the fault.

TYPICAL INDUSTRIAL SINGLE LINE

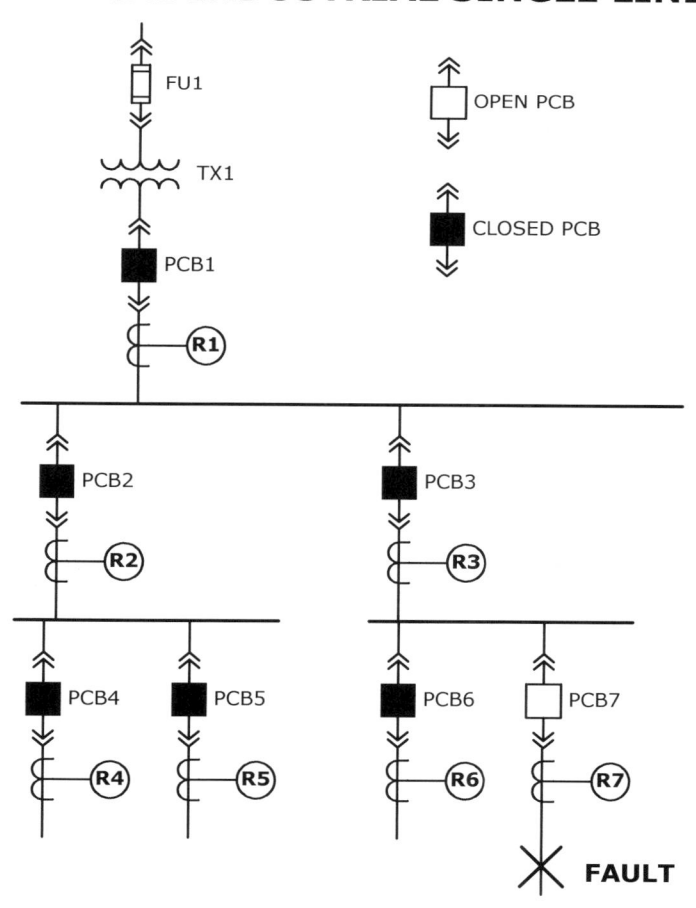

Figure 2-4: Protective Relay Coordination

71

D) Compensate for Normal System Fluctuations

Infrequent or short duration overloads and/or system fluctuations caused by motor starting, switching loads, sags, or swells could cause nuisance trips if time delays were not applied to protective relaying. For example, the inrush current for a typical induction motor can be up to 6-10x its full load rating (FLA). The pickup settings for protective relays is typically set to 125% of the full load rating (R2 pickup = 125A in our example). However, the induction motor can draw up to 300A (6 x FLA) while the motor starts and Relay R2 could monitor a total of 350A. The relay should not operate during motor starting, but it should operate if a fault is detected. A time delay of at least three seconds must be added to allow the motor to start as shown in Figure 2-5.

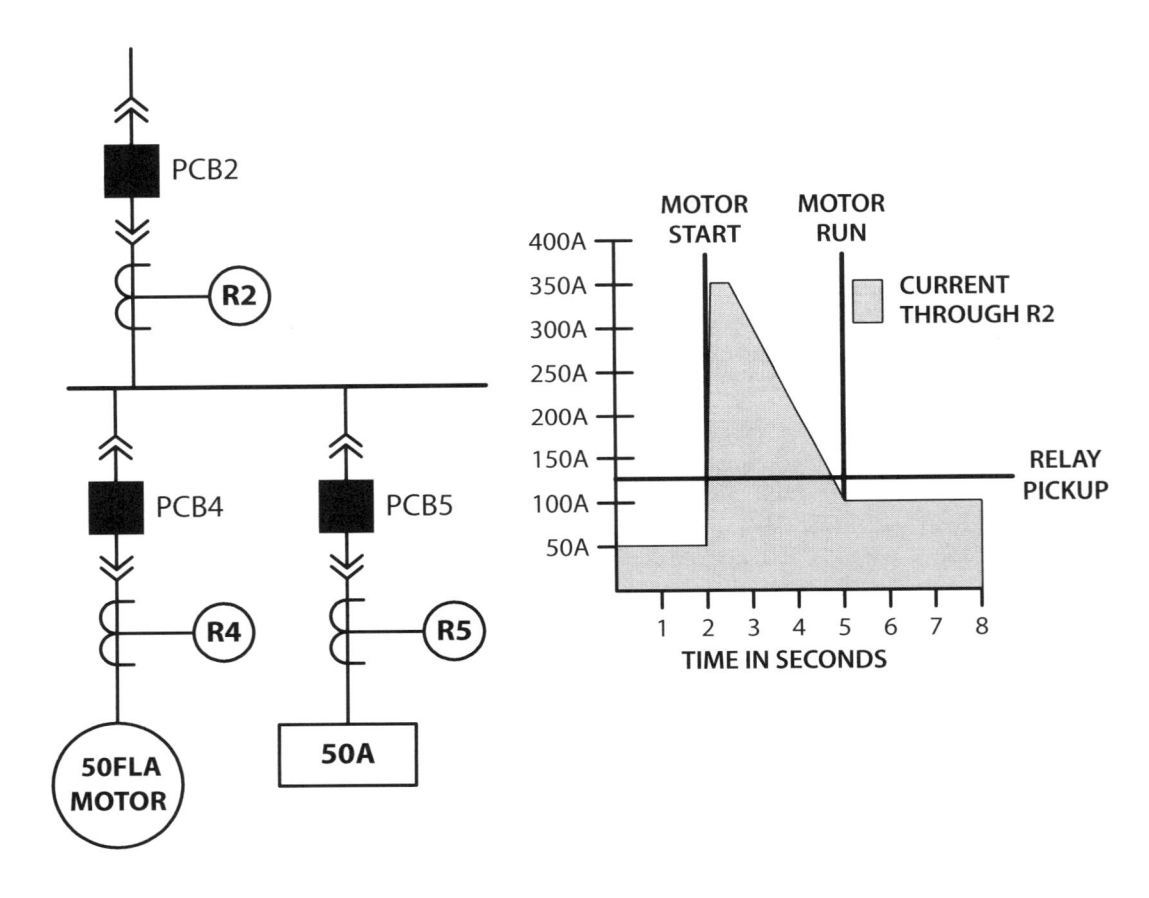

Figure 2-5: Compensation for Normal System Fluctuations

2. Time Coordination Curves (TCC) and Coordination

One of the key functions of a protective relay is to coordinate with upstream and downstream devices. Time Coordination Curves (TCC) were developed to help the engineer ensure that the relays in a system will coordinate with each other. Relay coordination should be determined during the fault and coordination study.

A) Fault Study or Short Circuit Study

A fault study is performed for new electrical systems to determine the maximum fault current at different locations within the system. Fault studies should also be performed on existing systems whenever there are significant changes in the electrical system. Fault studies are usually calculated for the four most common faults in symmetrical and asymmetrical quantities. The asymmetrical current exists during the first few cycles of a fault when the maximum disruption occurs due to DC offset at the location of the fault along with motor and generator initial fault contributions. Symmetrical currents are steady-state fault currents after the initial disruption has stabilized, usually after a few cycles. The four most common faults calculated are:

- Three-phase (3P)
- Line to Ground (SLG)
- Line to Line (LL)
- Line to Line to Ground (LLG)

The fault study uses all of the following information regarding an electrical system to create a mathematical model to calculate the maximum fault currents:

- Source impedance and maximum fault contribution (Usually the primary source of fault current)
- Transformer voltages and percent impedance (Usually the largest impedance in an electrical system)
- Cable sizes and lengths (Usually second largest impedance in an electrical system)
- Bus ampacity and maximum fault current withstand (To determine if equipment is rated for maximum fault currents)
- Generator impedance and maximum fault contribution (To determine generator fault contribution)
- Motor size (horsepower (hp) or MW) and impedance (Motors can be source of fault current during the first few cycles of a fault)

An example of a fault study is depicted in Figure 2-6:

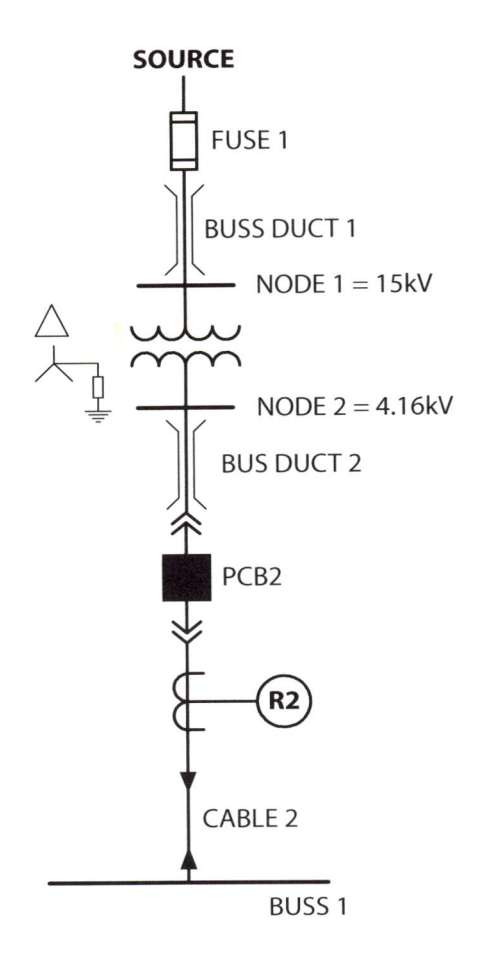

Node	Symmetrical Amps		Asymmetrical Amps	
	3-Phase	SLG	3-Phase	SLG
Source 15 kV	123,319	7	202,237	12
Node 1 15 kV	123,097	7	201,917	12
Node 2 4.16 kV	157,359	1006	235,777	1,006
Buss1 4.16 kV	145,561	1004	208,564	1,004

Figure 2-6: Fault Study Short Circuit Currents

Using the examples in Figure 2-6, we can interpret the following:

- The asymmetrical fault current is significantly higher than the symmetrical amps.
- There is relatively little difference between the Source and Node 1 currents because the fuse and bus duct, by design, are low impedance paths for fault current.
- There is almost zero line to ground (SLG) fault current on the primary of the transformer because the transformer primary is delta connected and there is no impedance path to ground. The zero sequence current flowing through the transformer secondary will circulate in the delta primary windings and will not leave the transformer windings.
- At first glance, there appears to be a rise in fault current through the transformer. However, the two fault currents are calculated at different voltages and we are comparing apples to oranges. We must decide on a reference voltage (base voltage) so that we can compare apples to apples. Using 15kV as the base voltage, we make the calculation in Figure 2-7 to convert the 4.16kV fault current to 15kV levels:

$$\frac{\text{Fault Voltage}}{\text{Base Voltage}} = \frac{\text{Base Amps}}{\text{Fault Amps}} \textbf{ TO } \frac{4.16\text{kV}}{15\text{kV}} = \frac{\text{Base Amps}}{157,359} \textbf{ TO }$$

$$\text{Base Amps} = \frac{4.16\text{kV} \times 157,359\text{A}}{15\text{kV}} = 43,640\text{A}$$

Figure 2-7: Calculation to Convert Current to Base Voltage

- After converting the Node 2 current to 15kV values, you can see a drop in fault current through the transformer. Transformer windings are typically the largest impedance to fault current in an electrical system.
- The current drop between Node 2 and the buss is very small but not as insignificant as the drop between the source and Node 1 because the cable between Node 2 and the buss has more resistance than the buss duct.
- The Line to Ground fault current is limited to 1000A by the neutral ground resistor and there is no difference between symmetrical and asymmetrical currents.

The fault current calculations should be compared to the equipment ratings to ensure that the switchgear, etc can withstand the maximum fault current. Switchgear and buss ducts should have "Bus Bracing" or "Short Circuit Current" ratings greater than the symmetrical short circuit calculation for the equipment location. In our example, Buss 1's rating should be greater than 145,561 Amps. If the rating were less than the short circuit current, the switchgear buss bars could shake free of their bracings and cause catastrophic damage to the switchgear and anything in the area.

Interrupting devices like fuses must be rated to interrupt the maximum fault current or they may not correctly isolate equipment during a fault. In our example, the fuse must be able to interrupt 123,319 Amps.

Other interrupting devices such as circuit breakers and motor starters should also be able to interrupt the short circuit current, but their relays can be set to operate less than the interrupting device's capabilities.

Protective relays must be set to operate below the maximum expected fault current or they would NEVER operate. Relay R2 pickup settings should be less than 145,561 Primary Amps.

B) Coordination Studies

Coordination studies are performed to ensure that all of the protective devices protect the equipment and coordinate with each other within an electrical system. You would think that this is a simple, cut and dried, task but coordination is more of an art than a science because there are nearly unlimited possibilities with today's protective relays.

i) Time Coordination Curves

The time coordination curve (TCC) drawing is used to display all related equipment and protective devices on one simple-to-understand drawing to ensure proper coordination. The TCC plots all device and equipment characteristics in relation to time and current on a log-log graph to ensure proper protection and coordination.

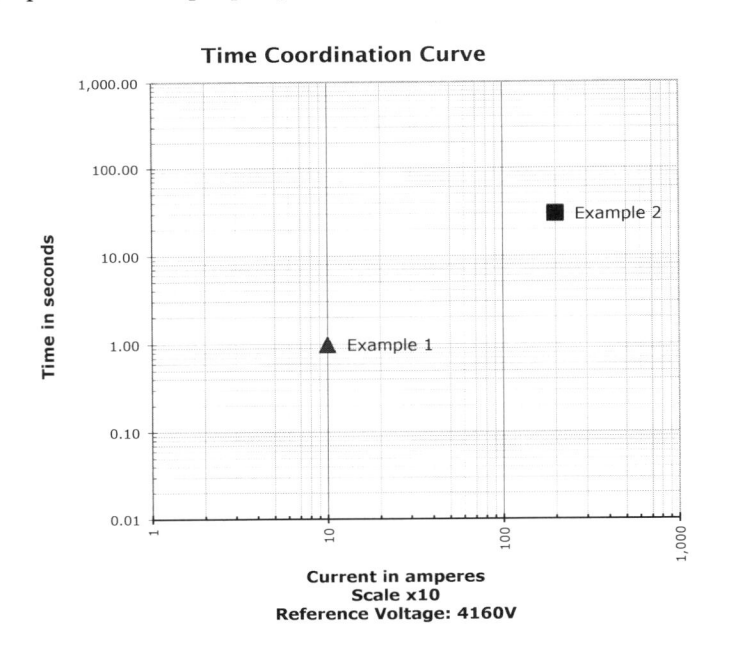

Figure 2-8: Example Time Coordination Curve (TCC)

The vertical or Y-axis represents time using a logarithmic scale and the major dividers will always be a factor of 10. The minor, unlabeled dividers split each major divider into 10 sections. For example, the dividers between 1 and 10 are 2,3,4,5,6,7,8, and 9. Example 1 is across from the 1.00 major divider and represents 1 second. Example 2 is across from the second minor divider above 10.00 and represents 30 seconds.

The horizontal or x-axis represents current using a logarithmic scale and the major dividers will also be a factor of 10. The current axis is a little more complicated because there may be an additional scaling factor listed. In this case any number on the x-axis is multiplied by 10 to determine the actual current because of the "Scale x10" listing in the axis labels. This scaling factor is usually listed in multiples of 10 using exponents. (10^0 = x10, 10^{-1} = x1, 10^2 = x100). Example 1 is in line with 10 on the x-axis but the final current represented by the graph would be 100 Amps due to the scaling factor. Example 2 is in line with 200 on the x-axis but the final current represented on the graph would be 2,000 Amps due to the scaling factor

It is important to note that the spaces in between dividers are not linear. For example, the 105 Amps would not be found at the midpoint between 100 and 110 but significantly to the right of midpoint in the similar relationship that 150 is to 100 and 200.

Coordination studies are often performed for and around transformers that change system voltages. Devices on the high and low side of a transformer must coordinate with each other, but it can be difficult to relate these devices to each other at different voltage levels because currents are also transformed with the voltages. This problem is overcome by relating all currents to one voltage reference when determining device coordination. The voltage reference is usually listed in the TCC margins and should be verified when interpreting results. Use the formula in Figure 2-9 to convert from and to the reference voltage:

$$\frac{\text{Actual Voltage}}{\text{Reference Voltage}} = \frac{\text{Reference Voltage Amps}}{\text{Actual Voltage Amps}}$$

$$\text{Actual Voltage Amps} = \frac{\text{Reference Voltage} \times \text{Reference Voltage Amps}}{\text{Actual Voltage}}$$

$$\text{Reference Voltage Amps} = \frac{\text{Actual Voltage} \times \text{Actual Voltage Amps}}{\text{Reference Voltage}}$$

Figure 2-9: Reference Voltage Conversions

ii) Damage Curves

Electrical equipment can withstand currents greater than their ratings for a period of time without sustaining damage. As the magnitude of overcurrent increases and/or the overload time increases, the equipment can be irreversibly damaged and service life reduced. A damage curve represents the maximum amount of current and time that equipment can withstand without damage. Every piece of equipment has a different damage characteristic and protective devices should operate before the current reaches the damage curve. Look at the Figures 2-10 and 2-11 for examples of damage curves:

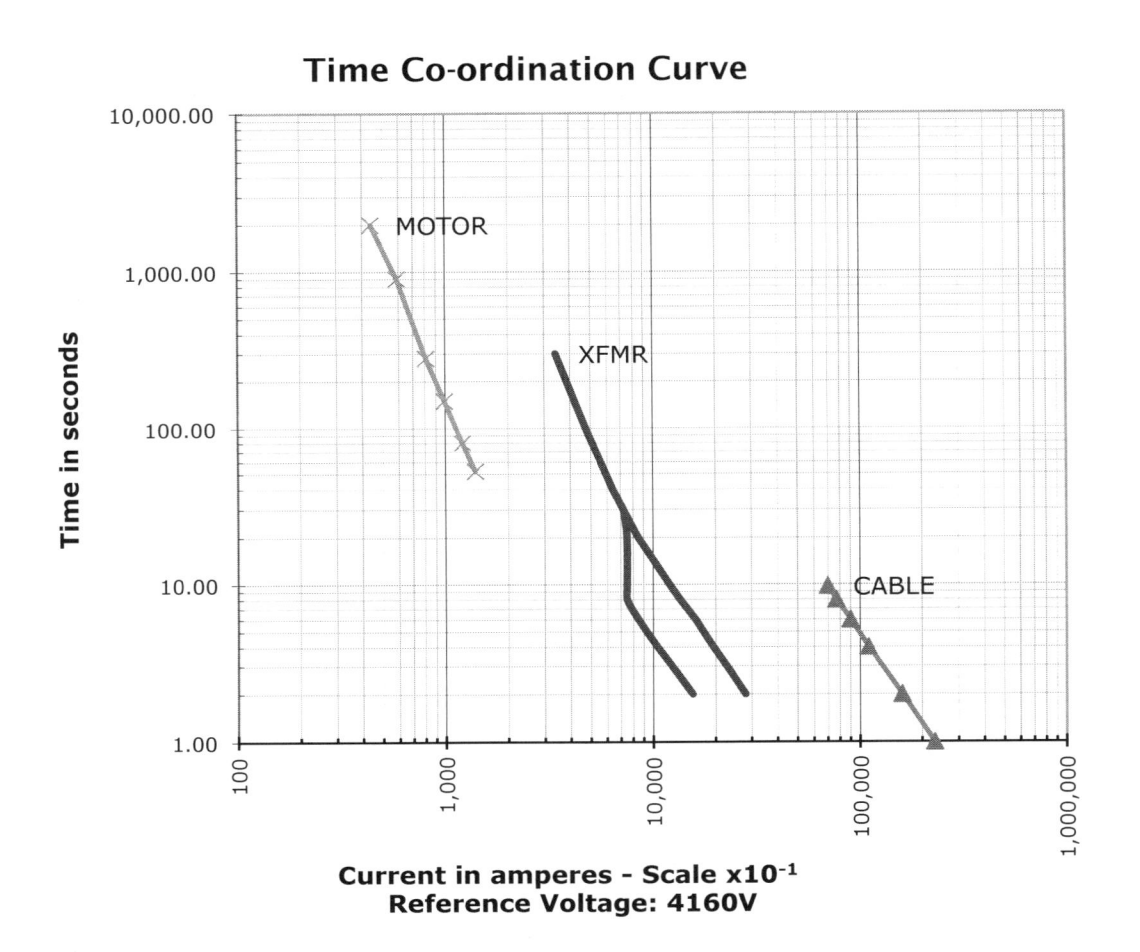

Figure 2-10: Damage Curves

Each line in Figure 2-10 represents the damage curve for a different piece of equipment. The equipment will be damaged if the current and/or time is above and on the right hand side of the damage curve. If the current is below and/or on the left hand side of the curve, the equipment will not be damaged. For example, the motor can withstand 50 Amps for about 1200 seconds without damage, but it will be damaged if 100 Amps flows for more than 160 seconds.

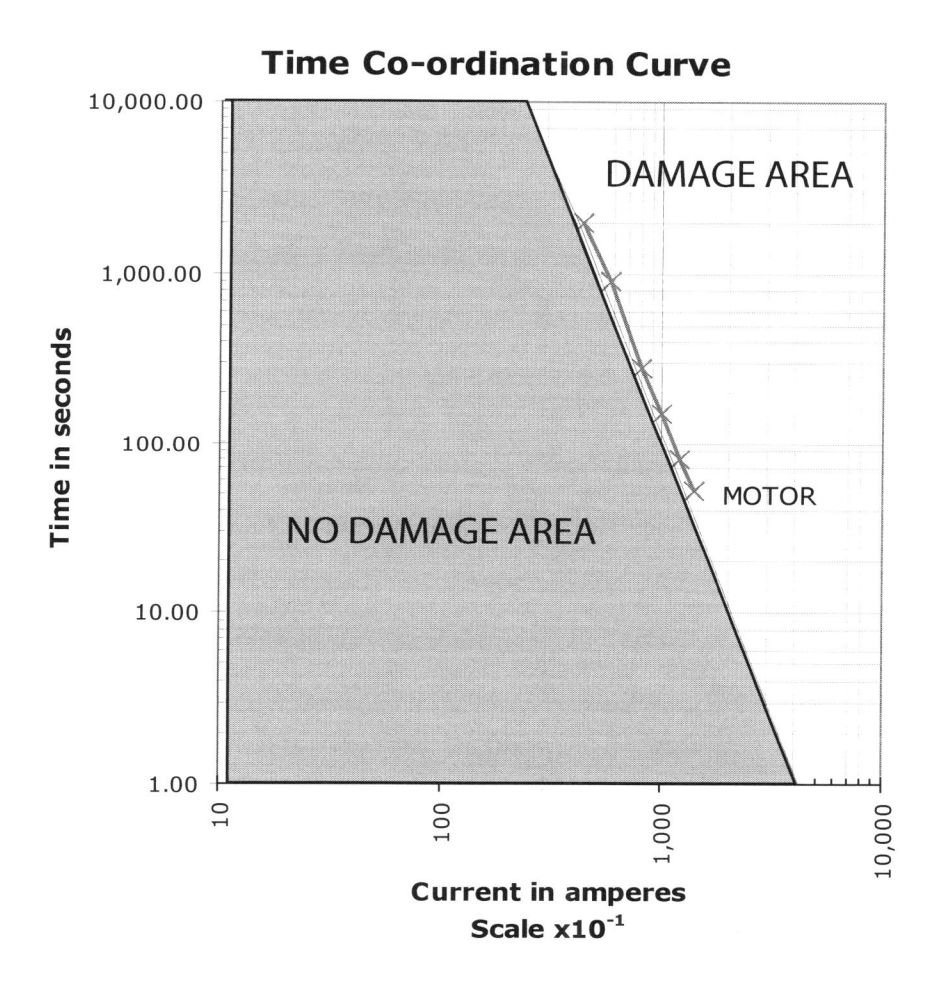

Figure 2-11: Damage Curve TCC

iii) Protective Device Curves

Protective devices, like relays and fuses, are selected and sized to operate before an overcurrent fault damages the equipment. The protective device characteristic is chosen to mimic the equipment damage curve to achieve a balance between the interests of protection and operation. If the designer leans toward protection, the relay may operate during normal operation. If the designer leans in favor of operation, the equipment may be damaged during overloads. Although this seems straightforward, the relay settings must also coordinate with other relays in the electrical system that can complicate matters. Some equipment, like transformers and motors, have inrush currents during energization or starting that must be accounted for as well. In the TCC curves in Figures 2-13, 2-14, and 2-15, we can see how combining time and instantaneous overcurrent elements can enhance equipment protection.

We can see in TCC Curve #1 that "R2 Time Overcurrent Relay Curve" is a good match for the cable damage curve. However, the two curves cross and the relay will only protect the cable from damage if the fault current is less than 16,000 Amps. This problem could be corrected by lowering the pickup or time-dial setting. In this instance, a downstream fuse prevents us from lowering either setting because the main feeder overcurrent protection would operate before the fuse. The entire bus could be isolated instead of the motor under fault. We can add instantaneous protection to the main overcurrent relay curve to ensure the cable is protected at all fault levels as shown in TCC Curve #2.

Adding the instantaneous overcurrent protection has added a new problem shown in TCC #2. The new instantaneous setting does not coordinate with the downstream fuse as shown where the two curves cross at the bottom of the graph. This would have been an acceptable compromise in the past because of protective relay limitations. Today we can add a small time delay to the instantaneous element as shown in TCC Curve #3 to correct the mis-coordination.

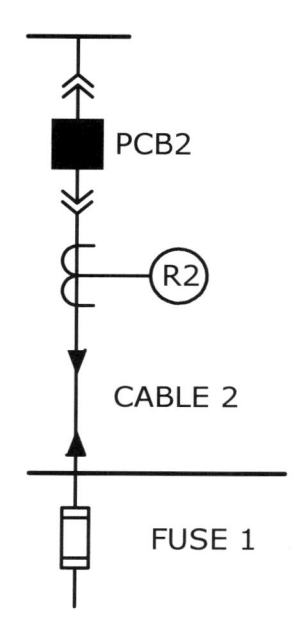

Figure 2-12: Single Line Drawing

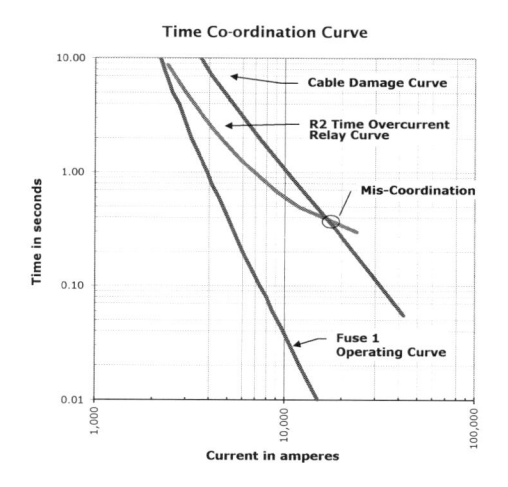

Figure 2-13: TCC Curve #1

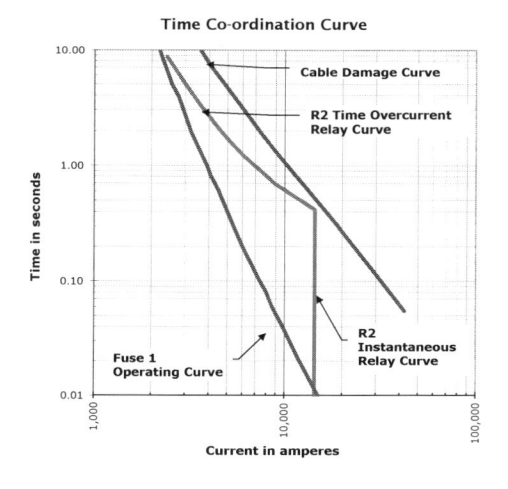

Figure 2-14: TCC Curve #2

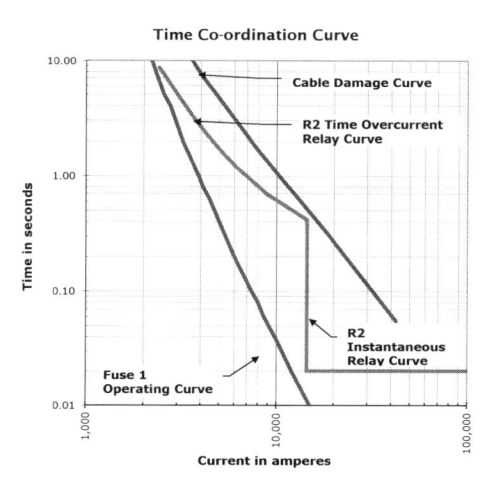

Figure 2-15: TCC Curve #3

Chapter 3

A Brief History of Protective Relays

1. Electromechanical Relays

Electromechanical relays are considered the simplest form of protective relays. Although these relays have very limited operating parameters, functions, and output schemes, they are the foundation for all relays to follow and can have very complicated mechanical operating systems. The creators of these relays were true geniuses as they were able to apply their knowledge of electrical systems and protection to create protective relays using magnetism, polarizing elements, and other mechanical devices that mimicked the characteristics they desired.

The simplest electromechanical relay is constructed with an input coil and a clapper contact. When the input signal (current or voltage) creates a magnetic field greater than the mechanical force holding the clapper open, the clapper closes to activate the appropriate control function. The relay's pickup is adjusted by changing the coil taps and/or varying the core material via an adjusting screw. The relay has no intentional time delay but has an inherent delay due to mechanical operating times.

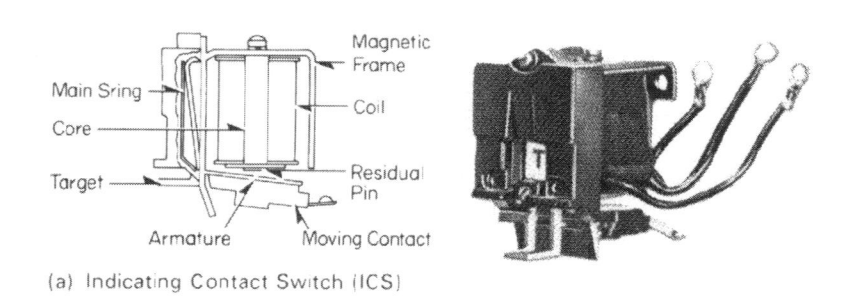

(a) Indicating Contact Switch (ICS)

Figure 3-1: Example of Clapper Style Relay

The next level of electromechanical relay incorporates an internal time delay using a rotating disk suspended between two poles of a magnet. When the input coil's magnetic strength is greater than the mechanical force holding the disk in the reset position, the disk will begin to turn toward the trip position. As the input signal (voltage or current) increases, the magnetic force increases, and the disk turns faster. The relay's pickup is adjusted by changing input coil taps and/or adjusting the holding spring tension. The time delay is altered by moving the starting position and varying the magnet strength around the disk. Time characteristics are preset by model.

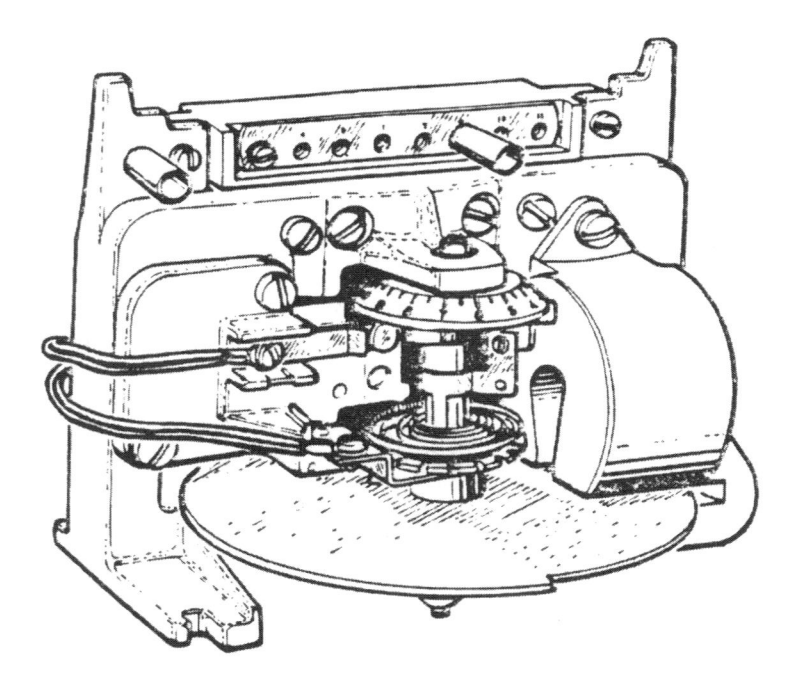

Figure 3-2: Typical Electromechanical Relay with Timing Disk

The next level of complexity included polarizing elements to determine the direction of current flow. This element is necessary for the following protective functions to operate correctly:

- Distance (21)
- Reverse Power (32)
- Loss of Field (40)
- Directional Overcurrent (67)

These relays used the components previously described but will not operate until the polarizing element detects that the current is flowing in the correct direction. Polarizing elements use resistors, capacitors, and comparator circuits to monitor current flow and operate a clapper style contact to shunt or block the protective functions accordingly.

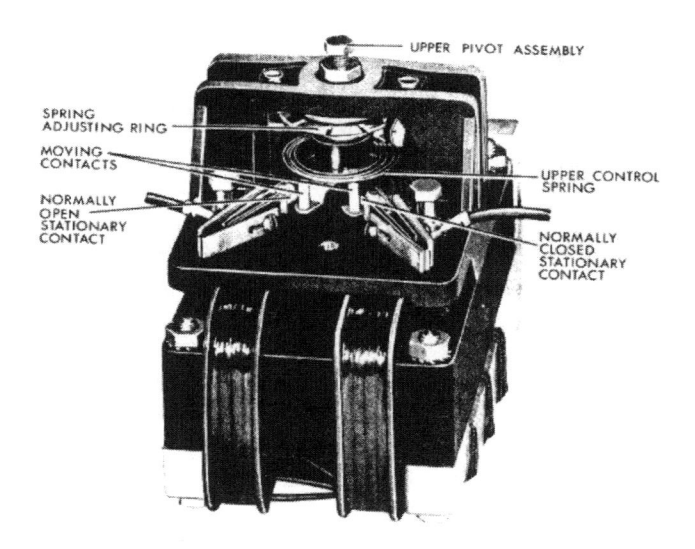

Figure 3-3: Example of an Electromechanical Relay Polarizing Element

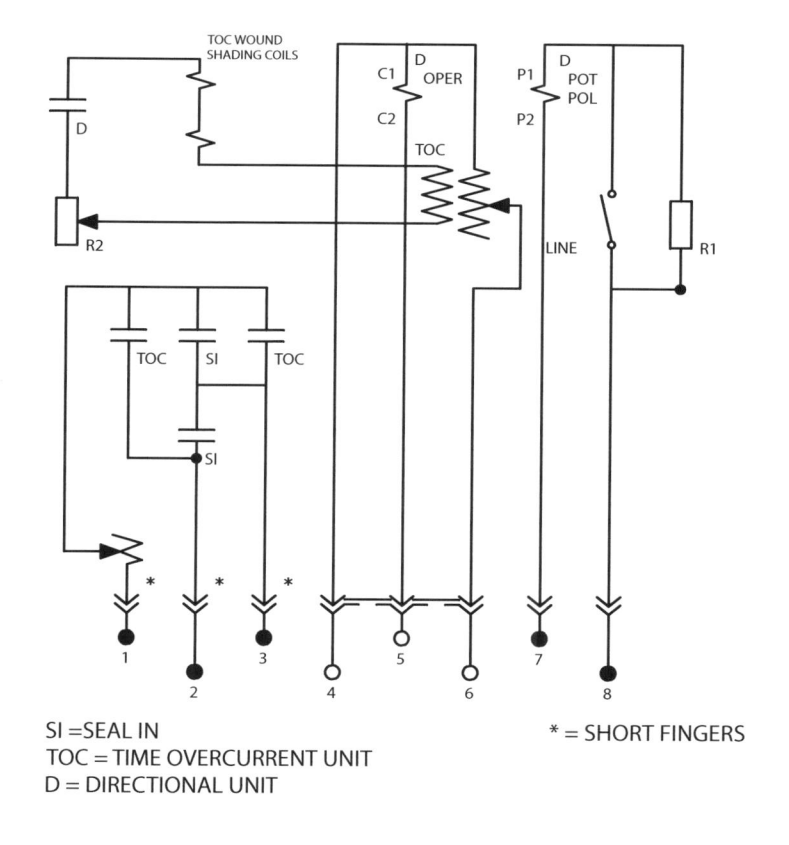

SI = SEAL IN
TOC = TIME OVERCURRENT UNIT
D = DIRECTIONAL UNIT

* = SHORT FINGERS

Figure 3-4: Typical Polarizing Element Electrical Schematic

As electromechanical relays are largely dependent on the interaction between magnets and mechanical parts, their primary problems are shared with all mechanical devices. Dirt, dust, corrosion, temperature, moisture, and nearby magnetic fields can affect relay operation. The magnetic relationship between devices can also deteriorate over time and cause the pickup and timing characteristics of the relays to change or drift without regular testing and maintenance.

These relays usually only have one or two output contacts and auxiliary devices are often required for more complex protection schemes. Figure 3-5 depicts a simple overcurrent protection scheme using electromechanical relays for one feeder. Notice that four relays are used to provide optimum protection and any single-phase relay can be removed for testing or maintenance without compromising the protection scheme.

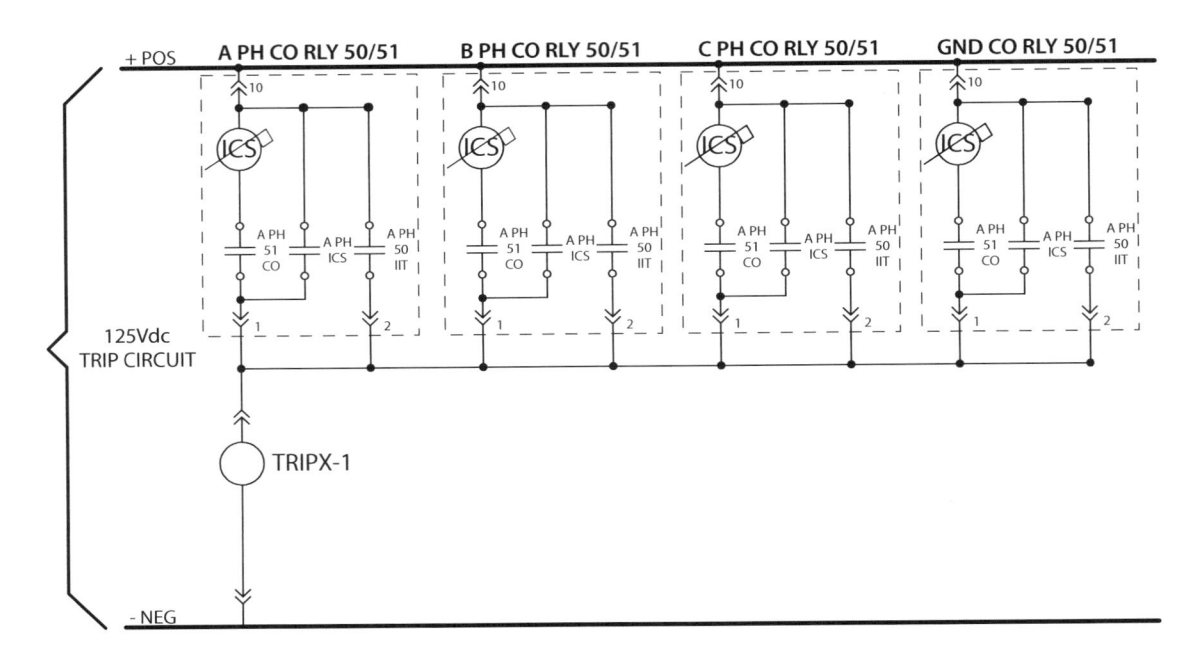

Figure 3-5: Typical Electromechanical Overcurrent Trip Schematic

2. Solid-State Relays

As technology progressed and electronic components shrunk in size, solid-state relays began to appear. The smaller, lighter, and cost-effective solid-state relays were designed to be direct replacements for the electromechanical relays. However, this generation of relays introduced new, unforeseen problems including; power supply failures and electronic component failures that prevented relays from operating, and sensitivity to harmonics that caused nuisance trips. Protective relays are the last line of defense during an electrical fault, and they must operate reliably. Unfortunately, early solid-state relays were often unreliable, and you will probably find many more electromechanical relays than solid-state relays in older installations.

Solid-state relays used electronic components to convert the analog inputs into very small voltages that were monitored by electronic components. Pickup and timing settings were adjusted via dip switches and/or dials. If a pickup was detected, a timer was initiated which caused an output relay to operate. Although the new electronics made the relays smaller, many models were made so they could be inserted directly into existing relay cases allowing upgrades without expensive retrofitting expenses. Early models were direct replacements with no additional benefits other than new technology, but later models were multi-phase or multi-function.

3. Microprocessor Based Relays

Microprocessor based relays are computers with preset programming using inputs from the analog-to-digital cards (converts CT and PT inputs into digital signals), digital inputs, communications, and some form of output contacts. The digital signals are analyzed by the microprocessor using algorithms (computer programs) to determine operational parameters, pickup, and timing based on settings provided by the end user.

The microprocessor relay, like all other computing devices, can only perform one task at a time. The microprocessor will analyze each line of computer code in predefined order until it reaches the end of the programming where it will begin analyzing from the beginning again. The relay scan time refers to the amount of time the relay takes to analyze the complete program once. A simplified program might operate as follows:

1. Start
2. Perform self-check
3. Record CT inputs
4. Record PT inputs
5. Record digital input status
6. Overcurrent pickup? If yes, start timer.
7. Instantaneous Pickup? If yes, start timer.
8. Any element for OUT101 On (1)? If yes, turn OUT101 on.
9. Any element for OUT102 On (1)? If yes, turn OUT102 on.
10. Back to Start

Microprocessor relays also evaluate the input signals to determine if the analog input signal is valid using complex analyses of the input signal waveforms. These evaluations can require significant portions of a waveform or multiple waveform cycles to properly evaluate the input signal which can increase the microprocessor relay's operating time.

Electrical faults must be detected and cleared by the relay and circuit breaker as quickly as possible because an electrical fault can create an incredible amount of damage in a few cycles. The microprocessor relay's response time is directly related to the amount of programming and its processor speed. Early microprocessor speeds were comparatively slow, but they were also simple with small programs. As each additional feature or level of complexity is added, the processor speed must be increased to compensate for the additional lines of computer code that must be processed or the relay response time will increase.

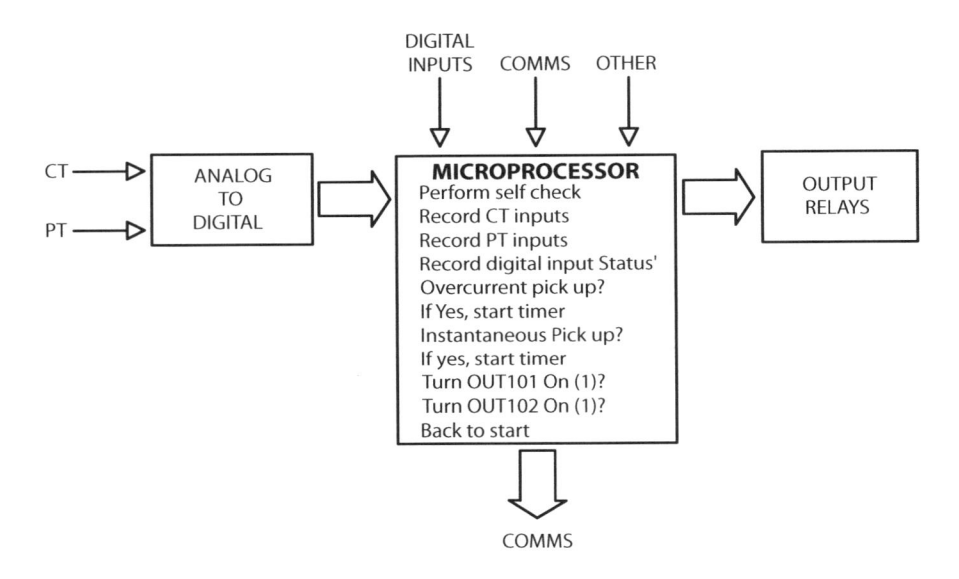

Figure 3-6: Simple Microprocessor Operation Flowchart

ANALOG TO
DIGITAL CONVERSION

MICROPROCESSOR

OUTPUT RELAYS

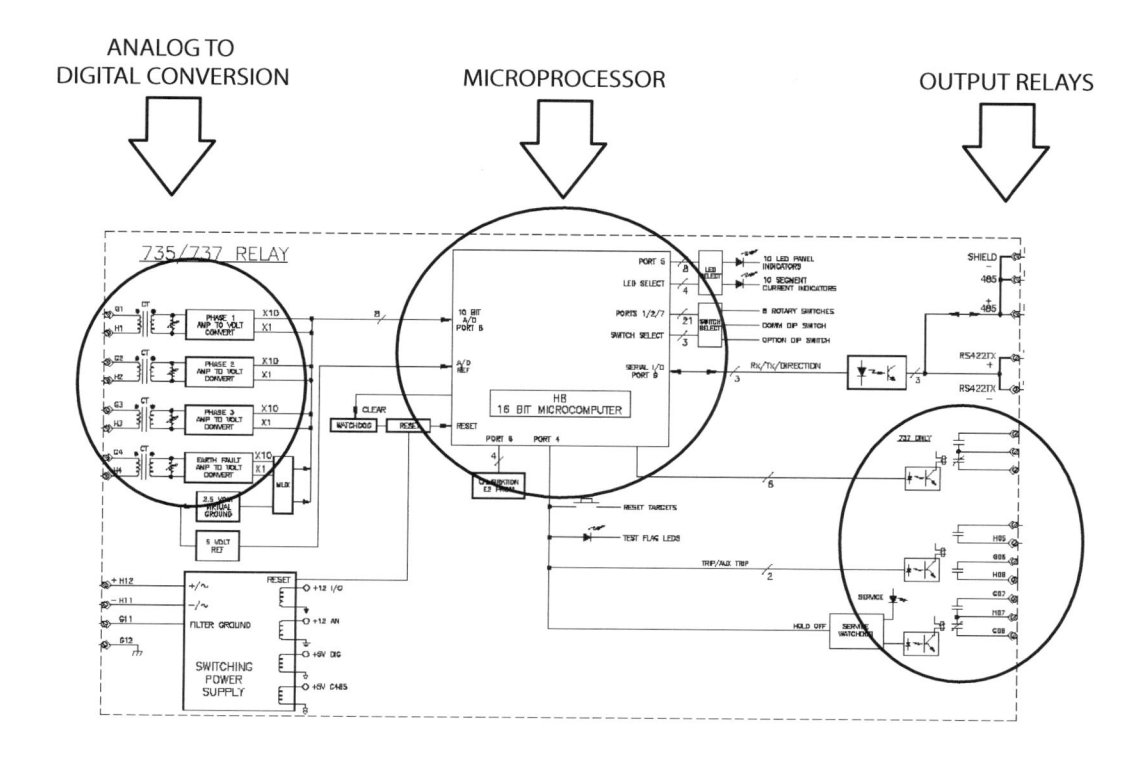

Figure 3-7: Simple Microprocessor Internal Schematic

A) Simple Digital Relays

Early microprocessor based relays were nothing more than direct replacements for electromechanical and solid-state relays. Most were simple multi-phase, single function relays with limited outputs. These relays were typically cheaper than comparable relays from previous generations and added additional benefits including:

- More sensitive settings
- Multiple time curve selections
- Metering functions
- Remote communications
- Self-test functions that monitored key components to operate an LED on the front display or operate an output contact
- Simple fault recording

These relays were relatively simple to install, set, and test as they had limited functions and limited contact configurations. The General Electric MIF/MIV Series or Multilin 735/737 are good examples of simple digital relays.

B) **Multi-Function Digital Relays**

As technology improved and microprocessors became faster and more powerful, manufacturers began to create relays with the all-in-one-box philosophy we see today. These relays were designed to provide all the protective functions for an application instead of a protective element as seen in previous relay generations. Instead of installing four overcurrent (50/51), three undervoltage (27), three overvoltage (59), two frequency (81), and one synchronizing (25) relays; just install one feeder management relay such as the SEL-351 or GE Multilin 750 relay to provide all these functions in one box for significantly less cost than any one of the previous relays. As a bonus, you also receive metering functions, a fault recorder, oscillography records, remote communication options, and additional protection functions you probably haven't even heard of. Because all the protective functions are processed by one microprocessor, individual elements become interlinked. For example, the distance relay functions are automatically blocked if a PT fuse failure is detected.

The all-in-one-box philosophy caused some problems as all protection was now supplied by one device and if that device failed for any reason, your equipment was left without protection. Periodic testing could easily be performed in the past with minimal risk or system disruption because only one element was tested at a time. Periodic testing with digital relays is a much more intrusive process as the protected device must either be shut down or left without protection during testing. Relay manufacturers downplay this problem by stating that periodic testing is not required because the relay element tests will not change over time because of the digital nature of microprocessor relays and the relays have many self-check features that will detect most problems. However, output cards, power supplies, input cards, and analog-to-digital converters can fail without warning or detection and leave equipment or the system without protection. Also, as everyone who uses a personal computer can attest, software can be prone to unexpected system crashes and digital relays are controlled by software.

As relays became more complex, relay settings became increasingly confusing. In previous generations a fault/coordination study was performed and the relay pickup and time dial settings were determined then applied to the relay in secondary amps. Today we can have multiple elements providing the same protection but now have to determine whether the pickup is in primary values, secondary values, or per unit. There can also be additional settings for even the simplest overcurrent (51) element including selecting the correct curve, voltage controlled or restraint functions, reset intervals, etc. Adding to the confusion is the concept of programmable outputs where each relay output contact could be initiated by any combination of protective elements and/or external inputs, and/or remote inputs via communications. These outputs are programmed with different setting interfaces based on the relay model or manufacturer with no standard for schematic drawings.

Multifunction digital relays have also added a new problem through software revisions. The computer software industry appears to be driven by the desire to add new features, improve operation, and correct bugs from previous versions. Relay programmers from some manufacturers are not immune from these tendencies. It is important to realize that each new revision changes the relay's programming and, therefore creates a brand new relay that must be tested after every revision change to ensure it will operate when required. In the past, the relay manufacturers and consumers, specifically the utility industry, extensively tested new relay models for months — simulating various conditions to ensure the relay was suitable for their systems. The testing today is either faster or less stringent as the relays are infinitely more complex, and new revisions or models disappear before some end users approve their replacements for use in their systems.

Examples of multifunction digital relays include the Schweitzer Laboratories product line (SEL), the GE Multilin SR series and the Beckwith M Series.

C) The Future of Protective Relays

A paradigm shift occurred when relay designers realized that all digital relays use the same components (analog-to-digital cards, input cards, output cards, microprocessor, and communication cards) and the only real differences between relays is programming. With this principle in mind, new relays are being produced that use interchangeable analog/digital input/output, communication, and microprocessor cards. Using this model, features can be added to existing relays by simply adding cards and uploading the correct software. These protective relays will have the same look and feel as their counterparts across the product line.

These relays will be infinitely configurable but will also be infinitely complex, requiring specialized knowledge to be able to operate and test. Also, software revisions will likely become more frequent. Another potential physical problem is also created if the modules are incorrectly ordered or installed.

Examples of this kind of relay include the Alstom M series, General Electric UR series, and the ABB REL series.

D) Digital Relay Considerations

While digital relays are more powerful, flexible, effective, and loaded with extras, they are also exponentially more complex than their predecessors and can easily be misapplied. As the relay accomplishes more functions within its programming, many of the protective functions within an electrical system become invisible and are poorly documented.

The electrical protection system in the past was summarized on a schematic drawing that was simple to understand and almost all of the necessary information was located in one place. If information was missing, you could look at the control cabinet wiring and trace the wires.

Today's protection system is locked within a box and poorly documented at best; this makes it nearly impossible for an operator or plant electrician to troubleshoot. It can even be difficult for a skilled relay technician due to communication issues, complex logic, cyber security, and difficult software. A one-character mistake can be catastrophic and can turn thousands of dollars' worth of protective relay into a very expensive Christmas decoration. Some problems may never be discovered because the relay may only need to operate one percent of the time and testing periods are rare due to manufacturer's claims that metering and self-diagnosis tests are all that's required for maintenance.

There is also a lack of standardization between engineering companies and relay manufacturers regarding how control, annunciation, and protection should be documented. While electromechanical or solid-state relay schemes would be well documented in the schematic diagrams for the site, few engineering companies provide an equivalent drawing of internal relay output logic and often reduce a complex logic scheme into "OUT 1/TRIP" or simply "OUT1." This kind of documentation is almost useless when trying to troubleshoot the cause of an outage, especially if the relay has a cryptic message on its display panel. An operator usually does not need the extra hassle of finding a laptop, RS-232 cable and/or RS-232 to RS-485 adapter, cable adapter, sorting through the various communication problems and menu layers only to determine that an overcurrent (51) element caused the trip. In the past, the operator could simply walk to the electrical panel and look for the relay target and read the label above the relay to determine what caused the trip.

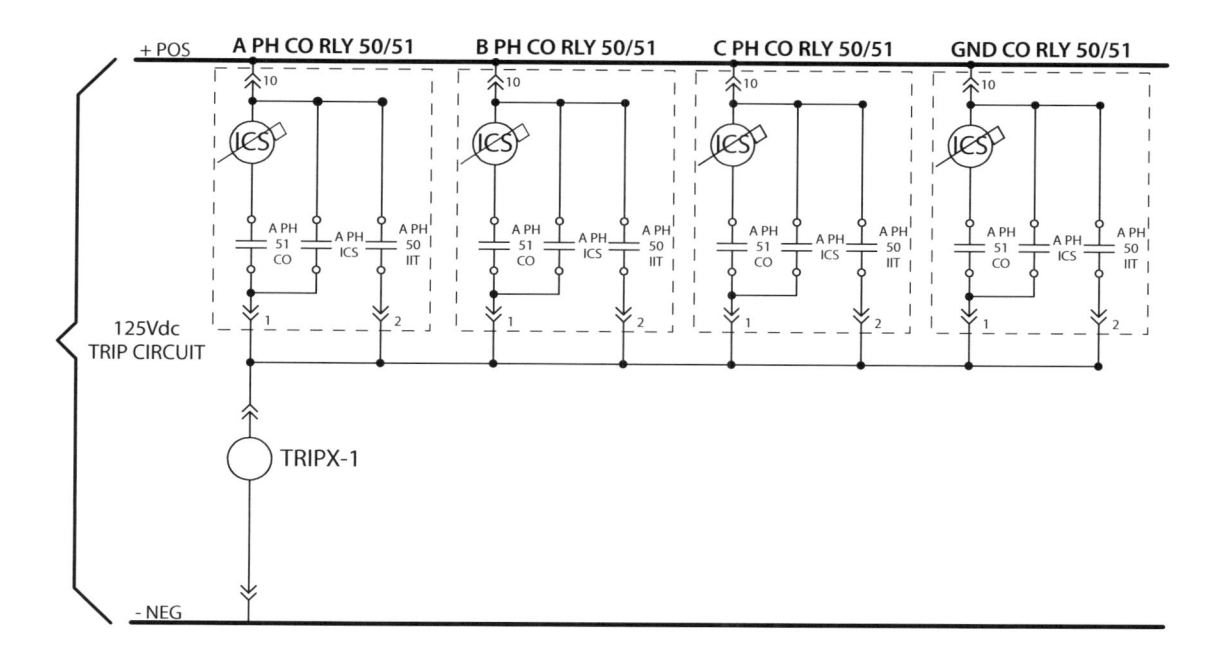

Figure 3-8: Typical Electromechanical Overcurrent Trip Schematic

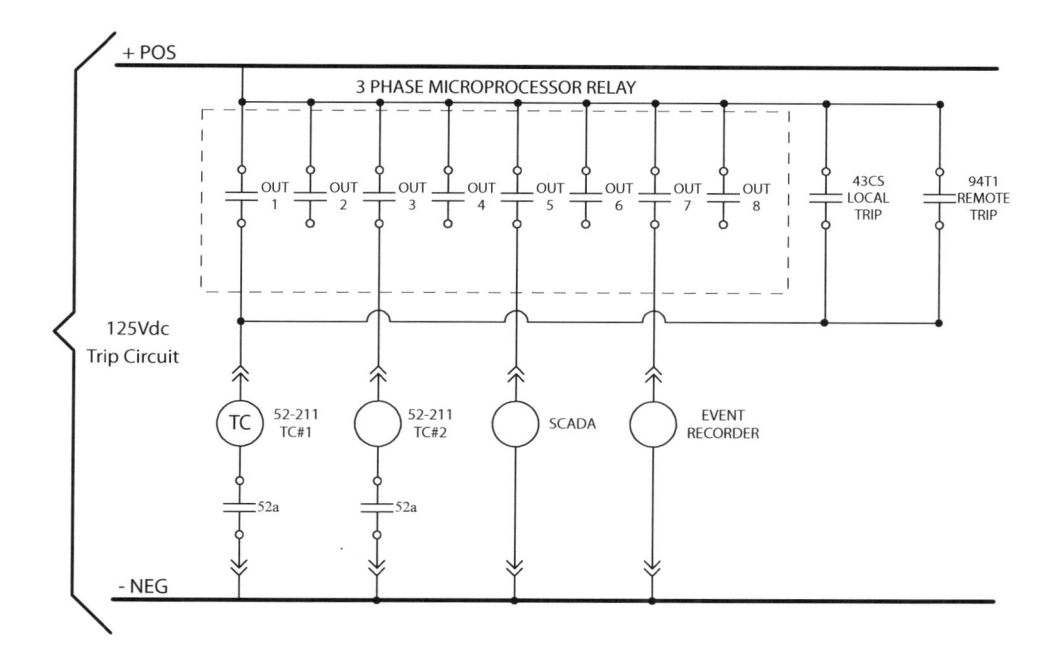

Figure 3-9: Typical Digital Relay Trip Schematic

Figure 3-10: Electromechanical Generator Protection Panel

Figure 3-11: Digital Generator Protection Panel

A properly applied microprocessor relay, on the other hand, will usually provide protection that is more sensitive and can be a godsend when you are looking for those elusive, quirky problems. As with all aspects of life, there are trade-offs for every benefit. The difference between electromechanical relays and digital relays can be compared to a carburetor vs fuel injection. The modern fuel injector is more efficient with less moving parts than a carburetor, but you would probably prefer a carburetor when you are broken down in the middle of the desert with no help in sight, because you could at least try something with the carburetor to get you home.

Chapter 4

Relay Testing Fundamentals

In the past, relay testing options were limited by the test equipment available and the simple (relatively speaking) electromechanical relays to be tested. Today's relays are highly sophisticated with an incredible number of settings that are tested with equally powerful test equipment and test systems. Complicated electromechanical relays required only a day to calibrate and test all functions. A digital relay could require weeks to test all of its functions. Relay testing is expensive and intelligent choices regarding what we test and how the test must be made. The best answers to these questions come from an evaluation of our goals and options.

1. Reasons for Relay Testing

Some of the reasons for relay testing include:

A) Type Testing

Type testing is a very extensive process performed by a manufacturer or end user that runs a relay through all of its paces. The manufacturer uses type testing to either prove a prospective relay model (or software revision), or as quality control for a recently manufactured relay. The end user, usually a utility or large corporation, can also perform type tests to ensure the relay will operate as promised and is acceptable for use in their system.

This kind of testing is very involved and, in the past, would take months to complete on complicated electromechanical relays. Every conceivable scenario that could be simulated was applied to the relay to evaluate its response under various conditions. Today, all of these scenarios are now stored as computer simulations that are replayed through advanced test equipment to prove the relay's performance in hours instead of weeks.

Type testing is very demanding and specific to manufacturer and/or end-user standards. Type testing should be performed before the relay is ordered and is not covered in this book. However, independent type testing is a very important part of a relay's life span and should be performed by end users before choosing a relay model.

B) Acceptance Testing

There are many different definitions of acceptance testing. For our purposes, acceptance testing ensures that the relay is the correct model with the correct features; is operating correctly; and has not been damaged in any way during transport. This kind of testing is limited to functional tests of the inputs, outputs, metering, communications, displays, and could also include pickup/timing tests at pre-defined values. Acceptance test procedures are often found in the manufacturer's supplied literature.

C) Commissioning

Commissioning and acceptance testing are often confused with each other but can be combined into one test process. Acceptance testing ensures that the relay is not damaged. Commissioning confirms that the relay's protective element and logic settings are appropriate to the application and will operate correctly. Acceptance tests are generic and commissioning is site specific.

Commissioning is the most important test in a digital relay's lifetime.

The tests recommended in this book are a combination of acceptance testing and commissioning and includes the following items:

- Functional tests of all digital inputs/outputs
- Metering tests of all current and voltage inputs
- Pickup and timing tests of all enabled elements
- Functional checks of all logic functions
- Front Panel display, target and LEDs

This battery of tests is a combination of acceptance testing and commissioning to prove:

- The relay components are not damaged and are acceptable for service
- The specific installation is set properly and operates as expected.
- The design meets the application requirements.
- The relay works in conjunction with the entire system.

D) Maintenance Testing

Maintenance tests are performed at specified intervals to ensure that the relay continues to operate correctly after it has been placed into service. In the past, an electromechanical relay was removed from service, cleaned, and fully tested using as-found settings to ensure that its functions had not drifted or connections had not become contaminated. These tests were necessary due to the inherent nature of electromechanical relays.

Today, removing a relay from service effectively disables all equipment protection in most applications. In addition, digital relay characteristics do not drift and internal self-check functions test for many errors. There is a heated debate in the industry regarding maintenance intervals and testing due to the inherent differences between relay generations.

In my opinion, the following tests should be performed annually on all digital relays:

- Perform relay self-test command, if available.
- Check metering while online and verify with external metering device, or check metering via secondary injection. (This test proves the analog-to-digital converters.)
- Verify settings match design criteria.
- Review event record data for anomalies or patterns.
- Verify all inputs from end devices.
- Verify all connected outputs via pulse/close command or via secondary injection. (Optionally, verify the complete logic output scheme via secondary injection.)

E) Troubleshooting

Troubleshooting is usually performed after a fault to determine why the relay operated or why it did not operate when it was supposed to. The first step in troubleshooting is to review the event recorder logs to find out what happened during the fault. Subsequent steps can include the following, depending on what you discover in the post-fault investigation.

- Change the relay settings accordingly.
- Change the event record or oscillography initiate commands.
- Re-test the relay.
- Test the relay's associated control schemes.
- Replay the event record through the relay or similar relay to see if the event can be replicated.

2. Relay Testing Equipment

Using the basic equipment listed, you will be able to apply all of the principles in this book and test over 90% of relay applications.

A) Test-Set

Your test-set should have a minimum of two current and two voltage channels with independent controls for phase angle, magnitude, and frequency. It is possible to simulate a balanced 3-phase system using two currents and voltages as shown in Figure 4-1. 3-phase voltage and currents are, of course, preferred. The test-set must be able to supply at least 25 Amps per channel and be able to parallel outputs for higher currents. A relay set capable of producing 75 Amps will be able to test most applications. Some other features required to test most applications include:

- Normally open/closed and/or voltage sensing input for timing
- At least one normally open/closed output contact to simulate breaker status
- Independent prefault and fault modes for dynamic testing

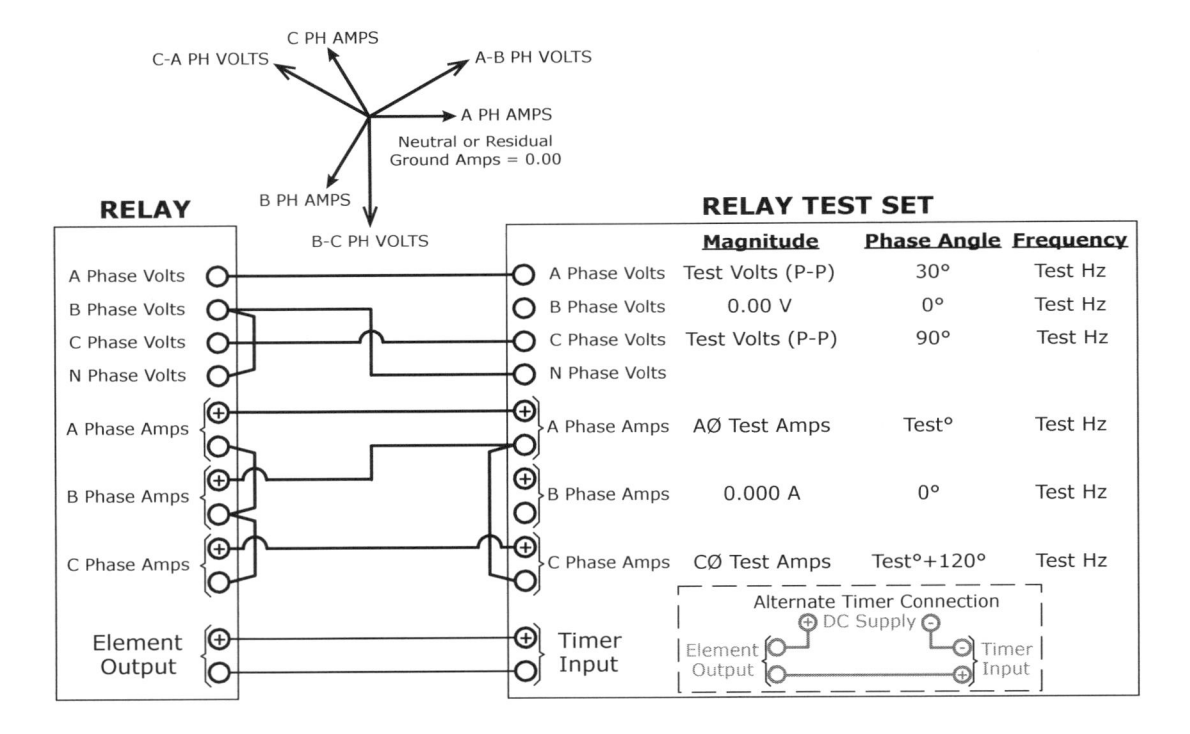

Figure 4-1: 3-Phase Relay Input Connection Using Two Sources

The perfect test-set features include:

- 4 independent voltage and 7 independent current sources
- 40 Amps (short-term) per phase in a 3-phase system
- The ability to control each channel independently, three phases together, and any two phases together
- True breaker-status simulation
- Number pad, thumbwheel, and computer control
- Communication port replication to allow one serial port to communicate with the test-set and relay
- Configurable timer for on/off delay, pulse timing, and multi-shot recloser timing

I have experienced many equipment malfunctions, computer failures, and communication nightmares using electrical testing equipment over the years and make it a policy to never buy a piece of test equipment that cannot be controlled manually in some fashion. By following this approach to equipment purchasing, I will never be held up waiting for a computer repair or delivery of the controlling software that was left in the office.

B) Test Leads

I usually carry the following list of test leads to minimize lead changing and wasted time:

- 8-foot banana-to-spade lug leads with red, black, blue, white, green, and yellow conductors rated for at least 30 Amps. (First choice for current connections)
- 12-foot banana-to-alligator leads with red, black, blue, white, green, and yellow conductors rated for at least 18 Amps. (For voltage connections or extensions)
- 8-foot banana-to-alligator leads with red, black, blue, white, green, and yellow conductors rated for at least 18 Amps. (For second current connections in differential circuits or for complicated input logic)
- 8-foot banana-to-banana leads with red, black, blue, white, green, and yellow conductors rated for at least 18 Amps. (Used as extensions)
- 12-foot banana-to-alligator with red and black conductors rated for at least 18 Amps. (For timer-stop input to relay set)
- 12-foot banana-to-alligator with blue and white conductors rated for at least 18 Amps to simulate contact input to relay.
- 12-foot banana-to-alligator with yellow and green conductors rated for at least 18 Amps to simulate a second contact input to the relay or as extension to reach behind panels.

C) Communication Supplies

Communication between relays is normally simple, and these basics will help you connect to nearly every relay:

- One, 10-foot 9-pin to 9-pin, male-to-female serial cable
- One male 9-pin gender bender
- One female 9-pin gender bender
- One null-modem adapter
- One, 25-foot, 9-pin to 9-pin, male-to-female serial cable
- One 25-foot Ethernet Cable
- One 25-foot crossover Ethernet Cable

If you expect problems, the following supplies can help you troubleshoot or overcome obstacles or connect to older equipment:

- Serial traffic monitor with lights to indicate communication paths. This proves both sides are trying to communicate.
- Serial breakout box that allows you to customize cable configurations with jumpers, or
- Customizable cable-end terminations for use with network cables.

3. Relay Testing Methods

This section will outline the evolution of relay testing to better understand the choices available to the relay tester when testing modern digital relays.

A) Electromechanical Relay Testing Techniques

Electromechanical relays operation was based on mechanics and magnetism and it was important to test all of the relay's characteristics to make sure that the relay was in tolerance. Various tests were applied to ensure all of the related parts were functioning correctly and, if the relay was not in tolerance; the relay resistors, capacitors, connections, and magnets were adjusted to bring the relay into tolerance. With enough patience, almost any relay could be adjusted to acceptable parameters.

Relay testing in the electromechanical age was very primitive due to the limitations of the test equipment available. The most advanced test equipment available to the average relay tester would include a variac for current output, another variac for voltage signals, a built in timer with contact sensing, and a phase shifter for more advanced applications. With this equipment, detailed test plans and connection diagrams, and a hefty dose of patience; the relay tester was able to test the pickup, timing, and characteristics of the electromechanical relays installed as well as most solid-state relays. Test plans and connections for currents and voltages often had very little resemblance to the actual operating connections because the limited test equipment could not create simulations of actual system conditions during a fault. Electromechanical relays were also built with inter-related components that needed to be isolated for calibration.

The following techniques were used when testing electrical-mechanical relays:

i) Steady State

Steady state testing is usually used for pickup tests. The injected current/voltage/ frequency is raised/lowered until the relay responds accordingly. Steady state testing can be replaced by jogging the injected value up/down until the relay responds.

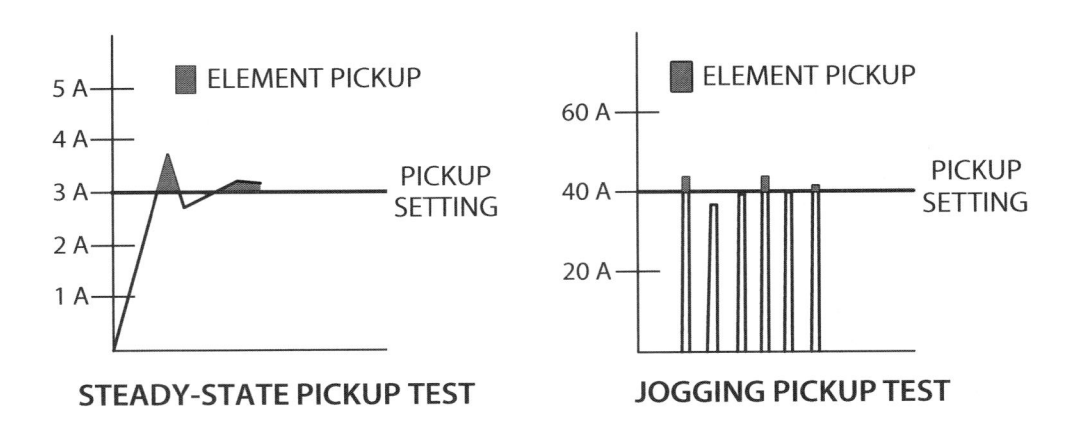

Figure 4-2: Steady State Pickup Testing

ii) Dynamic On/Off Testing

Dynamic on/off testing is the simplest form of fault simulation and was the first test used to determine timing. A fault condition is suddenly applied at the test value by closing a switch between the source and relay or activating a test-set's output.

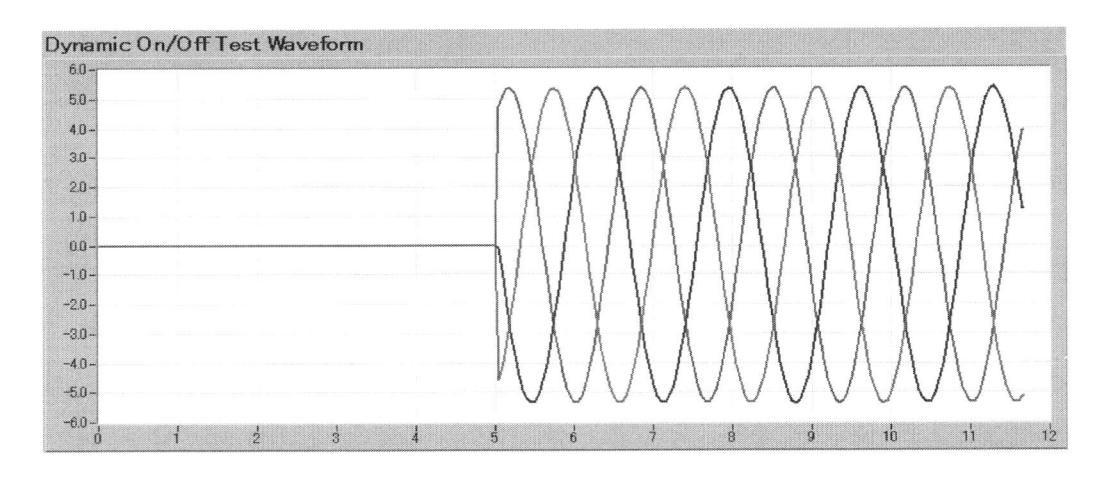

Figure 4-3: Dynamic On/Off Waveform

iii) Simple Dynamic State Testing

Some protective elements such as under-frequency (81) and undervoltage (27) require voltages and/or current before the fault condition is applied or the element will not operate correctly. Simple dynamic state testing uses prefault and/or post-fault values to allow the relay tester to obtain accurate time tests. A normal current/voltage/frequency applied to the relay suddenly changes to a fault value. The relay-response timer starts at the transition between prefault and fault, and the timer ends when the relay operates. Simple dynamic state testing can be performed manually with two sources separated by contacts or switches; applying nominal signals and suddenly ramping the signals to fault levels; or using different states such as prefault and fault modes.

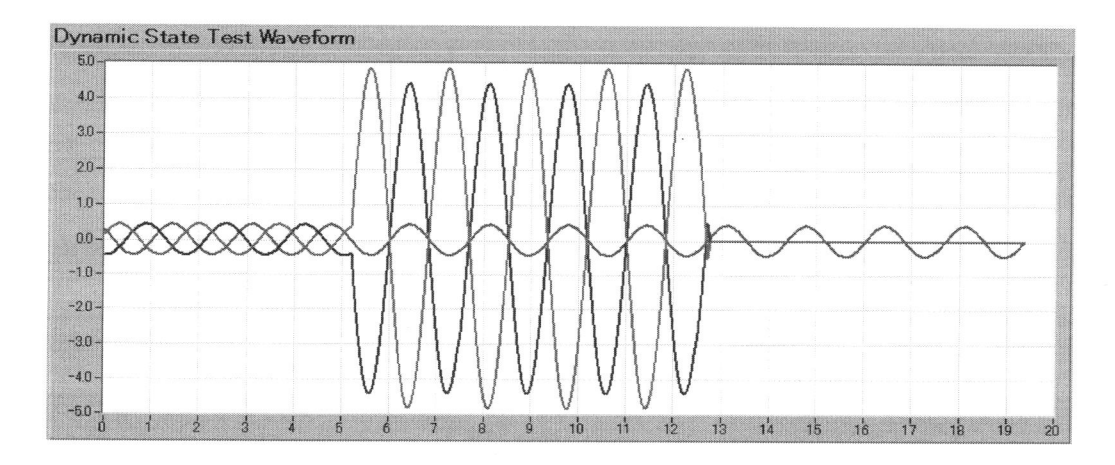

Figure 4-4: Simple Dynamic Test Waveform

B) Solid-State Relay Testing Techniques

Solid-state relays were primarily created to be direct replacements for electromechanical relays and the same test techniques were used for these relays. However, these solid-state relays were constructed with silicon chips, digital logic, and mathematical formulas instead of steel and magnetism so the test plans were the same but the results were often very different. When an electromechanical relay was found out of tolerance, there were resistors or springs to adjust. Solid-state relays did not have many adjustments besides the initial pickup setting and these relays either operated correctly or did not operate at all. Relay testers typically replaced entire cards instead of adjusting components when the relay failed.

Test equipment did not evolve much during this period and the typical test-set changed all of its analog displays to digital and made the previously described tests easier to perform.

C) **Microprocessor Relay Testing Techniques**

Simple microprocessor relays are almost identical in operation to the solid-state relays they replaced and the test techniques for these relays are identical to the techniques previously described.

Complex microprocessor relays included a large number of settings and interlinked elements which created confusion in the relay testing industry because a relay tester could spend an entire week testing one relay and barely scratch the surface of the relay's potential. The confusion increased when relay manufacturers claimed that relay testing was not required because the relay performed self-check functions and the end user would be informed if a problem occurred. Some manufacturers even argued that the relay could test itself by using its own fault recording feature to perform all timing tests. Eventually a consensus was reached where the relay tester would test all of the enabled features in the relay. Relay testers began modifying and combining their electromechanical test sheets to account for all of the different elements installed in one relay but the basic fundamentals of relay testing didn't change very much.

One of the first problems that a relay tester experiences when testing microprocessor relay elements is that different elements inside the relay often overlap. For example, an instantaneous (50) element set at 20A will operate first when trying to test a time-overcurrent (51) element at 6x (24A) its pickup setting (4A). The relay tester instinctively wants to isolate the element under test and usually changes the relay settings to set one output, preferably an unused one, to operate only if the element under test operates. Now they can perform that 6x test without interference from the 50-element. While these techniques will give the test technician a result for their test sheet, the very act of changing relay settings to get that result does not guarantee that the relay will operate correctly when required because the in-service relay settings and reactions have not been tested.

Relay testers often use the steady-state and simple-dynamic test procedures described previously to perform their element tests on microprocessor relays which create another problem. These complex relays are constantly monitoring their input signals to determine if those signals are valid. The steady-state and simple-dynamic test procedures are often considered invalid system conditions by the relay and the protection elements will not operate to prevent nuisance trips for a perceived malfunction. For example, if a relay tester tries to perform a standard electromechanical impedance test (21) on a digital relay, the relay will likely assume that there is a problem with a PT fuse and block the element; or the switch-on-to-fault (SOTF) setting could cause the relay to trip instantaneously. Relay testers who encountered this problem often disable those blocking signals to perform their tests and, hopefully, turned the blocking settings back on when they were complete. Again, the act of changing settings is fine if you need a number for a test sheet but will not guarantee that the relay will operate correctly when it is required.

Modern test equipment allows the relay tester to apply several different techniques to overcome any of the situations described above. When an instantaneous element operates before a time element timing test can operate, that is usually a good thing if the relay is programmed correctly. Instead of modifying the settings to get a result, the technician can modify the test plan to ensure that all time-element tests fall below the instantaneous pickup. If a ground element operates before the phase element you are trying to test, apply a realistic phase-to-phase or 3-phase test instead and the ground element will not operate. If the loss of protection element prevents a distance relay from operating, apply a balanced 3-phase voltage for a couple of seconds between each test to simulate real life conditions. If switch-on-to-fault operates whenever you apply the fault condition; use an output to simulate the breaker status, apply prefault current, or lower the fault voltage so that you can lower the fault current when testing impedance relays. All of these possibilities are easily applied with modern test equipment to make our test procedures more intelligent, realistic, and effective.

Modern test equipment also allows the following additional test methods.

i) Computer-Assisted Testing

Because modern test equipment is controlled by electronics, computer-assisted testing became available. Standard test techniques could be repetitive on relays that were functioning correctly. Computer programs were created that would ramp currents and voltages at fixed rates in an effort to make relay testing faster with more repeatable results because every test would be performed identically. Computer-assisted testing has evolved to the point where the software will:

- connect to the relay
- read the relay settings
- create a test plan based on the enabled settings
- modify the settings needed to isolate an element and prevent interference
- test the enabled elements
- restore the relay settings to as found values

By following the steps above, computer-assisted relay testing can replace the relay tester on a perfectly functioning relay and can theoretically perform the tests faster than a human relay tester can. This type of testing works extremely well when performing pickup and timing tests of digital relays because these relays are computer programs themselves.

However, it is very unlikely that the basic test procedures described by computer-assisted testing, whether initiated by computers or humans, will discover a problem with a digital relay. Most digital relay problems are caused by the settings engineer and not the relay. If a computer or human reads the settings from the relay and regurgitates them into the test plan, they will not realize that the engineer meant to enter a 0.50A pickup but actually applied 5.0A which will make the ground pickup larger than the phase pickup. Each element will operate as programmed and, when tested in isolation, will create excellent test results but may not be applied in the trip equation which was probably not tested by the automated program. An excellent relay technician could create additional tests to perform the extra steps necessary for a complete test; but will that technician be more capable or less capable as they rely more heavily on automation to perform their testing?

ii) State Simulation

State simulations allow the user to create dynamic tests where the test values change between each state to test the relay's reaction to changes in the power system. Multiple state simulations are typically required for more complex tests such as frequency load shedding, end-to-end tests, reclosing, breaker-fail, and the 5% under/over pickup technique described later in this chapter.

iii) Complex Dynamic State Testing

Complex dynamic state testing recognizes that all faults have a DC offset that is dependent on the fault incidence angle and the reactance/resistance ratio of the system. Changing the fault incidence angle changes the DC offset and severity of the fault and can significantly distort the sine wave of a fault as shown in Figure 4-5. This kind of test requires high-end test equipment to simulate the DC offset and fault incidence angle and may be required for high speed and/or more complex state-of-the-art relays.

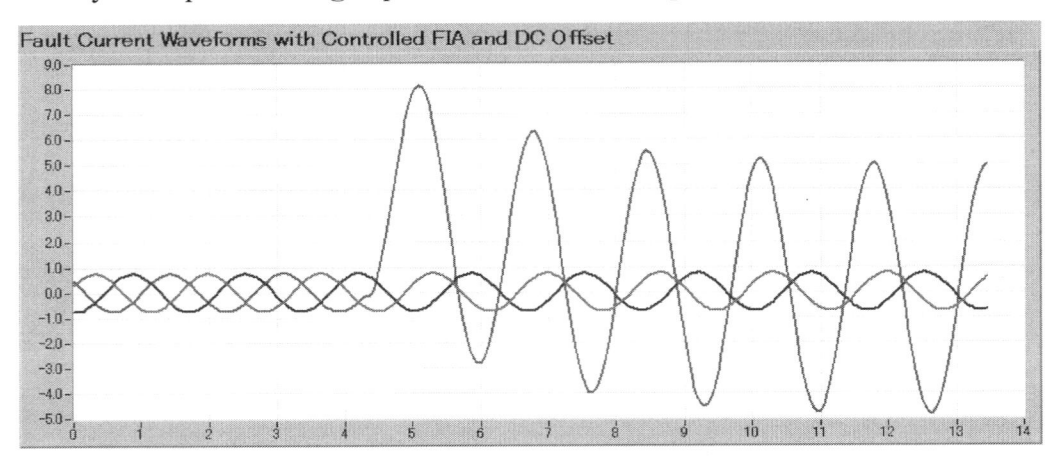

Figure 4-5: Complex Dynamic Waveform (Compliments of Manta Test Systems)

iv) Dynamic System Model Based Testing

Dynamic system model based testing uses a computer program to create a mathematical model of the system and create fault simulations based on the specific application. These modeled faults (or actual events recorded by a relay) are replayed through a sophisticated test-set to the relay.

Arguably, this is the ultimate test to prove an entire system as a whole. However, this test requires specialized knowledge of a system, complex computer programs, advanced test equipment, and a very complex test plan with many possibilities for error. My biggest concern regarding this kind of testing is the level of expertise necessary for a successful test. In the age of fast-track projects and corporate downsizing, many times the design engineer is barely able to provide settings in time for energization without relying on the designer to provide test cases as well.

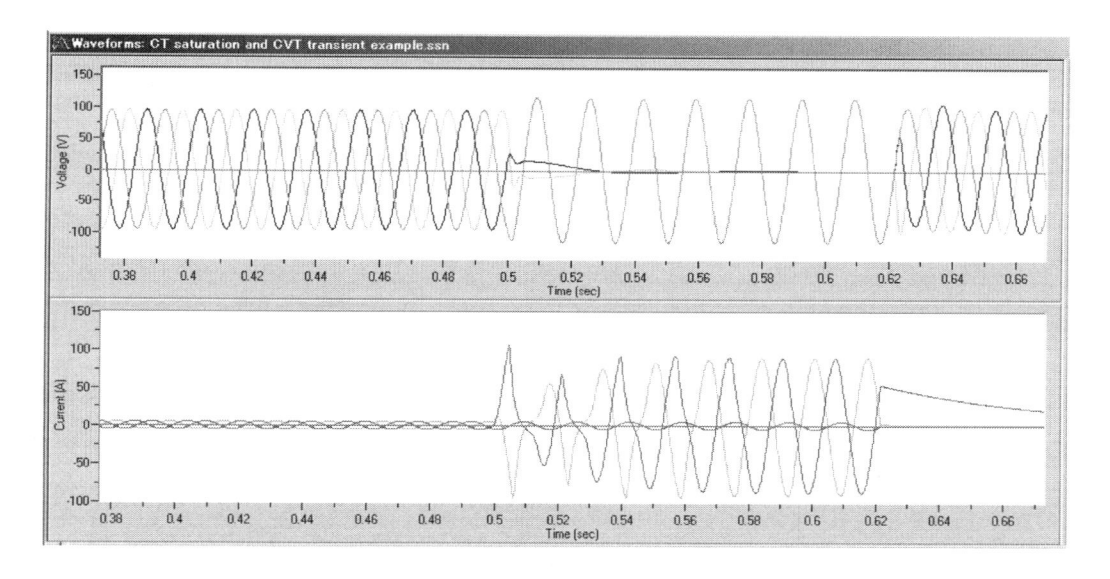

Figure 4-6: System Modeling Waveform (Compliments of Manta Test Systems)

v) End-to-End Testing

End to end testing is performed when two or more relays are connected together via communication channels to protect a transmission line. These relays can transfer status or metering information between each other to constantly monitor the transmission line in order to detect faults and isolate the faulted transmission line more quickly and reliably than single relay applications. The relays can communicate to each other through a wide range of possibilities including telecom equipment, fiber optic channels, or wave traps that isolate signals transferred over the transmission line.

Testing these complicated schemes in the past was limited to functional tests of the individual components with a simplified system test to prove that the basic functions were operating correctly. For example, a relay tester would configure and test the relay, configure and test the communication equipment, then inject a fault condition into the relay. A relay tester at the other location would verify that they received the signal and they would repeat the process at the remote end. This procedure tested the base components of the system but they often failed to detect problems that occurred with faults in real time. For example, a fault on parallel feeders could change direction in fractions of cycles when one breaker in the system operated that often caused the protection schemes to mis-operate.

Relay testers could only test one end at a time because there was no way to have two test-sets at remote locations start at exactly the same moment. If the test-sets do not provide coordinated currents and voltages with a fraction of a cycle, the protection scheme would detect a problem and mis-operate. Global Positioning System (GPS) technology uses satellites with precise clocks to communicate with equipment on earth which allowed test-set manufacturers to synchronize test-sets at remote distances. After the test-sets are synchronized, test plans could be created with simulated faults for each end of the transmission line. To perform a test, the relay testers synchronize their test-sets, load the same fault simulation with the values for their respective ends, set the test-sets to start at the exact same moment, and initiate the countdown. The test-sets will inject the fault into the relays simultaneously and they should respond as if the fault occurred on the line. The relays' reactions are analyzed and determined to be correct before proceeding to the next test. Any mis-operations are investigated to see where the problem originates and corrected.

End to end testing is typically considered to be the ultimate test of a system and should ideally perform using Dynamic System Model Testing to ensure that the system is tested with the most comprehensive test conditions. Simpler end-end schemes such as Line-Differential schemes can be tested using Simple Dynamic State testing.

4. Relay Test Procedures

A) Pickup Testing

There are several methods used to determine pickup, and we will review the most popular in order of preference. You must remember that we strive for minimum impact or changes when testing. If there is a method to determine pickup without changing a setting…use it!

i) Choose a Method to Detect Pickup

a) Front Display LEDs

Many relays will have LEDs on the front display that are predefined for pickup or can be programmed to light when an element picks up. Choose or program the correct LED and change the test-set output until the LED is fully lit. Compare the value to the manufacturer's specifications and record it on the test sheet.

- SEL relays often allow you to customize LED output configurations using the "TAR" or "TAR F" commands. Some SEL LEDs can only be controlled via the relay's front panel.
- Monitor Beckwith Electric relay pickup values by pressing and holding the reset button. The appropriate LED will light on pickup. Some elements share a single LED, and you must use alternate methods to determine pickup.
- GE UR relays allow you customize the front display LEDs using the "Product Set-up" "User Programmable LEDs" menu.
- Most of the GE/Multilin element pickups can be monitored via the pre-defined pickup LED.

b) Front Display Timer Indication

Some relays, including the Beckwith Electric models, have a menu item on the front panel display that shows the actual timer value in real time. After selecting the correct menu item, change the test-set output until the timer begins to count down. Compare the value to the manufacturer's specifications and record it on the test sheet.

c) Status Display via Communication

Some relays provide a real-time status display on an external computer or other device via communication. While this is the most unobtrusive method of pickup testing, the accuracy of this method is limited by the communication speed. Some relays will also slow down the communication speed during events that can further decrease accuracy.

Slowly change the test-set output and watch the display to get a feel for the time between updates. Change the relay test-set output at a slower rate than the update rate until the relay display operates. Compare the value to the manufacturer's specifications and record it on the test sheet.

d) Output Contact

This method requires you to assign an output contact to operate if the element picks up. Choose an unused output contact whenever possible and monitor the contact with your relay test-set or external meter. You can often hear the output contact operate, but you must be sure that you are listening to the correct relay. Another element in the control logic could be operating while you are performing your test.

If you are unable to assign a pickup element to an output contact, you can set the element time delay to zero. This method is obviously not a preferred method. Always test the time delay after the pickup to ensure the time delay was returned to the correct value.

ii) Pickup Test Procedure

After selecting a pickup method, apply the current/voltage/frequency necessary for pickup and make sure a pickup is indicated by whichever method you have selected. Slowly decrease the relay test-set output until the pickup indication disappears. Slowly increase the test-set output until pickup is indicated. If the test current/voltage is higher than the input rating, only apply test value for the minimum possible duration. See individual element testing for tips and tricks for individual elements.

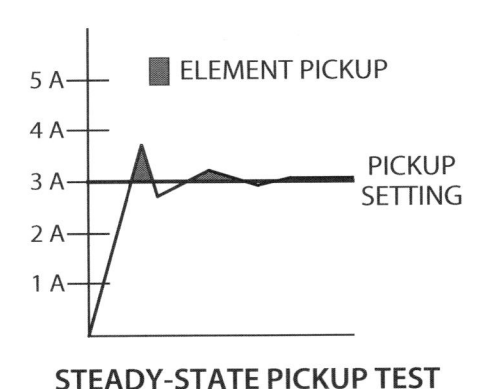

STEADY-STATE PICKUP TEST

Figure 4-7: Graph of Pickup Test

Calculate the percent error using the formula in Figure 4-8 and compare the results to the manufacturer's specifications. Record the pickup value and percent error on the test sheet.

$$\frac{\text{Actual Value - Expected Value}}{\text{Expected Value}} \times 100 = \text{percent error}$$

Figure 4-8: Percent Error Formula

Device Number	Function	Setpoint Ranges	Increment	Accuracy
(50)	50W1/50W2 Pickup #1, #2	1.0 to 100.0 A (0.2 to 20.0 A)	0.1 A	+/- 0.1 A or +/- 3% (+/- 0.02 A or +/- 3%)
	Trip TIme Response	Fixed 2 Cycles	-	+/- 2 Cycles

Figure 4-9: Beckwith Electric M-3310 Relay Element Specifications

For example, if we performed a pickup test for the 50-element on a relay and recorded a pickup value of 3.05A, the 50-element pickup setpoint is 3.0A. Using the formula in Figure 4-10 the percent error for this pickup is:

$$\frac{\text{Actual Value - Expected Value}}{\text{Expected Value}} \times 100 = \text{percent error}$$

$$\frac{3.05 - 3.0}{3.0} \times 100 = \text{percent error}$$

$$\frac{0.05}{3.0} \times 100 = \text{percent error}$$

1.66%

Figure 4-10: Example Pickup Percent Error Calculation

The percent error is within the specified "+/- 3% error" in the "Accuracy" column and is acceptable for service.

Always document setting changes and return the settings to their original values before proceeding!

B) Timing Tests

Timing tests apply a test input at a pre-defined value in the pickup region and measures the time difference from test initiation until the relay output-contact operates. This is the dynamic on/off testing method described earlier. Some elements like undervoltage (27) or under-frequency (81U) require prefault, non-zero values in order to operate correctly.

The output contact used to turn the timing set off is preferably the actual output contact used while in service. The timing test contact can be a spare output contact if another element interferes with the element timing, but the actual element output contacts must be verified as well. Some outputs are designed to operate at different speeds. Always use the high-speed output if a choice is available.

If the time delay is a constant value such as zero seconds or one second, apply the input at 110% of the pickup value and record the time delay. Determine if the manufacturer's specified time delays are in seconds, milliseconds, or cycles and record test results.

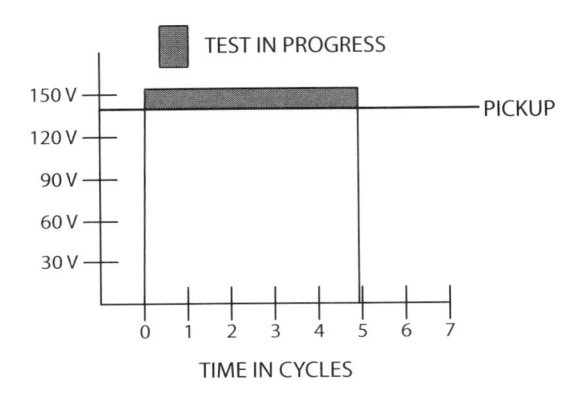

Figure 4-11: Simple Off/On Timing Test

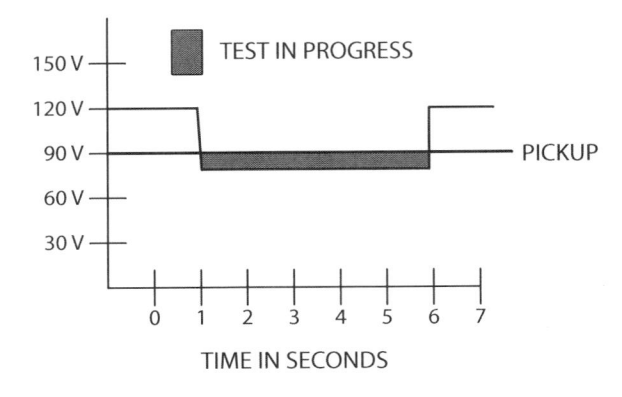

Figure 4-12: Dynamic On/Off Testing

Calculate the percent error using the percent error formula and compare the results to the manufacturer's specifications. Record the pickup value and percent error on the test sheet.

Device Number	Function	Setpoint Ranges	Increment	Accuracy
(51)	51W1 / 51W2 Pickup	0.50 to 12.00 A (0.10 to 2.4 A)	0.01 A	+/- 0.1 A or +/- 3% (+/- 0.02 A or +/- 3%)
	Characteristic Curve	Definite Time / Inverse / Very Inverse / Extremely Inverse / IEC Curves		
	Time Dial Setting	0.5 to 11.0 0.05 to 1.10 (IEC Curves)	0.1 0.01	+/- 3 Cycles or +/- 5%

Figure 4-13: Beckwith Electric M-3310 Relay Element Specifications

For example, a timing test result for the 51-element in Figure 4-13 was recorded as 0.8615 seconds and the manufacturer specified value is 0.8600 seconds. Using the formula for calculating percent error, we find a 0.17% error in our example as shown in Figure 4-14.

$$\frac{\text{Actual Value - Expected Value}}{\text{Expected Value}} \times 100 = \text{percent error}$$

$$\frac{0.8615 - 0.8600}{0.8600} \times 100 = \text{percent error}$$

$$\frac{0.0015}{0.8600} \times 100 = \text{percent error}$$

0.17 % Error

Figure 4-14: Example Timing Test Percent Error Calculation

The percent error is within the specified "+/- 5% error" from the "Accuracy" column and is acceptable for service.

When the digital relay time delay is zero or very small (less than two seconds), the actual measured time delay can be longer than expected. There is an inherent delay before the relay can detect a fault plus an additional delay between fault detection and relay output operation. Some relays use error checking features that can also increase the expected time delay because most dynamic faults involve sudden gaps in the current/voltage waveforms that would not occur during a real fault. These delays are very small (less than five cycles) and are insignificant with time delays greater than two seconds.

The first delay exists because the relay is a computer, and computers can only perform one task at a time. The relay evaluates each line of programming, one line at a time, until it reaches the end of the program, and then returns to the start to scan the entire program again. As more features are added, more program lines are added. Large program sizes are offset by faster processors that spend less time evaluating each line. If a fault occurs just after the relay processes the line of code that detects that particular fault, the relay has to run through the entire program one more time before the fault is detected. This time delay is usually a fraction of a cycle. The "Timer Accuracy" specifications in Figure 4-15 detail this time delay.

Instantaneous/Definite-Time Overcurrent Elements

Pickup Range:	OFF, 0.25 - 100.00 A, 0.01 A steps (5 A nominal) OFF, 0.05 - 20.00 A, 0.01 A steps (1 A nominal)
Steady State Pickup Accuracy:	+/- 0.05 A and +/-3% of Setting (5 A nominal) +/- 0.01 A and +/-3% of Setting (1 A nominal)
Transient Overreach:	< 5% of Pickup
Time Delay:	0.00 - 16,000.00 cycles, 0.25-cycle steps
Timer Accuracy:	+/- 0.25 cycle and +/-0.1% of setting

Figure 4-15: SEL-311C Relay Element Specifications

The second time delay occurs after the relay has detected the fault and issues the command to operate the output relays. There is another fraction of a cycle delay to evaluate what output contacts should operate and then the actual contact operation can add up to an additional cycle depending on relay manufacturer, model, etc. "Pickup Time" in the Figure 4-17 represents this delay for the specified relay.

Output Contacts:	30 A Make 6A continuous carry at 70 C; 4 A continuous carry at 85 C 50A for one second MOV Protected: 270 Vac, 360 Vdc, 40 J; Pickup Time: <5ms.

Figure 4-16: SEL-311C Output Contact Specifications

Your test-set also adds a minor delay to the test result as shown by the "Accuracy" specifications in Figure 4-17:

MANTA MTS-1710 TIME MEASUREMENT SPECIFICATIONS	
Auto ranging Scale:	0 – 99999 sec
Auto ranging Scale:	0 – 99999 cycles
Best Resolution:	0.1 ms / 0.1 cycles
	Two \ pulse timing mode
Accuracy:	0 – 9.9999 sec scale: +/-0.5ms +/- 1LS digit
	all other scales: +/- 0.005% +/- 1LS digit

Figure 4-17: Manta Test Systems MTS-1710 Technical Specifications

What does all this mean? With a time delay of zero, the time test result for a SEL 311C relay, using a Manta MTS-1710 test-set, could be as much as 28.0 ms or 1.68 cycles as shown in Figure 4-18:

MINIMUM TIME TEST RESULT	
Relay Operate Time:	0.25 cycles
Relay Timing Accuracy:	0.10 cycles (0.1% of setting, because setting is zero and next setting is 1 cycle, use 0.1 cycles)
Relay Operate Time:	0.30 cycles (< 5 ms)
Test-set :	0.03 cycles (+/-0.005%)
	1.00 cycles (+/- 1LS digit)
	1.68 cycles or 28 ms

Figure 4-18: 50-Element Maximum Expected Error

An SEL-311C relay instantaneous overcurrent element is set for one cycle and the timing test result was 2.53 cycles. That equals 153% error using the percent error formula and the relay appears to fail the timing test. But we determined that the relay/test-set combination could add 1.68 cycles to the time delay. The expected time (one cycle) plus the maximum "50-element expected Error" (1.68 cycles) equals 2.68 cycles. The test result is lower than the maximum expected time calculated in the previous sentence and, therefore, the relay passes the timing test.

C) Logic Testing

The relay testing methods described so far have limitations when applied to microprocessor relays and are more suited to acceptance testing because they only prove that the analog inputs (voltage and current signals) are operating correctly, at least one output is operating correctly, and the relay will do what it is programmed to do when elements are isolated. Logic testing attempts to apply a more holistic testing approach that tests the relay without changing relay settings and monitoring the contacts that will actually operate when the relay is in service.

There are some serious flaws when you use traditional test methods to perform commissioning tests. The goal of a commissioning test should be to ensure that the relay will operate correctly when applied to a specific application using the installed settings. This requires testing with the applied settings and ensuring the relay has been properly configured. Are you really performing a commissioning test of as-left settings when you are changing settings to test? If OUT101 is connected to your trip circuit and all of your testing is performed on OUT107, how do you know that OUT101 is operating correctly? Does your output logic equation include all of the enabled elements? Are all of the enabled elements in your trip equation?

Almost all problems found in the field with microprocessor relays have absolutely nothing to do with the actual relay and occur because of drawing and/or relay setting mistakes. Here are some examples of some common mistakes found during relay testing.

- A differential relay element tests correctly on all three phases when isolated but a GE T-60 relay's output setting is "XFMR PCNT DIFF OP A" which will only operate if an A-phase fault is detected. B and C differential protection is effectively disabled. The correct element was "XFMR PCNT DIFF OP". A one character mistake could have made a B or C-phase differential fault much worse than it could have been.

- The 50N1 (Instantaneous Overcurrent on I_N input) setting is in the trip equation but 50N1 is off in the element settings. 50G1 (Residual Instantaneous Overcurrent) is on in the element settings but missing from the trip equation. All ground protection is disabled.

- A generator step-up transformer differential element is to be disabled by the lower voltage starting breaker 52a signal when the generator is run up to speed. However, a 52b breaker signal is actually sent which disabled differential protection when the generator is online. The relay will never trip and could cause millions of dollars in damage and lost revenue for a year waiting for a replacement transformer.

None of the examples described above would have been discovered using traditional testing techniques. Several of these problems were found several years after the relay was placed into service during maintenance testing by a different relay tester.

Logic testing starts by comparing all of the onsite documentation to the settings and making sure all drawings and the relay settings match as described in *Chapter 17: Review the Application*. You can create your test plan based on the relay settings once all of the site documentation is reviewed. It is important to look at the settings objectively and look for inconsistencies inside the settings themselves. Look for impossible logic conditions and make sure that an element that is enabled and setup is also found in the output logic. Look for elements in the output logic that aren't turned on or set. If there are no obvious errors, note the logic for each output, including signals sent over communication channels and LED or front panel displays. Once you have a comprehensive list of all of the output logic, create a checklist for each output broken down into simple OR statements. For example, a simple SEL overcurrent relay might have the following settings.

- TRIP = 51P1T + 51N1T + 50P1 + 50N1
- (Trip Breaker) OUT101 = TRIP
- (Scada/Remote Trip Indication) OUT107 = TRIP
- (Front Panel Display) 52A = IN101, DP1 = 52A, DP_1 = Breaker Closed, DP_2 = Breaker Open

For those unfamiliar with SEL logic, *Chapter 16: Understanding Digital Logic* describes SEL and GE logic schemes in detail. Brief descriptions of the SEL codes above include:

- **51P1T**—Phase Inverse Time Overcurrent Trip
- **51N1T**—Neutral Inverse Time Overcurrent Trip
- **50P1**—Phase Instantaneous Overcurrent Pickup
- **50N1**—Neutral Instantaneous Overcurrent Pickup
- **DP_1**—Front Panel Display Point One
- **DP_2**—Front Panel Display Point Two

If you wish to combine traditional pickup and timing testing with logic testing, your test plan could look like the following test plans. It is important to note that you must apply the appropriate fault simulation to ensure the element operates during the timing test to prevent interference by other elements. If a phase-related element is enabled, a phase-to-phase or 3-phase test should be applied. If a ground related element is applied, a phase-to-ground test should be applied.

Test Plan #1

1. Perform 51P1T pickup test using steady state technique and use pickup LED/ Display/computer to determine pickup (recommended) or assign unused output for pickup indication (not recommended).

2. Perform 51P1T timing test at 2x pickup and use OUT101 for timer stop.

3. Perform 51P1T timing test at 2x pickup and use OUT107 for timer stop.

4. Perform 51P1T timing test at 4x pickup and use OUT101 for timer stop.

5. Perform 51P1T timing test at 4x pickup and use OUT107 for timer stop.

6. Perform 51P1T timing test at 6x pickup and use OUT101 for timer stop.

7. Perform 51P1T timing test at 6x pickup and use OUT107 for timer stop.

8. Perform 51N1T pickup test using steady state technique and use pickup LED/ Display/computer to determine pickup (recommended) or assign unused output for pickup indication (not recommended).

9. Perform 51N1T timing test at 2x pickup and use OUT101 for timer stop.

10. Perform 51N1T timing test at 2x pickup and use OUT107 for timer stop.

11. Perform 51N1T timing test at 4x pickup and use OUT101 for timer stop.

12. Perform 51N1T timing test at 4x pickup and use OUT107 for timer stop.

13. Perform 51N1T timing test at 6x pickup and use OUT101 for timer stop.

14. Perform 51N1T timing test at 6x pickup and use OUT107 for timer stop.

15. Perform 50P1 pickup test using steady state technique and use pickup LED/ Display/computer to determine pickup (recommended) or assign unused output for pickup indication (not recommended).

16. Perform 50P1 timing test at 1.1x pickup and use OUT101 for timer stop.

17. Perform 50P1 timing test at 1.1x pickup and use OUT107 for timer stop.

18. Perform 50N1 pickup test using steady state technique and use pickup LED/ Display/computer to determine pickup (recommended) or assign unused output for pickup indication (not recommended).

19. Perform 50N1 timing test at 1.1x pickup and use OUT101 for timer stop.

20. Perform 50N1 timing test at 1.1x pickup and use OUT107 for timer stop.

21. Check breaker status and compare to front panel display. (If breaker is open, then display should indicate open.)

22. Change breaker status and compare front panel display.

Test Plan #2 *(Streamlined using alternating outputs)*

1. Perform 51P1T pickup test using steady state technique and use pickup LED/ Display/computer to determine pickup (recommended) or assign unused output for pickup indication (not recommended).

2. Perform 51P1T timing test at 2x pickup and use OUT101 for timer stop.

3. Perform 51P1T timing test at 4x pickup and use OUT107 for timer stop.

4. Perform 51P1T timing test at 6x pickup and use OUT107 for timer stop.

5. Perform 51N1T pickup test using steady state technique and use pickup LED/ Display/computer to determine pickup (recommended) or assign unused output for pickup indication (not recommended).

6. Perform 51N1T timing test at 2x pickup and use OUT107 for timer stop.

7. Perform 51N1T timing test at 4x pickup and use OUT107 for timer stop.

8. Perform 51N1T timing test at 6x pickup and use OUT101 for timer stop.

9. Perform 50P1 pickup test using steady state technique and use pickup LED/ Display/computer to determine pickup (recommended) or assign unused output for pickup indication (not recommended).

10. Perform 50P1 timing test at 1.1x pickup and use OUT101 for timer stop.

11. Perform 50P1 timing test at 1.1x pickup and use OUT107 for timer stop.

12. Perform 50N1 pickup test using steady state technique and use pickup LED/ Display/computer to determine pickup (recommended) or assign unused output for pickup indication (not recommended).

13. Perform 50N1 timing test at 1.1x pickup and use OUT101 for timer stop.

14. Perform 50N1 timing test at 1.1x pickup and use OUT107 for timer stop.

15. Check breaker status and compare to front panel display. (If breaker is open, then display should indicate open.)

16. Change breaker status and compare front panel display.

Principles and Practice

Test Plan #3 *(Streamlined using multiple inputs and timers)*

1. Perform 51P1T pickup test using steady state technique and use pickup LED/Display/computer to determine pickup (recommended) or assign unused output for pickup indication (not recommended).
2. Perform 51P1T timing test at 2x pickup and verify OUT101 and OUT107 operates.
3. Perform 51P1T timing test at 4x pickup and verify OUT101 and OUT107 operates.
4. Perform 51P1T timing test at 6x pickup and verify OUT101 and OUT107 operates.
5. Perform 51N1T pickup test using steady state technique and use pickup LED/Display/computer to determine pickup (recommended) or assign unused output for pickup indication (not recommended).
6. Perform 51N1T timing test at 2x pickup and verify OUT101 and OUT107 operates.
7. Perform 51N1T timing test at 4x pickup and verify OUT101 and OUT107 operates.
8. Perform 51N1T timing test at 6x pickup and verify OUT101 and OUT107 operates.
9. Perform 50P1 pickup test using steady state technique and use pickup LED/Display/computer to determine pickup (recommended) or assign unused output for pickup indication (not recommended).
10. Perform 50P1 timing test at 1.1x pickup and verify OUT101 and OUT107 operates.
11. Perform 50N1 pickup test using steady state technique and use pickup LED/Display/computer to determine pickup (recommended) or assign unused output for pickup indication (not recommended).
12. Perform 50N1 timing test at 1.1x pickup and verify OUT101 and OUT107 operates.
13. Check breaker status and compare to front panel display. (If breaker is open, then display should indicate open.)
14. Change breaker status and compare front panel display.

Notice that we do not simulate the breaker when performing the logic test for the 52A (IN101) front panel display. You should always use the actual end device to prove input status and logic to make sure the actual device status contact:

- uses the correct status indication
- is connected correctly
- uses the correct input voltage. Different relays use an internally supplied voltage source or external source to determine input status. Some relays can use both methods and an easily be connected incorrectly.

Relay logic is often more complex than the previous example and more complicated logic schemes should be broken down to a simple OR statement. For example, a breaker-failure logic scheme could be written as:

- SV1 = (50P2 [0.5A] + 50N2 [0.5A]) * (SV1T [Seal-in] + TRIP [Initiate]) [Breaker-fail operate logic]
- SV1PU = 15 cycles [Breaker-failure Timer]
- OUT102 = SV1T [Breaker-fail Signal = Current is still flowing through the breaker 15 cycles after the trip signal is sent and will stay closed until the current is lower than 0.5A. Send trip signal to the next upstream breaker]

This logic can be broken down into the following logic equations:
- OUT102 = 50P2 * TRIP
- OUT102 = 50N2 * TRIP
- OUT102 = 50P2 * SV1T
- OUT102 = 50N2 * SV1T

For those unfamiliar with SEL logic, *Chapter 16: Understanding Digital Logic* describes SEL and GE logic schemes in detail. Brief descriptions of the SEL codes above include:
- **50P2**—Phase Instantaneous Overcurrent Pickup
- **50N2**—Neutral Instantaneous Overcurrent Pickup
- **SV1**—SELogic Variable #1
- **SV1PU**—SELogic Variable #1 Time Delay to Operate
- **SV1T**—SELogic Variable #1 Operated

Broken down into its base components, we can now test each of these equations using the following test plan.

Breaker-Fail (OUT102) Test Plan

1. OUT102 = 50P2 * TRIP. Perform 51P1T timing test at 2x pickup and set timer to start when OUT101 operates and to stop when OUT102 operates.

2. OUT102 = 50N2 * TRIP. Perform 51N1T timing test at 2x pickup and set timer to start when OUT101 operates and to stop when OUT102 operates.

3. OUT102 = 50P2 * SV1T. Perform 51P1T timing test at 2x pickup. Do not stop the test after OUT101 and OUT102 operates. Lower fault current below 51P1 pickup setting but greater than 50P2 setting. OUT101 should open but OUT102 should still be closed. Lower fault current below 50P2 setting. Both outputs should now be open.

4. OUT102 = 50N2 * SV1T. Perform 51N1T timing test at 2x pickup. Do not stop the test after OUT101 and OUT102 operates. Lower fault current below 51N1 pickup setting but greater than 50N2 setting. OUT101 should open but OUT102 should still be closed. Lower fault current below 50N2 setting. Both outputs should now be open.

Applying logic testing will not find every problem but it will allow the relay tester to feel reasonably confident that the relay has been set correctly, there are no obvious logic errors, and the relay will operate when required and is connected properly.

D) Combining Pickup, Timing Tests, and Logic Testing

Modern test equipment uses a minimum of three voltage and three current outputs with the ability to independently vary the phase angles between any of the outputs. With this equipment, you can use different states to create more complicated dynamic tests which can make relay testing more realistic, effective, and efficient.

A microprocessor relay element does not fall out of calibration…it either works correctly or it doesn't. Using this principle, the pickup and timing test can be combined into one test. The simplest element to use as an example is the instantaneous overcurrent (50) element. If the applied current is greater than the pickup setting, the element will operate. If our example 50-element pickup setting is 25A, the element will not operate if the current is less than 25A, and will operate instantaneously if the current is greater than 25A in an ideal world. The test plan to test the 50-element in one test is shown on the following chart:

PREFAULT	FAULT 1	FAULT 2
Nominal Current (4.0A) for 2 seconds	24.99A for 1 second	25.00A Start timer Stop timer when relay output operates. Time should be instantaneous.

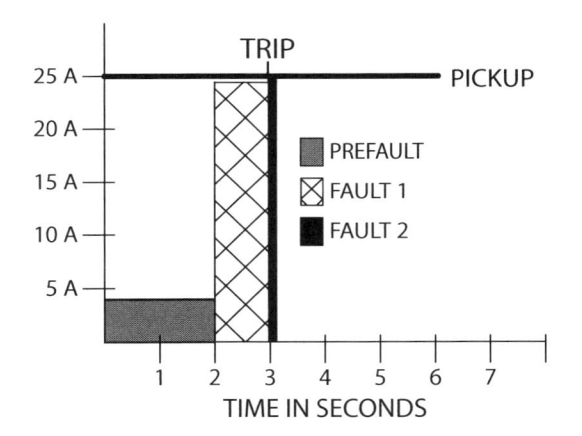

Figure 4-19: 50-Element Ideal Combination Test

We do not live in an ideal world and this test would probably fail due to accuracy errors of the relay and the test-set. In digital relays and modern test equipment, the combined error is usually less than 5%. We can modify our test-set to allow for the inherent error in relay testing using the following chart:

PREFAULT	FAULT 1	FAULT 2
Nominal Current (4.0A) for 2 seconds	23.75A for 1 second (25A - 5% = 25 - (25 x 0.05) = 25 - 1.25 = 23.75A)	26.25A (25A + 5% = 25 + (25 x 0.05) = 25 + 1.25 = 26.25A) Start timer Stop timer when relay output operates. Time should be less than 5 cycles

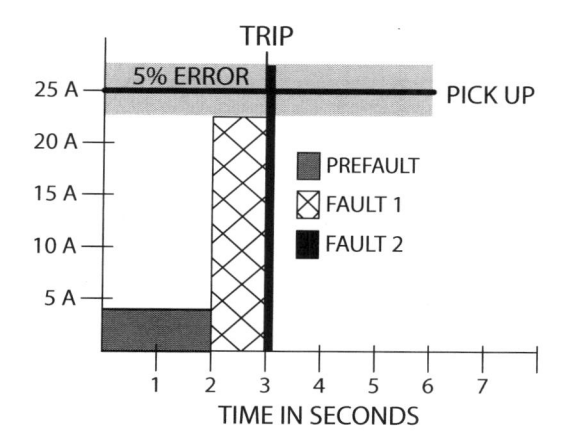

Figure 4-20: 50-Element Combination Test

This method will work for all relay elements, including elements with time delays such as the time overcurrent (51) element.

Multiple timing tests were performed on electromechanical relays to ensure the relay magnets and mechanical time dials were in the proper location. If the relay timing did not match the manufacturer's curve, a magnet or time dial setting was adjusted to bring the relay into tolerance. There are no possible adjustments to a microprocessor relay so the timing will be correct, or the relay is set wrong and it will be incorrect. The following test plan will allow the user to test the pickup and timing of a microprocessor based 51-element with a pickup of 5A.

PREFAULT	FAULT 1	FAULT 2
Nominal Current (4.0A) for 2 seconds	4.75A for 5 second (5A - 5% = 5 - (5 x 0.05) = 5 - 0.25 = 4.75A)	10.0A (2x nominal pickup) Start timer Stop timer when relay output operates. Time should be expected result for 2x pickup +/- 5%
Nominal Current (4.0A) for 2 seconds	4.75A for 5 seconds (5A - 5% = 5 - (5 x 0.05) = 5 - 0.25 = 4.75A)	20.0A (4x nominal pickup) Start timer Stop timer when relay output operates. Time should be expected result for 4x pickup +/- 5%
Nominal Current (4.0A) for 2 seconds	4.75A for 5 seconds (5A - 5% = 5 - (5 x 0.05) = 5 - 0.25 = 4.75A)	30.0A (6x nominal pickup) Start timer Stop timer when relay output operates. Time should be expected result for 6x pickup +/- 5%

You should notice that the timing test current was equal to the multiple of current without adding 5% to compensate for test-set and relay error. The 5% error is used when comparing the time in fault…not the applied current. If there was a problem with the applied settings, the measured time delay would be significantly shorter than the expected time delay to indicate the problem.

Another problem would become evident if you were testing this relay if the 50 and 51-elements were assigned to the same trip coil. The 50-element (25A) would operate when the relay tester tried to perform the 6x timing test (30A). The relay tester could isolate the 51-element to another relay output to perform their 6x test without interference…or they could step back and review their procedure. Two different issues come into play when protective elements overlap.

- The microprocessor relay's 51-element does not have any possible adjustments, so is a third test really necessary to prove the characteristic curve if two other tests are successful? If the third test is required, we can change the test current of the third test. The primary reason for choosing whole numbers for 51-element tests in electromechanical relays is that it is easier to determine the expected result on the graph using whole numbers. Most relay testers are using the formulas to determine expected values when testing microprocessor relays, so changing the third test current to 24A or 4.8x pickup should be no problem when calculating the expected result.

- The second issue in play is commissioning vs. acceptance testing. Testing without changing settings proves that the settings have been applied correctly. If 30 Amps are applied to the relay in service, would the 51 or 50 element operate? Is there any advantage to testing an element at a test point where it will never operate in service?

This technique works for all protective elements including more complicated relay elements such as distance protection (21). In fact, if you are unable to apply this test procedure and achieve a successful test, there is probably something wrong with the relay settings.

21-elements use the measured impedance and angle between the current and voltage to detect a fault on a transmission line or other electrical apparatus as described in *Chapter 15: Line Distance (21) Element Testing*. The most typically applied characteristic is a MHO circle. If the measured impedance falls within the circle, the 21-element operates after a pre-set time delay. Our Zone 2 element is set at 3.4Ω @ 87° with a 20 cycle time delay as shown in the following figure.

We can start the test by applying a prefault state using nominal conditions with an impedance near the x-axis, far away from the circle. We then apply the Fault 1 impedance just outside of the circle followed by the Fault 2 impedance applied just inside the circle. Don't forget that there is also a 5% error to account for and that is shown by the shaded band around the original circle. If the measured time between Fault 2 and the relay output is 20 cycles +/- the relay tolerance, the test is successful. You could use the same technique to plot the entire circle, but it is extremely unlikely you will find a problem with the relay's programming which should have been tested in the Type Testing process.

PREFAULT	FAULT 1	FAULT 2
Voltage (69.28V)	Voltage (30.0V)	Voltage (30.0V)
Current (3.0A @ -30°)	Current (8.40A) @ -87°	Current (9.29A) @ -87°
(23.09Ω @ 30°)	(3.57Ω @ 87°)	(3.23Ω @ -87°)
for 2 seconds	for 0.5 seconds	Start timer
		Stop timer relay output operates.
		Time should be 20 cycles +/- 5%

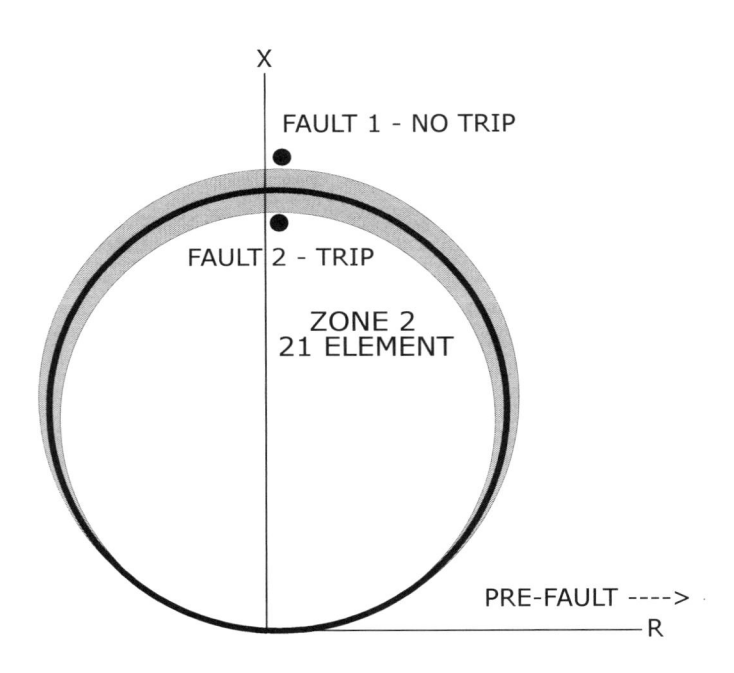

Figure 4-21: 21-Element Combination Test

This test technique can also be applied to overlapping zones of protection. Our example relay has a Zone 1 distance protection element set at 1.5Ω @ -87° with no intentional delay. We can modify the previous test plan for the new impedance to use the following test parameters:

PREFAULT	FAULT 1	FAULT 2
Voltage (69.28V)	Voltage (20.0V)	Voltage (30.0V)
Current (3.0A) @ -30°	Current (12.73A) @ -87°	Current (14.08A) @ -87°
(23.09Ω @ 30°)	(1.57Ω @ 87°)	(1.42Ω @ -87°)
for 2 seconds	for 1 seconds	Start timer
		Stop timer relay output operates.
		Time should be less than 5 cycles

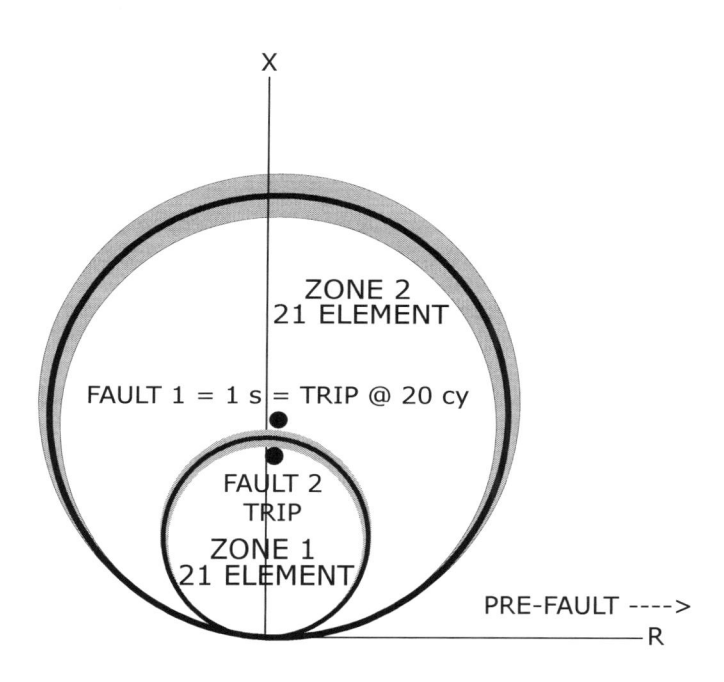

Figure 4-22: 21-Element Dual Zone Combination Test

The previous test would fail because the Fault 1 impedance falls within the Zone 2 circle and Zone 2 will trip in 20 cycles, which is a shorter time than the Fault 1 duration. A simple modification of the Fault 1 time to a value greater than the Zone 1 time but less than the Zone 2 time delay will make this a practical test. Change the Fault 1 time delay to 10 cycles (Zone 1 time (5 cycles) < Fault 1 time < (Zone 2 time (20 cycles) - Zone 1 time (5 cycles)) = 5 cycles < Fault 1 time < 15 cycles = Fault 1 time = 10 cycles).

PREFAULT	FAULT 1	FAULT 2
Voltage (69.28V)	Voltage (20.0V)	Voltage (30.0V)
Current (3.0A) @ -30°	Current (12.73A) @ -87°	Current (14.08A) @ -87°
(23.09Ω @ 30°)	(1.57Ω @ 87°)	(1.42Ω @ -87°)
for 2 seconds	for 10 cycles	Start timer
		Stop timer relay output operates.
		Time should be less than 5 cycles

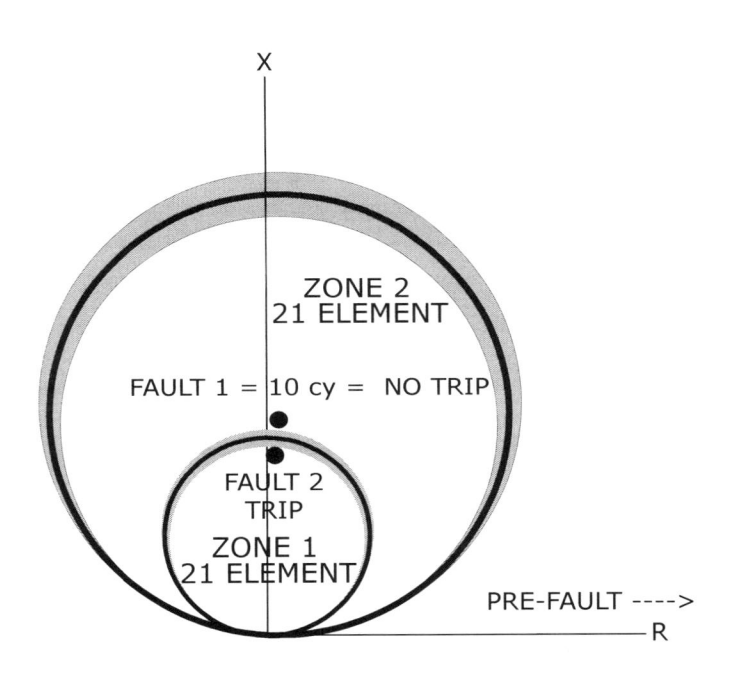

Figure 4-23: 21-Element Dual Zone Combination Test

Adding logic testing to this step is as simple as connecting all of the relay outputs used by the relay settings and monitoring all contacts during the test. If OUT101 and OUT107 are supposed to operate when a 50-element operates, make sure both elements operate during the 50-element test. Or, you could perform the test twice using each output for timing.

When this test technique is applied to the relay in the Logic Testing example, the test plan has fewer steps and is more comprehensive.

Relay Settings

1. SV1PU = 15 cycles [Breaker-failure Timer]

2. TRIP = 51P1T + 51N1T + 50P1 + 50N1

3. SV1 = (50P2 + 50N2) *(SV1T + TRIP) [Breaker-fail operate logic]

4. OUT101 = TRIP [Trip Breaker]

5. OUT102 = SV1T [Trip upstream breaker]

6. OUT107 = TRIP [Scada/Remote Trip Indication]

7. 52A = IN101

8. DP1 = 52A [Front Panel Display]

9. DP_1 = Breaker Closed

10. DP_2 = Breaker Open

Test Plan

1. Perform 51P1T combination test at 2x pickup and verify OUT101 and OUT107 operates.

2. Perform 51P1T combination test at 4x pickup and verify OUT101 and OUT107 operates.

3. Perform 51N1T combination test at 2x pickup and verify OUT101 and OUT107 operates.

4. Perform 51N1T combination test at 4x pickup and verify OUT101 and OUT107 operates.

5. Perform 50P1 combination test at 1.1x pickup and verify OUT101 and OUT107 operates.

6. Perform 50N1 combination test at 1.1x pickup and verify OUT101 and OUT107 operates.

7. Perform 51P1T combination test at 2x pickup and set timer to start when OUT101 operates and stop the timer and test when OUT102 operates.

8. Perform 51N1T combination test at 2x pickup and set timer to start when OUT101 operates and stop the timer and test when OUT102 operates.

9. Perform 51N1T combination test at 2x pickup and set timer to start when OUT101 operates and stop the timer OUT102 operates. Do not stop the test when OUT102 operates. Switch to an additional state with the neutral current set at 0.525A and verify OUT102 is still in the trip state. Switch to an additional state with the neutral current set at 0.475A and verify that OUT102 has changed state and create a timer to start when you enter the last state and stop when the contact opens. Compare the second timer to the SV1DO setting.

10. Perform 51P1T combination test at 2x pickup and set timer to start when OUT101 operates and stop the timer when OUT102 operates. Do not stop the test when OUT102 operates. Switch to an additional state with the phase current set at 0.525A and verify OUT102 is still in the trip state. Switch to an additional state with the phase current set at 0.475A and verify that OUT102 has changed state and create a timer to start when you enter the last state and stop when the contact opens. Compare the second timer to the SV1DO setting.

11. Check breaker status and compare to front panel display. (If breaker is open, then display should indicate open.)

12. Change breaker status and compare front panel display.

This technique, when applied correctly:

- Can be faster than traditional testing techniques
- Is more efficient than traditional testing techniques
- Is more comprehensive because it tests the pickup, timing, and logic in one step
- Provides true commissioning results because no settings are changed and tests are more realistic
- Can make maintenance testing simple if the tests are saved and re-played at maintenance intervals

E) **System Testing**

Logic testing combined with dynamic testing is a very powerful and effective test method when applied by an experienced relay tester who has a good understanding of the relay elements and the system the relay protection is applied to. However, there is a fatal flaw when performing relay testing based on supplied setting files…do the settings match the engineer's intent? As mentioned before, a modern microprocessor relay will perform the tasks that it is instructed to perform and cannot determine if the engineer has understood the relay's operation or not. This problem was coined Garbage in = Garbage Out when computers were first implemented in society but we appear to become far more trusting as computers became part of our daily life. Neither the relay tester nor the relay can determine whether the pickup setting is intended to be 0.5 instead of the 5.0 Amps the engineer accidently applied unless they have the engineer's notes or a coordination study or it is an obvious error. Testing a relay to its applied settings with no comparison to intent or common sense will almost always create a successful test unless there are gross mechanical or setting and test plan errors. A relay's mechanical problems can be more easily detected by simply applying voltage and current and performing a meter test followed by exercising each digital input and output. Gross setting errors can be detected by a combination of dynamic and logic testing. But what happens when the logic is too complex to decipher like this real world example of a capacitor control circuit:

Opening the Capacitor Bank

SV8 = (RB15 * !LT6 + PB7 * LT5) * LT10 + RB13 * !LT10 + /SV3T

Closing the Capacitor Bank Switches

SV9 = (RB16 * !LT6 + PB8 * LT5) * LT10 * SV10T + RB14 * !LT10 * SV10T

This logic doesn't look that complicated until you realize that any word bit that begins with RB is logic from another device that has over 100 lines of programming. If the logic was expanded to represent just what is inside this one relay, it would look like:

Opening the Capacitor Bank

SV8 = (RB15 * ! (PB10 * !LT6 * (!LT5 * PB5)) + PB7 * (!LT5 * PB5)) * (!LT10 * (PB6 * LT5 + RB12 * !LT6)) + RB13 * ! (!LT10 * (PB6 * LT5 + RB12 * !LT6)) + /3P27 * !50L * 52A

Closing the Capacitor Bank Switches

SV9 = (RB16 * ! (PB10 * !LT6 * (!LT5 * PB5)) + PB8 * (!LT5 * PB5)) * (!LT10 * (PB6 * LT5 + RB12 * !LT6)) * SV10T + RB14 * ! (!LT10 * (PB6 * LT5 + RB12 * !LT6)) * IN104

It turns out that testing this logic was quite simple after the engineer was contacted. This logic translates into the following bullet points:

Closing the Capacitor Bank Switches

1. If the capacitor switches are open and the capacitor control is in "Manual", close the capacitor switches when the "Close Capacitor" button is pushed.

2. If the capacitor switches are open and the capacitor control is in "Auto", the capacitor switches will close if:

    a. the phase-to-phase voltage is below 6.84 kV.
    b. there is a lagging power factor and the load is above 10 MW.
    c. there is a leading power factor between 0.99 and 1.00 and the load is above 15 MW.

Opening the Capacitor Bank

1. If the capacitor switches are closed and the capacitor control is in "Manual", open the capacitor switches when the "Open Capacitor" button is pushed.

2. If the capacitor switches are closed and the circuit breaker opens, open the capacitor switches.

3. If the capacitor switches are closed and the capacitor control is in "Auto", the capacitor will close if:

    a. the phase-to-phase voltage is above 7.74 kV.
    b. there is a leading power factor of 0.96 or less.

A very complex logic equation was translated into simple, easy to simulate conditions and all of the settings worked perfectly. If the logic had been tested without understanding the engineer's intent, it could take hours or days to reverse engineer and unless something obvious went wrong, no errors would have been detected because the relay would perform as programmed.

The ideal testing scenario would occur if the settings engineer created a description of operation that will allow us to test the relay based on their intent instead of their settings. An example of the setting engineer's description of operation for our example relay could include:

1. The Zone 2 element should operate OUT101 and OUT107 at 3.4Ω @ 87° after a 20 cycle delay.

2. The Zone 1 element should operate OUT101 and OUT107 at 1.5Ω @ -87° in less than 5 cycles.

3. The 51PT element should operate at OUT101 and OUT107 at 10A in 10.4 seconds.

4. The 51PT element should operate at OUT101 and OUT107 at 20A in 7.3 seconds.

5. The 51NT element should operate at OUT101 and OUT107 at 2A in 7.4 seconds during a phase-to-ground fault.

6. The 51NT element should operate at OUT101 and OUT107 at 4A in 3.6 seconds during a phase-to-ground fault.

7. The 50PT element should operate at OUT101 and OUT107 at 25A in less than 5 cycles.

8. The 50NT element should operate at OUT101 and OUT107 at 6A in less than 5 cycles during a phase-to-ground fault.

9. OUT102 will operate if the breaker remains closed 15 cycles after a trip signal is sent. The breaker is considered closed if the phase or residual current is greater than 0.5A.

With these instructions, we can build a test plan similar to the test plans described in this section to ensure that the relay will perform as the relay engineer intended instead of regurgitating what the relay was programmed to do. It is important that this description be created when the engineer designs the system and the relay's intended operation is fresh in their mind.

F) Dynamic System-Model Testing

Dynamic system model based testing uses a computer program to create a mathematical model of the electrical system and create fault simulations based on the specific application. These modeled faults (or actual events recorded by a relay) are replayed through a sophisticated relay test-set to the relay and, if performed correctly, is the ultimate test to prove an entire protection system as a whole. Dynamic System Model based testing can also provide more realism by creating waveforms that can incorporate real system conditions such as DC offset, transients, or CCVT distortions as shown in the example waveform in Figure 4-24.

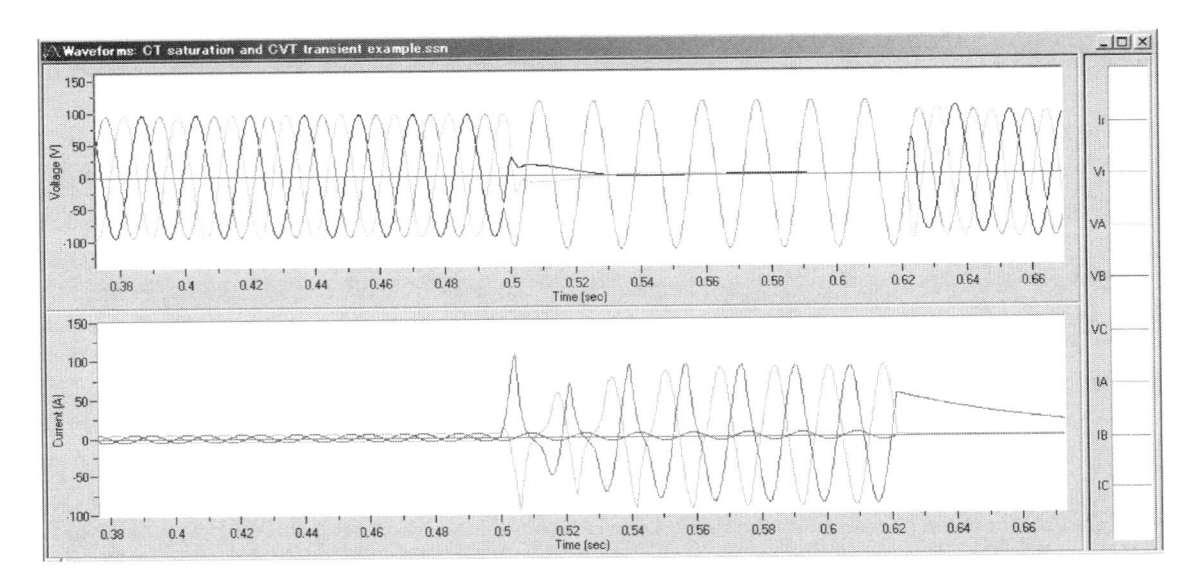

Figure 4-24: System Modeling Waveform (Compliments of Manta Test Systems)

This test is typically limited to type testing or end-end testing because it requires specialized knowledge of a system, complex computer programs, advanced test equipment, and a very complex test plan with many possibilities for error.

Chapter 5

Test Sheets and Documentation

After all the testing is complete, you want a complete record of all of your activities to show that you actually tested the relay and justify those huge relay-testing bills. Your final documentation should be comprehensive, and you should be able to answer any questions regarding the relay; its as-left, final settings; and the test results without leaving your office.

Of course, everyone has their own personal opinion of what a test sheet should look like, and it is difficult, if not impossible, to get two technicians to agree on font, size, or even what computer program is best for its design. Test sheets can be specifically designed for relay models or be generic for all relays or functions. They can automatically perform most of the percent error, pickup, and timing calculations necessary for relay testing. In fact, some test sheets can be as comprehensive as the higher end computer controlled test software recently introduced to increase efficiency.

After a lot of experimentation with various programs including Microsoft Access™, Microsoft Word™, IMSI Form Tool™, and Omni-Forms™, we determined that Microsoft Excel™ provided the most flexibility and the best calculating functions to speed up data entry and actual testing. Test sheets were created for individual relay models and most of the calculations for pickup, timing, and percent error are automatically calculated based on the settings that are entered into the test sheet. Unused functions are removed from the final test sheet and test blocks are copied from existing test sheet templates to create a new relay model test-sheet template.

Regardless of your personal preferences regarding test sheets, all test sheets should include the following information:

1. Your Company Name and Logo

Everyone has too much paper on his or her desk and placing your name and logo on the top of every page will make it easier for the final recipient to know where it belongs without much thought. This can also be a great promotional tool for you and your company because your test sheet is professional, detailed, and you did a great job.

The page number and number of pages are very important in a header or footer in case there is a paper explosion and a person needs to collect the test sheets in order.

It is a good idea to put the location and equipment description in the header or footer as well so that a person can easily find a specific test sheet by flipping through one of the corners instead of stopping to read each test sheet for specifics.

Figure 5-1: Example Test Sheet Header

2. Project Details

It is always a good idea to add some additional information to track and organize test sheets such as:

- Client name
- Internal project number or contract number
- Project Location
- Test equipment used and date of calibration
- The date of testing
- Your initials or name
- A space for external approval in case it is required
- The kind of testing you will be performing

SCHWEITZER LABORATORIES SEL-311C TEST FORM			
PROJECT #:	DATE:		
CLIENT:	TESTED BY:		
LOCATION:	APPROVED BY:		
TEST SET:	TESTING STD:	☐ ACCEPT	☐ MAINT

Figure 5-2: Example Test Sheet with Project Details

3. Nameplate Data

Write down all of the information from the relay nameplate. Also, include the relay's panel location and site designation. For example, a relay could be located in "Alexis Substation Panel 7A" with designation "LINE 1 PROT." Enter this information into the test sheet for tracking purposes. Example nameplate information includes:

- **Type**—Relay series or application.
- **Model Number**—for comparison to specifications and reference when contacting technical support or ordering a replacement.
- **Serial Number**—to keep track of relays and reference when contacting technical support.
- **Control Power**—The power supply voltage rating. Compare this value to the application control voltage before energizing the relay.
- **Nominal CT Secondary Current**—The relay nominal current rating should match the CT nominal rating, which is typically 5A.
- **Part Number**—for comparison to specifications and reference when contacting technical support or ordering a replacement.
- **Logical Input**—The digital input voltage rating if supplied by an external source. Compare this value to application voltages.
- **Rotation**—Expected phase rotation. Some relays have a preset phase rotation of clockwise or counter-clockwise as listed on the nameplate. Some relays have settings to change rotation. The system rotation should match the relay rotation or some protective elements may not operate correctly.
- **Frequency**—Expected input frequency. Some relays must have the nameplate rated frequency while others have a nominal frequency setpoint. The frequency nameplate or setting should match the system frequency.
- **Nominal Secondary Voltage**—The nominal system voltage should be equal to or less than this value. Always check to see if the specified rating is line-line or line-ground voltages.

NAMEPLATE DATA			
DESIGNATION:		LOCATION:	
TYPE:		PART NUMBER:	
MODEL NUMBER:		LOGIC INPUT:	
SERIAL NUMBER:		ROTATION:	
CONTROL POWER:		FREQUENCY:	
		MAXIMUM VOLTAGE:	
CT SEC NOMINAL:		NOMINAL VOLTAGE:	
PHASE CT RATIO:		PHASE VT RATIO:	
CT RATIO DRAWING:		VT RATIO DRAWING:	

Figure 5-3: Example Test Sheet with Nameplate Data

4. CT and PT Ratios

Document the CT and PT ratios with source drawing numbers and compare the ratios to the final relay settings.

5. Comments and Notes

Any problems or concerns you notice while testing are documented for later review. Also, this protects you should something happen in the future after a client has ignored your recommendations.

Document all temporary changes to ensure that all changes are returned to their proper values. Any discrepancies between the final settings and the supplied or design settings are documented with explanations to provide a paper trail.

Keep these notes on the first page to make them stand out before the end-user gets sick of flipping through pages. Also, it is a good idea to add two reminders to change the relay's date and time, and reset the event recorder when testing is finished.

NOTES
ALL OF THE PROBLEMS AND DIFFERENCES FOUND ARE DETAILED HERE
NOTE WHETHER THE PROBLEM IS OUTSTANDING OR WAS CORRECTED

Figure 5-4: Example Test Sheet with Notes

6. Metering Test Data

Verify the relay's metering data to ensure you have your test-set connected correctly and to prove that the analog-to-digital converters are operating correctly. It is easiest to apply 3-phase rated voltage and current and record all of the available information. If the relay monitors VARs or Watts, vary the 3-phase phase angles and record the Watts and VARs at the various test angles as shown in Figure 5-5:

METERING						
CURRENT (AMPS)						
SEC INJ INPUT	A PH	B PH	C PH	MFG	% ERROR	
PHASE ANGLE						
SEC INJ INPUT		In		MFG	% ERROR	
@ 0 degrees				In RATIO=120		
VOLTAGE (VOLTS)						
SEC INJ INPUT	A PH	B PH	C PH (kV)	MFG (kV)	% ERROR	
PHASE ANGLE						
3 PHASE METERING						
POWER (MW)				**VARS (MVAR)**		
SEC INJ INPUT	3 PH (MW)	MFG (MW)	%ERROR	SEC INJ INPUT	3 PH	MFG	%ERROR
COMMENTS:						
RESULTS ACCEPTABLE:		☐ YES		☐ NO		☐ SEE NOTES

Figure 5-5: Example Test Sheet with Metering

7. Input / Output Tests

All input and output functionality (even inputs/outputs not used) should be verified to ensure the relay is fully functional. Inputs in use should be verified with their end devices. Inputs not in use should be verified with control power when applicable, or rated test voltage as a last resort. Not all inputs use external voltages, be careful before applying any voltages. Outputs in use are verified during pickup, timing, and logic-verification tests to operate their end devices. Outputs not in use can be verified with an ohmmeter or relay test-set input and a pulse command from the relay or other simulation necessary to close the output contact.

INPUT/OUTPUT CHECKS									
OUT101	OUT102	OUT103	OUT104	OUT105	OUT106	OUT107	ALARM		
IN101	IN102	IN103	IN104	IN105	IN106	IN107			
COMMENTS:									
RESULTS ACCEPTABLE:			☐ YES		☐ NO			☐ SEE NOTES	

Figure 5-6: Example Test Sheet with Input / Output Verification

8. Element Test Results

The most important part of the test sheet is the test results section. Separate the test results into sections to organize the tests into categories. The results should be easily understood and include the following information.

- **Settings**—the settings applied during test should be readily available to allow the reviewer to confirm or calculate the expected values. Most of our test sheets automatically calculate expected values based on the entered settings to add incentive for the additional data entry. The reviewer could search for the appropriate setting in the attached setting sheets, but it is in your best interest to keep things simple for the reviewer and yourself.

- **Pickup Values**—You never want the reviewer to assume which unit of measure actually applies. Pickup test results should be included for every test on every phase tested. Clearly define the unit of measure. Pickup test results can be in ohms (Ω), amperes (A or Amps), Volts (V or Volts), kilovolts (kV), Frequency (Hz) or in degrees (°). Verify and document blocking elements internal to the element-under-test such as loss-of-potential or blocking inputs. Blocking elements added in the user-defined logic scheme will be confirmed later.

- **Timing Test Results**—Every timing test is performed with some input (Volts, Amps, Frequency, Impedance, etc.) and this value must be clearly documented with the result. The reviewer is unable to verify your results without knowing the applied input. Clearly document the actual timing results with the units clearly defined in seconds (s) or cycles (c).

- **Manufacturer's Expected Result**—All test results (pickup and timing) must include the manufacturer's specified values as per the applied settings. Compare this value to your test results to determine if the relay is acceptable for service. A perfect test sheet would also include the manufacturer's tolerances, but you will be hard-pressed to fit all of this information on one test sheet. Manufacturer's expected results from different phases with the same expected result could be entered as one value assuming that it is easily understood which expected result relates to which test results.

- **Percent Error**—The percent error for every test result should be calculated and evaluated against manufacturer's tolerances. Percent error calculations or difference between test and tolerances should be placed as close as possible to the test value to ease understanding.

- **Notes**—Detail any problems, necessary changes, re-configurations, concerns, and questions as notes for each element section. Summarize critical items on the first or last page for easy reference. Examples of notes include "50-element blocked for timing test" or "Loss-of-potential block tested" or "Setting does not coordinate with downstream device." If another setting was blocked to allow a test, always add a note stating that the element has been returned to service.

- **Evaluation**—A non-technical person should be able to quickly flip through your test sheets and see your evaluation of the results. In addition, adding an evaluation after each section allows a reviewer to home-in on trouble areas to speed up the review process. Good test sheets use a "Results Acceptable" check box to indicate test result evaluations.

3 PHASE DISTANCE TEST RESULTS (Ohms)															
Z1P - PU			Z2P - PU			Z3P - PU			Z4P - PU						
Z1PANG			Z2PANG			Z3PANG			Z4PANG						
Z1PD -TD			Z2PD - TD			Z3PD - TD			Z4PD - TD						
DIR1			DIR2			DIR3			DIR4						
50PP1			50PP2			50PP3			50PP4						
TEST ANGLE		**ZONE 1 TESTS**			**ZONE 2 TESTS**			**ZONE 3 TESTS**			**ZONE 4 TESTS**			
F	R	M1P	MFG	%ERR	M2P	MFG	%ERR	M3P	MFG	%ERR	M4P	MFG	%ERR	
-50,0	130,0													
-30,0	150,0													
0,0	180,0													
30,0	210,0													
112,5	292,5													
PICK UP		50PP1	MFG	%ERR	50L	MFG	%ERR	50PP3	MFG	%ERR	50PP4	MFG	%ERR	
50PP-AMPS														
TIMING TESTS (in cycles)														
MULT		M1PT	MFG	%ERR	M2PT	MFG	%ERR	M3P	MFG	%ERR	M4PT	MFG	%ERR	
1.2 * PU														

COMMENTS:

RESULTS ACCEPTABLE:　　　　☐ YES　　　　☐ NO　　　　☐ SEE NOTES

Figure 5-7: Example Test Sheet for Element Pickup Tests

PHASE OVERCURRENT TEST RESULTS							
51P - PICK UP		50P1 - PICK UP					
51P - CURVE		50P1 - TIME					
51P - TIME DIAL		50P2 - PICK UP					
51P - RESET		50P2 - TIME					
TEST	A PHASE PU	B PHASE PU	C PHASE PU		MFG	% ERROR	
51P PICKUP							
50P1 PICKUP							
50P2 PICKUP							
51P TIMING TESTS (in seconds)							
MULT	AMPS	A PH TRIP	B PH TRIP	C PH TRIP	MFG	% ERROR	
2	0,00						
3	0,00						
4	0,00						
RESET					0,00		
67P1T TIMING TESTS (in cycles)							
MULT	AMPS	A PH TRIP	B PH TRIP	C PH TRIP	MFG	% ERROR	
1,1							
50P2 TIMING TESTS (in cycles)							
MULT	AMPS	A PH TRIP	B PH TRIP	C PH TRIP	MFG	% ERROR	
1,1							
COMMENTS:							
RESULTS ACCEPTABLE:		☐ YES		☐ NO		☐ SEE NOTES	

Figure 5-8: Example Test Sheet #2 for Element Pickup Test

9. Element Characteristics

Some testing specifications require you to prove the element characteristics. Examples of this testing includes, for example, that the impedance relay characteristic circle is in the correct quadrant, or the offset on a loss-of-field element is correctly set for positive or negative. All test data should be added to the test sheet but graphs always make you look good and can make evaluation a breeze. Characteristic graphs should be constructed to easily identify different phases or elements and include a curve of the manufacturer's expected values based on the settings. If you have test results for similar elements, they should be collected on one graph unless the values are too different and the graph becomes distorted or illegible.

3 PHASE DISTANCE TEST RESULTS (Ohms)												
Z1P - PU	2,4	Z2P - PU	4,7	Z3P - PU	5	Z4P - PU						
Z1PANG	79	Z2PANG	79,00	Z3PANG	259,00	Z4PANG						
Z1PD -TD	0,00	Z2PD - TD	20,00	Z3PD - TD	60,00	Z4PD - TD						
DIR1	F	DIR2	F	DIR3	R	DIR4						
50PP1	2,00	50PP2	2,00	50PP3	1,00	50PP4						

TEST ANGLE		ZONE 1 TESTS			ZONE 2 TESTS			ZONE 3 TESTS			ZONE 4 TESTS		
F	R	M1P	MFG	%ERR	M2P	MFG	%ERR	M3P	MFG	%ERR	M4P	MFG	%ERR
29,0	209,0	1,55	1,54	0,47	3,04	3,02	0,63	-3,224	-3,21	0,31			
49,0	229,0	2,09	2,08	0,56	4,08	4,07	0,24	-4,337	-4,33	0,16			
79,0	259,0	2,41	2,40	0,42	4,71	4,70	0,21	-5,02	-5,00	0,40			
109,0	289,0	2,08	2,08	0,07	4,04	4,07	-0,74	-4,33	-4,33	0,02			
139,0	300,0	1,21	1,20	0,83	2,36	2,35	0,43	-3,79	-3,77	0,53			
PICK UP		50PP1	MFG	%ERR	50L	MFG	%ERR	50PP3	MFG	%ERR	50PP4	MFG	%ERR
50PP-AMPS		2,05	2,0	2,50	2,03	2,0	1,50	0,99	1,0	-1,00			

TIMING TESTS (in cycles)												
MULT		M1PT	MFG	%ERR	M2PT	MFG	%ERR	M3P	MFG	%ERR	M4PT	MFG	%ERR
1.2 * PU		3,5	3,00	OK	20,3	20,00	1,50	60,05	60,00	0,08			

COMMENTS:

RESULTS ACCEPTABLE: ☑ YES ☐ NO ☐ SEE NOTES

Figure 5-9: Test Sheet Example Element

3 PHASE DISTANCE TEST RESULTS (Ohms)

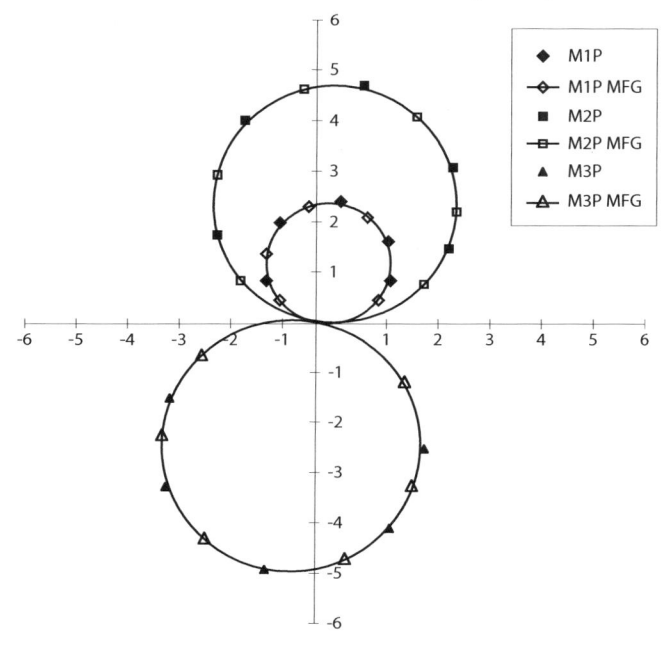

Figure 5-10: Example Test Sheet with Element Characteristics

10. Final Output Checks

This is the most important test you can perform on a digital relay. This test is performed after all other testing is complete, the relay settings have been re-checked to make sure that no further corrections or changes are necessary, and that the relay is ready for service. Each element or combination of elements in an output logic scheme is simulated and the output operation is verified. This test replaces the relay wiring function testing of the past that the relay logic replaced.

There are many ways to document this test. If no other testing was performed, the current/voltage/frequency test values should be recorded to prove the correct element was applied to verify the correct output. If pickup and timing tests were performed, a list of applied elements for each output is satisfactory. An example report of output tests looks like Figure 5-11:

FINAL OUTPUT CHECKS							
OUT101	OUT102	OUT103	OUT104	OUT105	OUT106	OUT107	
"TRIP"	"SV1T"	TRIP	"SV2T"				
M1P	IN102		IN102*50P1*12cy				
Z1G			IN102*50G1*12cy				
M2PT			TRIP*50P1*12cy				
Z2GT			TRIP*50G1*12cy				
OC			SV2*50P1				
(M2P+Z2G)*RMB1A			SV2*50G1				
COMMENTS:		UNABLE TO TEST HIGHLIGHTED ITEMS - COMMUNICATION NOT ENABLED					
RESULTS ACCEPTABLE:		✓ YES		☐ NO		☐ SEE NOTES	

Figure 5-11: Example Test Sheet with Final Output Check

11. Test Sheet Template

VALENCE ELECTRICAL TRAINING SERVICES		EQ. DESG:
		LOCATION:
		SYSTEM:

SCHWEITZER LABORATORIES SEL-311C TEST FORM

PROJECT #:		DATE:	
CLIENT:		TESTED BY:	
LOCATION:		APPROVED BY:	
TEST SET:		TESTING STD:	☐ ACCEPT ☐ MAINT

NAMEPLATE DATA

DESIGNATION:		LOCATION:	
TYPE:		PART NUMBER:	
MODEL NUMBER:		LOGIC INPUT:	
SERIAL NUMBER:		ROTATION:	
CONTROL POWER:		FREQUENCY:	
		MAXIMUM VOLTAGE:	
CT SEC NOMINAL:		NOMINAL VOLTAGE:	
PHASE CT RATIO:		PHASE VT RATIO:	
CT RATIO DRAWING:		VT RATIO DRAWING:	

NOTES

INPUT/OUTPUT CHECKS

OUT101	OUT102	OUT103	OUT104	OUT105	OUT106	OUT107	ALARM		
IN101	IN102	IN103	IN104	IN105	IN106	IN107			

COMMENTS:

RESULTS ACCEPTABLE: ☐ YES   ☐ NO   ☐ SEE NOTES

METERING							
CURRENT (AMPS)							
SEC INJ INPUT	A PH		B PH	C PH	MFG	% ERROR	
PHASE ANGLE							
SEC INJ INPUT		In			MFG	% ERROR	
@ 0 degrees					In RATIO=120		
VOLTAGE (VOLTS)							
SEC INJ INPUT	A PH		B PH	C PH (kV)	MFG (kV)	% ERROR	
PHASE ANGLE							
SEC INJ INPUT		Vs (kV)			MFG (kV)	% ERROR	
@ 0 degrees					Vs RATIO = 60		

3 PHASE METERING							
FREQUENCY (Hz)				**POWER FACTOR**			
SEC INJ INPUT	3 PH (Hz)	MFG (Hz)	%ERROR	SEC INJ INPUT	3 PH	MFG	%ERROR
57.50		57.50		30 Degrees Lag		0.866	
60.00		60.00		0 degrees		1.000	
62.50		62.50		30 Degrees Lead		-0.866	
POWER (MW)				**VARS (MVAR)**			
SEC INJ INPUT	3 PH (MW)	MFG (MW)	%ERROR	SEC INJ INPUT	3 PH	MFG	%ERROR

COMMENTS:

RESULTS ACCEPTABLE:

3 PHASE DISTANCE TEST RESULTS (Ohms)													
Z1P - PU		Z2P - PU		Z3P - PU		Z4P - PU							
Z1PANG		Z2PANG		Z3PANG		Z4PANG							
Z1PD -TD		Z2PD - TD		Z3PD - TD		Z4PD - TD							
DIR1		DIR2		DIR3		DIR4							
50PP1		50PP2		50PP3		50PP4							

TEST ANGLE		ZONE 1 TESTS			ZONE 2 TESTS			ZONE 3 TESTS			ZONE 4 TESTS		
F	R	M1P	MFG	%ERR	M2P	MFG	%ERR	M3P	MFG	%ERR	M4P	MFG	%ERR
-50,0	130,0												
-30,0	150,0												
0,0	180,0												
30,0	210,0												
112,5	292,5												
PICK UP		50PP1	MFG	%ERR	50L	MFG	%ERR	50PP3	MFG	%ERR	50PP4	MFG	%ERR
50PP-AMPS													

TIMING TESTS (in cycles)													
MULT		M1PT	MFG	%ERR	M2PT	MFG	%ERR	M3P	MFG	%ERR	M4PT	MFG	%ERR
1.2 * PU													

COMMENTS:

RESULTS ACCEPTABLE: ☐ YES   ☐ NO   ☐ SEE NOTES

PHASE TO PHASE ZONE 1 DISTANCE TEST RESULTS

	PICK UP TEST RESULTS (OHMS)			EXPECTED RESULTS (OHMS)			
TEST ANGLE	A-B PHASE	B-C PHASE	C-A PHASE	MFG	% ERROR		
-50							
-30							
0							
30							
50							
	A-B PHASE	B-C PHASE	C-A PHASE	MFG	% ERROR		
50PP1							

TIMING TESTS (in cycles)

	A-B PHASE	B-C PHASE	C-A PHASE	MFG	% ERROR		
1.2 * PU							

COMMENTS:

RESULTS ACCEPTABLE: ☐ YES   ☐ NO   ☐ SEE NOTES

VOLTAGE AND SYNCHRONIZING ELEMENTS

SINGLE PHASE VOLTAGE

ELEMENT	27A	27B	27C	MFG	% ERROR		
27P				10,000			

THREE PHASE VOLTAGE

ELEMENT	PICKUP (V)	MFG (V)	%ERROR	NOTES		
27SP		10,00		27S		
59V1P		40,00		POSITIVE SEQUENCE OVERVOLTAGE		
59SP		50,00		59S		

25 SYNCHRONIZING

ELEMENT	PICKUP (VPH)	PICKUP (VS)	MFG (V)	%ERR (VP)	%ERR (VS)	NOTES
25VLO			53,00			
25VHI			79,00			
25SF			0,10			
25ANG1			15,00			
25ANG2			-15,00			

COMMENTS:

RESULTS ACCEPTABLE: ☐ YES   ☐ NO   ☐ SEE NOTES

79 RECLOSER				
RECLOSE TIMERS (cycles)				
TIMER	TIME	MFG	%ERROR	NOTES
79RSLD		120,00		TIME FROM !52A TO 52a - MONITOR **79LO**
79RS		3000,00		TIME FROM 52a & FAULT THEN !52a & NO-FAULT TO 52A
79O1		60,00		TIME FROM FAULT & 52A TO **CLOSE**
CFD		60,00		TIME FROM FAULT & 52A TO **79LO** MINUS 79O1
RECLOSE LOGIC				
TIMER	LOGIC		NOTES	
79CLS			CLOSE SUPERVISION	
79DLS			DRIVE TO LAST SHOT	
79DTL			DRIVE TO LOCKOUT	
79RI			RECLOSE INITIATE	
79RIS			RECLOSE INITIATE SUPERVISION	
79STL			STALL OPEN INTERVAL TIMING	

COMMENTS:

RESULTS ACCEPTABLE: ☐ YES ☐ NO ☐ SEE NOTES

FINAL OUTPUT CHECKS								
OUT101	OUT102	OUT103	OUT104	OUT105	OUT106	OUT107	TMB1B	TMB1B
							TMB1A	TMB1A

COMMENTS:

RESULTS ACCEPTABLE: ☐ YES ☐ NO ☐ SEE NOTES

12. Final Report

The final report should document all of your test results, comments, and a final copy of the relay settings to allow the project manager to review the results and final settings. The following items should be included in every test report.

A) Cover Letter

The cover letter should describe the project, provide a brief history about it, and (most importantly) list all your comments. This letter summarizes all of the test sheets and should be written with non-electrical personnel in mind. Ideally, you should be able to review this document years from now and be able speak with authority about your activities. Any comments should be clearly explained with a brief history of any actions taken regarding the comment, and its status at the time of the letter. Organize comments in order of importance and by relay or relay type if the same comment applies to multiple relays. An example comment is: "The current transformer ratio on Drawing A and the supplied relay settings did not match. The design engineer was contacted and the correct ratio of 600:5 was applied to the relay settings and confirmed in the field. No further action is required."

This is where you justify your work on the project and why your client's money was well spent.

B) Test Sheet

The test sheet should clearly show all of your test results, the expected result from the manufacturer's manual, and whether the result is acceptable as discussed previously in this chapter.

C) Final Settings

The final, as-left settings should be documented at the end of your test sheet. A digital copy should also be saved and all relay settings for a project should be made available to the client or design engineer for review and their final documentation. Setting files should be in the relay's native software and in a universal format such as word processor or pdf™ file to allow the design engineer to make changes, if required, and allow anyone else to review the settings without special software.

Chapter 6

Testing Overvoltage (59) Protection

1. Application

Higher than rated voltages stress the electrical insulation of equipment and cause it to deteriorate. The effects of overvoltages are cumulative and may cause in-service failures over time. Overvoltage (59-element) protection is applied to protect the equipment and will operate if the measured voltage rises above the 59-element pickup setting. 59-elements almost always incorporate time delays to prevent nuisance tripping caused by transients or swells in the system voltage.

While the actual Overvoltage (59) protection element is relatively simple, it can be difficult to determine the correct voltage application. The voltage element settings are often related to the nominal line voltage setting of the relay. The nominal potential transformer (PT) secondary voltage could be line-line (L-L) or line-ground (L-G) voltages depending on the system, number of PTs, and/or the PT connection as discussed in the *Instrument Transformers* section starting on page 34.

After you have determined whether the relay measures phase-to-phase (L-L) or phase-to-neutral (L-G) voltages, you should review the relay's 59-element and relay nominal voltage settings and make sure that they are correct. For example, if a relay is connected to a system with two PTs, the nominal voltage is likely to be between 115-120V using phase-to-phase voltages. A 59-element setting below 115V will likely cause nuisance trips using this configuration.

Sometimes a 59-element will be applied to monitor breaker status or to determine whether a bus or line is energized. These applications will have lower voltage settings (approximately 90V L-L) and are used in control applications.

2. Settings

The most common settings used in 59-elements are explained below:

A) Enable Setting

Many relays allow the user to enable or disable settings. Make sure that the element is ON or the relay may even prevent you from entering settings. If the element is not used, the setting should be disabled or OFF to prevent confusion.

B) Pickup

This setting determines when the relay will start timing. Different relay models use different methods to set the actual pickup. Make sure you determine whether Line-to-Line or Line-to-Neutral voltages are selected in the relay. The most common pickup setting definitions are:

- **Secondary Voltage**—Pickup = Setting
- **Multiple of Nominal or Per Unit (P.U.)**—If the relay has a nominal voltage setting, the pickup could be a multiple of that nominal voltage setting. Or it could be a multiple of the nominal PT secondary if a nominal PT secondary setting exists in the relay.
  Pickup = Setting x Nominal Volts, **OR**
  Pickup = Settings x Nominal PT Secondary Setting
- **Primary Volts**—There must be a setting for PT ratio if this setting style exists. Check the PT ratio from the drawings and make sure that the drawings match the settings.
  Pickup = Setting / PT Ratio, **OR**
  Pickup = Setting * PT secondary / PT primary

C) Time Curve Selection

59-element timing can have a fixed time (definite time) or an inverse curve that will cause the element to operate faster as the overvoltage magnitude increases. This setting determines which characteristic applies. DO NOT assume that a time-delay setting with a "definite" time-curve selection is the actual expected element time. Some "definite-time" settings are actually inverse-time curves. Check the manufacturer's literature.

D) Time Delay

The time-delay setting for the 59-element can be a fixed-time delay that determines how long (in seconds or cycles) the relay will wait to trip after the pickup has been detected.

59-elements can also be inverse-time curves and this setting simulates the time-dial setting of an electromechanical relay. ANSI curves usually have a time delay between 1 and 10 and IEC time delay settings are typically between 0 and 1.

E) Reset

Electromechanical 59-element relay timing was controlled by a mechanical disc that would rotate if the voltage was higher than the pickup setting. If the voltage dropped below the pickup value, the disc would rotate back to the reset position. The disc would move to the reset position faster in relation to a smaller voltage input.

Some digital relays simulate the reset delay using a curve that is directly proportional to the voltage in order to closely match the electromechanical relay reset times. Other relays have a preset time delay or user defined reset delay that should be set to a time after any related electromechanical discs. Some relays will reset immediately after the voltage drops below the pickup voltage.

F) Required Phases

Some relays allow the user to determine what combination of overvoltages must be present in order to start the 59-element timer.

- **Any Phase**—Any phase or any combination of measured voltages that rise above the 59-element pickup setpoint will start the timer.
- **Any Two Phases**—At least two measured voltages must be above the 59-element pickup setpoint before the timer will start.
- **All Three Phases**—All measured voltages must be above the 59-element pickup setpoint before the timer will start.

3. Pickup Testing

Overvoltage pickup testing is very simple. Apply nominal voltage, check for correct metering values, and raise the voltage until pickup operation is detected. It is often easier to adjust all 3Ø voltages simultaneously, and remove voltage leads from the test-set for the phases not under test rather than adjust one phase at a time. Use the following settings for a GE/Multilin SR750 relay for the steps below.

	SETTING	TEST RESULT
VT Connection Type	Wye	
Nominal VT Secondary Voltage	69.28V	
VT Ratio	35:1	
Overvoltage Pickup	1.1 x VT	76.32V
Time Dial	3.0	3.05s
Phases Required	Any Two	

Figure 6-1: Example 59-Element Settings and Test Results with Phase-to-Neutral Voltages

	SETTING	TEST RESULT
VT Connection Type	Delta	
Nominal VT Secondary Voltage	120.0V	
VT Ratio	35:1	
Overvoltage Pickup	1.1 x VT	132.2 V
Time Dial	3.0	3.05s
Phases Required	Any Phase	Any Phase

Figure 6-2: Example 59-Element Settings and Test Results with Phase-to-Phase Voltages

When the VT connection type is set to Wye in the SR-750 relay, all voltages are phase-to-neutral. Phase-to-phase values are required when the VT connection is in Delta. Use the following substitution chart to modify the procedures below for the different PT configurations.

	WYE	DELTA
AØ	A-N	A-B
BØ	B-N	B-C
CØ	C-N	C-A

Figure 6-3: Substitution Chart for 59-Procedures

A) Test-Set Connections

Use the following test connections for the different PT choices. Remember that the "Magnitudes" are phase-to-neutral values that must be multiplied by $\sqrt{3}$ to obtain phase-to-phase values.

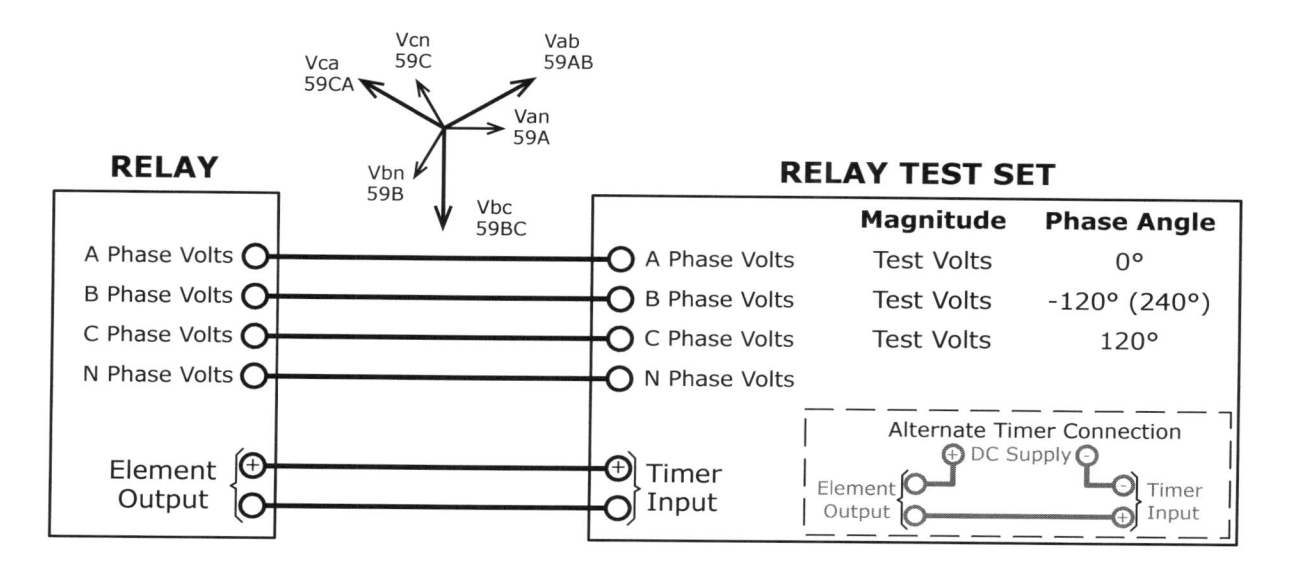

Figure 6-4: Simple Phase-to-Neutral Overvoltage Connections

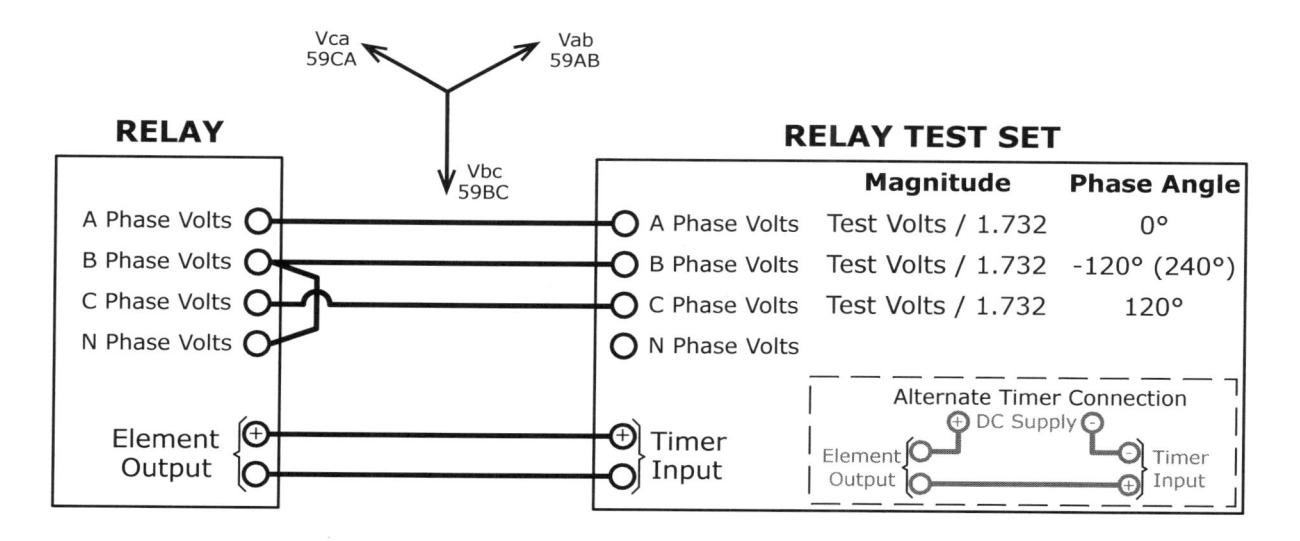

Figure 6-5: Simple Phase-to-Phase Overvoltage Connections

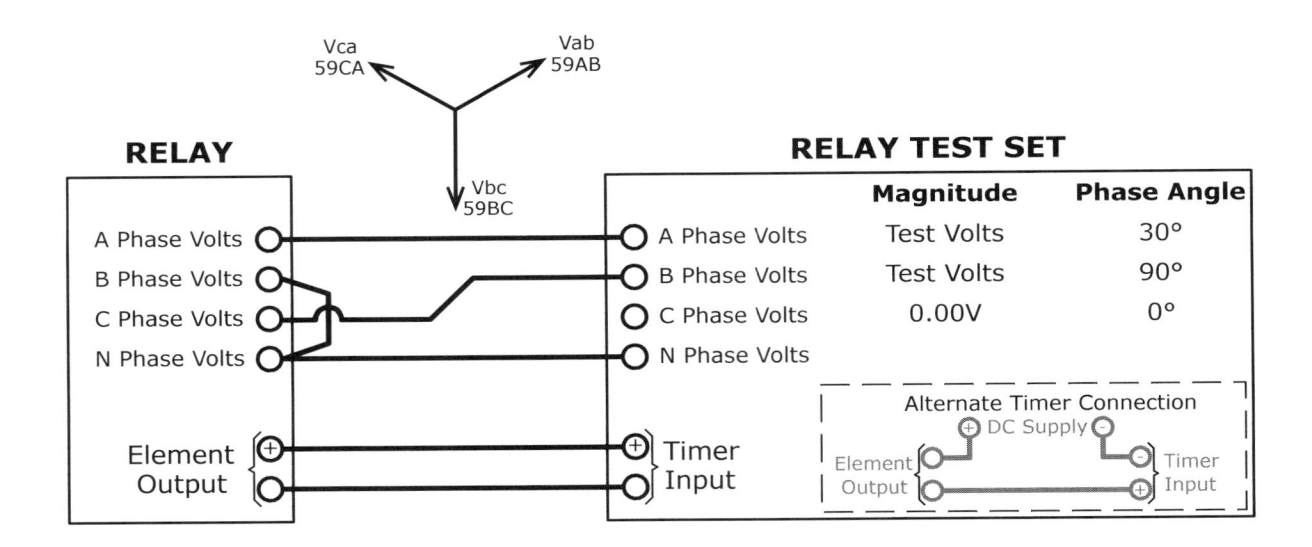

Figure 6-6: Phase-to-Phase Overvoltage Connections with 2 Voltage Sources

B) 3-Phase Voltage Pickup Test Procedure

- Determine how you will monitor pickup and set the relay accordingly, if required. (Pickup indication by LED, output contact, front panel display, etc. See the *Relay Test Procedures* section starting on page 109 for details.)
- Apply 3Ø rated voltage and check metering values.
- Apply 3Ø voltage 5% higher than the pickup setting and verify pickup indication is on.
- Slowly lower all 3Ø voltages simultaneously until the pickup indication is off. Slowly raise voltage until pickup indication is fully on. Record pickup values on test sheet.
- While the pickup indication is on, slowly lower any single-phase voltage below the pickup value and ensure the pickup drops out. This proves all three phases must be above the pickup level for the 59-element to operate.

C) More than One Phase Voltage Pickup Test Procedure

- Determine how you will monitor pickup and set the relay accordingly, if required. (Pickup indication by LED, output contact, front panel display, etc. See the *Relay Test Procedures* section starting on page 109 for details.)
- Apply rated 3-phase voltage and check metering values.
- Increase AØ voltage 5% above pickup (76.55V = V_{Rated} * Pickup * 1.05= 66.28V *1.1 * 1.05) and verify that pickup does not operate.
- Increase BØ voltage 5% above pickup and verify that the pickup indication operates. This proves that two voltages must be above the pickup setting before the 59-element will operate.

- Slowly lower the AØ voltage until the pickup indication is off. Slowly raise voltage until pickup indication is fully on. Record pickup values on test sheet. Set AØ voltage to nominal.
- Raise CØ voltage 5% above pickup and verify that the pickup indication operates.
- Slowly lower the BØ voltage until the pickup indication is off. Slowly raise voltage until pickup indication is fully on. Record pickup values on test sheet. Set BØ voltage to nominal.
- Raise AØ voltage 5% above pickup and verify that the pickup indication operates.
- Slowly lower the CØ voltage until the pickup indication is off. Slowly raise voltage until pickup indication is fully on. Record pickup values on test sheet.

D) Any Phase Pickup Test Procedure

- Determine how you will monitor pickup and set the relay accordingly, if required. (Pickup indication by LED, output contact, front panel display, etc. See the *Relay Test Procedures* section starting on page 109 for details.)
- Apply rated 3-phase voltage and check metering values.
- Increase AØ voltage 5% above pickup (138.06V = V_{Rated} * Pickup * 1.05= 120V *1.1 * 1.05) and verify that the pickup indication operates.
- Slowly lower the AØ voltage until the pickup indication is off. Slowly raise voltage until pickup indication is fully on. Record pickup values on test sheet and return AØ voltage to rated voltage.
- Repeat for BØ and CØ.

E) Evaluate Results

Before we can evaluate the test results, we must determine manufacturer's expectations and tolerances. Use the following specifications from a GE/Multilin SR750 relay to determine expected values for our examples.

BUS AND LINE VOLTAGE

Accuracy (0 to 40°C)	+/- 0.25% of full scale (10 to 130 V). (For open delta, the calculated phase has errors 2 times those shown above.)

VOLTAGE

Accuracy	+/- 0.25% of full scale

OVERVOLTAGE

Level Accuracy:	Per voltage input
Timing Accuracy:	+/- 100ms

Figure 6-7: GE/Multilin SR-750 Overvoltage Relay Specifications

Using "Bus and Line Voltage/Accuracy = +/- 0.25%" in Figure 6-7 we can determine that the allowable error is +/- 0.325V (0.0025 * 130V). In our example, the difference between the pickup setting (132.0V) and the test result (132.2V) is 0.2V (132.2V - 132.0V). The difference is within the manufacturer's tolerance and we have a successful pickup test. We could also calculate the percent error as shown below.

$$\frac{\text{Actual Value} - \text{Expected Value}}{\text{Expected Value}} \times 100 = \text{Percent Error}$$

$$\frac{132.2 - 132.0}{132.0} \times 100 = \text{Percent Error}$$

$$\frac{0.2}{132.0} \times 100 = \text{Percent Error}$$

$$0.15\% \text{ Error}$$

4. Timing Tests

59-element timing tests are very straightforward. If the time delay is fixed like the SR-750 relay in our example, apply 110% of the pickup voltage and measure the time between test start and output contact operation. The time-test result is compared to the setting and manufacturer's tolerances to make sure it is acceptable for service.

If the time delay is an inverse curve, perform the timing test by applying a multiple of pickup voltage and measure the time between the test start and output contact operation. Repeat the test for at least one other point to verify the correct curve has been applied.

A) Timing Test Procedure with Definite Time Delay and All Three Phases Required

- Determine which output the 59-element operates and connect the test-set-timing input to the relay output contact.
- Set the 3Ø fault voltage 10% higher than the pickup setting. The test for our example would be performed at 145.2V (V_{Rated} * Pickup * 1.10 = 120V *1.1 * 1.10). Set your test-set to stop when the timing input operates and to record the time delay from test start to stop.
- Apply rated voltage. Apply test voltage, ensure timing input operates, and note the time on your test sheet. Compare the test time to the settings to ensure timing is correct.
- Review relay targets to ensure the correct element and phases are displayed.

B) Timing Test Procedure with Definite Time Delay and Two Phases Required

- Determine which output the 59-element operates and connect the test-set-timing input to the relay output contact.

- Set the AØ and BØ fault voltage 10% higher than the pickup setting. The test for our example would be performed at 83.8V (V_{Rated} * Pickup * 1.10 = 69.28V *1.1 * 1.10). Set your test-set to stop when the timing input operates and to record the time delay from test start to stop.
- Apply rated voltage. Apply test voltage, ensure timing input operates, and note the time on your test sheet. Compare the test time to the settings to ensure timing is correct.
- Review relay targets to ensure the correct element and phases are displayed.
- Repeat the steps above for BØ-CØ and CØ-AØ.

C) Timing Test Procedure with Definite Time Delay and Any Phase Required

- Determine which output the 59-element operates and connect the test-set-timing input to the relay output contact.
- Set the AØ fault voltage 10% higher than the pickup setting. (V_{Rated} * Pickup * 1.10) Set your test-set to stop when the timing input operates and to record the time delay from test start to stop.
- Apply rated voltage. Apply test voltage, ensure timing input operates, and note the time on your test sheet. Compare the test time to the 59-element settings to ensure timing is correct.
- Review relay targets to ensure the correct element and phases are displayed.
- Repeat the steps above for BØ and CØ.

D) Timing Test Procedure with Inverse Time Delay and All Three Phases Required

- Determine which output the 59-element operates and connect the test-set-timing input to the relay output contact.
- Pick the first test point from manufacturer's curve. (Typically in percent of pickup) Set the 3Ø fault voltage at the test point. The first test for our example at 110% pickup would be performed at 145.2V (V_{Rated} * Pickup * 1.10 = 120V *1.1 * 1.10). Set your test-set to stop when the timing input operates and to record the time delay from test start to stop.
- Apply rated voltage. Apply test voltage, ensure timing input operates, and note the time on your test sheet. Compare the test time to the 59-element timing curve or formula to ensure timing is correct.
- Review relay targets to ensure the correct element and phases are displayed.
- Perform second test at another point on the manufacturer's timing curve. (E.g. 120%= 120V * 1.1 * 1.2= 158.4V)
- Apply rated voltage. Apply test voltage, ensure timing input operates, and note the time on your test sheet. Compare the test time to the 59-element timing curve or formula to ensure timing is correct.
- Review relay targets to ensure the correct element and phases are displayed.

E) Timing Test Procedure with Inverse Time Delay and Two Phases Required

- Determine which output the 59-element operates and connect the test-set-timing input to the relay output contact.
- Pick the first test point from manufacturer's curve. (Typically in percent of pickup) Set the AØ and BØ fault voltage at the test point. The first test for our example at 110% pickup would be performed at 83.8V (V_{Rated} * Pickup * 1.10 = 69.28V *1.1 * 1.10). Set your test-set to stop when the timing input operates and to record the time delay from test start to stop.
- Apply rated voltage. Apply test voltage, ensure timing input operates, and note the time on your test sheet. Compare the test time to the 59-element settings to ensure timing is correct.
- Review relay targets to ensure the correct element and phases are displayed.
- Perform second test at another point on the manufacturer's timing curve. (E.g. 120%= 120V * 1.1 * 1.2= 158.4V)
- Review relay targets to ensure the correct element and phases are displayed.
- Repeat the steps above for BØ and CØ.

F) Timing Test Procedure with Inverse Time Delay and Any Phases Required

- Determine which output the 59-element operates and connect the test-set-timing input to the relay output contact.
- Pick the first test point from manufacturer's curve. (Typically in percent of pickup) Set the AØ fault voltage at the test point. The first test for our example at 110% pickup would be performed at 145.2V (V_{Rated} * Pickup * 1.10 = 120V *1.1 * 1.10). Set your test-set to stop when the timing input operates and to record the time delay from test start to stop.
- Apply rated voltage. Apply test voltage, ensure timing input operates, and note the time on your test sheet. Compare the test time to the 59-element settings to ensure timing is correct.
- Review relay targets to ensure the correct element and phases are displayed.
- Perform second test at another point on the manufacturer's timing curve. (E.g. 120%= 120V * 1.1 * 1.2= 158.4V)
- Review relay targets to ensure the correct element and phases are displayed.
- Repeat the steps above for BØ and CØ.

G) Evaluate Test Results for Definite Time Settings

Before we can evaluate the test results, we must determine manufacturer's expectations and tolerances. Use the following specifications from a GE/Multilin SR750 relay to determine expected values.

BUS AND LINE VOLTAGE

Accuracy (0 to 40°C)	+/- 0.25% of full scale (10 to 130 V). (For open delta, the calculated phase has errors 2 times those shown above.)

VOLTAGE

Accuracy	+/- 0.25% of full scale

OVERVOLTAGE

Level Accuracy:	Per voltage input
Timing Accuracy:	+/- 100ms

Figure 6-8: GE/Multilin SR-750 Overvoltage Relay Specifications

The difference between the time delay setpoint and the test result in our examples is 0.05s or 50ms (Test Result—Setting = 3.05s—3.00s). Our example is a successful test because the difference between the setting and test is within manufacturer's tolerances using "Overvoltage/Level Accuracy = +/- 100ms".

Principles and Practice

H) Evaluate Test Results for Inverse Time Settings

There are two methods available when an inverse curve is selected for time delays. The first method is less accurate and uses the manufacturer's supplied curve to determine the expected time. The manufacturer's curve will plot a series of curves with the voltage magnitude on the x-axis and time on the y-axis. The voltage magnitude should be plotted in multiples or percent of the pickup setting and the time scale may be plotted using a logarithmic scale. Use the following steps and Figure 6-9 to determine the expected time delay on a manufacturer's supplied curve:

- Find the multiple or percent of pickup on the x-axis. (110% for our example)
- Draw an imaginary vertical line from that point until the line crosses the curve that matches the time delay setting. If your setting is not plotted on the curve, you must extrapolate the correct curve by picking a point between the two nearest curves. Remember that the logarithmic scale is not linear. For example, the arrow on the y-axis between 1 and 10 marks 5 seconds which would be near the midpoint between 1 and 10 on a linear scale but is significantly closer to 10 on a logarithmic scale.
- Draw a horizontal line from the end of your vertical line across to the y-axis. The expected time is the point where your imaginary line crosses the y-axis.

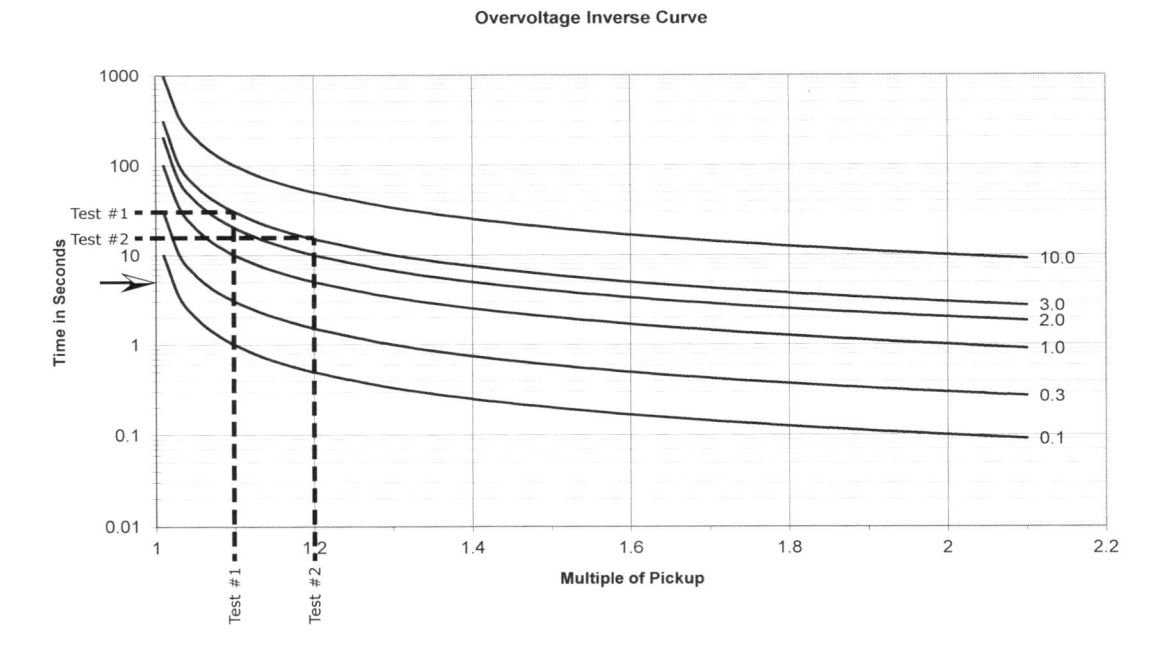

Figure 6-9: Inverse Curve for Overvoltage Protection

Using the steps above, our expected time delays are 30 seconds at 110% of pickup and approximately 13.5 seconds at 120%. This method can be inaccurate based on the extrapolation necessary, especially with logarithmic scales.

162

The second method uses a manufacturer supplied formula for inverse elements and is obviously the preferred method to determine the expected time delay because it is the most accurate. The formula for the Multilin SR-489 inverse curve is:

$$T = \frac{D}{\left(\dfrac{V}{V_{pickup}} - 1 \right)}$$

T = **Operating Time**

D = **Undervoltage Delay Setting**
(D = 0.00 operates instantaneously)

V = **Secondary Voltage Applied to the Relay**

V_{pickup} = **Pickup Level**

Use the formula to calculate the expected time delay. The results should be the same as the results obtained by the graph method, but the formula results will always be more accurate and are preferred.

$$T = \frac{3.0}{\left(\dfrac{145.2}{120 \times 1.1} - 1 \right)} = 30s \qquad T = \frac{3.0}{\left(\dfrac{158.4}{120 \times 1.1} - 1 \right)} = 15.00s$$

The formulas can be simplified by replacing V/V_{pickup} with the actual multiple of pickup you will be using for the test.

$$T = \frac{3.0}{(1.1 - 1)} = 30s \qquad T = \frac{3.0}{(1.2 - 1)} = 15.00s$$

Now that we have an accurate expected time, we can use the manufacturer's specifications to evaluate the test results.

OVERVOLTAGE

Pickup Accuracy: +/- 0.5% of Full Scale

Timing Accuracy: +/- 100ms or +/- 0.5% of Total Time

Figure 6-10: GE/Multilin SR-489 Overvoltage Relay Specifications

Our test results should be within the following parameters to be acceptable for service:

Test at 110% = 30 seconds	Test at 120% = 15 seconds
30s * +/- 0.5% = 29.85 to 30.15 s	15s * +/- 0.5% = 14.925—15.075 s
30s +/- 100ms = 39.9 to 30.1 s	15s +/- 100ms = 14.9 to 15.1 s
Use 29.85 to 30.15s (+/- 0.5%)	**Use 14.9 to 15.1s (+/-100ms)**

5. Tips and Tricks to Overcome Common Obstacles

The following tips or tricks may help you overcome the most common 59-element testing obstacles.

- Before you start, apply voltage at a lower value and perform a meter command to make sure your test-set is actually producing voltage and the PT ratio is set correctly.
- More sophisticated relays may require positive sequence voltage. Check system settings and connections for correct phase sequences.
- It is often easier to set 3Ø voltages and remove leads instead of changing test-set settings for multiple tests.
- Applying prefault nominal voltages can increase timing accuracy.
- Fuse failure (60) may block voltage tests. Check your configuration settings for fuse failure protection.

Chapter 7

Undervoltage (27) Protection Testing

1. Application

Equipment operating at lower than nominal voltages could overheat due to the increased amperage necessary to produce the same amount of power at the lower voltage. Undervoltage protection (27-element) is used to protect equipment from thermal stress created from lower than rated voltages. 27-elements almost always incorporate time delays to prevent nuisance tripping caused by transients or sags in the system voltage.

Undervoltage protection (27) is a little more complicated than overvoltage (59) protection described in the previous chapter because dynamic testing is necessary for accurate test results. It can also be difficult to determine the correct application. The undervoltage settings are often related to the nominal voltage setting of the relay, which could be line-line or line-ground voltages depending on the system, number of potential transformers (PTs), and/or the PT connection as discussed in the *Instrument Transformers* section starting on page 34.

After you have determined whether the relay measures phase-to-phase or phase-to-neutral voltages, you should review the relay's 27-element and nominal voltage settings to make sure that they are correct. For example; if a relay is connected to a system with two PTs, the nominal voltage is likely to be between 115-120V and would be phase-to-phase voltages. A 27-element setting above 110V will likely cause nuisance trips.

Sometimes a 27-element will be applied to monitor breaker status or to determine whether a bus or line is de-energized. These applications will have lower voltage settings (approximately 30V) and are used for control applications.

Undervoltage (27) protection is the opposite of overvoltage protection. When the protected equipment is de-energized, the input voltage should fall below the undervoltage setting and cause an undervoltage trip. The trip indication from the 27-element can be a nuisance for operators; and even prevent the breaker from closing in some control schemes.

Relay voltage inputs are supplied by PTs and an open circuit PT fuse could cause a 27-element to trip. Would you want your entire plant to shut down because someone accidentally touched a PT circuit? There are many methods for dealing with these problems and some are automatically applied by more sophisticated relays. You may:

- Interlock 27-element operation with a breaker signal to block tripping unless the breaker is closed.
- Interlock 27-element operation with a small 50-element setting to block tripping unless a preset amount of current flows.
- Only allow 27-element operation within a voltage window between pickup and minimum voltage. For example; if the voltage window is between 30 and 90V. The 27-element will not operate if the voltage is above 90V or below 30V.
- Interlock 27-element operation with a loss-of-fuse or PT Fuse Failure protection to block tripping if a PT fuse opens.
- Block 27-element operation if the positive sequence voltage is less than a predefined value to prevent nuisance trips if a PT fuse opens.

The test voltage must be initially higher than the 27-element pickup setpoint or the 27-element will always be on. Therefore, some kind of prefault voltage must be applied for timing tests.

The input voltage is continuously turned off and on during relay testing and 27-elements often interfere with tests or are just a plain nuisance. I often disable 27-element while testing generator protection and save it until the very last test after it is re-enabled. Remember, disabled elements should be tested AFTER the settings have been re-enabled.

2. Settings

A) Enable Setting

Many relays allow the user to enable or disable settings. Make sure that the element is ON or the relay may prevent you from entering settings. If the element is not used, the setting should be disabled or OFF to prevent confusion.

B) Pickup

This setting determines when the relay will start timing. Different relay models use different methods to set the actual pickup. Make sure you determine whether Line-to-Line or Line-to-Neutral voltages are selected in the relay. The most common pickup setting definitions are:

- **Secondary Voltage**—Pickup = Setting
- **Multiple of Nominal or Per Unit (P.U.)**—If the relay has a nominal voltage setting, it could be a multiple of the nominal voltage as defined in the relay settings or it could be a multiple of the nominal PT secondary if a nominal PT secondary setting exists in the relay.

  Pickup = Setting x Nominal Volts, **OR**

  Pickup = Setting x Nominal PT secondary setting
- **Primary Volts**—There must be a PT ratio setting if this style exists. Check the PT ratio from the drawings and check to make sure that the drawing matches the settings.

  Pickup = Setting / PT Ratio, **OR**

  Pickup = Setting * PT secondary / PT primary

C) Time Curve Selection

27-element timing can be a fixed time (definite time), instantaneous, or an inverse curve where the time delay is relative to the severity of undervoltage. This setting determines which characteristic applies. DO NOT assume that the time delay setting with a "definite" time curve setting is the actual expected element time. Some "definite time" settings are actually curves that can vary the time delay with magnitude of voltage. Check the manufacturer's literature.

D) Time Delay

The time delay setting for the 27-element can be a fixed time delay that determines how long (in seconds or cycles) the relay will wait to trip after the pickup has been detected.

27-elements can also be inverse time curves and this setting simulates the time dial setting on an electromechanical relay. ANSI curves usually have a time delay between 1 and 10 and IEC time delay settings are typically between 0 and 1.

E) Reset

Electromechanical 27-element relay timing was controlled by a mechanical disc that would rotate if the voltage was lower than the pickup setting. If the voltage rose above the pickup value, the disc would rotate back to the reset position.

Some digital relays simulate the reset delay using a linear curve that is directly proportional to the voltage in order to closely match the electromechanical relay reset times. Other relays have a preset time delay or user defined reset delay that should be set to a time after any related electromechanical discs.

Some relays will reset immediately after the voltage rises above the 27-element pickup setting.

F) Required Phases

Some relays allow the user to determine what combination of undervoltages must be present in order to start the 27-element timer.

- **Any Phase**—Any phase or any combination of measured voltages below the 27-element pickup setpoint will start the timer.
- **Any Two Phases**—At least two measured voltages must be below the 27-element pickup setpoint before the timer will start.
- **All Three Phases**—All measured voltages must be below the 27-element pickup setpoint before the timer will start.

G) Mode

Some relays allow the user to determine whether phase-to-phase or phase-to-neutral voltages are used for 27-element measurements.

H) Minimum Voltage

If the voltage drops below this setpoint, the 27-element will not operate. This setting will prevent nuisance trips during normal switching or maintenance procedures.

I) Blocking Signal

This setting would be a breaker status contact or a loss of fuse signal from another element to prevent nuisance trips.

3. Pickup Testing

Undervoltage pickup testing is very simple. Apply nominal voltage, check for correct metering values, and lower the voltage until you detect pickup operation.

Use the following settings for a GE D60 relay for the steps below.

	SETTING	TEST RESULT
VT Connection Type	Wye	
Nominal VT Secondary Voltage	69.28V	
VT Ratio	35:1	
PHASE UV1 MODE:	Line-Ground	
PHASE UV1 PICKUP:	0.95pu	65.75V
PHASE UV1 CURVE:	Inverse	
PHASE UV1 DELAY:	2.0	20.2s / 5.05s
PHASE UV1 MINIMUM VOLTAGE:	30.0V	30.1V

Figure 7-1: 27-Element Example Settings and Test Results

When the UV1 mode is set to line-ground in the D-60 relay, all voltages are phase-to-neutral. Phase-to-phase values are required when the VT connection is in delta. Use the following substitution chart to modify the procedures below for the different PT configurations.

	WYE	DELTA
AØ	A-N	A-B
BØ	B-N	B-C
CØ	C-N	C-A

Figure 7-2: Substitution Chart for 27-Procedures

A) Test-Set Connections

Use the following test connections for the different PT choices. Remember that the magnitudes are phase-to-neutral values. Multiply them by $\sqrt{3}$ to obtain phase-to-phase values.

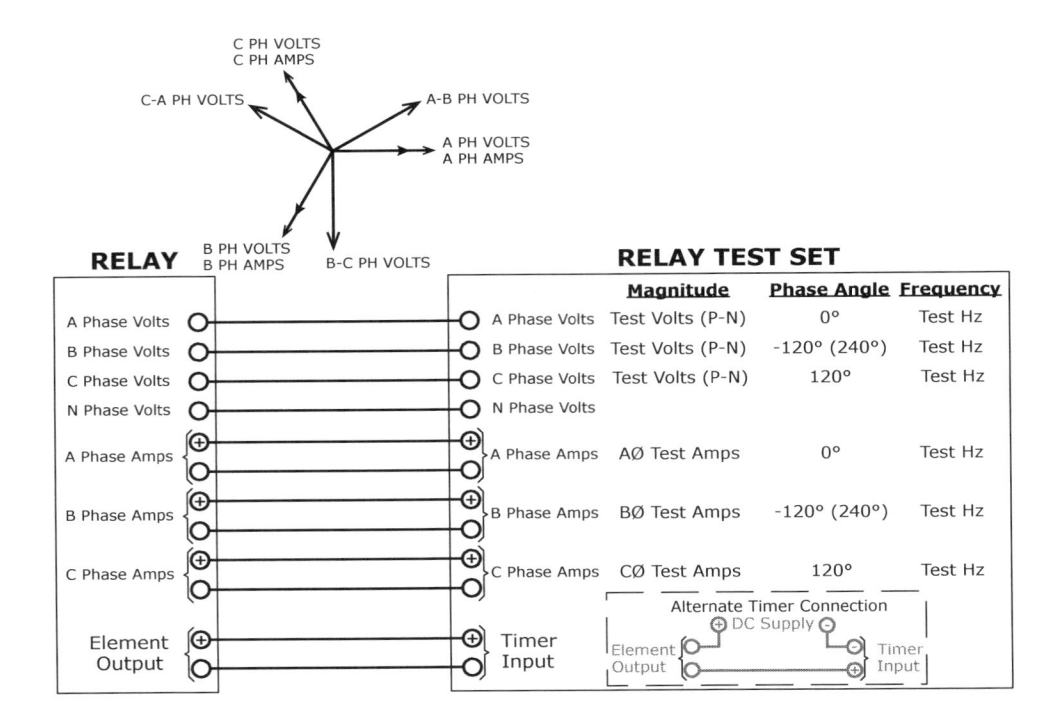

Figure 7-3: Simple Phase-to-Neutral Undervoltage Connections

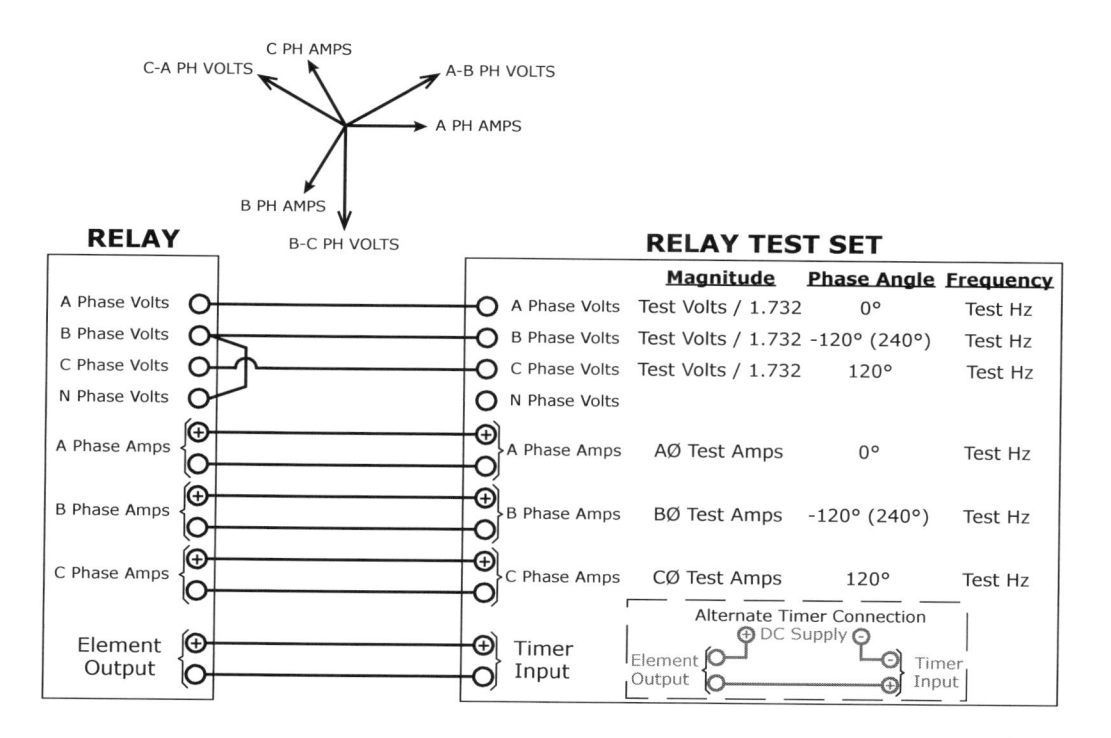

Figure 7-4: Simple Phase-to-Phase Undervoltage Connections

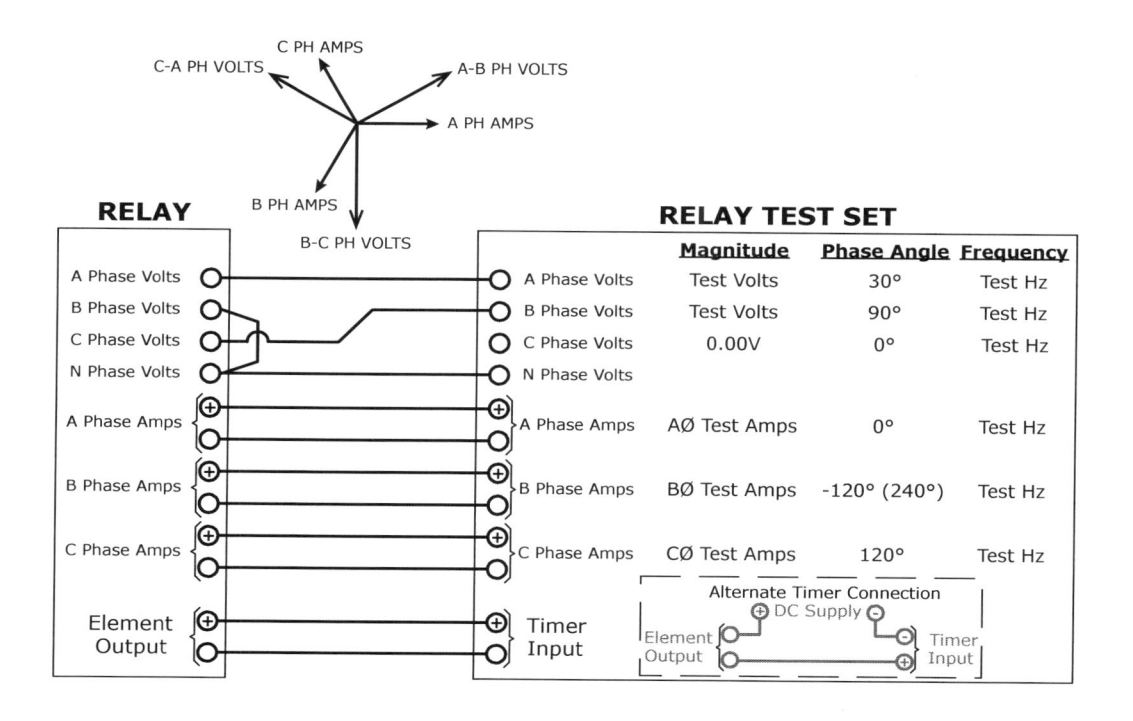

Figure 7-5: Phase-to-Phase Undervoltage Connections with 2 Voltage Sources

B) **Pickup Test Procedure If the Relay Requires 3-Phase Voltage to Operate**

- Determine how you will monitor pickup and set the relay accordingly, if required. (Pickup indication by LED, output contact, front panel display, etc. See the *Relay Test Procedures* section starting on page 109 for details.)
- Determine if a breaker status contact is used to disable 27-element protection and ensure it is in the correct state.
- Determine if input current is required for 27-element operation and apply nominal 3Ø current as per the wiring diagrams in Figures 2-3, 2-4, and 2-5.
- Apply rated 3Ø voltage (69.28V) and check metering values.
- Apply 3Ø voltage 5% lower than the pickup setting (62.53V = V_{Rated} * Pickup * 0.95= 69.28V *0.95 * 0.95) and verify pickup indication is on.
- Slowly raise the 3Ø voltages simultaneously until the pickup indication turns off. Slowly lower the 3Ø voltages simultaneously until pickup indication is fully on. Record pickup values on your test sheet.
- Raise one voltage above the pickup level. The pickup indication should turn off. This proves that all three voltages must be below the pickup setting to operate. Set the voltage below the pickup value and ensure pickup indication is back on.
- If current was required for 27-element operation, turn current off and make sure pickup indication turns off. Turn current back on.
- If a breaker status or blocking input disables the 27-element, reverse the input state and make sure pickup indication turns off.
- If a minimum voltage setpoint exists, continue lowering voltage until minimum voltage setpoint is reached and ensure pickup indication turns off. Record this value on your test sheet. Raise the voltage until the pickup indication is back on.

C) **Pickup Test Procedure If Relay Requires More Than One-Phase Voltage to Operate**

- Determine how you will monitor pickup and set the relay accordingly, if required. (Pickup indication by LED, output contact, front panel display, etc. See the *Relay Test Procedures* section starting on page 109 for details.)
- Determine if a breaker status contact is used to disable 27-element protection and ensure it is in the correct state.
- Determine if input current is required for 27-element operation and apply nominal 3Ø current as per the wiring diagrams in Figures 2-3, 2-4, and 2-5.
- Apply rated 3Ø voltage and check metering values.
- Lower AØ voltage 5% below pickup (62.53V = V_{Rated} * Pickup * 0.95= 69.28V *0.95 * 0.95) and verify that pickup does not operate.
- Lower BØ voltage 5% below pickup and verify that the pickup indication operates.
- Slowly raise the AØ voltage until the pickup indication turns off. Slowly lower voltage until pickup indication is fully on. Record pickup values on test sheet. Set AØ voltage to nominal.

- Lower CØ voltage 5% below pickup and verify that the pickup indication operates.
- Slowly raise the BØ voltage until the pickup indication turns off. Slowly lower voltage until pickup indication is fully on. Record pickup values on test sheet. Set BØ voltage to nominal.
- Lower AØ voltage 5% below pickup and verify that the pickup indication operates.
- Slowly raise the CØ voltage until the pickup indication turns off. Slowly lower voltage until pickup indication is fully on. Record pickup values on test sheet.
- If current was required for 27-element operation, turn current off and make sure pickup indication turns off. Turn current back on.
- If a breaker status or blocking input disables the 27-element, reverse the input state and make sure pickup indication turns off.
- If a minimum voltage setpoint exists, continue lowering voltage until minimum voltage setpoint is reached and ensure pickup indication turns off. This may require all three voltages to be below the minimum pickup setting. Raise the voltages until the pickup indication is back on.

D) Pickup Test Procedure for Any Phase Pickup

- Determine how you will monitor pickup and set the relay accordingly, if required. (Pickup indication by LED, output contact, front panel display, etc. See the *Relay Test Procedures* section starting on page 109 for details.)
- Determine if a breaker status contact is used to disable 27-element protection and ensure it is in the correct state.
- Determine if input current is required for 27-element operation and apply nominal 3Ø current as per the wiring diagrams in Figures 2-3, 2-4, and 2-5.
- Apply rated 3Ø voltage and check metering values.
- Lower AØ voltage 5% below pickup ($62.53V = V_{Rated}$ * Pickup * 0.95 = 69.28V *0.95 * 0.95) and verify that the pickup indication operates.
- Slowly raise the AØ voltage until the pickup indication turns off. Slowly lower voltage until pickup indication is fully on. Record pickup values on test sheet and return AØ to rated voltage.
- If current was required for 27-element operation, turn current off and make sure pickup indication turns off. Turn current back on.
- If a breaker status or blocking input disables the 27-element, reverse the input state and make sure pickup indication turns off.
- If a minimum voltage setpoint exists, continue lowering voltage until minimum voltage setpoint is reached and ensure the pickup indication turns off. This may require all three voltages to be below the minimum pickup setting. Raise the voltages until the pickup indication is back on.
- Repeat for BØ and CØ.

E) Evaluate Results

Before we can evaluate the test results, we must determine manufacturer's expectations and tolerances. Use the following specifications from a GE D-60 relay to determine expected values.

UNDERVOLTAGE

Voltage:	Phasor only
Level Accuracy:	+/- 0.5% of reading from 10 to 208 V
Timing Accuracy:	Operate @ <0.90 x Pickup
	+/- 3.5% of operate time or +/- 4 ms

Figure 7-6: GE D-60 Undervoltage Relay Specifications

Using "Level Accuracy = +/- 0.5%" we can determine that the allowable error is +/- 0.329V (0.005 * 65.82V). In our example, the difference between the pickup setting (65.82V) and the test result (65.75V) is 0.066V (65.82V—65.75V). The difference is within the manufacturer's tolerance and we have a successful pickup test. We could also calculate the percent error as follows.

$$\frac{\text{Actual Value - Expected Value}}{\text{Expected Value}} \times 100 = \text{percent error}$$

$$\frac{65.75 - 65.82}{65.82} \times 100 = \text{percent error}$$

$$\frac{0.066V}{65.82V} \times 100 = \text{percent error}$$

0.11% Error

4. Timing Tests

If the time delay uses a definite time curve, apply 90% of the pickup voltage and measure the time between test start and output contact operation. The time delay is compared to the setting and manufacturer's tolerances to make sure they match.

If the time delay is an inverse curve, perform the timing test by applying a multiple of pickup voltage and measure the time between the test start and output contact operation. Repeat the test for at least one other point to verify the correct curve has been applied.

Remember that prefault voltage higher than the pickup setting is required for successful 27-element testing!

A) Timing Test Procedure with Definite Time Delay and All Three Phases Required

- Determine which output the 27-element operates and connect timing input to the output.
- Determine if a breaker status contact is used to disable 27-element protection and ensure it is in the correct state.
- Determine if input current is required for 27-element operation and apply nominal 3Ø current as per the wiring diagrams in Figures 2-3, 2-4, and 2-5.
- Set the prefault voltage to nominal 3Ø voltage.
- Set the 3Ø fault voltage 10% lower than the pickup setting. The test for example would be performed at 59.23V (V_{Rated} * Pickup * 0.90= 69.28V *0.95 * 0.90). Set your test-set to stop when the timing input operates and to record the time delay from test start to stop.
- Apply prefault test voltage. Apply fault voltage and ensure timing input operates. Note the time on your test sheet. Compare the test time to the 27-element settings to ensure timing is correct.
- Lower the test voltage to any value below the first test and above the minimum voltage setting, if one exists. Apply prefault, and fault voltages. The time delay should be the same.
- Review relay targets to ensure the correct element operated.

B) Timing Test Procedure with Definite Time Delay and Two Phases Required

- Determine which output the 27-element operates and connect timing input to the output.
- Determine if a breaker status contact is used to disable 27-element protection and ensure it is in the correct state.
- Determine if input current is required for 27-element operation and apply nominal 3Ø current as per the wiring diagrams in Figures 2-3, 2-4, and 2-5.
- Set the AØ and BØ fault voltage 10% lower than the pickup setting. The test for our example would be performed at 59.23V (V_{Rated} * Pickup * 0.90= 69.28V *0.95 * 0.90). Set your test-set to stop when the timing input operates and to record the time delay from test start to stop.
- Apply prefault test voltage. Apply fault voltage and ensure timing input operates. Note the time on your test sheet. Compare the test time to the 27-element settings to ensure timing is correct.
- Review relay targets to ensure the correct element operated.
- Lower the test voltage to any value below the first test and above the minimum voltage setting, if one exists. Apply prefault, and fault voltages. The time delay should be the same.
- Repeat the steps above for BØ-CØ and CØ-AØ.

C) Timing Test Procedure with Definite Time Delay and Any Phase Required

- Determine which output the 27-element operates and connect timing input to the output.
- Determine if a breaker status contact is used to disable 27-element protection and ensure it is in the correct state.
- Determine if input current is required for 27-element operation and apply nominal 3Ø current as per the wiring diagrams in Figures 2-3, 2-4, and 2-5.
- Set the AØ fault voltage 10% lower than the pickup setting 59.23V (V_{Rated} * Pickup * 0.90= 69.28V *0.95 * 0.90). Set your test-set to stop when the timing input operates and to record the time delay from test start to stop.
- Apply prefault test voltage. Apply fault voltage and ensure timing input operates. Note the time on your test sheet. Compare the test time to the 27-element settings to ensure timing is correct.
- Review relay targets to ensure the correct element operated.
- Lower the test voltage to any value below the first test and above the minimum voltage setting, if one exists. Apply prefault, and fault voltages. The time delay should be the same.
- Repeat the steps above for BØ and CØ.

D) Timing Test Procedure with Inverse Time Delay and All Three Phases Required

- Determine which output the 27-element operates and connect timing input to the output.
- Determine if a breaker status contact is used to disable 27-element protection and ensure it is in the correct state.
- Determine if input current is required for 27-element operation and apply nominal 3Ø current as per the wiring diagrams in Figures 2-3, 2-4, and 2-5.
- Pick first test point from manufacturer's curve. (Typically in percent of pickup) Set the 3Ø fault voltage at the test point. The first test for our example at 90% pickup would be performed at 59.23V (V_{Rated} * Pickup * 0.90= 69.28V *0.95 * 0.90). Set your test-set to stop when the timing input operates and to record the time delay from test start to stop.
- Apply prefault test voltage. Apply fault voltage and ensure timing input operates. Note the time on your test sheet. Compare the test time to the 27-element settings to ensure timing is correct.
- Review relay targets to ensure the correct element operated.
- Perform second test at another point on the manufacturer's timing curve. (E.g. 60%= 69.28V * 0.95 * 0.6 = 39.49V)
- Apply prefault test voltage. Apply test voltage, ensure timing input operates, and note the time on your test sheet. Compare the test time to the 27-element timing curve or formula to ensure timing is correct.
- Review relay targets to ensure the correct element operated.

E) Timing Test Procedure with Inverse Time Delay and Two Phases Required

- Determine which output the 27-element operates and connect timing input to the output.
- Determine if a breaker status contact is used to disable 27-element protection and ensure it is in the correct state.
- Determine if input current is required for 27-element operation and apply nominal 3Ø current as per the wiring diagrams in Figures 2-3, 2-4, and 2-5.
- Pick first test point from manufacturer's curve. (Typically in percent of pickup) Set the AØ and BØ fault voltage at the test point. The first test for our example at 90% pickup would be performed at 59.23V (V_{Rated} * Pickup * 0.90= 69.28V *0.95 * 0.90) Set your test-set to stop when the timing input operates and to record the time delay from test start to stop.
- Apply prefault test voltage. Apply test voltage, ensure timing input operates, and note the time on your test sheet. Compare the test time to the 27-element settings to ensure timing is correct.
- Review relay targets to ensure the correct element operated.
- Perform second test at another point on the manufacturer's timing curve. (E.g. 60%= 69.28V * 0.95 * 0.6 = 39.49V)
- Apply prefault test voltage. Apply test voltage, ensure timing input operates, and note the time on your test sheet. Compare the test time to the 27-element timing curve or formula to ensure timing is correct.
- Review relay targets to ensure the correct element operated.
- Repeat the steps above for BØ-CØ and CØ-AØ.

F) Timing Test Procedure with Inverse Time Delay and Any Phase Required

- Determine which output the 27-element operates and connect timing input to the output.
- Determine if a breaker status contact is used to disable 27-element protection and ensure it is in the correct state.
- Determine if input current is required for 27-element operation and apply nominal 3Ø current as per the wiring diagrams in Figures 2-3, 2-4, and 2-5.
- Pick first test point from manufacturer's curve. (Typically in percent of pickup) Set the AØ fault voltage at the test point. The first test for our example at 90% pickup would be performed at 59.23V (V_{Rated} * Pickup * 0.90= 69.28V *0.95 * 0.90). Set your test-set to stop when the timing input operates and to record the time delay from test start to stop.
- Apply prefault test voltage. Apply test voltage, ensure timing input operates, and note the time on your test sheet. Compare the test time to the 27-element settings to ensure timing is correct.
- Review relay targets to ensure the correct element operated.

- Perform second test at another point on the manufacturer's timing curve. (E.g. 60%= 69.28V * 0.95 * 0.6 = 39.49V)
- Apply prefault test voltage. Apply test voltage, ensure timing input operates, and note the time on your test sheet. Compare the test time to the 27-element timing curve or formula to ensure timing is correct.
- Review relay targets to ensure the correct element operated.
- Repeat the steps above for BØ and CØ.

G) Evaluate Test Results with Definite Time Settings

Before we can evaluate the test results, we must determine manufacturer's expectations and tolerances. Use the following specifications from a GE D-60 relay to determine expected values.

UNDERVOLTAGE

Voltage:	Phasor only
Level Accuracy:	+/- 0.5% of reading from 10 to 208 V
Timing Accuracy:	Operate @ <0.90 x Pickup
	+/- 3.5% of operate time or +/- 4 ms

Figure 7-7: GE D-60 Undervoltage Relay Specifications

The test time should equal the time delay setting within 3.5% error or within 4ms of the specified time. For example, if the time setting was 20.0s and the measured time delay was 20.2 seconds, we could use the percent error calculation below to determine a 1% error which is within the manufacturer's tolerances. You would use the 4ms criteria for time delay settings less than 1.14 seconds because the measured time difference could be greater than 3.5% error but less than 4ms.

$$\frac{\text{Actual Value - Expected Value}}{\text{Expected Value}} \times 100 = \text{percent error}$$

$$\frac{20.2s - 20.00s}{20.00s} \times 100 = \text{percent error}$$

$$\frac{0.20s}{20.00s} \times 100 = \text{percent error}$$

$$1.0\% \text{ Error}$$

H) **Evaluate Test Results with Inverse Time Settings**

If "inverse curve" is defined, there are two different methods to calculate the time delay. The first method uses the manufacturer's supplied curve as shown in Figure 7-8. Our example uses a time delay line from the graph. If the setting is not drawn on the graph, move on to the second method. Find the first test voltage (90%) on the x-axis, remembering to read the x-axis scale. The scale is usually percent of pickup. Make a straight line up to the curve that represents the time delay setting (D=2.0). Make a horizontal line from the intersect point over to the Y-axis. The intersection at the Y-axis is the time delay (20.0s). Repeat for the second test point. (5.0s).

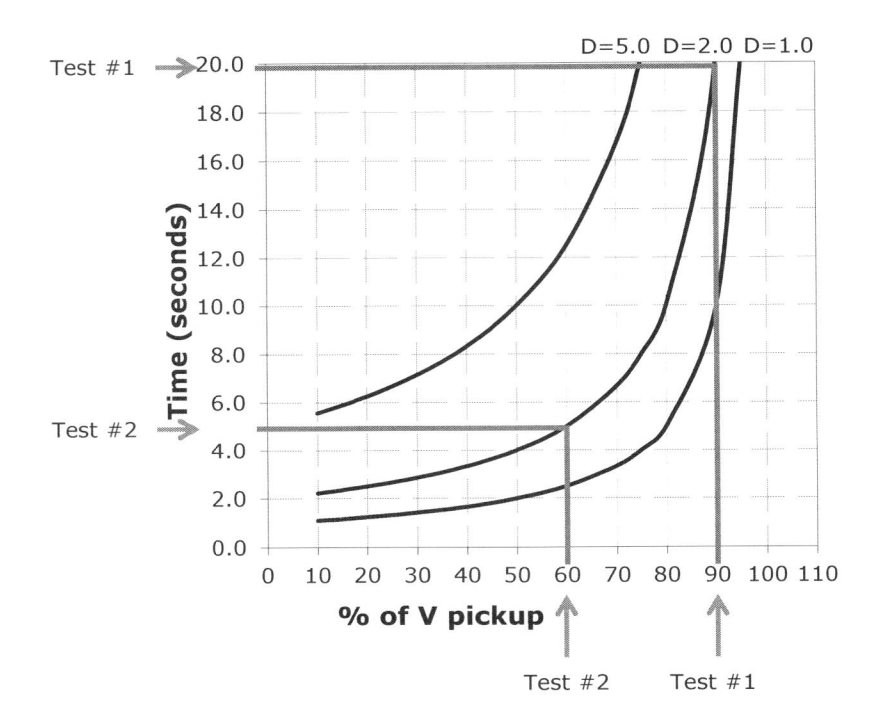

Figure 7-8: Undervoltage Inverse Curve

The second method uses a manufacturer supplied formula for inverse elements and, in this case, the formula is:

$$T = \frac{D}{\left(1 - \dfrac{V}{V_{pickup}}\right)}$$

T = Operating time

D = Undervoltage Delay Setting

(D = 0.00 operates instantaneously)

V = Secondary Voltage Applied to the Relay

V_{pickup} = Pickup Level

Use the formula to calculate the expected time delay. The results should be the same as the results obtained by the graph method but the formula results will always be more accurate and are preferred.

$$T = \frac{2.0}{\left(1 - \frac{59.23}{65.82}\right)} = 19.975s$$

$$T = \frac{2.0}{\left(1 - \frac{39.49}{65.82}\right)} = 5.00s$$

With the expected values, you can calculate percent error using the following formulas:

$$\frac{\text{Actual Value - Expected Value}}{\text{Expected Value}} \times 100 = \text{percent error}$$

$$\frac{20.2s - 20.00s}{20.00s} \times 100 = \text{percent error}$$

$$\frac{0.20s}{20.00s} \times 100 = \text{percent error}$$

1.0% Error

$$\frac{\text{Actual Value - Expected Value}}{\text{Expected Value}} \times 100 = \text{percent error}$$

$$\frac{5.08s - 5.00s}{5.00s} \times 100 = \text{percent error}$$

$$\frac{0.08s}{5.00s} \times 100 = \text{percent error}$$

1.6% Error

5. Tips and Tricks to Overcome Common Obstacles

The following tips or tricks may help you overcome the most common 27-element testing obstacles.

- Before you start, apply nominal voltage and perform a meter command to make sure your test-set is actually producing voltage.
- More sophisticated relays may require positive sequence voltage. Check system settings and connections for correct phase sequences.
- Does an external input affect operation?
- Do you need current for element to operate?
- Make sure the fuse failure element is not energized.
- Always use prefault nominal voltages before applying fault voltages.
- Are all voltages above the minimum voltage?
- Is the 27-element enabled?

Chapter 8

Over/Under Frequency (81) Protection Testing

1. Application

Frequency protective devices monitor the system frequency (usually via the voltage inputs) and will operate if the system frequency exceeds the setpoint for over-frequency (81O) protection. The under-frequency (81U) element will operate if the frequency drops below the setpoint.

Frequency protection is most commonly found in generator applications. Gas and steam turbines are particularly susceptible to damage when the generator runs at lower than normal frequencies. Over-frequency operation does not typically cause generator problems and any adverse effects of over-frequency operation occur in the prime mover of the generator. Mechanical over-speed protection is usually applied to protect the prime-mover and over-frequency protection is for back-up or control purposes only.

Generator protection is not the only factor to be considered when applying generator frequency protection. All major electrical systems in North America have guidelines for generator frequency protection to help maintain the stability of the electrical grid. The electrical system becomes more powerful and stable as more generators are added to it but that capacity must be available when a major fault on the grid creates a significant drop in the system-wide frequency. If the generator's frequency protection is too sensitive and removes the generator from the grid prematurely, it can cause a cascading failure across the entire electrical system. If you think of the electrical system as a clothesline stretched between two trees, each generator adds an additional strand to the clothesline to make it stronger. If a large, wet blanket (a fault) is suddenly thrown on the clothesline and it is thick enough to hold the weight, there will be some swings but it will stabilize eventually. If strands start breaking (generators separating from the grid), the clothesline may snap.

Generator frequency protection settings, when connected to the North American electrical grid, must meet the minimum criteria specified by the North American Electrical Reliability Council (www.nerc.com/regional) or by the associated region the generator is installed in. The following frequency standards are from the Western Electricity Coordinating Council (WECC) which will be similar to all NERC regions in an attempt to maintain electrical grid stability. The generator frequency settings must be longer than the specified setting in each range or the generator may not be able to legally synchronize to the grid. Remember that these settings are only for protection assigned to trip the generator offline. Alarm settings do not need to meet these criteria.

UNDER-FREQUENCY LIMIT (HZ)	OVER-FREQUENCY LIMIT (HZ)	MINIMUM TIME
60.0—59.5	60.0—60.5	Must Not Trip
59.4—58.5	60.6—61.5	3 minutes
58.4—57.9	61.6—61.7	30 seconds
57.8—57.4		7.5 seconds
57.3—56.9		45 cycles
56.8—56.5		7.2 cycles
Less than 56.5	Greater than 61.7	Instantaneous

Figure 8-1: Generator Trip Frequency Requirements

Over-frequency protection can also be used as over-speed back-up protection should the prime-mover protection fail to operate. This protection is possible because the frequency is directly related to the generator speed.

A dip in system frequency indicates a system overload in isolated systems supplied by on-site or emergency generators. Small or short fluctuations in frequency are normal as the generator's governor compensates for the changing load but a consistently low frequency indicates that the load is greater than the generator's capacity as the rotor begins to slip or the prime mover bogs down. Load-shedding is applied in these situations to disconnect non-essential loads from the overloaded systems to try to keep the essential systems on-line. Frequency relays (or setpoints) are installed throughout the system and the least important loads will have more sensitive frequency protection.

2. Settings

The most common settings used in 81-elements are explained below:

A) Enable Setting

Many relays allow the user to enable or disable settings. Make sure that the element is ON or the relay may prevent you from entering settings. If the element is not used, the setting should be disabled or OFF to prevent confusion.

B) Pickup

This setting determines when the relay will start timing and is always in Hertz. Some relays use specific under and over-frequency elements and the frequency setpoint should be below the nominal frequency for under-frequency elements and over the nominal frequency for over-frequency elements. A few relays automatically determine over and under frequency depending on whether the setpoint is above or below the nominal frequency. The nominal frequency could be fixed and written on the nameplate or a setting inside the relay.

C) Time Delay

The time delay setting for the 81-element is a fixed time delay that will determine how long (in seconds or cycles) the relay will wait to trip after the pickup has been detected. It is important to understand how the relay counts cycles and how your test-set counts cycles. SEL relays will count cycles based on the frequency applied to the relay while Beckwith Electric relays count cycles based on the system nominal (60 Hz in North America). See the *Frequency* section starting on page 5 for details.

D) Current Sensing

Most relays use the input voltage as the default source for frequency. Some relays can use current for frequency detection and will switch to current inputs if no voltages are connected or installed. A few relays can measure both signals simultaneously and this setting determines if current is used to determine frequency.

E) Minimum Operating Current

This setting determines how much current must be flowing into the relay before the frequency element will operate. This setting is used to prevent signal noise from influencing the value at low current levels and can also be used to disable frequency protection when the equipment is offline (the equipment is assumed to be offline if no current flows). The setting can be a fixed secondary current, a percentage of nominal secondary amps (CT secondary maximum rating = 5A), or a percentage of the nominal current as defined by a nominal current setpoint inside the relay.

F) Minimum Operating Voltage / Cutoff Voltage

Small voltage inputs created by signal noise when equipment is de-energized can cause nuisance tripping or prevent re-energization of equipment. Some relays automatically disable frequency protection at a preset voltage limit as defined in the manufacturer's bulletin. Other relays allow the user to define when frequency protection is enabled to prevent operation for noise or generator run up. The voltage setting can be a fixed secondary voltage, a percentage of nominal secondary volts, or a percentage of the nominal voltage as defined by a nominal voltage setpoint inside the relay. Make sure you know what the nominal voltage should be when reviewing this setting.

G) Under-Frequency Block

This setting can be a minimum voltage setting as described above or can be a digital input or other blocking element such as loss-of-fuse protection to prevent frequency protection under special conditions.

H) Block Under-Frequency from Online

There is usually an abrupt change in generator frequency when the generator initially synchronizes with the grid. This time setting determines how long frequency protection is disabled after the generator circuit breaker is closed to prevent nuisance trips as the generator synchronizes with the grid. This setting interacts with the generator relay's online status and should be tested to ensure it works correctly or frequency protection can be permanently disabled.

3. Pickup Testing

Frequency pickup testing is very simple and usually very accurate. Apply nominal voltage (or current if the relay uses current for frequency monitoring) and frequencies, check for correct metering values, and raise/lower the frequency until pickup operation is detected.

Use the following settings from a SEL-300G relay for the steps below.

SETTING DESCRIPTION	SETTING	TEST RESULT
VT Connection Type	Delta	
Nominal Frequency	60 Hz	
VNOM (Nominal VT Secondary Voltage)	115	
PTR (VT Ratio)	120:1	
27B81P (Cutoff Voltage)	60	59.8 V
81D1P (Pickup)	58	57.990 Hz
81D1D (Time Delay)	5	5.035 s
81D2P (Pickup)	62	62.02 Hz
81D2D (Time Delay)	5	5.038 s

Figure 8-2: 81-Element Example Settings and Test Results

A) Test-Set Connections

Use the following test connections for the different PT choices. Remember that the "Magnitudes" are phase-to-neutral values and they must be multiplied by √3 to obtain phase-to-phase values.

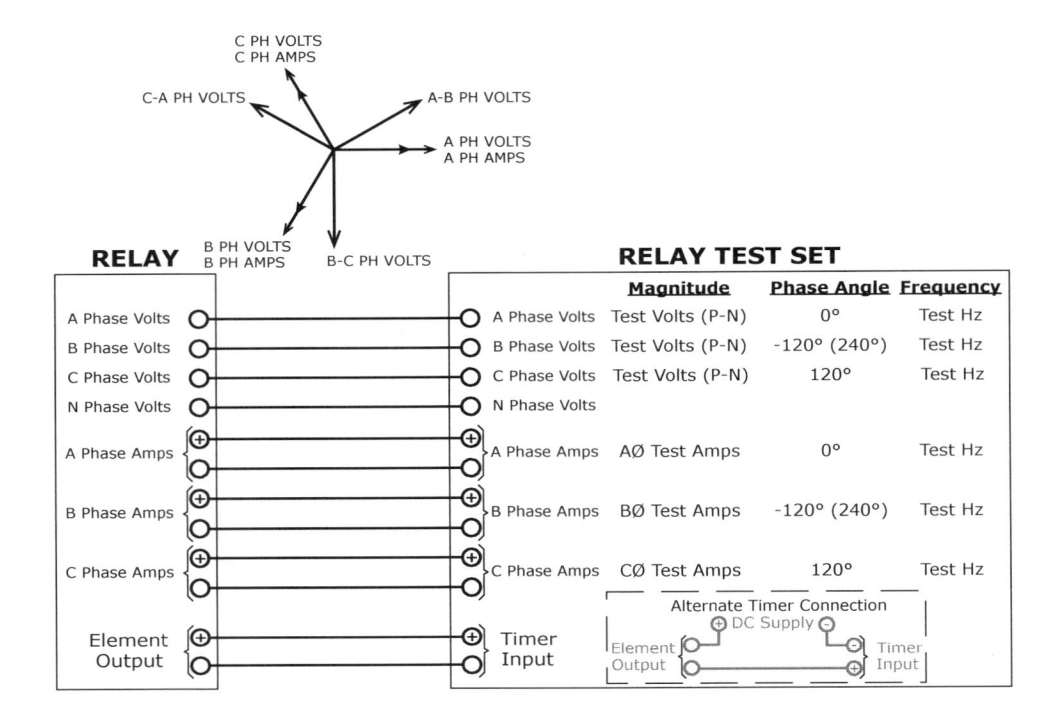

Figure 8-3: Simple Phase-to-Neutral Frequency Test Connections

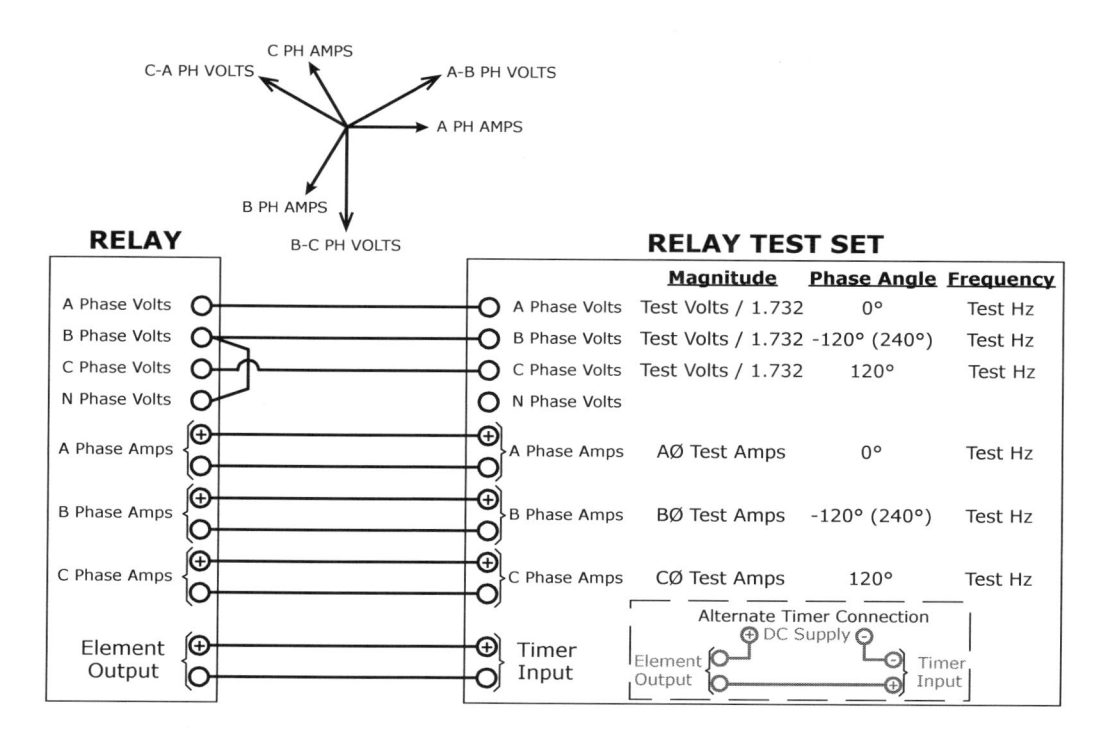

Figure 8-4: Simple Phase-to-Phase Overvoltage Connections

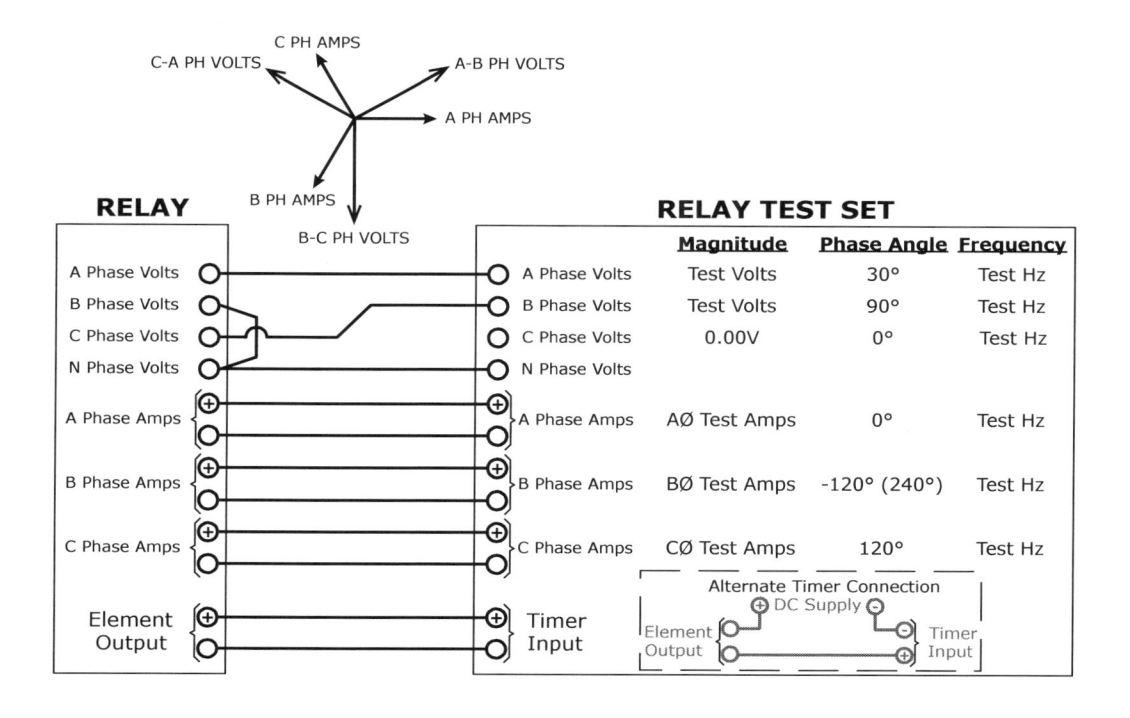

Figure 8-5: Phase-to-Phase Overvoltage Connections with 2 Voltage Sources

B) Under-Frequency Pickup Test Procedure

- Determine how you will monitor pickup and set relay accordingly, if required. (Pickup indication by LED, output contact, front panel display, etc. See the *Relay Test Procedures* section starting on page 109 for details.)
- Apply rated 3-phase voltage and frequency and check metering values.
- Lower frequency to value 0.2 Hz lower than the pickup setting (81D1P = 58—0.2 = 57.8) and verify pickup indication is on.
- Slowly raise the frequency until the pickup indication is off. Slowly lower the frequency until pickup indication is fully on. Record pickup values on test sheet.
- If undervoltage cutoff settings exist, set the frequency 0.2 Hz lower than the pickup setting and check for pickup indication. Lower 3Ø voltages 5% below the undervoltage cutoff setting (27B81P = 60.0 / 1.05 = 57.14V) and ensure that the 81-element pickup indication drops out. Slowly raise 3Ø voltages until the 81-element pickup is on. Slowly lower 3Ø voltages until 81-element pickup drops out and record this value on your test sheet.

C) Over-Frequency Pickup Test Procedure

- Determine how you will monitor pickup and set relay accordingly, if required. (Pickup indication by LED, output contact, front panel display, etc. See the *Relay Test Procedures* section starting on page 109 for details.)
- Apply rated 3-phase voltage and frequency and check metering values.
- Raise frequency to value 0.2 Hz higher than the pickup setting (81D2P = 62 + 0.2 = 62.2) and verify pickup indication is on.
- Slowly lower the frequency until the pickup indication is off. Slowly raise the frequency until pickup indication is fully on. Record pickup values on test sheet.
- If undervoltage cutoff settings exist, test its pickup. Set the frequency 0.2 Hz higher than the pickup setting and check for pickup indication. Lower 3Ø voltages 5% below the undervoltage cutoff setting (27B81P = 60.0 / 1.05 = 57.14V) and ensure that the 81-element pickup indication drops out. Slowly raise 3Ø voltages until the 81-element pickup is on. Slowly lower 3Ø voltages until 81-element pickup drops out and record this value on your test sheet.

D) Evaluate Results

Before we can evaluate the test results, we must determine manufacturer's expectations and tolerances. Use the following specifications from a SEL-300G relay to determine expected values.

DEFINITE-TIME UNDER/OVERFREQUENCY ELEMENTS (81):	
Frequency:	20-70 Hz, 0.01 Hz Steps
Pickup Time:	32ms
Delay Accuracy:	+/- 0.1%, +/-4.2ms at 60 Hz
Supervisory 27:	0-150V, +/- 5%, +/- 0.1V
METERING ACCURACY	
Voltages:	+/-0.1%
OUTPUT CONTACTS	
Pickup Time:	<5ms

Figure 8-6: SEL-300G Frequency Specifications

Using the data from Figure 8-2, the expected pickup was 58 Hz and the actual pickup test result was 57.99 Hz. We can pass this test result as it is within the metering resolution as shown in Figure 8-6 (20-70 Hz, 0.01 Hz Steps). We could also calculate the percent error as shown below.

$$\frac{\text{Actual Value - Expected Value}}{\text{Expected Value}} \times 100 = \text{percent error}$$

$$\frac{57.99 - 58.00}{58.00} \times 100 = \text{percent error}$$

$$\frac{0.01}{58.00} \times 100 = \text{percent error}$$

$$0.017\,\%$$

The undervoltage cutoff test results are also evaluated using the data from Figure 8-2 with a setting of 60.00V and test result 59.8V. The manufacturer's tolerances from Figure 8-6 indicate that the expected tolerance is +/- 5% with an additional metering tolerance of +/- 0.1% for a total of 5.1%. Using the percent error formula below, the measured percent error is 0.33% and is well below the specified relay tolerance for error.

$$\frac{\text{Actual Value - Expected Value}}{\text{Expected Value}} \times 100 = \text{percent error}$$

$$\frac{59.8 - 60.00}{60.00} \times 100 = \text{percent error}$$

$$\frac{0.2}{60.00} \times 100 = \text{percent error}$$

$$0.33\%$$

4. Timing Tests

81-element timing tests can be straightforward. Start with nominal frequency and apply +/- 0.5 Hz of the pickup frequency. Measure the time between test start and output contact operation. The measured time delay is compared to the setting with manufacturer's tolerances to make sure the test results are acceptable. However, different relays and test-sets calculate frequencies in different ways. For example, SEL relays will count cycles based on the frequency applied to the relay while Beckwith Electric relays count cycles based on the system nominal (60 Hz in North America). See the *Frequency* section starting on page 5 for details. Be sure you understand how the relay and your test-set counts cycles, or set your timer to measure seconds and convert from seconds to cycles to be sure.

A) Timing Test Procedure

- Determine which output the 81-element operates and connect the test-set timing input to the relay's output.
- Set the 3Ø fault frequency 0.5 Hz higher/lower than the pickup setting. The tests for our examples would be performed at 57.5 Hz (81D1D = 58.0 - 0.5 = 57.5 Hz) and 62.5 Hz (81D2D = 62.0 + 0.5 = 62.5 Hz). Set your test-set to stop when the timing input operates and to record the time delay from test start to stop.
- Apply nominal voltage and frequency. Change to test frequency at nominal voltage, ensure the timing input operates, and note the time on your test sheet. Compare the test time to the 81-element settings to ensure timing is correct.
- Review relay targets to ensure the correct element operated.

B) Evaluate Test Results

Using Figure 8-2, we can compare the timing test results to the factory expected results. The under frequency timing test result is 5.035s with a time delay setting of 5.00s. The expected time delay is calculated using the specifications listed in Figure 8-6 that indicate that there can be additional delays up to +/- 0.1% and +/- 4.2ms (from Delay Accuracy) plus an additional 5ms (from Output Contacts) for the contact to close. It can also take up to 32ms (from Pickup Time) for the relay to detect pickup which must also be added to the equation. The pickup time (32ms); plus delay accuracy ((0.1% * 5s) + 4.2ms); and additional output contact pickup time (5ms); are all added to the element setpoint (5s). All of these delays allow for a maximum measured time delay of 5.0462s (0.032 + 0.005 + 0.0042 + 0.005 + 5.00). Our test result (5.035 s) is lower than the expected number and the test result is acceptable for service.

5. Tips and Tricks to Overcome Common Obstacles

The following tips or tricks may help you overcome the most common 81-element testing obstacles.

- Before you start, apply nominal voltage & frequency and perform a meter command to make sure your test-set is actually producing the correct voltage and frequency.
- More sophisticated relays may require positive sequence voltage. Check system settings and connections for correct phase sequences.
- Applying prefault nominal voltages can increase timing accuracy.
- Frequency pickup tests are very accurate but timing tests can vary up to 10% depending on the relay model. Some relay models require the frequency to ramp to the test value instead of the sudden jump between pre-test and test values normally produced during dynamic testing.
- Your relay set may measure the time delay timing based on the test frequency and indicate an incorrect time delay at frequencies other than nominal. Use time delay results in seconds and multiply the result by 60 to obtain the correct time delay measurement in cycles.
- Make sure that any blocking functions (Voltage / Current) are in the correct state to allow 81-element operation.

Chapter 9

Instantaneous Overcurrent (50) Element Testing

1. Application

Although the official designation of the 50-element is "instantaneous overcurrent," a time delay is often added to transform it into a definite-time overcurrent element. A 50-element will operate if the current is greater than the pickup setpoint for longer than the time delay setting. When the instantaneous overcurrent element is used for phase overcurrent protection, it is labeled with the standard IEEE designation "50." Ground or neutral instantaneous overcurrent elements can have the designations 50N or 50G depending on the relay manufacturer and/or relay model.

The 50-element can be used independently or in conjunction with time overcurrent (51) functions. When used in a grounding scheme, typically all feeders have identical pickup and time delay settings. The main breaker would have a slightly higher setting and/or longer time delay to ensure that a ground fault on a feeder will be isolated by the feeder breaker before the main breaker operates. An example 50-element ground protection scheme is shown in the following figures.

The 50-element protective curve looks like an "L" on a Time Coordination Curve (TCC, See the *Time Coordination Curves (TCC) and Coordination* section starting on page 73 for details.) The element will operate if the measured current is on the right side of the vertical line for longer than the time indicated by the horizontal line of the protective curve in Figure 9-2. In this example, a feeder ground fault greater than 10 Amps must last longer than one second before the 50-element will operate. The main breaker protection will operate if any ground fault is greater than 15 Amps for longer than two seconds.

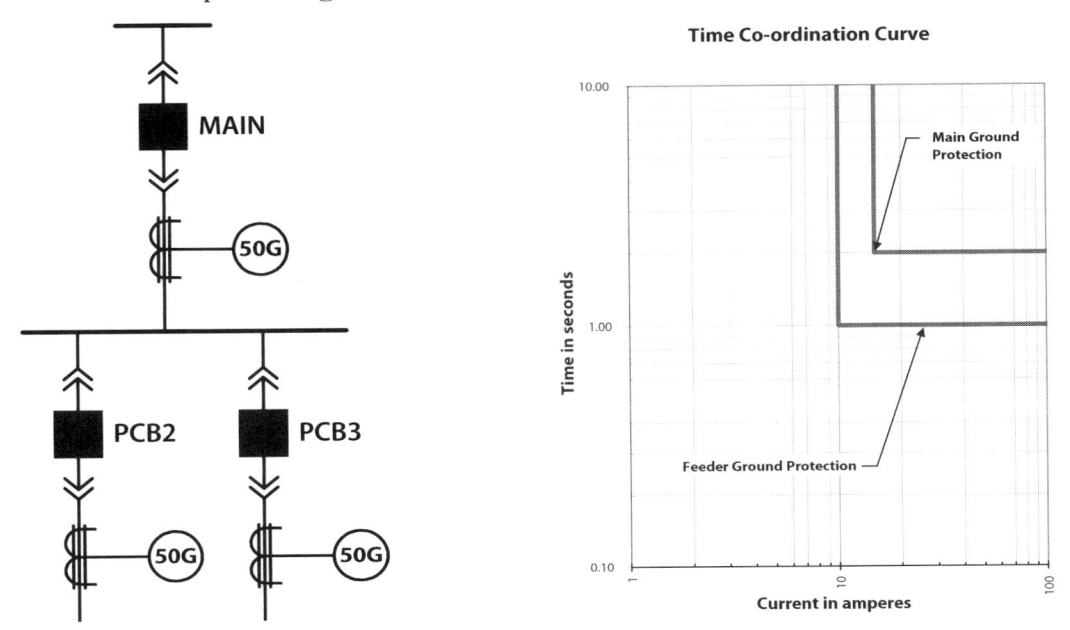

Figure 9-1: Ground Fault Protection Single-Line Drawing

Figure 9-2: Ground Protection TCC

The 50-element can also be applied in conjunction with inverse-time overcurrent elements (51) to better protect equipment during high-current faults. The amount of damage created during a fault can be directly related to the amount and duration of fault current. To limit equipment damage, the relay should operate faster during high fault currents.

The following figures display how the 50-element can enhance equipment protection as well as coordination with other devices. In Figure 9-3, the time overcurrent (51) relay curve intersects the cable damage curve and, therefore, does not provide 100% protection for the cable. The cable is only 100% protected if its damage curve is completely above the protection curve. Adding a 50-element to the time overcurrent element will provide 100% cable protection as shown in Figure 9-4. However, the addition of the 50-element creates a mis-coordination between the R2 relay and downstream Fuse 1 because the two curves now cross. The relay will operate before the fuse when the relay curve is below and to the left of the fuse curve. This problem can be solved by adding a slight time delay of 0.03 seconds, which will coordinate with the downstream fuse as shown in Figure 9-5.

If we wanted to provide the best protection for the cable and fully utilize the available options of most relays, we could add a second 50-element with no intentional time delay set with a pickup setting higher than the maximum fuse current. This is shown in Figure 9-6. Adding another 50-element will cause the relay to trip sooner at higher currents and will hopefully reduce the amount of damage caused by fault.

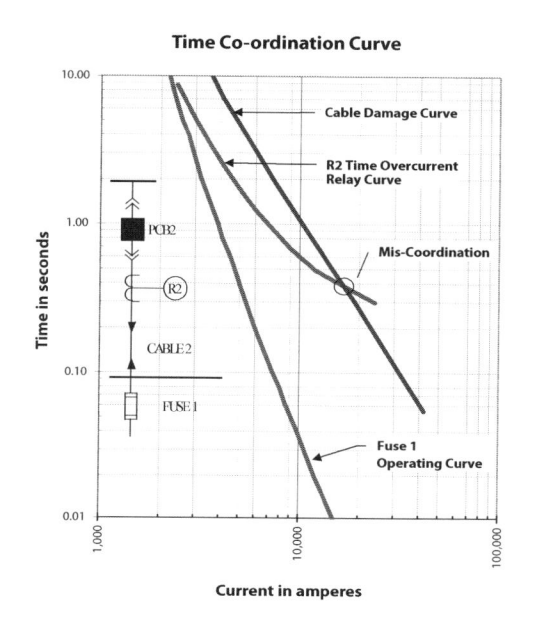

Figure 9-3: 50/51 TCC #1

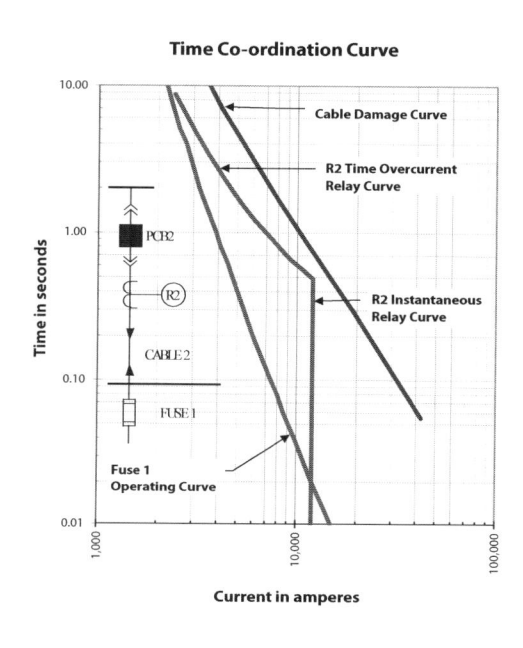

Figure 9-4: 50/51 TCC #2

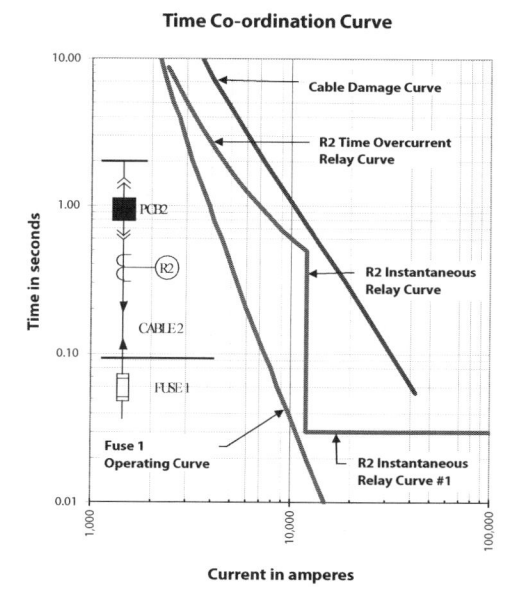

Figure 9-5: 50/51 TCC #3

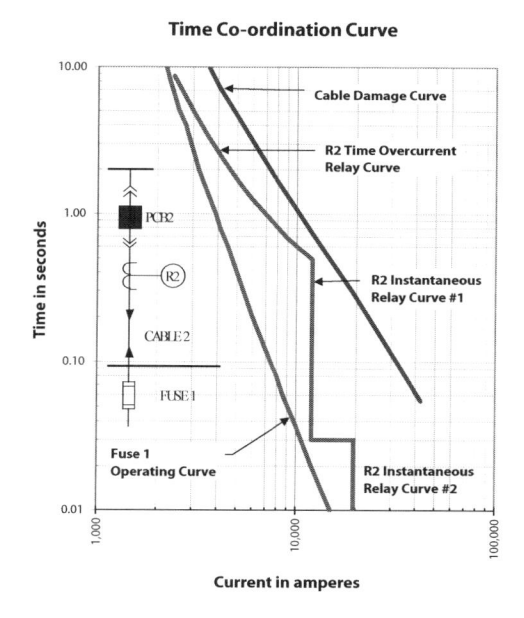

Figure 9-6: 50/51 TCC #4

50-elements can also be used to determine if the downstream equipment is operating and/or the circuit breaker or motor starter is closed. When used in this fashion, the 50-element is set very low, at some level below the minimum expected operating current. If the current flow exceeds the 50-element setpoint, the circuit breaker is considered closed because there would be no current flow if the circuit breaker was open. This method of breaker status indication will also detect flashovers or insulation breakdown inside the circuit breaker that would not be detected by a 52a or b contact and is often used in breaker-failure (50BF) or inadvertent energization (50/27) protection.

2. Settings

The most common settings used in 50-elements are explained below:

A) Enable Setting

Many relays allow the user to enable or disable settings. Make sure that the element is ON or the relay may prevent you from entering settings. If the element is not used, the setting should be disabled or OFF to prevent confusion.

B) Pickup

This setting determines when the relay will start timing. Different relay models use different methods to set the actual pickup and the most common methods are:

- **Secondary Amps**—the simplest unit. Pickup Amps = Setting
- **Per Unit (P.U.)**—This method can only exist if the relay settings include nominal current, watts, or VA. This setting could be a multiple of the nominal current as defined or calculated. If no such setting exists, it could be a multiple of the nominal CT (5A) secondary or a multiple of the 51-element pickup setting.
  Pickup = Setting x Nominal Amps, **OR**
  Pickup = Setting x Watts / (nominal voltage x √3 x power factor) OR
  Pickup = Setting x VA / (nominal voltage x √3), OR
  Pickup = Setting x CT secondary (typically 5 Amps)
  Pickup = Setting x 51-element Pickup
- **Primary Amps**—There must be a setting for CT ratio if this setting style exists. Check the CT ratio from the drawings to make sure that the drawings match the settings.
  Pickup = Setting / CT Ratio, **OR**
  Pickup = Setting * CT secondary / CT primary

C) Time Delay

The time delay setting for the 50-element is a fixed-time delay that determines how long the relay will wait to trip after the pickup has been detected. This setting is set in cycles, milli-seconds, or seconds.

3. Pickup Testing

Instantaneous overcurrent testing is theoretically simple. Apply a current into the appropriate input and increase it until you observe pickup indication. However, the actual application can be frustrating and require some imagination. High currents are usually involved and the relay could be damaged during testing. Most protective relay current inputs are rated for a maximum of 10 continuous Amps. Any input current greater than 10 Amps must be applied for the minimum amount of time possible to prevent damage. It's not a good feeling when you apply too much current for too long and get that slight smell of burning insulation, quickly followed by smoke billowing from the relay.

Instantaneous elements often interfere with time-overcurrent (51) testing and many relay testers turn the 50-element off during 51-element testing. This practice may be required by the testing specification but is NOT recommended when testing micro-processor relays. If the 50-element is disabled, it MUST be tested AFTER the 51-element tests are complete and the 50-element has been enabled. The opposite problem could occur because the 51-element function can interfere with the instantaneous pickup tests. Do NOT turn off the time-overcurrent (51) element to determine instantaneous pickup.

Before you begin testing, write down the pickup and time settings, and then calculate the pickup current. Make sure that you know which unit is used. Some relays use secondary Amps for time-overcurrent (51) and multiples of that pickup for 50-elements. Use the formulas described in the "Settings" section of this chapter to determine what the pickup actually is.

Now that you have determined the pickup and time delay settings, convert the current to primary values using the following formulas:
- Primary Current = Secondary Pickup Current * CT ratio, **OR**
- Primary Current = Secondary Pickup Current * CT Primary / CT Secondary.

It is extremely unlikely that you will find a microprocessor relay out of calibration. We should be performing these tests to check relay operation, verify the settings have been correctly interpreted by the design engineer, and that the settings were entered into the relay correctly. Using the relay settings to create the test plan only proves relay operation. If we want to check all of the reasons for testing, we should check the primary values and time delays against the coordination study and make sure they match. Make sure the supplied TCC curves are at the correct voltage levels as discussed in the Time Coordination Curves (TCC) and Coordination section starting on page 73. Use the voltage conversions discussed in the previously mentioned section if necessary. If you do not have the coordination study, quickly check that the upstream 50-element setting is higher and the downstream 50-element setting is lower than the relay under test.

The interrupting device (circuit breaker, etc.) must be rated to operate at the 50-element pickup level or it may not be able to clear the fault once a trip signal is initiated. Check the interrupting rating of the switchgear and circuit breaker or other disconnecting means. Make sure the equipment interrupting rating is greater than the setting.

Look in the short circuit study and determine the maximum fault level at the switchgear. The maximum fault level should be higher than the 50-element setpoint. If it's not, question the setting because the 50-element will likely never operate because there is not enough fault current available. If no coordination study is provided, look at the next upstream transformer and use the following formula to determine the maximum fault current that could flow through the transformer. The setting should be less than this value:

Maximum Fault Current = Transformer VA / (System Voltage * %Z)

A) Test-Set Connections

Because of the high currents involved with 50-element testing, you may need to try some of the alternative test-set connections shown below. Some technicians carry an older test-set when their modern test-sets are unable to reach the 50-element test levels.

You can prove the element is applied correctly by temporarily lowering the setting, but only use this method as a last resort. In the past, there have been some relay models that did not operate when secondary currents exceeded 100A even though the relay allowed settings larger than 100A. If the testers who discovered this had not tested at the higher fault current levels, it would never have been discovered.

Residual ground (externally connected or internally calculated) and negative sequence elements often interfere with 50-element tests. This problem can be overcome as shown in the following figures if your test-set is powerful or flexible enough. There will be some instances where the residual and negative sequence setting will have to be disabled but, **disabling settings is a last resort** and should only be undertaken if all other possibilities have been exhausted. All disabled elements must be tested **AFTER** the instantaneous element tests have been performed.

Connections are shown for AØ related tests. Simply rotate connections or test-set settings to perform BØ and CØ related tests. Simple phasor diagrams are shown above each connection to help you visualize the actual input currents.

If your test-set experiences problems during the test, even though the output is within its theoretical capabilities, you may need to connect two or more test leads in parallel for the phase **AND** neutral connections to lower the lead resistance. If this doesn't work, try connecting directly to the relay terminals as the circuit impedance may be more than your test-set can handle.

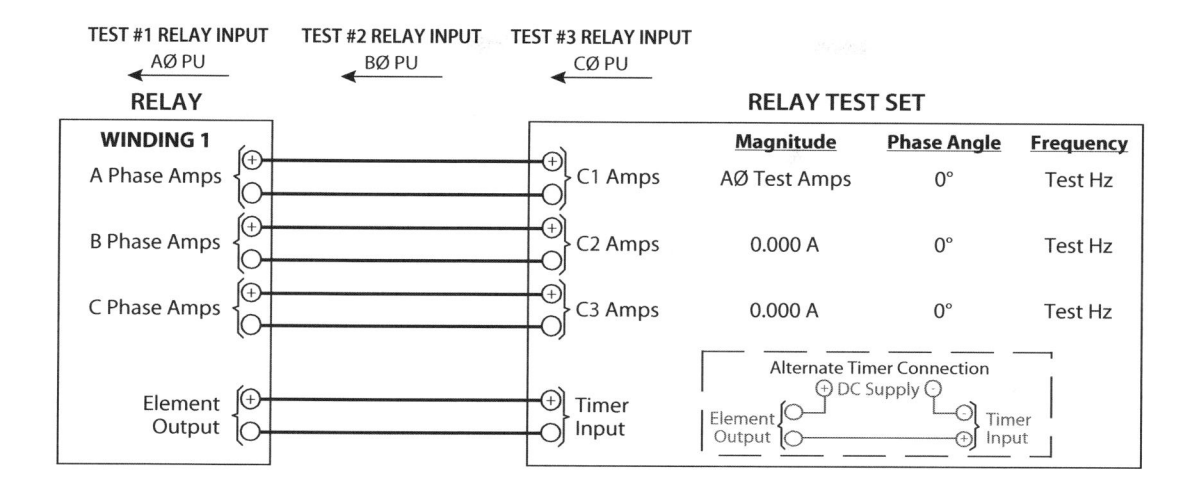

Figure 9-7: Simple Instantaneous Overcurrent Connections

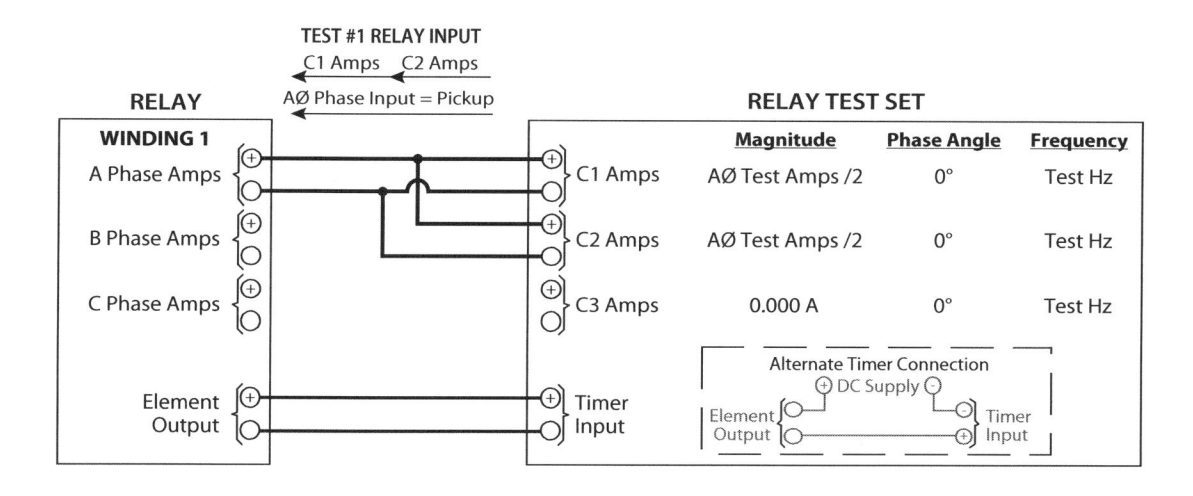

Figure 9-8: High Current Connections #1

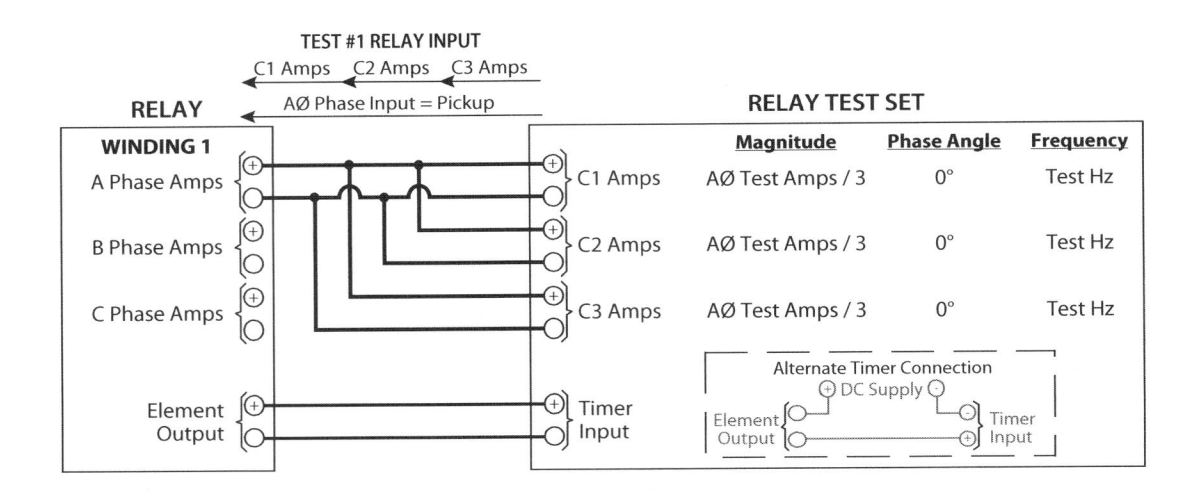

Figure 9-9: High Current Connections #2

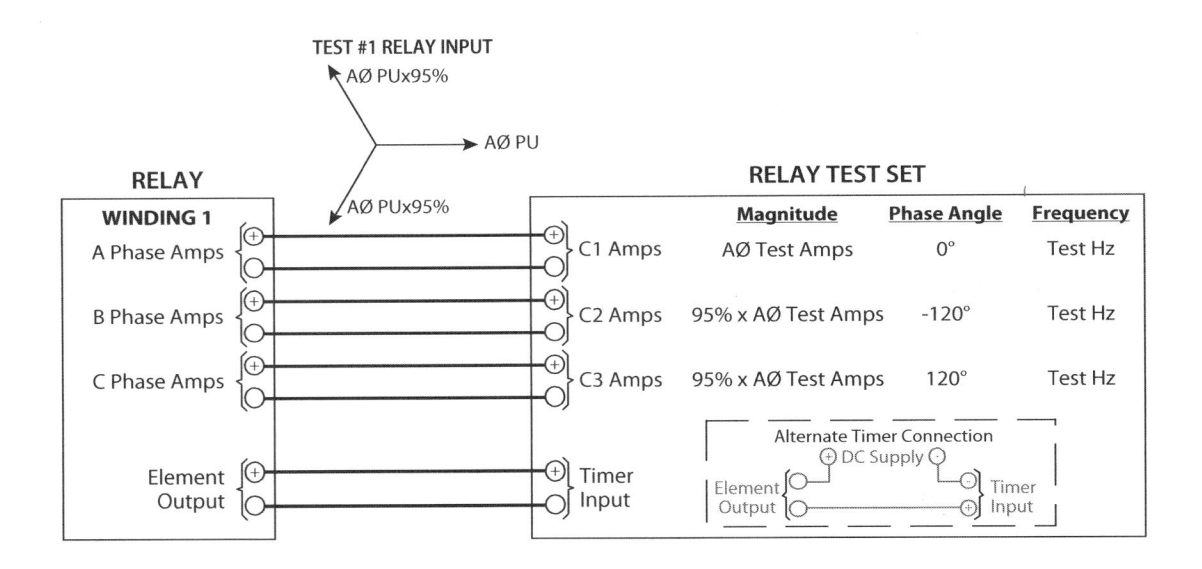

Figure 9-10: Neutral or Residual Ground Bypass Connection

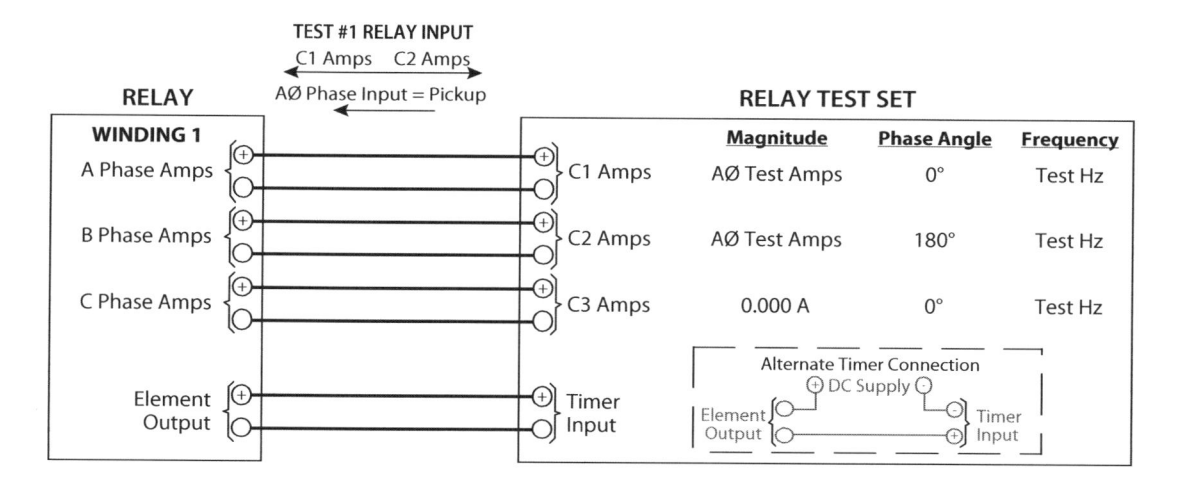

Figure 9-11: Neutral or Residual Ground Bypass Connection via Ø-Ø Connection

B) Pickup Test Procedure if Pickup is Less Than 10 Amps

Use the following steps to perform a pickup test if the setting is less than 10 secondary Amps:

- Determine how you will monitor pickup and set the relay accordingly, if required. (Pickup indication by LED, output contact, front panel display, etc. See the *Relay Test Procedures* section starting on page 109 for details.)
- Set the fault current 5% higher than the pickup setting. For example, 8.40 Amps for an element with an 8.00 Amp setpoint. Make sure pickup indication operates.
- Slowly lower the current until the pickup indication is off. Slowly raise current until pickup indication is fully on. Chattering contacts or LEDs are not considered pickup. Record pickup values on test sheet. The following figure displays the pickup procedure.

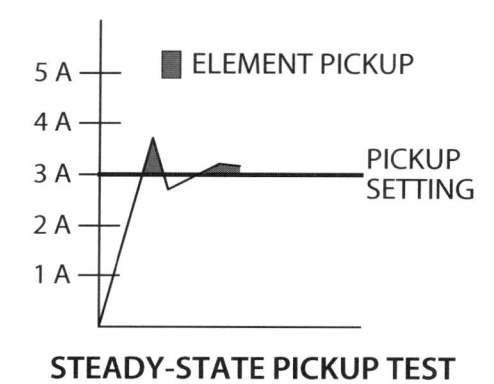

STEADY-STATE PICKUP TEST

Figure 9-12: Pickup Test Graph

C) Pickup Test Procedure if Pickup is Greater Than 10 Amps

Use the following steps to determine pickup if the setting is greater than 10 secondary Amps:

- Check the maximum per-phase output of the test-set, and use the appropriate connection shown in Figures 1-7 to 1-11. For example, if the 50-element pickup is 35A and your test-set's maximum output is 25 Amps per phase; use "High Current Connections #1." If the pickup setting is greater than 50 Amps, use "High Current Connections #2." If the pickup is higher than 75A (3x25A), you will have to use another test-set or temporarily lower the setting. Remember, setting changes are a last resort.
- Determine how you will monitor pickup and set the relay accordingly, if required. (Pickup indication by LED, output contact, front panel display, etc. See the *Relay Test Procedures* section starting on page 109 for details.)
- Set the fault current 5% higher than the pickup setting. For example, set the fault current at 42.0 Amps for an element with a 40.0 Amp setpoint. Apply current for a moment, and make sure the pickup indication operates. If pickup does not operate, check connections and settings and run the test again until the pickup indication operates.
- Set the fault current 5% lower than the pickup setting. Apply current for a moment and watch to make sure the pickup indication does not operate. Increase and momentarily apply current in equal steps until pickup is indicated. If large steps were used, reduce the amount of current per step around the pickup setting. See the following figure for a graph of this pickup method.

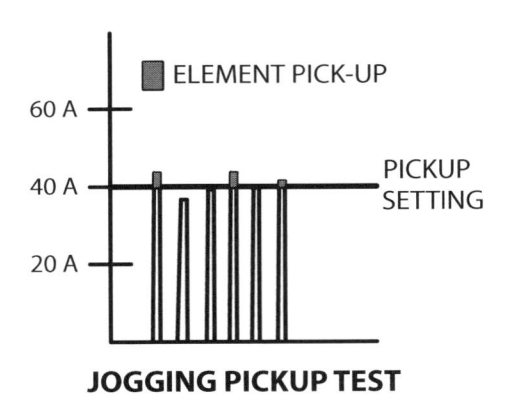

Figure 9-13: Pickup Test Graph via Jogging Method

D) Test Procedure to Avoid Setting Changes and Interference

It can be easier and more practical to test 50-elements without changing settings or disabling elements. The 50-element time delay setting is usually very small. The 50-element should trip before the time overcurrent (51) at the 50-element pickup level. The following procedure allows 50-element pickup testing without changing settings.

- Determine which output the 50-element trips and connect timing input to the relay output.
- Check the maximum per-phase output of the test-set and use the appropriate connection from Figures 1-7 to 1–11 in this chapter. For example, if the 50-element pickup is 35A and your test-set can only output 25 Amps per phase; use "High Current Connections #1." If the pickup setting is greater than 50 Amps, use "High Current Connections #2." If the pickup is higher than 75A (3x25A), you will have to use another test-set or temporarily lower the setting. Remember, setting changes are a last resort.
- Set the fault current 5% higher than the pickup setting. For example, set the fault current at 42.0 Amps for an element with a 40.0 Amp setpoint. Set your test-set to stop when the timing input operates and to record the time delay from test start to stop. Apply test current and ensure the relay output stopped the test and note the test time. Compare the test time to the 50-element time delay setting to ensure timing is correct. Review relay targets to ensure the correct element operated.
- Set the fault current 5% lower than the pickup setting. Apply test current and watch for timing input operation. If the relay does not operate after the 50-element time delay, stop the test manually. If the timing input operates, ensure the time delay is longer than the 50-element and review targets to ensure the 50-element did not operate. Increase and apply current in increasing steps until the 50-element time delay is observed. If large steps were used, lower the current below the pickup setting and use smaller steps to achieve better resolution.

4. Timing Tests

There is often a time delay applied to the 50-element protection even though the 50-element is defined as instantaneous overcurrent protection. Timing tests should always be performed even if time delay is not assigned.

50-element timing tests are performed by applying 110 % of pickup current (or any value above pickup) to the relay and measuring the time between the start of the test and relay operation. The start command could be an external trigger, a preset time, or a push button on the relay set. The stop command should be an actual output contact from the relay because that is what would happen under real-life conditions.

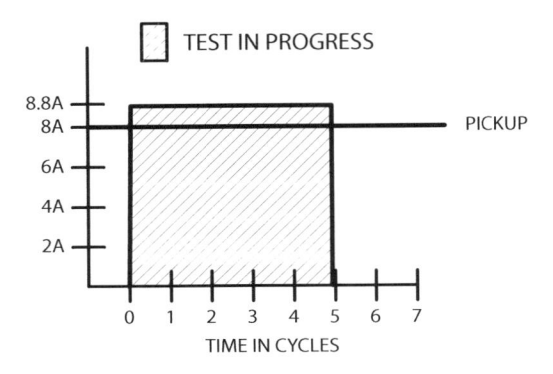

Figure 9-14: 50-Element Timing Test

When the 50-element time delay is zero or very small (less than 2 seconds), the actual measured time delay can be longer than expected. There is an inherent delay before the relay can detect a fault plus an additional delay between fault detection and output relay operation. These delays are very small (less than 5 cycles) and are insignificant with time delays greater than 2 seconds.

The first delay exists because the relay is constantly analyzing the input data to determine if it is valid and this analysis takes a fraction of a cycle. The relay cannot determine the magnitude of the input signal until it has enough of the waveform to analyze and determine the rms or peak current or voltage. The relay is also a computer and computers can only perform one task at a time. If a fault occurs just after the relay processes the line of code that detects that particular fault, the relay has to run through the entire program one more time before the fault is detected. All of these delays usually require a fair portion of a cycle to complete. The "Operate Time" and "Timer Accuracy" specifications in the following figure detail this time delay.

PHASE / NEUTRAL / GROUND IOC	
CURRENT:	**PHASOR ONLY**
Pickup Level:	0.000 pu to 30.000pu in steps of .001 pu
Dropout Level:	97% to 98% of Pickup
Level Accuracy:	+/- 0.5% of reading or +/- 1% of rated (Whichever is greater) from 0.1 to 2.0 x CT ration +/- 1.5% of reading > 2.0 x CT rating < 2%
Overreach:	< 2%
Pickup Delay:	0.00 to 600.00 in steps of 0.01 s
Reset Delay:	0.00 to 600.00 in steps of 0.01 s
Operate Time:	< 20 ms @ 3 x Pickup @ 60Hz
Timing Accuracy:	Operate @ 1.5 x Pickup +/- 3% or +/- 4ms (whichever is greater)

Figure 9-15: GE D-60 Relay Overcurrent Technical Specifications

The second time delay occurs after the relay has detected the fault and issues the command to operate the output relays. There is another fraction of a cycle delay to evaluate what output contacts should operate and then the actual contact operation can add up to an additional cycle depending on relay manufacturer, model, etc. "Operate Time" in the following figure represents this delay for the specified relay.

FORM-C AND CRITICAL FAILURE RELAY OUTPUTS	
Make and Carry for 0.2 sec:	10A
Carry Continuous:	6A
Break @ L/R of 40ms:	0.1 ADC max
Operate Time:	< 8ms
Contact material:	Silver Alloy

Figure 9-16: GE D-60 Relay Output Contact Technical Specifications

Your test-set can also add a small time delay to the test result as shown by the "Accuracy" specification of the following figure:

MANTA 1710 TIME MEASUREMENT SPECIFICATIONS	
Auto ranging Scale:	0—99999 sec
Auto ranging Scale:	0—99999 cycles
Best Resolution:	0.1ms / 0.1 cycles
	Two wire pulse timing mode
Accuracy:	0—9.9999 sec scale: +/-0.5ms +/- 1LS digit
	all other scales: +/- 0.005% +/- 1 digit

Figure 9-17: Manta Test Systems M-1710 Technical Specifications

What does all this mean? With a time delay of zero, the time test result for a GE D-60 relay, using a Manta M-1710 test-set, could be as much as 32.6ms or 1.956 cycles as shown in the following figure:

MINIMUM TIME TEST RESULT	
Relay Operate Time:	< 20 ms
Relay Timing Accuracy:	+/- 4ms
Relay Operate Time:	< 8 ms
Test-Set :	+/-0.5ms
	+/- 1LS digit (0.1 ms)
	32.6ms or 1.956 cycles

Figure 9-18: 50-Element Minimum Pickup

A) Timing Test Procedure

- Determine which output the 50-element trips and connect the test-set timing input to the relay-output.
- Check the maximum per-phase output of the test-set and use the appropriate connection from Figures 1-7 to 1-11. For example, if the 50-element pickup is 35A and your test-set can only output 25 Amps per phase; use "High Current Connections #1." If the pickup setting is greater than 50 Amps, use "High Current Connections #2." If the pickup is higher than 75A (3x25A), you will have to use another test-set or temporarily lower the setting. Remember, setting changes are a last resort.
- Set the fault current 10% higher than the pickup setting. For example, set the fault current at 44.0 Amps for an element with a 40.0 Amp setpoint. Set your test-set to stop when the timing input operates and to record the time delay from test start to stop.
- Apply test current and ensure timing input operates and note the time on your test sheet. Compare the test time to the 50-element timing to ensure timing is correct.
- Review relay targets to ensure the correct element operated.
- Repeat for other two phases.

5. Residual Neutral Instantaneous Overcurrent Protection

Residual neutral overcurrent protection is typically set well below phase overcurrent values. In these cases, follow the previous steps but apply current in one phase at a time. It is good practice to perform pickup tests on A-phase and timing tests on B-phase and C-phases to make sure the relay uses all three phases to calculate residual current.

If the phase overcurrent settings interfere with residual testing or the pickup results are not as accurate as they should be, connect the relay and test-set as shown earlier in Figure 9-7, but apply all 3-phase currents simultaneously at the same phase angle. The magnitude of each phase should be one-third of the test current. Some relay models need currents through all three phases to accurately calculate residual current.

6. Tips and Tricks to Overcome Common Obstacles

The following tips or tricks may help you overcome the most common obstacles.

- Before you start, apply current at a lower value and review the relay's measured values to make sure your test-set is actually producing an output and your connections are correct.
- If the element does not operate, watch the metering during the test if possible.
- Check to make sure your settings are correct.
- Make sure you are connected to the correct output.
- Check the output connections by pulsing the output and watching the relay input.
- Some relay test-set-inputs are polarity sensitive. If the connections look good, try reversing the leads.
- Have any of your test leads fallen off?
- If you are paralleling more than one relay output, do all channels have the same phase angle?
- Check for settings like "Any Two Phases" (Any two phases must be above the pickup to operate) or "All Three Phases" (All three phases must be higher than the pickup to operate) or "Any Phase" (Any phase above pickup operates element).

- If you need more than one phase to operate the 50-element but your test-set only has enough VA for one phase, put two or more phases in series as shown below:

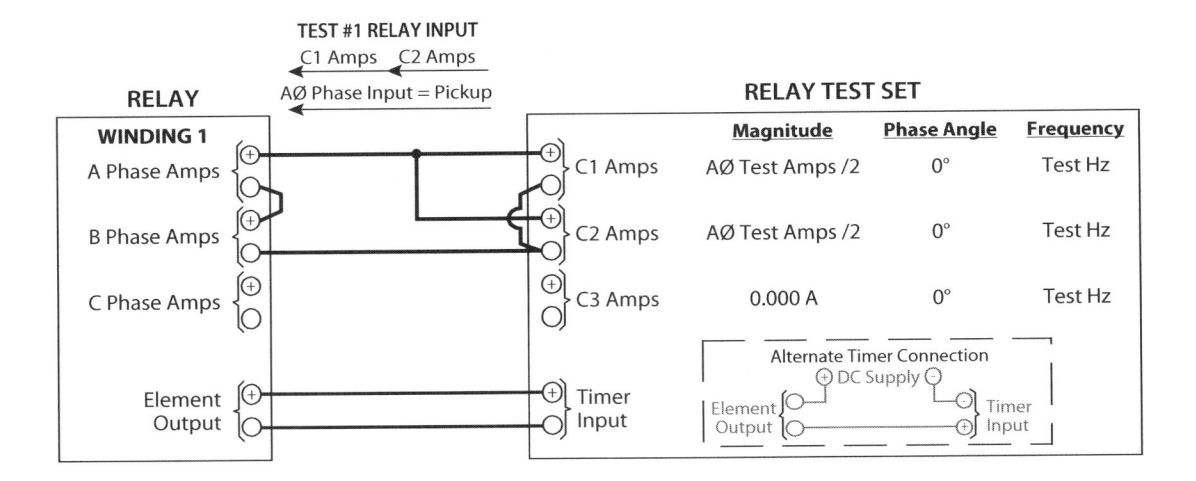

Figure 9-19: 50-Element Alternate Relay Connection

- Check for blocking inputs.
- Does the relay need breaker status or other input to operate?

Sometimes neutral or residual ground protection is applied and this protection is inevitably set lower than the phase elements. These elements trip first before the phase element operates and can be a nuisance at best. The following solutions can help overcome this obstacle:

- Perform tests using 3-phase, balanced inputs as shown in the "Neutral or Residual Ground Bypass Connection." Residual current will be zero.
- Perform tests with three phase inputs; put two phases slightly below the pickup and slowly raise one phase at a time until pickup is indicated.
- Apply a phase-to-phase fault by applying equal current to any two phases with the current applied 180° from each other as per Figure 9-11. For example, a 25 Amp pickup could be tested by applying 25 Amps @ 0° in Phase A-N and 25 Amps @ 180° into Phase B-N.

Chapter 10

Time Overcurrent (51) Element Testing

1. Application

The 51-element is the most common protective element applied in electrical systems. It uses an inverse curved characteristic and will operate more quickly as the fault magnitude increases. There are many different styles of curves in use and each style mimics a different damage characteristic. The most common curve characteristics used in North America are usually described as ANSI or U.S. curves. Some relays allow you to select European curves that are usually described as IEC curves. All of these curves are mathematical models of electromechanical relays to allow coordination between different generations of relays. General Electric used special curves for their IAC electromechanical relay line and some relays also have these curves available. Custom curves could also be available to create specific protection curves unique to an individual piece of equipment (like motors), but this feature is seldom used.

Examples of the different styles of curves with identical settings are shown in Figure 10-1. Notice that the x-axis values represent a multiple of the element's pickup setting. This is typical so that all curves can be plotted without site-specific values. Most manufacturers display their curves in multiples of pickup or its equivalents—"percent of pickup" or "I/Ipkp".

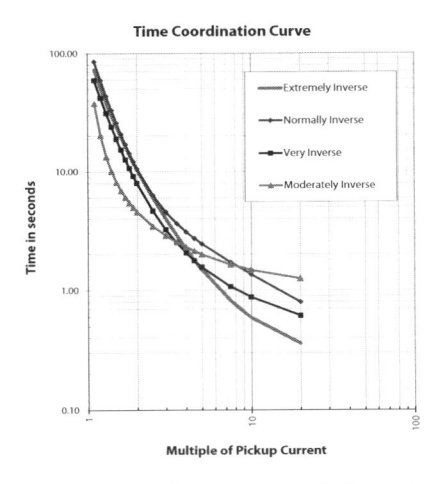

Figure 10-1: 51-Element North American Curves

Figure 10-2: 51-Element IEC European Curves

After the appropriate curve style is chosen, 51-elements typically have two primary settings, pickup and timing. The pickup setting changes the starting point of the curve. As the pickup setting increases, the curve moves from left to right as shown in Figure 10-3 which depicts an ANSI Extremely Inverse (EI) curve with different pickup values. Figure 10-4 shows how the curve moves vertically as the time dial setting is increased using the same ANSI Extremely Inverse curve with different time dials.

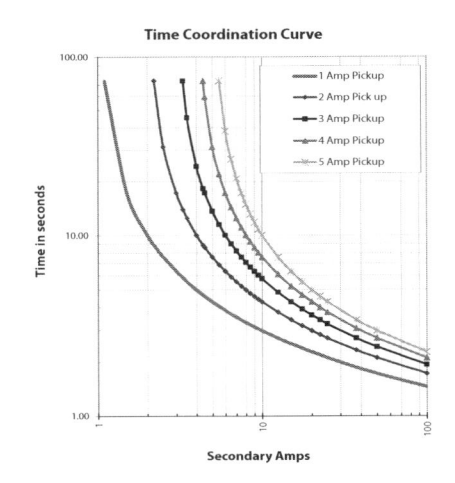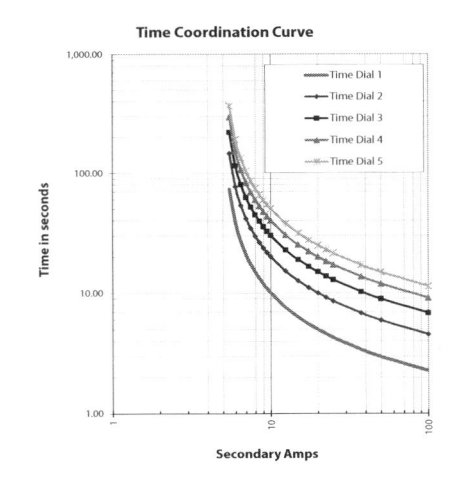

Figure 10-3: ANSI Extremely Inverse with Different Pickup Settings

Figure 10-4: ANSI Extremely Inverse with Different Timing Settings

2. Settings

Typical settings for 51-elements are described below:

A) Enable Setting

Many relays allow the user to enable or disable settings. Make sure that the element is ON/Enabled or the relay may prevent you from entering settings. If the element is not used, the setting should be disabled or OFF to prevent confusion.

B) Pickup

This setting determines when the relay will start timing. Different relay models use different methods to set the actual pickup and the most common methods are:

- **Secondary Amps**—the simplest unit. Pickup Amps = setting
- **Per Unit (P.U.)**—This setting could be a multiple of the nominal current as defined or calculated if the relay has setpoints for nominal current, Watts, or VA. It could also be a multiple of the nominal CT secondary.
  Pickup = Setting x Nominal Amps, **OR**
  Pickup = Setting x Watts / (nominal voltage x $\sqrt{3}$ x power factor), **OR**
  Pickup = Setting x VA / (nominal voltage x $\sqrt{3}$), **OR**
  Pickup = Settings x CT Secondary (typically 5 Amps)
- **Primary Amps**—There must be a setting for CT ratio if this setting style exists. Check the CT ratio from the drawings and make sure that the drawing matches the settings.
  Pickup = Setting / CT Ratio, **OR**
  Pickup = Setting * CT secondary / CT primary

C) Curve

This setting chooses which curve will be used for timing. Be very careful to select the correct curve as there can be subtle differences between curve descriptions. Compare the curve selection to the coordination study to ensure the correct curve is selected.

D) Time Dial/Multiplier

This setting simulates the time dial setting on an electromechanical relay and sets the time delay between pickup and operation. ANSI curves usually have a time delay between 1 and 10. IEC time dial setting are typically between 0 and 1.

E) Reset

Electromechanical 51-element relay timing was controlled using a mechanical disc that would rotate if the current was higher than the pickup setting. If the current dropped below the pickup value, the disc would slowly rotate back to the reset position. The disc speed in the trip and reset directions are directly related to the amount of current flowing through the relay.

Some digital relays simulate this reset delay using a linear curve that is directly proportional to the current to closely match the electromechanical relays. Other relays have a preset time delay or user defined reset delay that should be set to allow any electromechanical discs to reset for proper coordination between devices.

3. Pickup Testing

Time-overcurrent pickup testing is theoretically simple. Apply current into the appropriate input and increase until you observe the pickup indication. However, the actual application can be complicated and requires some imagination because 51-element testing can overlap with neutral overcurrent protection and timing tests can interfere with 50-element testing. Look in the tips and tricks heading of this section for ways to avoid interference from other elements.

Start by writing down all settings related to the 51-element and calculate what the pickup current should be using the formulas described in the previous section.

Now that you have determined the pickup and time delay settings, convert the current to primary values using the following formulas:

Primary Current = Secondary Pickup Current * CT ratio, **OR**

Primary Current = Secondary Pickup Current * CT Primary / CT Secondary.

It is extremely unlikely that you will find a microprocessor relay out of calibration. We perform these tests to check relay operation and verify that the engineer has correctly interpreted the settings. Check the primary values and time delays against the coordination study and make sure they match. Make sure the supplied TCC curves are at the correct voltage levels as discussed in the *Time Coordination Curves (TCC) and Coordination* section starting on page 73. If you do not have the coordination study, quickly check that the upstream relay 51-element pickup and timing settings are higher and the downstream relay 51-elements settings are lower. Use voltage conversions discussed in the *Time Coordination Curves (TCC) and Coordination* section if necessary.

A) Test-Set Connections

Because timing tests can require large test currents, you may need to try some of the alternative test-set connections shown in Figures 2-5 to 2-9.

Residual ground (externally connected or internally calculated) and negative sequence elements often interfere with 51-element tests. Overcome these problems by using one of the test connections in the following figures, if your test-set is powerful or flexible enough. There will be some instances where the residual and/or negative sequence settings need to be disabled. If you apply realistic voltages and currents for the element you are testing, you may want to ask the design engineer to review the settings for errors because overlapping elements with realistic test quantities often indicate setting errors. Only disable settings as a last resort after all other possibilities are exhausted. Test all disabled elements **AFTER** the 51-element tests are performed.

The following five figures have three separate current outputs. Some test-sets have only one neutral current terminal and the connections are easily modified to work by imagining the non-polarity connections have been jumpered together. Connections are shown for A-phase related tests. Simply rotate connections or test-set settings to perform B-phase and C-phase related tests. Simple phasor diagrams are shown above each connection to help you visualize the actual input currents.

Test-set output errors during tests can occur even though the output current is within the test-set's specifications because of high impedance connections. You may need to connect two test leads in parallel for the phase **AND** neutral connections to lower the lead resistance. If this does not work, try connecting directly to the relay terminals because the circuit impedance may be too great.

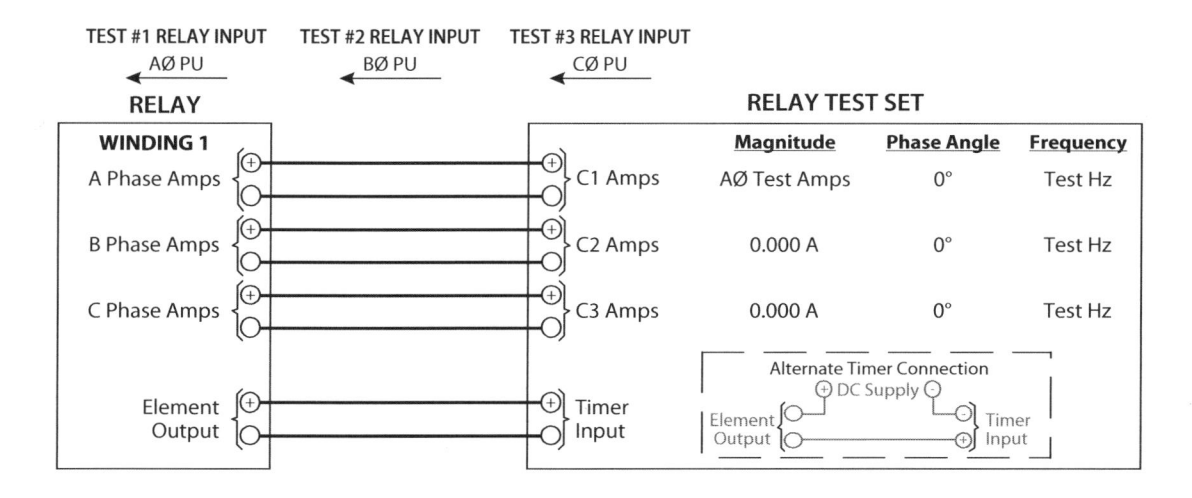

Figure 10-5: Simple Time Overcurrent Connections

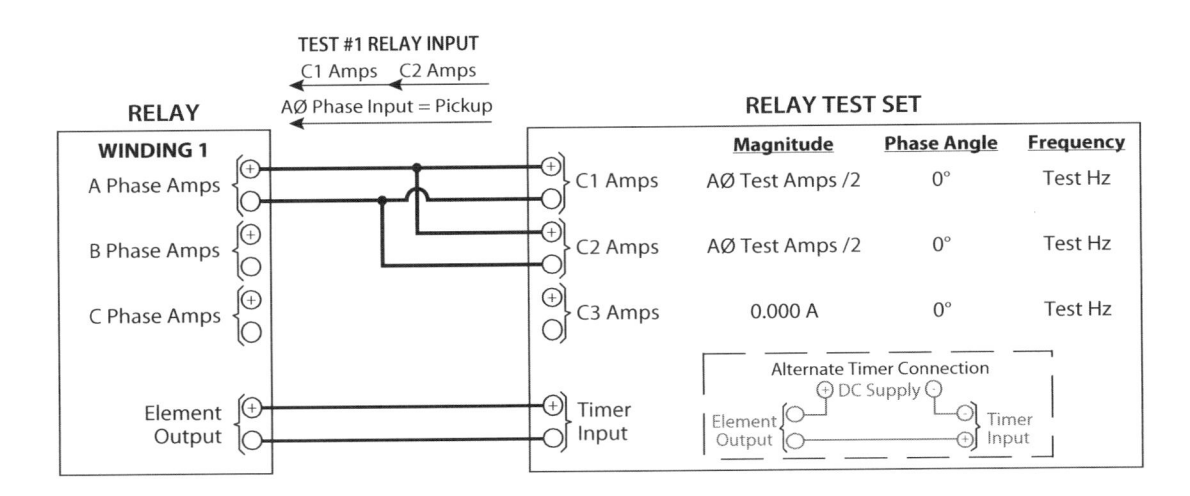

Figure 10-6: High Current Connections #1

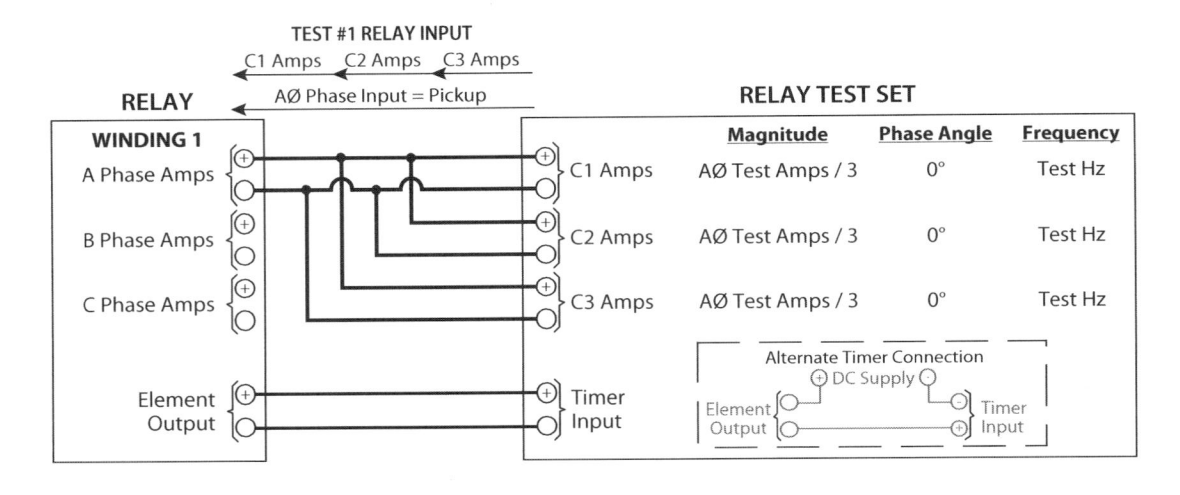

Figure 10-7: High Current Connections #2

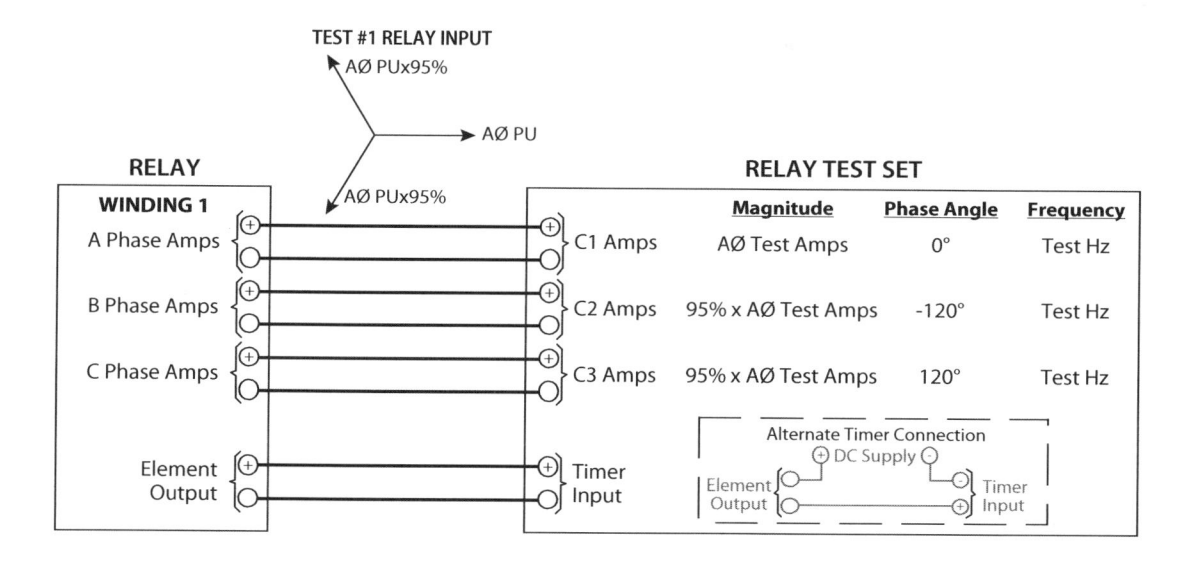

Figure 10-8: Neutral or Residual Ground Bypass Connection

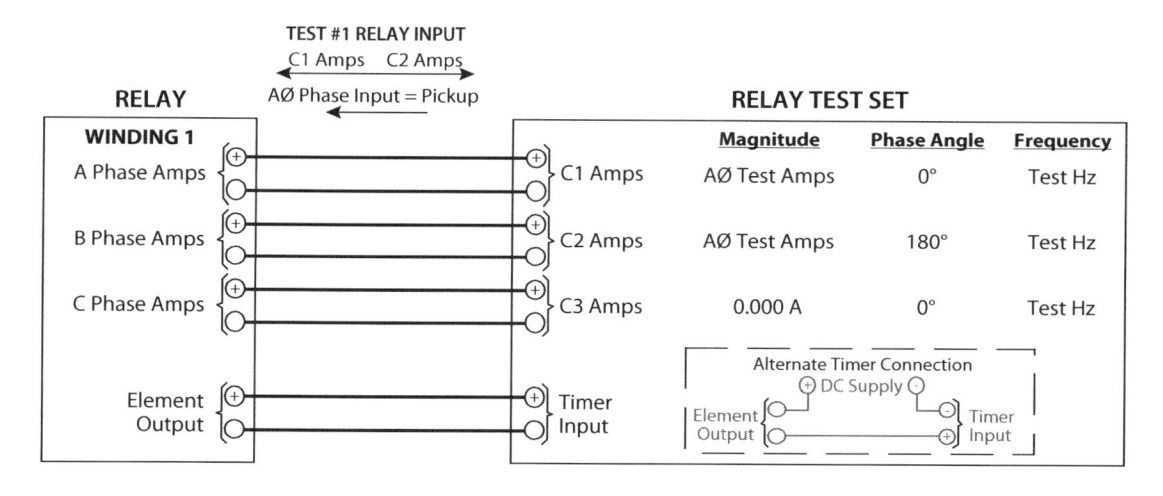

Figure 10-9: Neutral or Residual Ground Bypass Connection via Ø-Ø Connection

B) Pickup Test Procedure

Use the following steps to determine pickup. If the pickup is greater than 6 Amps, consult the design engineer to ensure the setting is correct. If the setting is correct, review the pickup procedure described in *Chapter 1: Instantaneous Overcurrent (50) Element Testing*.

- Determine how you will monitor pickup and set the relay accordingly, if required. (Pickup indication by LED, output contact, front panel display, etc. See the *Relay Test Procedures* section starting on page 109 for details.)

- Set the fault current 5% higher than the pickup setting. For example, set the fault current at 3.15 Amps for an element with a 3.00 Amp setpoint. Make sure pickup indication operates.
- Slowly lower the current until the pickup indication is off. Slowly raise current until pickup indication is fully on. (Chattering contacts or LEDs are not considered pickup) Record the pickup values on your test sheet. The following graph displays the pickup procedure.

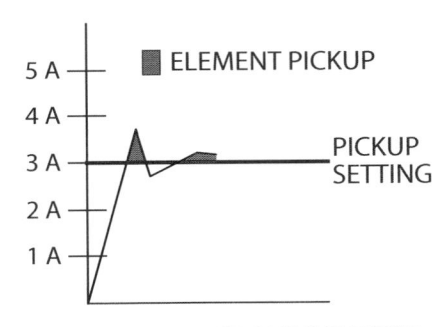

STEADY-STATE PICKUP TEST

Figure 10-10: Pickup Test Graph

- Compare the pickup test result to the manufacturer's specifications and calculate the percent error as shown in the following example:

In our example, the 51-element pickup setting is 4.00A and the measured pickup was 4.02A. Looking at the "Currents" specifications in "Metering Accuracy," we see that the acceptable metering error is 0.05A. The difference between the test result and setting is 0.02A (4.02A - 4.00A). We can immediately consider it acceptable as it falls within the metering accuracy specifications. The result is also acceptable as shown in the "Steady State Pickup Accuracy" specification in the "Time-Overcurrent Elements" section of this chapter.

Metering Accuracy

Voltages	Va, Vb, Vc, Vs	+/- 0.67 V Secondary
Currents	Ia, Ib, Ic, In	+/-0.05 A secondary (5 A nominal) +/-0.01 A secondary (1 A nominal)

Time-Overcurrent Elements

Pickup Range:	OFF, 0.50 -16.00 A, 0.01 A steps (5 A nominal) OFF, 0.10 - 3.20 A, 0.01 A steps (1 A nominal)
Steady State Pickup Accuracy:	+/- 0.05 A and +/-3% of Setting (5 A nominal) +/- 0.01 A and +/-3% of Setting (1 A nominal)
Time Dial Range:	0.50 - 15.00, 0.01 steps (US) 0.05 - 1.00, 0.01 steps (IEC)
Curve Timing Accuracy:	+/- 1.50 cycle and +/-4% of curve time for current between 2 and 30 multiple of pickup

Figure 10-11: SEL-311C 51 Time Overcurrent Specifications

Using these two sections, we can calculate the manufacturer's allowable percent error. The allowable percent error is 5.5% as shown in the calculation below. The test error of 0.5% is below 5.5% and the test result is acceptable.

Expected =	[Metering Accuracy]	+	[51-Element pickup Accuracy]	+	[Pickup Setting]
Expected =	0.05A	+	0.05A + (4.00A * 3%)	+	4.00A
Expected =	0.05A	+	0.05A + 0.12A	+	4.00A
Expected =			**4.22A**		

$$\frac{\text{Actual Value - Expected Value}}{\text{Expected Value}} \times 100 = \text{percent error}$$

$$\frac{4.22\,A - 4.00\,A}{4.0\,A} \times 100 = \text{percent error}$$

$$\textbf{5.5 \%}$$

$$\frac{\text{Actual Value - Expected Value}}{\text{Expected Value}} \times 100 = \text{percent error}$$

$$\frac{4.02\,A - 4.00\,A}{4.0\,A} \times 100 = \text{percent error}$$

$$\textbf{0.5 \%}$$

- Repeat the pickup test for the other affected phases.

4. Timing Tests

It is unlikely that you will discover a problem with the actual relay timing. Timing tests are performed to ensure the settings have been entered and interpreted correctly by the design engineer which is why you should use the coordination study instead of relay settings whenever possible. There can be small differences between settings that can easily be missed during a setting review. A minimum of two 51-element timing tests should be performed; one on each side of the bend in the curve to ensure the correct curve is selected. ANSI curves can have very similar time delays at different multiples as shown below. Everyone has their own preferences, but I prefer to perform timing tests at 2x, 4x, and 6x the pickup.

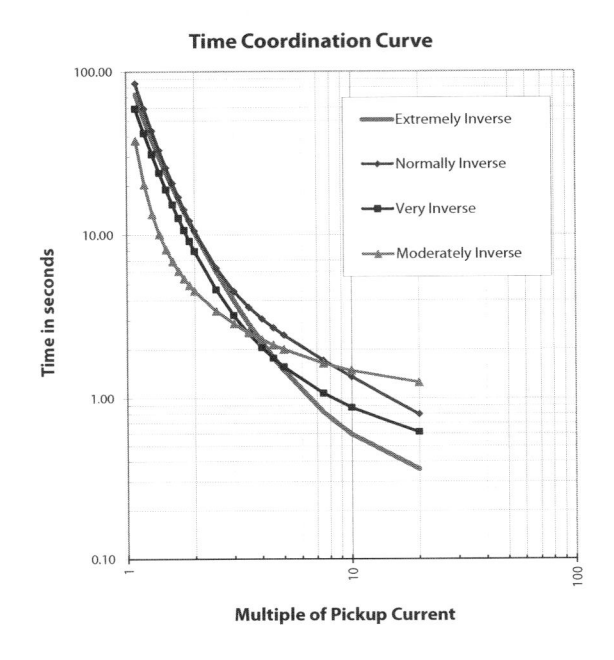

Figure 10-12: 51-Element North American Curves

51-element timing tests are performed by applying a multiple of the pickup current to the relay and measuring the time between the start of the test and relay operation as shown in Figure 10-13. The start command could be an external trigger, a preset time, or a push button on the test-set. The stop command should be an relay-output-contact because that is what would happen if an actual fault or overload condition existed. You should use the actual output contact assigned to trip when in service to ensure the relay will operate the correct output when in service and minimize setting changes.

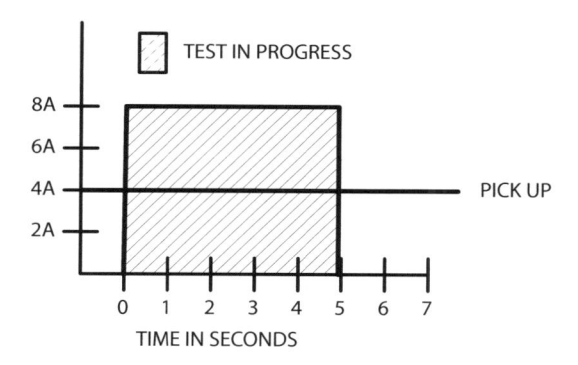

Figure 10-13: 51-Element Timing Test

Figuring out what the timing test result should be can often be the hardest part of 51-element timing tests when using relay settings instead of the coordination study. Manufacturers can display the curve characteristics as a formula, a curve, and/or a chart. We will discuss each method in detail using various relay models all set with the following example settings:

- Pickup = 3.5A
- Curve = Extremely Inverse
- Time Dial = 4.50

A) **Using Formulas to Determine Time Delay**

Before we get started, the 51-element curve setting must be confirmed because our example states "Extremely Inverse" but there are two extremely inverse curves to choose from in an SEL relay. Assume we have contacted the design engineer who instructed us to use curve "U4." Always use the "tp" equation in the following figure when calculating expected time in an SEL relay. We will use the "U.S. Extremely Inverse Curve: U4" to calculate the expected time delay. We will perform timing tests at 2x, 4x, and 6x the pickup current. "M" is the multiple of pickup of the test in this relay and is **NOT** the actual test current. TD is the time dial setting which, in our case, is 4.5.

tp = operating time in seconds

tr = electromechanical induction-disk emulation reset time in seconds
 (if electromechanical reset setting is made)

TD = time dial setting

M = applied multiple of pickup current
[for operating time (tp), M>1; for reset time (tr), M<=1].

U.S. Moderately Inverse Curve: U1	**U.S. Inverse Curve: U2**
$tp = TD * [0.0226 + 0.0104/(M^{0.02}-1)]$	$tp = TD * [0.180 + 5.95/(M^2-1)]$
$tr = TD * [1.08/(1-M^2))$	$tr = TD * [5.95/1-M^2)]$
U.S. Very Inverse Curve: U3	**U.S. Moderately Inverse Curve: U4**
$tp = TD * [0.0963 + 3.88/(M^2-1)]$	$tp = TD * [0.0352 + 5.67/(M^2-1)]$
$tr = TD * [3.88/(1-M^2)]$	$tr = TD * [5.67/(1-M^2)]$
U.S. Short Term Inverse Curve: U5	
$tp = TD * [0.00262 + 0.00342/(M^{0.02}-1)]$	
$tr = TD * [0.323/(1-M^2)]$	
I.E.C. Class A Curve (Standard Inverse): C1	**I.E.C. Class B Curve (Very Inverse): C2**
$tp = TD * [0.14/(M^{0.02}-1)]$	$tp = TD * [13.5/(M-1)]$
$tr = TD * [13.5/(1-M^2)]$	$tr = TD * [47.3/(1-M^2)]$
I.E.C. Class C Curve (Standard Inverse): C3	**I.E.C. Long-Time Inverse Curve: C4**
$tp = TD * [80.0/(M^2-1)]$	$tp = TD * [120.0/(M-1)]$
$tr = TD * [80.0/(1-M^2)]$	$tr = TD * [120/(1-M)]$
I.E.C. Short-Time Inverse Curve: C5	
$tp = TD * [0.05/(M^{0.04}-1)]$	
$tr = TD * [4.85/(1-M^2)]$	

Figure 10-14: 51-Element SEL-311C Timing Curve Characteristic Formulas

If you do not have a fancy calculator that allows you to perform the calculation as one formula, you must break the calculation into steps. The following breakdown should work on most calculators.

Time @ 2x Pickup

- Test Current = M * Pickup Setting = 2 * 3.5A = 7.0A
- **Step 1**: $(M^2-1) = (2^2-1) = $ **3.0**
- **Step 2**: 5.67 / Step 1 = 5.67 / 3.0 = **1.89**
- **Step 3**: 0.0352 + Step 2 = 0.0352 + 1.89 = **1.9252**
- **Step 4**: TD * Step 3 = 4.5 * 1.9252 = **8.6634 seconds**

Time @ 4x Pickup

- Test Current = M * Pickup Setting = 4 * 3.5A = 14.0A
- **Step 1**: $(M^2-1) = (4^2-1) = $ **15.0**
- **Step 2**: 5.67 / Step 1 = 5.67 / 15.0 = **0.378**
- **Step 3**: 0.0352 + Step 2 = 0.0352 + 0.378 = **0.4132**
- **Step 4**: TD * Step 3 = 4.5 * 0.4132 = **1.8594 seconds**

Time @ 6x Pickup

- Test Current = M * Pickup Setting = 6 * 3.5A = 21.0A
- **Step 1**: $(M^2-1) = (6^2-1) = $ **35.0**
- **Step 2**: 5.67 / Step 1 = 5.67 / 35.0 = **0.162**
- **Step 3**: 0.0352 + Step 2 = 0.0352 + 0.162 = **0.1972**
- **Step 4**: TD * Step 3 = 4.5 * 0.1972 = **0.8874 seconds**

I use Microsoft Excel® for my test sheets and use the following formula to calculate the expected time delay where "TD" and "M" reference the cell for the appropriate setting.

- =(TD*(0.0352+(5.67/(POWER(M,2)-1))

B) Using Graphs to Determine Time Delay

You can also determine the expected time delay using the manufacturer's supplied time characteristic curves using the following steps:

- Locate the correct Time Coordination Curve.
- Find the line associated with the time dial setting. (If Time Dial Setting is a fraction of a whole number, round the time dial to the lower number.) Counting from the bottom, determine the time dial line number. (Time Dial 4.0 is the 5th highest line in our example.)
- Locate the vertical line associated to the first timing test multiple. (2x in our example)
- Follow the vertical line up and count the TD lines until you reach the target TD line. (5th highest in our example). If the Time Dial setting is a whole number, mark the intersection between the target TD line and the vertical Line. If the Time Dial Setting is a fraction, approximate the fraction between lines. (0.5 between 4 and 5 in our example) Remember that it is a logarithmic graph and the scale is logarithmic.
- Follow the previous mark using a straight edge to the time axis and record the time.
- Repeat for all test points.

Obviously this method is not as accurate as the formula method because we obtained the following results from the two different methods:

TEST	EXPECTED TIME VIA GRAPH	EXPECTED TIME VIA FORMULA
2x PU	9.00 s	8.6634 s
4x PU	1.80 s	1.8594 s
6x PU	0.89 s	0.8874 s

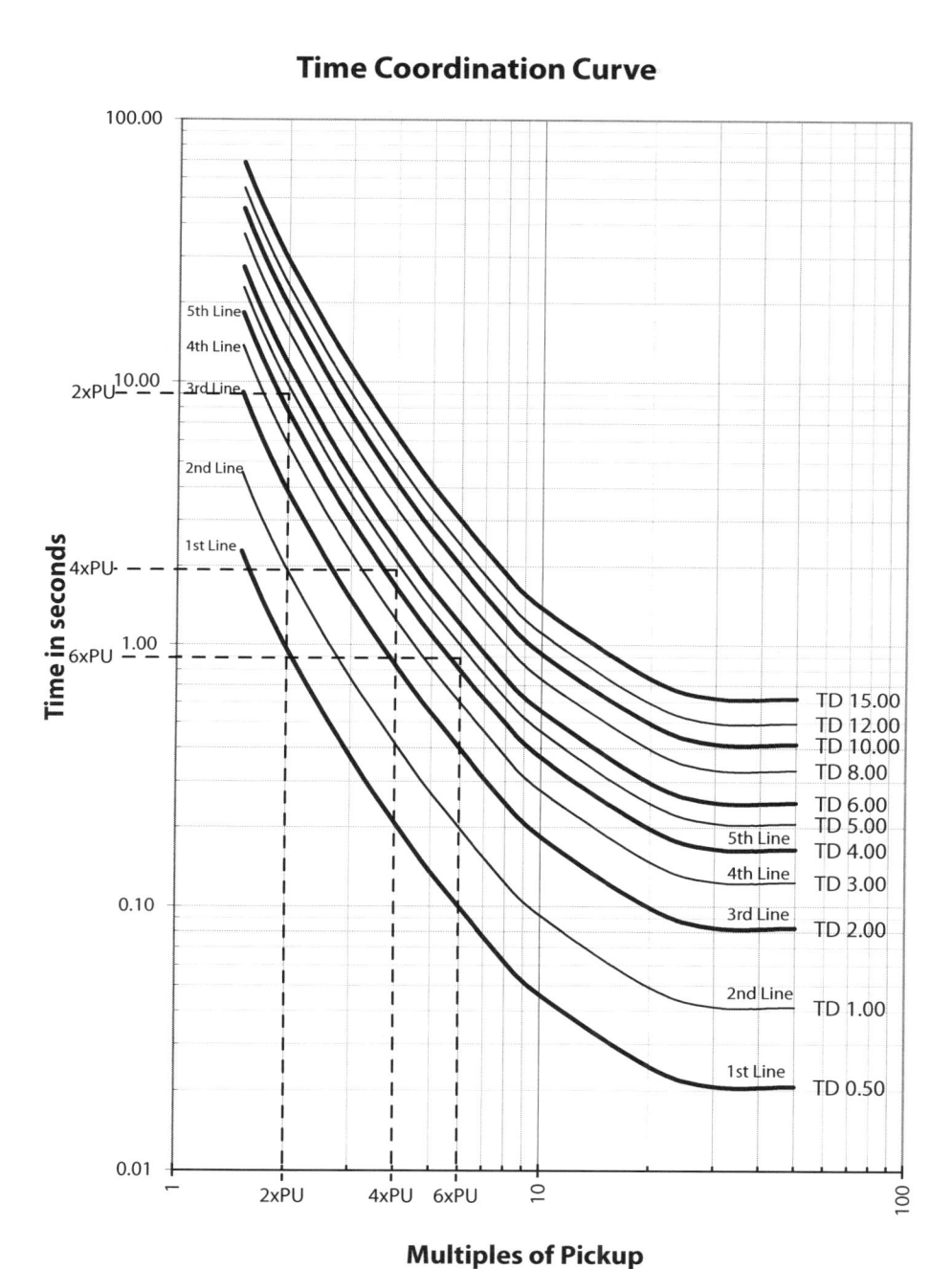

Figure 10-15: 51-Element Example Time Coordination Curve

C) Using Tables to Determine Time Delay

The third method for displaying 51-element timing is the table method as shown in Figure 10-17. Determine 51-element timing using the following steps:

- Find the correct table and correct curve section. ("Table 5-11: IEEE Curve Trip Times / IEEE Extremely Inverse" in our example)
- Find the column associated with the test value. (2.0)
- Find the row associated with the Time Dial (Multiplier) and record the expected time. If our Time Dial (Mulitplier) was 4.0; our 2x time delay would be 38.087s, our 4x time delay would be 8.007s, and our 6x time delay would be 3.710S.

MULTIPLIER	Current (I/I_{pickup})							
	1.5	2.0	3.0	4.0	5.0	6.0	7.0	8.0
IEEE EXTREMELY INVERSE								
0.5	11.341	4.761	1.823	1.001	0.648	0.464	0.355	0.285
1.0	22.682	9.522	3.647	2.002	1.297	0.927	0.709	0569
2.0	45.363	19.043	7.293	4.003	2.593	1.855	1.418	1.139
4.0	90.727	38.087	14.587	8.007	5.187	3.710	2.837	2.277
6.0	136.090	57.130	21.880	12.010	7.780	5.564	4.255	3.416
8.0	181.454	76.174	29.174	16.014	10.374	7.419	5.674	4.555
10.0	226.817	95.217	36.467	20.017	12.967	9.274	7.092	5.693

Figure 10-16: 51-Element Timing for GE D-60

- If the time dial is not shown in the multiplier column as per our example, find the two closest values in the appropriate column (38.087 & 57.130) and calculate the expected time using the formula:

$$\text{Time} = \left[\frac{\text{Time Dial} - \text{Min Multiplier}}{\text{Max Multiplier} - \text{Min Multiplier}} \times (\text{Max Time Delay} - \text{Min Time Delay}) \right] + \text{Min Time Delay}$$

Figure 10-17: 51-Element Time Delay Calculation for Table Method

$$\text{Time} = \left[\frac{4.5 - 4.0}{6.0 - 4.0} \times (57.130 - 38.087) \right] + 38.087 = [0.25 \times 19.043] + 38.087 = 42.85\text{s}$$

- Repeat for the other test points.

The GE relay also provides formulas to calculate the expected time delay and we can use the following calculation to check our results.

$$T = TDM \times \left[\frac{A}{\left(\dfrac{I}{I_{pickup}} \right)^{p} - 1} + B \right] = 4.5 \times \left[\frac{28.2}{\left(\dfrac{7.0}{3.5} \right)^{2} - 1} + 0.1217 \right]$$

$$T = 4.5 \times \left[\frac{28.2}{(2)^{2} - 1} + 0.1217 \right] = 4.5 \times \left[\frac{28.2}{4 - 1} + 0.1217 \right]$$

$$T = 4.5 \times \left[\frac{28.2}{3} + 0.1217 \right] = 4.5 \times [9.4 + 0.1217] = 4.5 \times [9.5217]$$

$$T = 42.848 \text{ seconds}$$

D) Timing Test Procedure

- Determine which output the 51-element trips and connect the test-set timing input to the relay output.
- Check the maximum per-phase output of the test-set and use the appropriate connection from the previous section. For example, if the 51-element timing test is 35A and your test-set can only produce 25 Amps per phase; use "High Current Connections #1." If the pickup setting is greater than 50 Amps, use "High Current Connections #2." If the pickup is higher than 75A (3x25A), you will have to use another test-set or temporarily lower the setting. Remember, setting changes are a last resort.
- Set the fault current to the 51-element test current. Set your test-set to stop when the timing input operates and to record the time delay from test start to stop.
- Apply test current, ensure timing input operation, and note the time on your test sheet.
- Compare the test time to the 51-element timing calculations to ensure timing is correct.
- Repeat for other two phases.
- Set test current to 2nd test level.
- Apply test current, ensure timing input operation, and note the time on your test sheet.
- Compare the test time to the 51-element timing calculations to ensure timing is correct.
- Repeat for other two phases.
- Set test current to 3rd test level.
- Apply test current, ensure timing input operation, and note the time on your test sheet.
- Compare the test time to the 51-element timing calculations to ensure timing is correct.
- Repeat for other two phases.

5. Reset Tests

Exact reset time testing can be performed using complicated test plans but a simple verification is usually all that is required. The test procedure to measure the exact reset time is highly dependent on your test-set and we will concentrate on reset verification instead of an actual reset time measurement.

A) Reset Test Procedure

- Set up and perform a normal 51-element timing test.
- Immediately after the test is complete, immediately perform the test again.
- The time delay should be significantly smaller than the original timing test if the reset feature is enabled.
- Wait for a time longer than the reset delay and perform the test again. The timing result should be close to the first time delay.

6. Residual Neutral Time Overcurrent Protection

Residual neutral overcurrent protection calculates the unbalance current between all three phase values and is usually an easy test because it should be set well below phase overcurrent values. In these cases, follow the steps above applying current in one phase at a time. It is good practice to perform pickup tests on A-phase and timing tests on B-phase and C-phases to make sure the relay uses all three phases to calculate residual current.

If the phase overcurrent settings interfere with residual testing, or the pickup results are not as accurate as they should be, connect the relay and test-set as shown earlier in Figure 10-9. Apply all three phase currents simultaneously at the same phase angle. The magnitude of each phase should be one-third of the test current. Some relay models need currents on all three phases to accurately calculate residual current.

7. Tips and Tricks to Overcome Common Obstacles

The following tips or tricks may help you overcome the most common obstacles.

- Before you start, apply current at a lower value and perform a meter command to make sure your test-set is actually producing an output and your connections are correct.
- If the element does not operate, watch the metering during the test if possible.
- Check to make sure your settings are correct.
- Make sure you are connected to the correct output.
- Check the output connections by pulsing the output and watching the test-set input.
- Some relay test-set inputs are polarity sensitive. If the connections look good, try reversing the leads.

- Have any of your test leads fallen off?
- If you are paralleling more than one relay output, do all channels have the same phase angle?
- Check for settings like "Any Two Phases" (Any two phases must be above the pickup to operate) or "All Three Phases" (All three phases must be higher than the pickup to operate) or "Any Phase" (Any phase above pickup operates element).
- If you need more than one phase to operate the 50-element, but your test-set only has enough VA for one phase, put two or more phases in series as shown below:

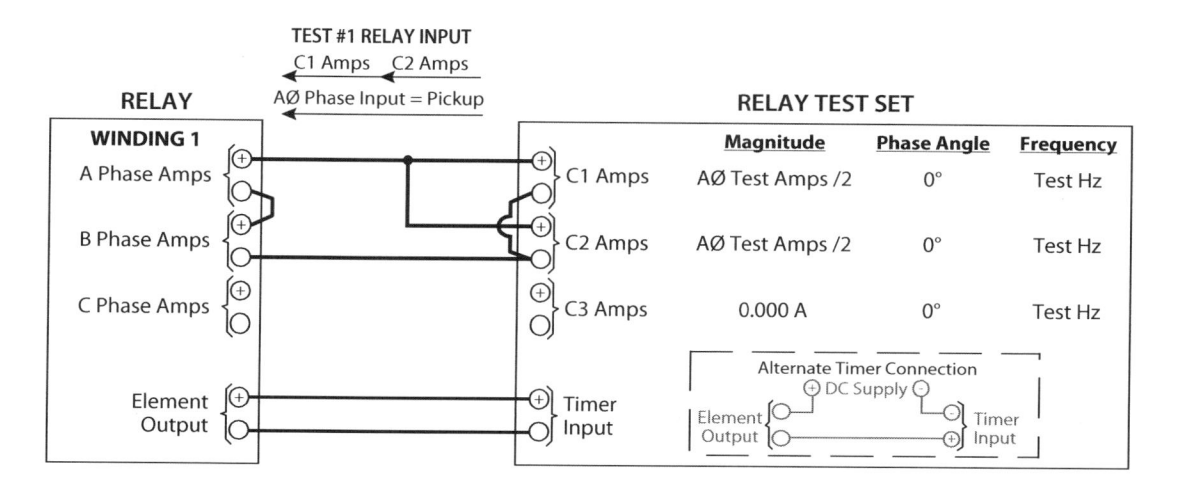

Figure 10-18: 51-Element Alternate Relay Connection

- Check for blocking inputs.
- Does the relay need breaker status or other input to operate?

Sometimes neutral or residual ground protection is applied. This protection is inevitably set lower than the phase settings. These elements trip first before the phase protection operates and can be a nuisance at best. The following solutions can help overcome this obstacle:

- Perform tests using 3-phase current with equal magnitudes, 120° apart. Residual current will be zero.
- Perform tests with three phase inputs; put two phases slightly below the pickup and slowly raise one phase at a time until pickup is indicated as shown in Figure 10-8.
- Performing phase-to-phase tests as shown in Figure 10-9 will also eliminate zero sequence interference.

Chapter 11

Directional Overcurrent (67) Element Testing

1. Application

Instantaneous overcurrent (50) and time overcurrent (51) protection can protect equipment from overloads and short circuits, but there are situations where their ability to protect a system is limited. Many applications exist where the direction of current is not fixed due to multiple sources and/or parallel feeders that prevent simple overcurrent protection from adequately protecting the electrical system. Directional overcurrent (67) protection only operates if the current flows in a pre-defined direction to provide more selective and sensitive protection.

Previous generations of directional overcurrent protection were limited by the construction materials available at the time and used different connections to determine the operating direction and sensitivity. Some of their operating parameters were based on power elements. Operating times varied depending on the magnitude of current and the measured phase angle compared to the maximum torque angle (MTA). Modern directional (67) elements operate like standard overcurrent elements (50 or 51) with a switch that turns the protection on if the current flows in the trip direction.

The most common directional overcurrent applications are listed below and described in the next few sections.

A) Parallel Feeders
B) Transmission Line Ground Protection
C) Power Flow

A) **Parallel Feeders**

Directional Overcurrent elements are typically installed in substations or other distribution systems to create zones of protection as described in the Zones of Protection section starting on page 66. In Figure 11-1, two parallel lines feed a load to provide system stability and all relays are set with typical 51-element overcurrent protection. A fault on either transmission line could cause them both to trip offline because the relay that detects the most fault current (R1) will operate first and any combination of relays R3, R4, or R2 could also operate depending on their settings. If R3 or R4 operates during this fault, both lines will be offline, and the load will be de-energized due to a fault on one line...the opposite result intended by parallel lines.

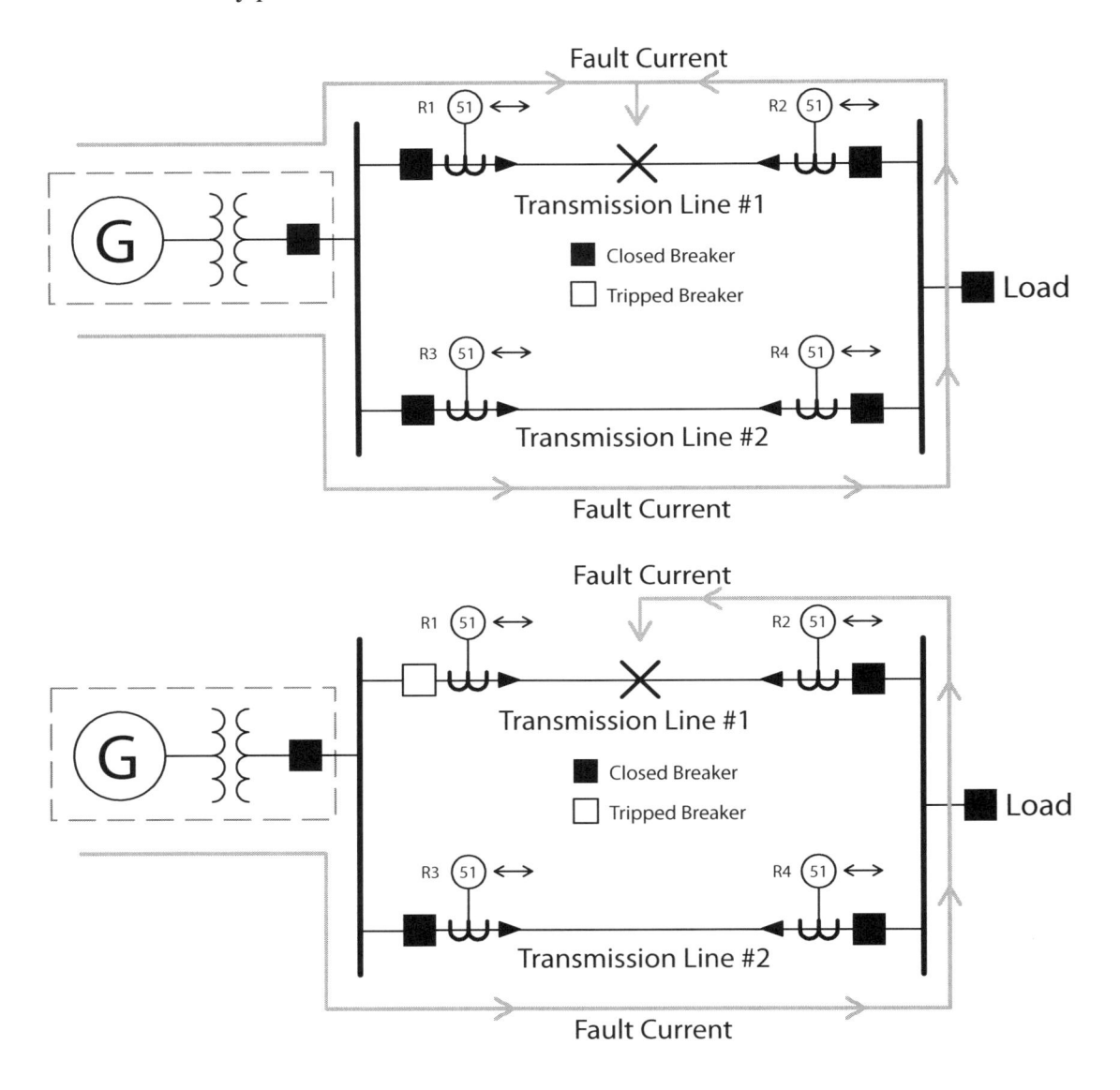

Figure 11-1: Parallel Transmission Lines with Standard Overcurrent Protection

In Figure 11-2, the load side relays (R2 and R4) use directional overcurrent protection that will only operate if the current flows toward the transmission line. These relays also have smaller overcurrent pickup settings to make them more sensitive to faults on the transmission line they are designed to protect. A fault on one transmission line will be isolated with no interruption of service because the directional Relay R2 will operate first (due to its lower pickup setting) and R1 will trip shortly thereafter as the fault current can only flow through R1 after R2 operates.

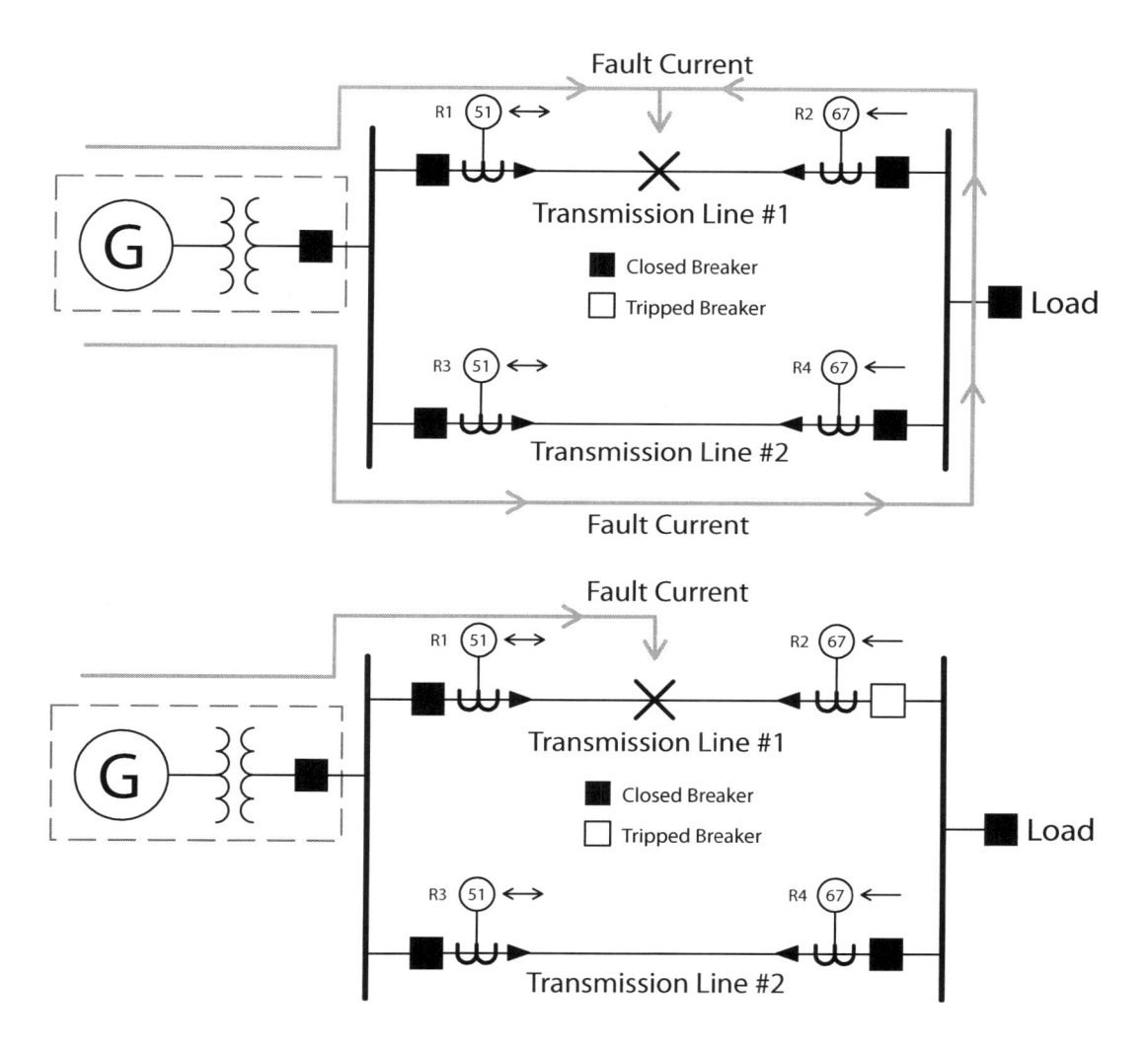

Figure 11-2: Parallel Transmission Lines with Directional Overcurrent Protection

B) Transmission Line Ground Protection

Directional overcurrent protection is often applied to transmission lines as backup protection for ground faults that may not be detected by impedance relays. These relays are set to provide ground fault protection on the transmission line only. Any fault behind the 67-element will be ignored by the feeder relays and another protective element will operate to protect the bus as shown in Figure 11-3.

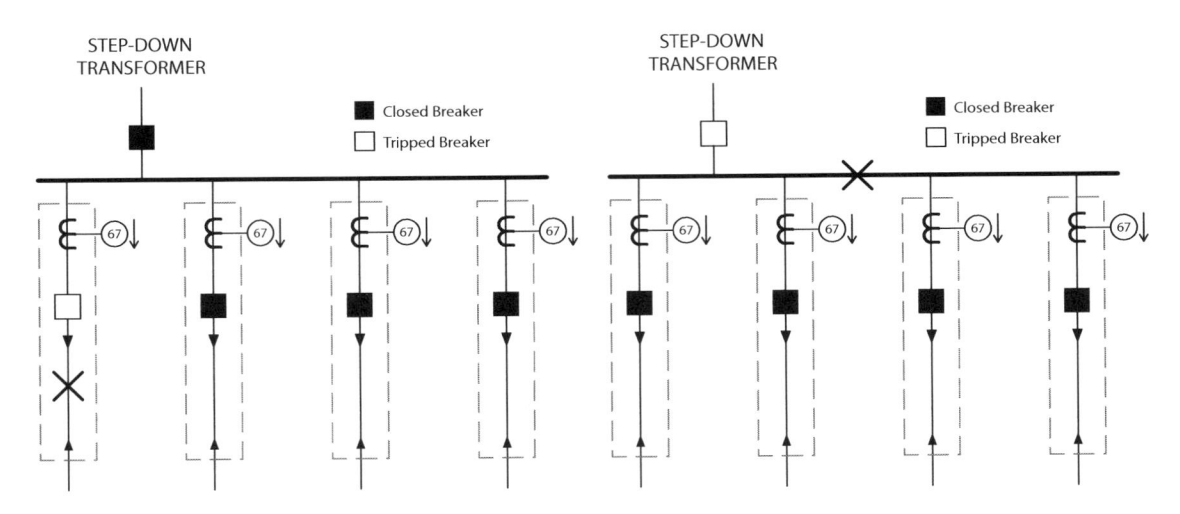

Figure 11-3: Directional Ground Overcurrent Protection for Transmission Lines

C) Power Flow

Directional relays can be found in facilities with their own on-site generating capabilities and a stand-by utility feed. Directional overcurrent protection is applied to prevent the plant from supplying power to the utility. If current flows from the plant to the utility, the Utility circuit breaker will open.

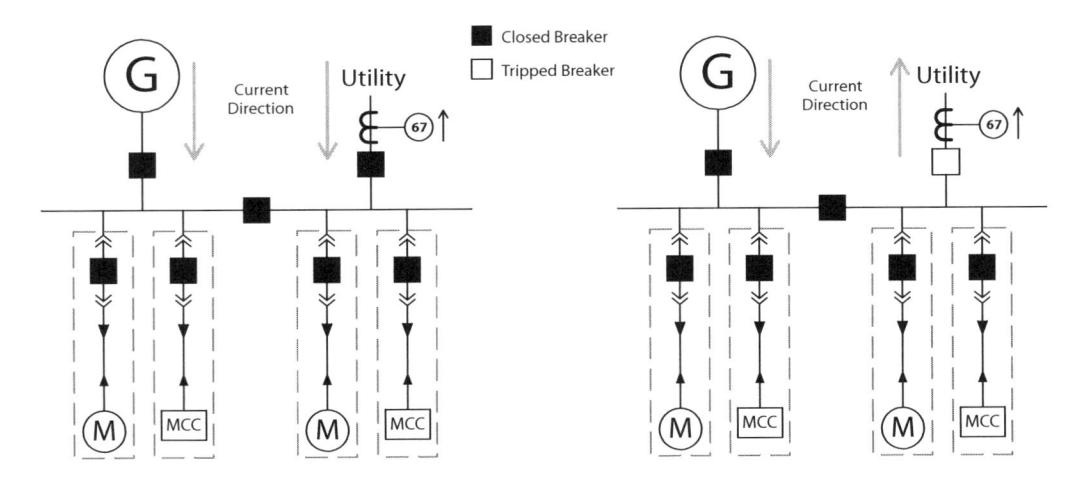

Figure 11-4: Directional Overcurrent Protection in an Industrial Application

2. Operation

Electromechanical relays used magnetism to produce torque in the polarizing element and were designed to operate when the measured current and voltage were a specific phase angle apart. These relays used the non-faulted voltages for polarizing signals. Figure 11-5 demonstrates all of the available phasors that the relay designer could use to obtain the desired Maximum Torque Angle (MTA). The maximum torque angle is the defining point for directional control and was typically fixed at 90° or 60° in electromechanical relays. A typical configuration (60°) used I_A current phasors referenced to the V_{BC} voltage to detect an A-phase fault; I_B referenced to V_{CA} to detect a B-phase fault, and I_C referenced to V_{AB} to detect a C-phase fault. Figure 11-6 displays the operating characteristic of an A-phase relay with a MTA of 60°. The dotted line is drawn 90° from the MTA to indicate the zero torque line which is the transition between the forward and reverse directions. Any A-phase current phasor below or to the right of the dotted line is flowing in the positive direction and will cause the relay to trip if the current exceeds the pickup setting. Current phasors above and to the left of the dotted line flow in the reverse direction and will never trip.

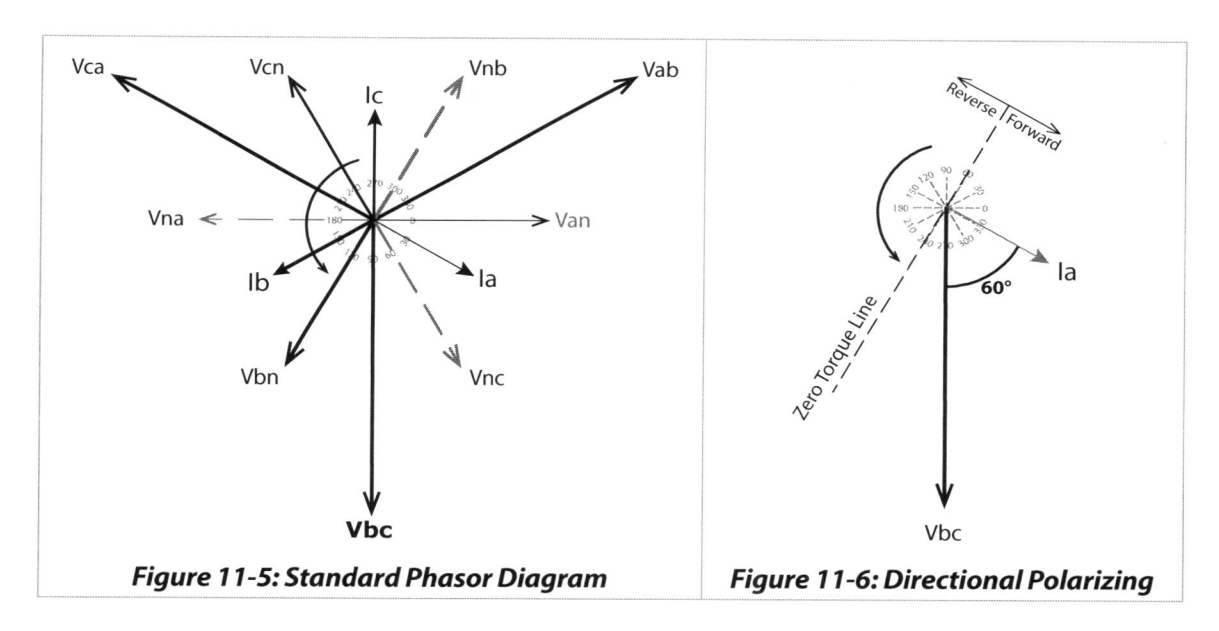

Figure 11-5: Standard Phasor Diagram   **Figure 11-6: Directional Polarizing**

Modern directional overcurrent protection starts with a polarizing element acting as an internal switch that turns the overcurrent element on or off. If the current flows in the trip direction, the overcurrent protection operates as a standard overcurrent (50 or 51) element. If the current flows in the reverse direction, the directional element does not turn "on" and blocks the element from operating. The polarizing element can be an integral part of the directional-overcurrent (67) protection or it can be a separate element used to block or permit an independent 50/51 element. Basic polarizing elements use two separate signals (Voltage and/or Current) and compare the phase angle between the two signals to determine direction.

Traditionally, a phase-to-phase voltage was compared to a line current in the polarizing element but this configuration can cause nuisance trips under certain fault conditions. Some modern relays can use any voltage, current, or sequence component as the polarizing source to ensure reliable directional operation. Some relays even use a prefault value for directional control or the relay can choose the best option from a list of choices depending on the type of fault.

3. Settings

Typical settings for 67-elements are described below:

A) Enable Setting

Many relays allow the user to enable or disable settings. Make sure that the element is ON/Enabled or the relay may prevent you from entering settings. If the element is not used, the setting should be disabled or OFF to prevent confusion.

B) Pickup

This setting determines when the relay will start timing if the current flows in the correct direction. Different relay models use different methods to set the actual pickup and the most common methods are:

- **Secondary Amps**—the simplest unit. Pickup Amps = setting
- **Per Unit (P.U.)**—This setting could be a multiple of the nominal current as defined or calculated if the relay has setpoints for nominal current, Watts, or VA. It could also be a multiple of the nominal CT secondary.
  Pickup = Setting x Nominal Amps, **OR**
  Pickup = Setting x Watts / (nominal voltage x $\sqrt{3}$ x power factor), **OR**
  Pickup = Setting x VA / (nominal voltage x $\sqrt{3}$), **OR**
  Pickup = Settings x CT Secondary (typically 5 Amps)
- **Primary Amps**—There must be a setting for CT ratio if this setting style exists. Check the CT ratio from the drawings and make sure that the drawing matches the settings.
  Pickup = Setting / CT Ratio, **OR**
  Pickup = Setting * CT secondary / CT primary

C) Curve

This setting chooses which curve will be used for timing. Be very careful to select the correct curve as there can be subtle differences between curve descriptions. Compare the curve selection to the coordination study to ensure the correct curve is selected.

D) Time Dial/Multiplier

This setting simulates the time dial setting on an electromechanical relay to determine the time delay between pickup and operation in conjunction with the selected curve. ANSI curves usually have a time delay between 1 and 10. IEC time setting are typically between 0 and 1.

E) Reset

Electromechanical 51-element relay timing was controlled using a mechanical disc that would rotate if the current was higher than the pickup setting. If the current dropped below the pickup value, the disc would rotate back to the reset position.

Some digital relays simulate the reset delay using a linear curve that is directly proportional to the current to closely match the electromechanical relays. Other relays have a preset time delay or user-defined reset delay that should be set to allow any related electromechanical discs to reset for proper coordination between devices.

F) Phase Directional MTA (Maximum Torque Angle)

This setting determines the maximum torque angle to be used by the directional element. It is set in degrees and sets the angle between the polarizing value and the measured current as shown in Figure 11-6. Be sure that you know whether phase angle leads or lags the polarizing element. Most General Electric relays use angle measurements that lag. Make sure you understand which value is used for polarizing.

Some relays use directional overcurrent settings to block rather than enable overcurrent protection. Review the relay's instruction manual to determine whether overcurrent protection is blocked or enabled at the maximum torque angle. Compare this to the system drawings to make sure that the correct setting has been applied.

G) Phase Directional Relays

This setting determines which output relay(s), if any, will operate when the current flows in the pre-determined direction.

H) Minimum Polarizing Voltage

This setting is used to ensure that the polarizing voltage reference is large enough to provide the correct reference angles when required. This is automatically set in some relays and exists to prevent nuisance trips. If the PT fuses to the relay were not installed and this setting was not applied, any induced voltages or noise could provide an incorrect reference for the directional element.

I) Block OC When Voltage Memory Expires

Some of the most severe faults will cause the voltage to collapse to near zero which will not provide a valid phase-voltage signal for the polarizing element. The relay constantly records the system voltages to use the prefault voltage as a reference when the fault voltages are too low. This setting will allow the directional-overcurrent element to operate until the memory time-delay expires.

J) Directional Signal Source

Some relays can have multiple voltage or current sources and this setting determines which CT/PT input to use for the directional element reference.

K) Directional Block

This setting is a logic function and if the logic applied is true, the directional element will be blocked and not operate.

L) Directional Target

Some relays allow you to define what front panel LED or message will be displayed on the relay front panel. This setting determines what display indication, if any, will operate if the element operates.

M) Directional Events

Some relays allow you to define what events will appear in the event recorder. This setting determines if any directional events will be recorded.

N) Directional Order

Some relays allow you to define multiple sources for a directional reference. This setting determines which order that the relay will use to determine a valid directional reference. For example, a 3-phase fault will not create much zero sequence voltage and the relay could switch to the next reference source if it determined that a zero-sequence voltage was not adequate.

4. Pickup Testing

Directional overcurrent (67) pickup testing is essentially the same as traditional overcurrent pickup testing once you are sure that the current is flowing in the correct direction. Write down all settings related to the 67-element and calculate what the pickup current should be using the formulas described in the previous section.

Check the primary values and time delays against the coordination study and make sure they match. Pay close attention to your application and ensure that you know the normal flow of current. If the relay uses a maximum torque angle setting, draw the nominal vectors for your application, and then draw the maximum torque angle vectors and compare them. Make sure the correct direction has been set based on the application.

The single line drawing in Figure 11-7 depicts the substation portion of a generating plant. The lines with arrows indicate the normal flow of current which flows from the generator out Line #2, and to the Station Service Transformer. The most common application for directional overcurrent protection is found by the 67N element in relay 6. This relay is a line protection relay that is designed to protect the transmission line.

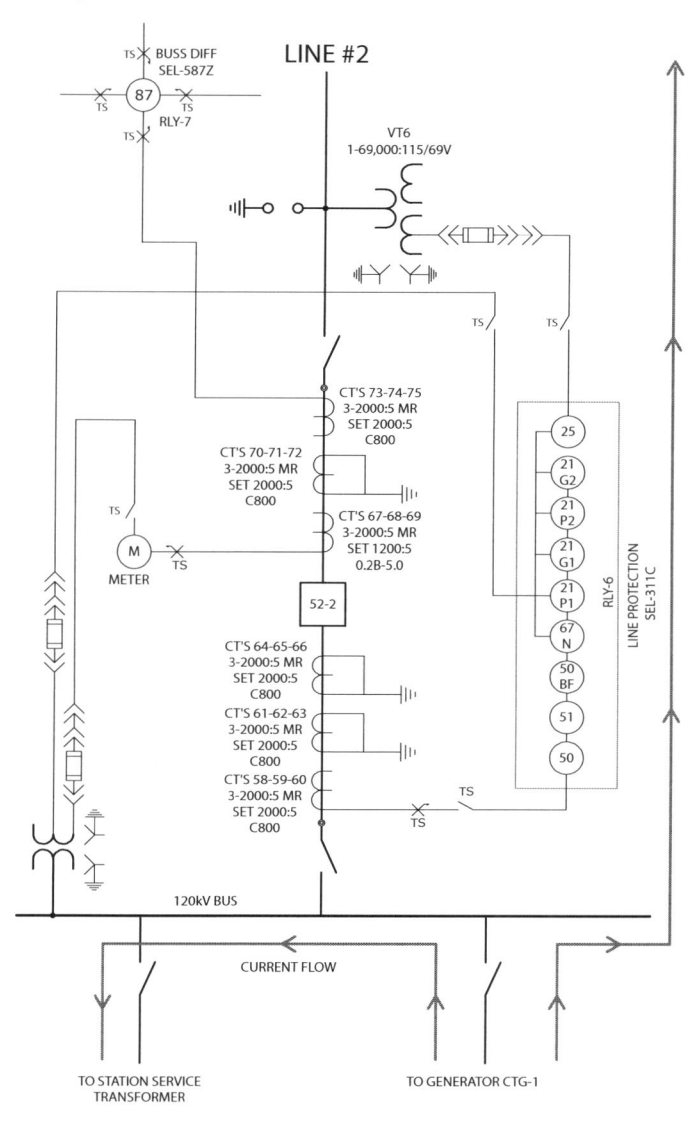

Figure 11-7: Example Single-Line Drawing

A standard phasor diagram for this relay would be depicted by Figure 11-8. All SEL relay directional elements are based on a general direction instead of a fixed angle so, the 67N directional element should be set in the forward direction to provide line protection and ignore faults inside the generator or station service transformers.

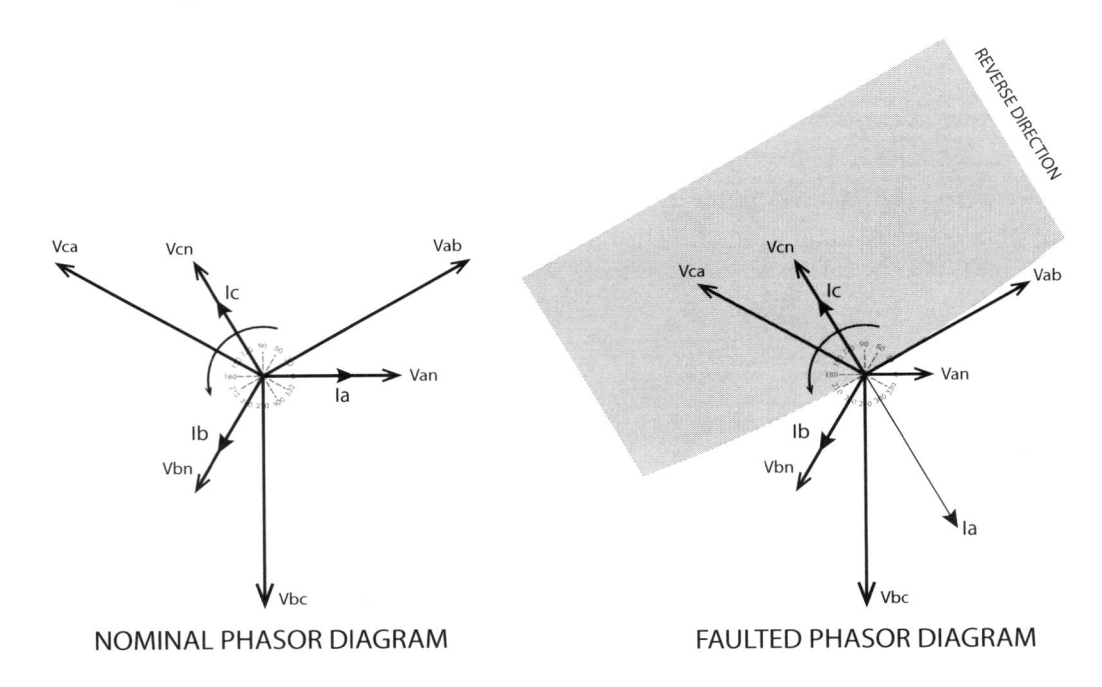

Figure 11-8: Typical Directional Polarizing Using SEL Relays

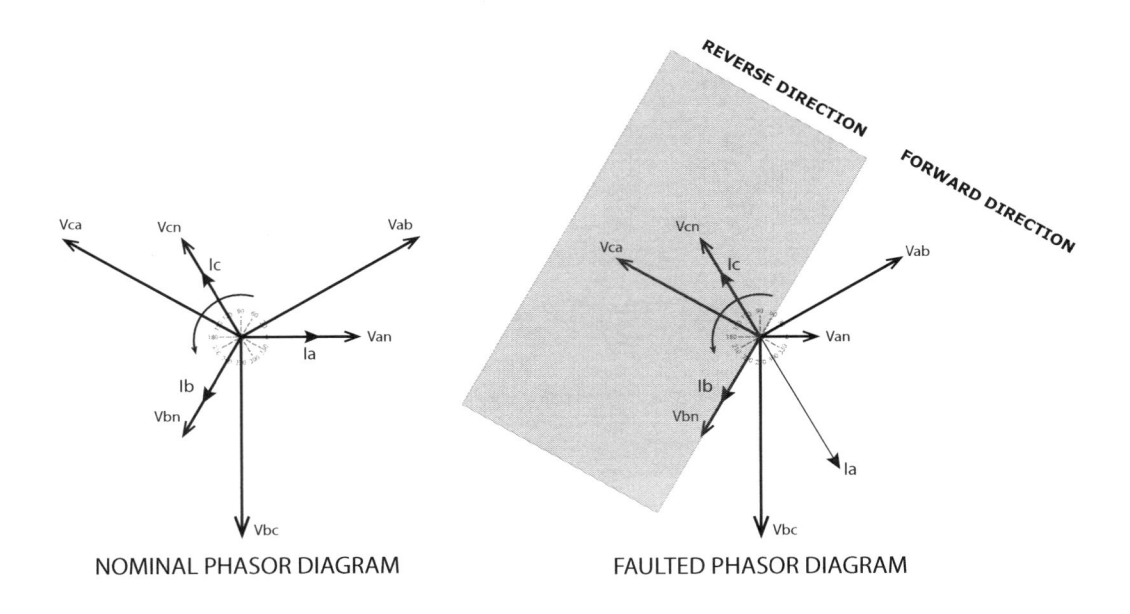

Figure 11-9: Directional Polarizing Using GE Relays and a 60° MTA Setting

A) Test-Set Connections

Because directional overcurrent protection is highly dependent on correct current and voltage connections, it is extremely important that your test-set connections match the application's 3-line drawings. Use the following figures to correctly simulate the current and voltage connections.

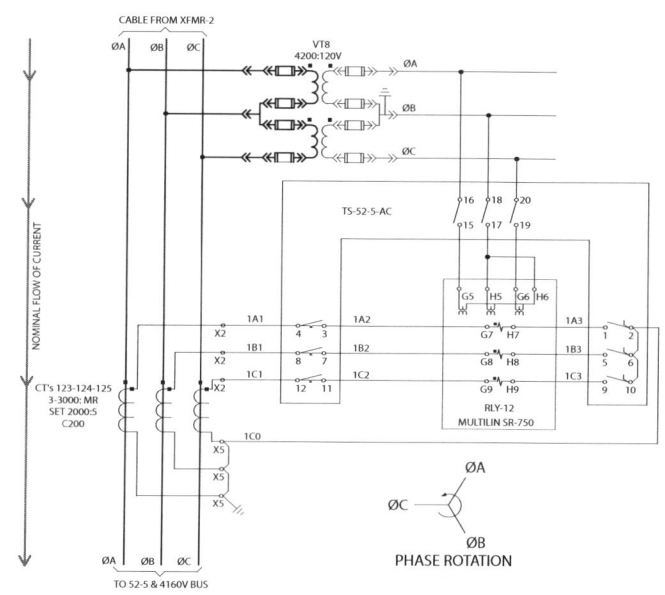

Figure 11-10: 3-Line Drawing for Example Test-Set Connection

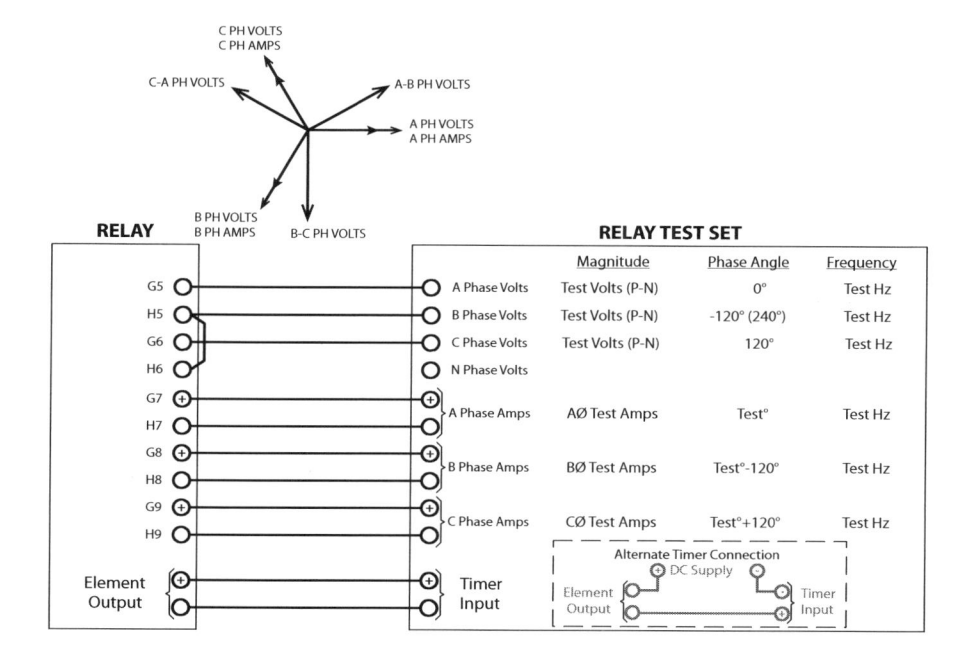

Figure 11-11: Directional Overcurrent Test-Set Connections

241

B) Determine Maximum Torque Angle in GE Relays

The first step to any test procedure is determining what the expected value is. The settings important to directional control in an SR-750 are:

- PHASE TIME OC 1 FUNCTION = Trip
- PHASE TIME OC 1 PICKUP = 1 x CT
- PHASE TIME OC 1 DIRECTION = Forward
- PHASE INST OC 1 FUNCTION = N/A
- PHASE INST OC 1 PICKUP = N/A
- PHASE INST OC 1 DIRECTION = N/A
- PHASE DIRECTIONAL FUNCTION = Control
- PHASE DIRECTIONAL MTA = 30° Lead
- MIN POLARIZING VOLTAGE = 0.05 x VT
- BLOCK OC WHEN VOLT MEM EXPIRES = Disabled

The settings above also include the example settings we will use for our test. To determine what to use as our reference, we can use the following chart from the SR-750/760 Feeder Management Relay Instruction Manual.

QUANTITY	OPERATING CURRENT	POLARIZING VOLTAGE	
		A-B-C PHASE SEQUENCE	A-C-B PHASE SEQUENCE
Phase A	Ia	Vbc	Vcb
Phase B	Ib	Vca	Vac
Phase C	Ic	Vab	Vba

We will use A-B-C phase sequence for our example and draw all of our normal phasors as shown in Figure 11-12. If we want to test Phase A, we can remove all phasors except Ia and Vbc and draw the MTA at 30° as per the PHASE DIRECTIONAL MTA setting. The operating range will be 90° from the MTA in both directions and the relay operates in the forward direction as per the PHASE TIME OC 1 DIRECTION setting. Figure 11-13 depicts the operating characteristic for Phase A.

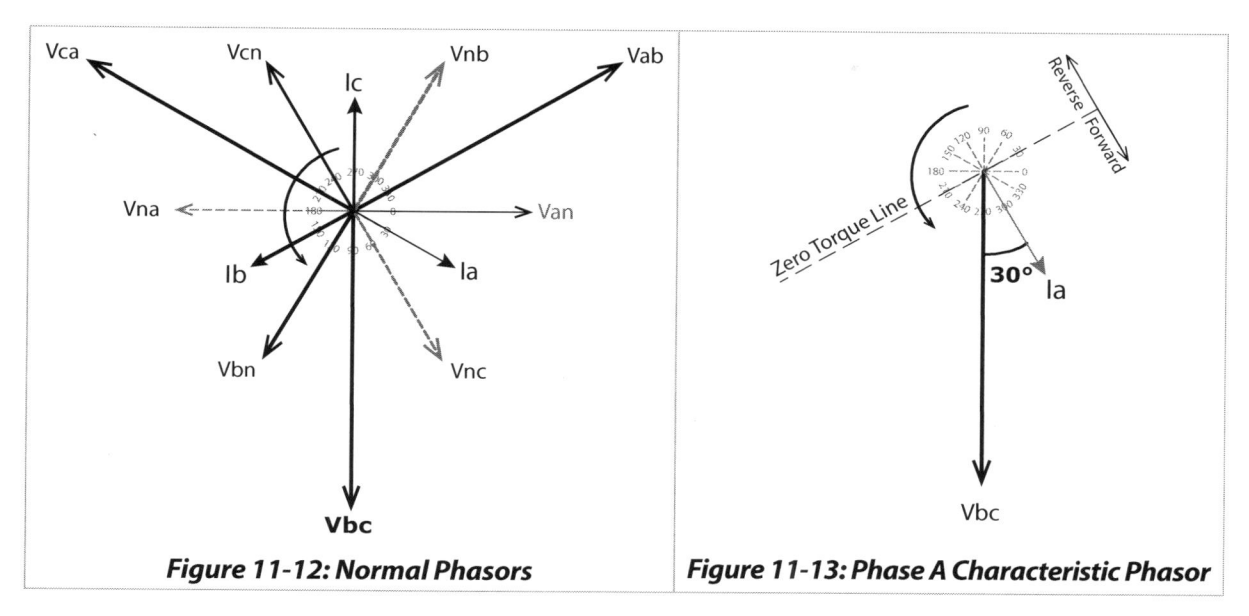

Figure 11-12: Normal Phasors   **Figure 11-13: Phase A Characteristic Phasor**

To test the MTA of this element, choose your method of monitoring pickup as described in the *Pickup Testing* section starting on page 109 and follow these steps.

a. Apply 3-phase balanced voltages and A-phase current above the pickup setting as per the following settings:

- Van = Nominal voltage @ 0°
- Vbn = Nominal Voltage @ -120°
- Vcn = Nominal Voltage @ 120°
- Ia = 125% of pickup current @ 0° (1.25 * 5A = 6.25A)
- Ib = 0A
- Ic = 0A

b. Adjust the Ia phase angle in the positive direction until the pickup indication drops out. This should happen at approximately 30°. Adjust the phase angle until pickup is indicated and record the pickup value (30.3°).

c. Adjust the Ia phase angle to 220° (-140°). The pickup should still be illuminated.

d. Adjust the Ia phase angle in the negative direction (clockwise) until the pickup indication drops out at approximately 210° (-150°). Adjust the phase angle into the positive direction until pickup is indicated and record the pickup result. (-150.3°)

e. Take the average of the two values ([30.3 + -150.3] / 2 = -60°) to find the measured MTA and compare to the calculated MTA (Vbc @ -90° + PHASE DIRECTIONAL MTA = 30° Lead = -90° + 30° = -60°).

f. Repeat for Ib and Ic.

C) Quick and Easy Directional Overcurrent Test Procedures

With enough time and the right equipment, it is possible to test every aspect of the 67-element protection with detailed test results for MTA, operating characteristic, memory drop-out, polarizing memory, etc. However, testing a 67-element in accordance with the applied settings will never fail on a relay that is operating correctly. Some relays, such as SEL models, do not have user defined characteristics and operate dynamically based on actual operating conditions. Depending on the relay, the 67-element can be rather complex and confusing for the design engineer which could cause setting errors. A more efficient test for the 67-element would be a functional test of its operation based on the application or engineer's intent. Use the following procedure to test nearly every relay application to ensure it will operate correctly when placed into service instead of simply testing the applied settings:

a. Contact the design engineer and determine whether the element should operate in the forward or reverse direction. Determine if there are any special conditions that must occur before the directional element will operate for complicated installations such as wind farms, etc. If you cannot contact the design engineer, review the drawings to determine the correct tripping direction. Use the settings as the basis for your tests as a last resort.

b. Once you have determined the correct tripping direction, simulate a line-to-ground fault using the following test-set settings.

PREFAULT	FAULT
Van = Nominal voltage @ 0°	Van = 85% of Nominal voltage @ 0°
Vbn = Nominal Voltage @ -120°	Vbn = Nominal Voltage @ -120°
Vcn = Nominal Voltage @ 120°	Vcn = Nominal Voltage @ 120°
Ia = 0A	Ia = 125% of pickup current @ (-60° if trip direction
Ib = 0A	is forward, 120° if trip direction is reverse)
Ic = 0A	Ib = 0A
	Ic = 0A

c. Determine how you will monitor the pickup as described in the *Relay Test Procedures* section starting on page 109.

d. Apply the prefault currents and voltages.

e. Apply the fault currents and voltages. The 67-element pickup indication should be on.

f. Reverse the A-phase current phase angle by 180° (-60° + 180° = 120°). The pickup indication should turn off. Change the A-phase current back to the original fault angle. The pickup indication should be on.

g. Slowly lower the A-phase current until the pickup indication is off. Slowly raise the A-phase current until the pickup indication is fully on. This is the 67-element pickup.

h. You can determine the MTA at this point, if you wish, by rotating the A-phase current angle in either direction until the pickup indication turns off, reverse direction and record the angle that the 67-element picks up again. Rotate the phase angle to the opposite side and repeat. The MTA can be determined using the following formula: (MTA = 1st angle pickup - [(1st angle pickup - 2nd angle pickup) / 2] If we use the previous GE relay example with a 30° MTA setting (-60° MTA), the first angle pickup would be 30° and the second angle pickup would be -150°. Using our formula: MTA = 30° - [(30° - -150°)/2] = 30° - (180°/2) = 30° + -90° = -60° .

i. You can also test other functions such as minimum polarizing voltage by simulating the condition you wish to test. For minimum polarizing voltage, apply a current 125% greater than the pickup settings and change the fault voltage magnitudes to a value below the pickup level (you may need to multiply your voltage magnitudes by 1.732 to account for differences in phase-to-phase and phase-to-neutral setting/application differences), the 67-element should not pickup. Increase all three phase-voltages until the 67-element picks up. This is the minimum polarizing voltage pickup.

j. You can also test polarizing memory by applying nominal prefault voltages and changing all of the fault voltages to zero. Apply prefault values with nominal voltages. Apply the fault values with zero voltage. If the 67-element operates, polarizing memory is operating. If the element does not operate, polarizing memory is not enabled or operating. If the 67-element picks up and then drops out while the fault is being applied, the polarizing memory has expired. You can time this value as well.

k. Repeat the tests on B-phase and C-phases with the following fault settings. Prefault values will stay the same.

B-PHASE FAULT	C-PHASE FAULT
Van = Nominal voltage @ 0°	Van = Nominal voltage @ 0°
Vbn = 85% of Nominal voltage @ -120°	Vbn = Nominal Voltage @ -120°
Vcn = Nominal Voltage @ 120°	Vcn = 85% of Nominal voltage @ 120°
Ia = 0A	Ia = 0A
Ib = 125% of pickup current @ (180° if trip direction is forward, 0° if trip direction is reverse)	Ib = 0A
Ic = 0A	Ic = 125% of pickup current @ (60° if trip direction is forward, -120° if trip direction is reverse)

5. Timing Test Procedures

The timing test procedure for directional overcurrent elements is identical to the procedure described in the earlier "Time Overcurrent (51) Protection Testing" or "Instantaneous Overcurrent (50) Protection Testing" chapters of this publication once the correct direction has been applied. Please review those chapters for detailed timing test procedures and ensure that the correct direction is applied for tests.

6. Tips and Tricks to Overcome Common Obstacles

The following tips or tricks may help you overcome the most common obstacles.

- Apply prefault currents and voltages and perform a metering test.
- All of the examples have been applied for A-B-C or counter-clockwise rotation with 90° in the upper quadrants and -90° in the lower quadrants or the phasor diagram. Adjust the angles accordingly if you use different rotation or references.
- Different relay manufacturers have different phasor references. Make sure you understand the manufacturer's phasor references. For example, GE relay phasors use a lagging reference; SEL relays use a leading reference. 30° displayed on a GE relay is -30° on an SEL relay.
- Some relays use sequence components to determine direction. Applying a P-N fault will create positive, negative, and zero sequence components if the faulted phase-voltage is less than nominal and the faulted current is greater than the pickup setting is the best option for simple directional testing. If the element does not operate, try lowering the fault voltage for the corresponding high current to create a larger reference signal.
- Make sure the current under test is greater than the pickup and is not at unity power factor.
- SEL relays that have manual directional settings can use impedance blinders that may prevent normal directional operation. Ask the design engineer to provide specific test parameters.
- Is the direction element turned on?
- Is the directional element applied to the overcurrent element?

Chapter 12

Simple Percent Differential (87) Element Testing

Differential protection (87) is applied to protect equipment with high replacement costs and/or long replacement/repair times, or a group of equipment which is integral to the electrical system. 87-elements are selective and will only operate when a fault occurs within a specified zone of protection with no inherent time delay. There are several methods to apply differential protection but they are all designed to achieve the same basic principal; trip for faults within the zone of protection and ignore faults outside of the zone of protection. The zone of protection is defined by strategic placement of current transformers (CTs) that measure the current entering and exiting the zone. If the difference between currents is larger than the 87-element setpoint, the relay will trip. Percent differential protection is the most commonly applied differential element and this chapter will lay the foundation for understanding most differential protection schemes.

This chapter only discusses the simplified differential protection typically found in generator, motor, or other applications where the voltage and CT ratios are the same on both sides of the protected device. All descriptions are made on a single-phase basis and you should keep in mind that identical configurations exist for the other two phases in a three-phase system.

We will discuss complex differential protection typically installed for transformers in the next chapter.

1. Application

Differential protection operates on the principle that any current entering the protected zone must equal the current leaving the protected zone, and that a difference between the two currents is caused by an internal fault. Differential CTs are often strategically installed to provide overlapping protection of circuit breakers and, therefore, a generator differential scheme may include other equipment such as PTs and circuit breakers.

The following figure displays a simplified version of differential protection that would typically be applied to a generator or bus. The equipment being protected in this example would be connected between CT1 and CT2 which defines the differential zone of protection. The current flows from left to right in this system and enters the polarity mark of CT1. The CT secondary current follows the circuit and flows through the Iop coil from the bottom to top. Simultaneously, the current enters the non-polarity mark of CT2 and the CT secondary current flows through the Iop coil from top to bottom. If the CTs are operating perfectly and have the same CT ratio, the net Iop current would be zero amps because the two currents cancel each other out. Early differential elements were simple overcurrent devices and the Iop coil would have a relatively low setting.

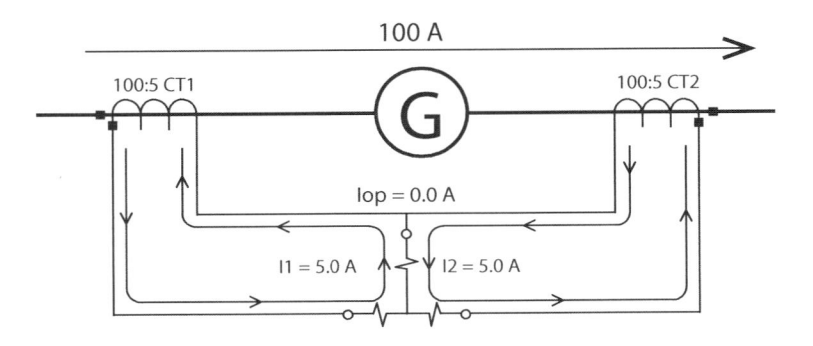

Figure 12-1: Simple Differential Protection

The next figure displays an external fault in an ideal world. The current entering the zone of protection equals the current leaving the zone and cancel each other out just like the first example. If the Iop coil was a simple overcurrent element, nothing would happen because zero amps flow though the Iop coil.

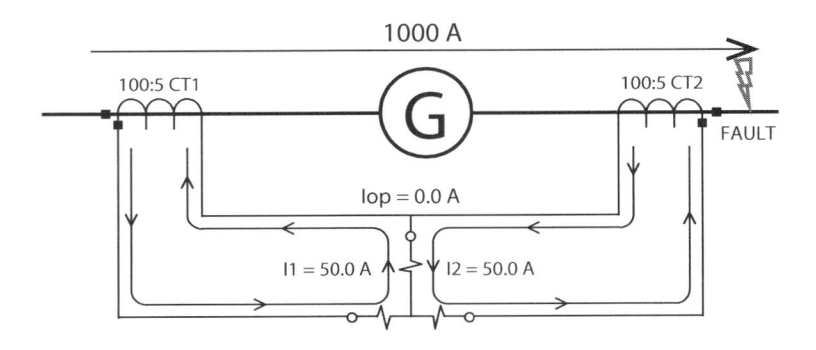

Figure 12-2: Simple Differential Protection with External Fault

The next figure shows a fault in the opposite direction. The current-in equals the current-out and cancel each other out. The Iop element does not operate.

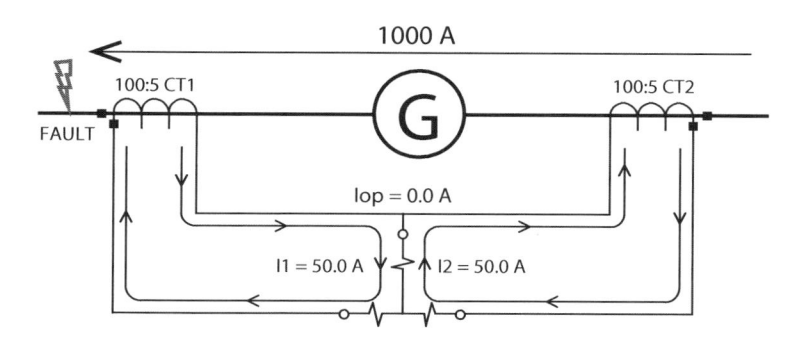

Figure 12-3: Simple Differential Protection with External Fault 2

The next figure displays an internal fault with Source 1 only. In this scenario, current flows into CT1 and the secondary current flows through Iop, bottom to top. There is no current flowing through CT2 and, therefore, nothing to cancel the CT1 current. If the CT1 current is greater than the Iop setting, the Iop element will trip.

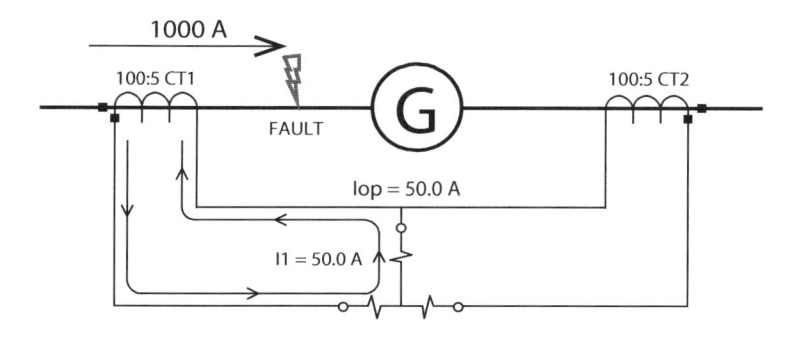

Figure 12-4: Simple Differential Protection with Internal Fault

The next figure displays an internal fault with sources on both ends. Current flows into CT1 and its secondary current flows through Iop from bottom to top. Current flows through CT2 and its secondary current flows through Iop from bottom to top also. In this case the currents add together instead of canceling each other out and the Iop element will trip if the combined currents are larger than the Iop trip setting.

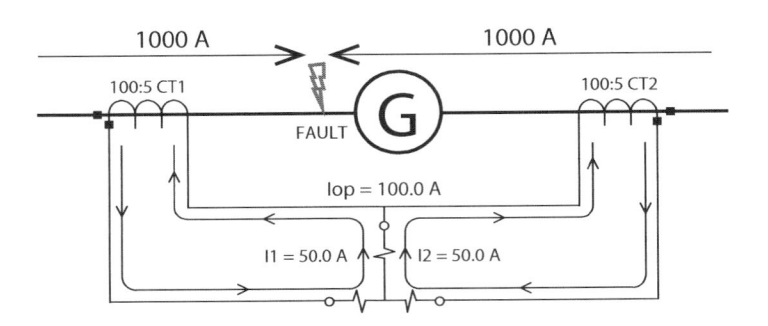

Figure 12-5: Simple Differential Protection with Internal Fault 2

In an ideal world, differential current is a simple overcurrent relay connected between two or more CTs as shown in previous figures. Unfortunately, we do not live in an ideal world and no two CTs will produce exactly the same output current even if the primary currents are identical. In fact, protection CTs typically have a 10% accuracy class which can cause problems with the protection scheme described previously. The news gets worse when you also consider that CT accuracy can jump to 20% when an asymmetrical fault is considered. The difference between CT operating characteristics is called CT mismatch and the effects of CT mismatch are displayed in the following figures.

This is a worst case scenario with nominal current. 100A flows through CT1, and the secondary current with -10% error equals 4.5A $(5.0A - [5.0A \times 0.10])$ flowing through Iop from bottom to top. 100A flows through CT2, and the secondary current with +10% error equals 5.5A $(5.0A + [5.0A \times 0.10])$ flowing through Iop from top to bottom. The difference between the two CT secondaries is 1.0A. The Iop element must be set larger than 1.0A or the relay will trip under normal load conditions assuming a worst case CT mismatch.

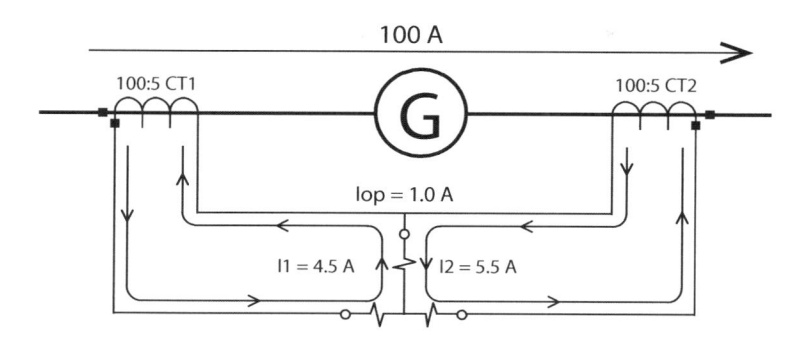

Figure 12-6: Simple Differential Protection with Worst Case CT Error

The next figure displays an external fault with the worst case CT mismatch. As in our previous example, the CT1 secondary current with -10% error is 45A flowing through the Iop from bottom to top. The CT2 secondary current with +10% error is 55A flowing through Iop from top to bottom. The differential current is 10A and the Iop element would incorrectly trip for a fault outside the zone using the previously defined setting of 1.0A. This element would have to be set greater than 10A to prevent nuisance trips if a fault occurs outside the zone of protection. This setting is very high and any internal fault would need to cause greater than 200A (2x the rated current for a single source system) or 100A (dual source system) of fault current to operate the differential element. This setting does not appear to limit equipment damage very well and a new system was developed to provide more sensitive protection and prevent nuisance trips for external faults.

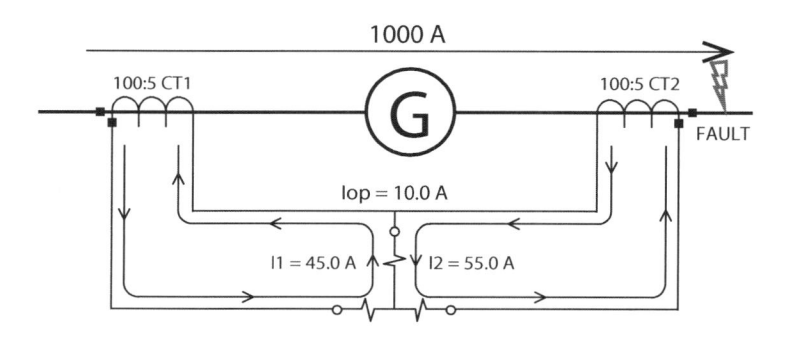

Figure 12-7: Simple Differential Protection with Worst Case CT Error and External Fault

This new system adds a restraint coil (Ir) along with the existing operate coil (Iop) in the circuit as shown in Figure 12-8. This Ir coil provides counter-force in electromechanical relays as shown in Figure 12-9 which pulls the trip contacts apart. The Ir coil force is directly related to the average of the I1 and I2 currents. The Iop coil attempts to pull the trip contacts together and the coil force is directly related to the Iop current. The two coils are designed so the Iop coil will be able to close the contacts if the ratio of Iop to Ir exceeds the relay's slope setting. Therefore, any slope setting is a ratio of Iop to Ir. We will define current flowing through the Iop coil as "operate current" and current flowing through the Ir current as the "restraint current" for the rest of this book.

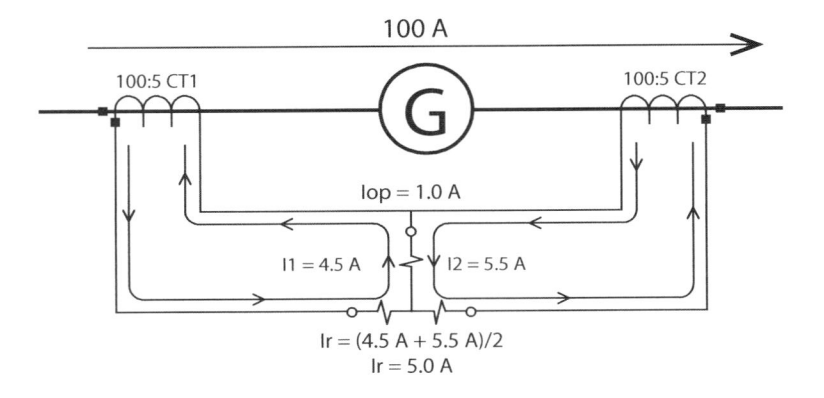

Figure 12-8: Percentage Differential Protection Schematic

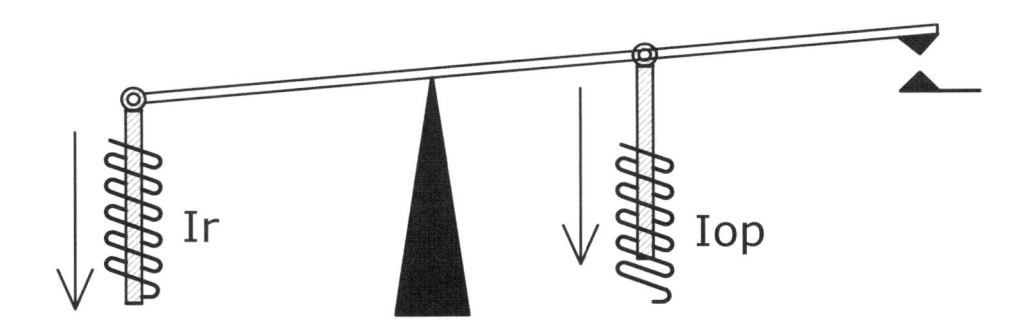

Figure 12-9: Percentage Differential Protection Operating Mechanism

With this new system, the designer chooses a slope setting instead of a fixed current and the relay will adjust the pickup setting to the system parameters. Our example slope setting will be 25% for the following figures which repeat the previous fault simulations with the new design.

Figure 12-10 displays normal operating conditions. The CT1 secondary current (4.5A) flows through the Iop coil as before. The CT2 current (5.5A) flows through Iop in opposition and the 1.0A differential current energizes the Iop coil which tries to close the trip contact. The trip contact does not close because one-half of the restraint coil has 4.5A of CT1 current and the other half has 5.5A of CT2 current. The average of these two currents creates the Ir force holding the contacts open. In this case the average of the two CT currents is 5.0A. The ratio of operate coil current and restraint coil current is 20% ($\frac{10A}{50A} = 0.2, 0.2 \times 100 = 20\%$). The pickup setting of this relay is 25% so the restraint coil is applying more force than the operate coil and the relay will not trip.

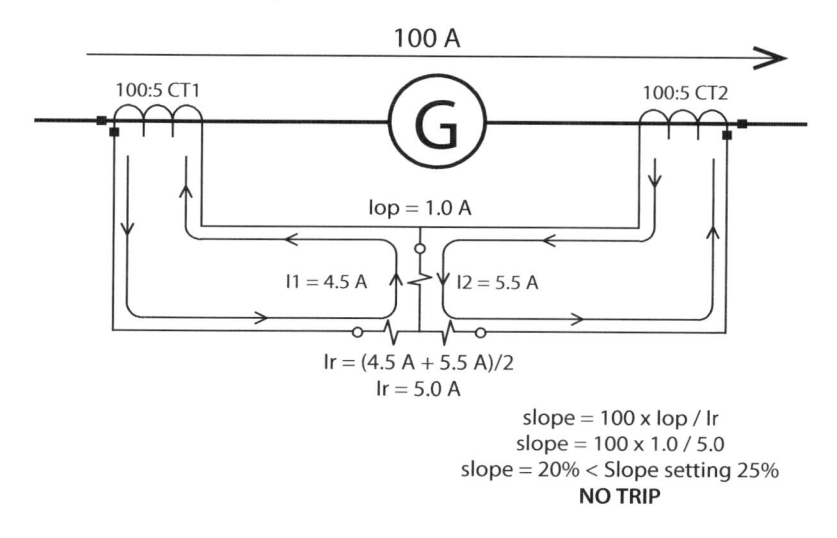

Figure 12-10: Percentage Differential Protection and External Faults

The next figure displays an internal fault with only one source. The relay trips in this scenario due to the 200% slope.

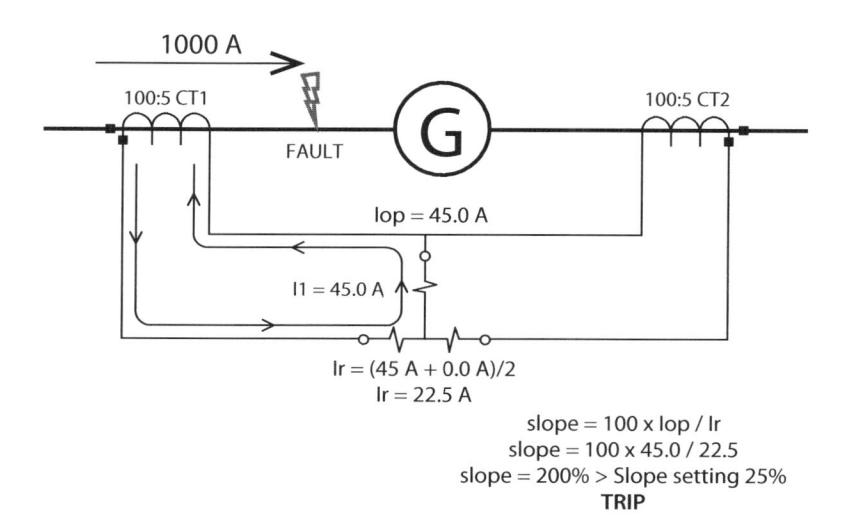

Figure 12-11: Percentage Differential Protection and Internal Faults

The next figure displays an internal fault with two sources. The relay trips in this scenario due to the 200% slope.

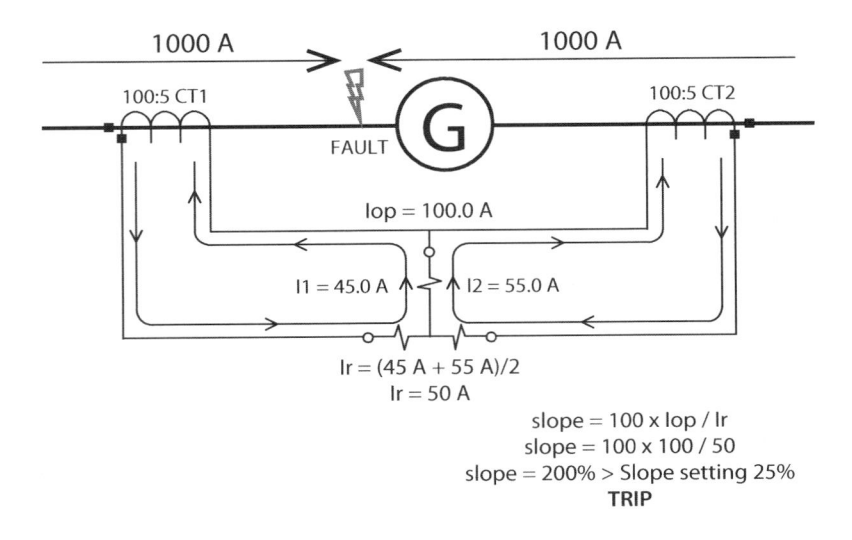

Figure 12-12: Percentage Differential Protection and Internal Faults 2

The characteristic curve of this element is shown in the following figure. The relay will trip if the ratio of Iop to Ir is plotted above the line and will not trip if the ratio falls below the line.

Differential Protection 25% Characteristic Curve

Figure 12-13: Percentage Differential Protection Characteristic Curve

Most relays have a Tap setting that defines the normal operating current based on the rated load of the protected equipment, the primary voltage, and the CT ratio. For example, imagine a 3-phase, 4160V, 600 kW generator. Using the standard 3-phase power formula, we can determine that the rated primary current is 83.27A.

$$Power = P\text{-}P\ Volts \times Amps \times \sqrt{3}$$

$$Amps = \frac{600000W}{\left(4160V \times \sqrt{3}\right)} = 83.27A$$

The secondary current at full load using 100:5 CTs would be 4.16A

$$\frac{CT_{SEC}}{CT_{PRI}} = \frac{Amps_{SEC}}{Amps_{PRI}}$$

$$\frac{5A}{100A} = \frac{Amps_{SEC}}{83.27A}$$

$$Amps_{SEC} = \frac{5A \times 83.27A}{100A} = 4.16A$$

We can combine the two formulas to determine the calculated Tap setting in one step.

$$Amps_{SEC} = \frac{CT_{SEC} \times Power}{P\text{-}P\ Volts \times \sqrt{3} \times CT_{PRI}}$$

The Tap setting could also be considered the per-unit current of the protected device and, if the Tap settings exist, is the basis for all differential calculations.

If we were to zoom in on the origin of the graph in Figure 12-13, we would see that it takes very little operate current to trip this element at low power levels. Any induced current, noise, or excitation current could cause this element to trip under these normal conditions. One common situation occurs when the transformer is energized with no connected load. There will be some excitation current on the primary side of the transformer and no current on the secondary side. This is the definition of a differential fault, but it is a normal condition, and we do not want the 87-element to operate under normal conditions.

A new setting called Minimum Pickup is introduced to ensure that the 87-element will trip under fault conditions and ignore normal mismatch at low current levels. This setting sets the minimum amount of current that must be present before the 87-element will operate. This setting is sometimes in secondary amps for simple differential applications, but it is more commonly set as a percentage or multiple of the Tap setting and is typically set at 0.3 times the Tap setting.

The differential characteristic curve changes when the Minimum Pickup setting is applied as shown in Figure 12-14. Notice the flat line between 0 and 1.00 Restraint Current at the beginning of the curve which represents the Minimum Pickup setting (0.3 x Tap current). Also notice that the Operate and Restraint currents are plotted as Multiples of Tap and not amps.

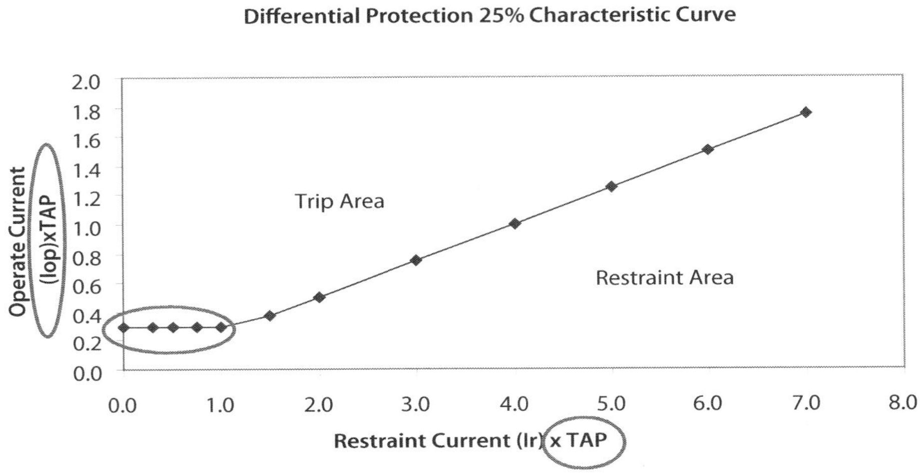

Differential Protection 25% Characteristic Curve

Figure 12-14: Percentage Differential Protection Characteristic Curve with Minimum Pickup

As the fault current increases, the chance of CT saturation or other problems occurring dramatically increase which can cause up to a 20% error during asymmetrical faults. Because digital relays use programming instead of components for their pickup evaluations, we can add a second slope setting to be used when the restraint current exceeds a manufacturer or user-defined level of current. This second slope has a higher setting, and the differential protection will be less sensitive during high-level external faults.

The transition between the two slope settings is called the breakpoint or knee and is pre-defined by the manufacturer in some relays or user-defined in others. The breakpoint is set in multiples of Tap. The differential characteristic curve with a Minimum Pickup and two slope settings is shown in the following figure:

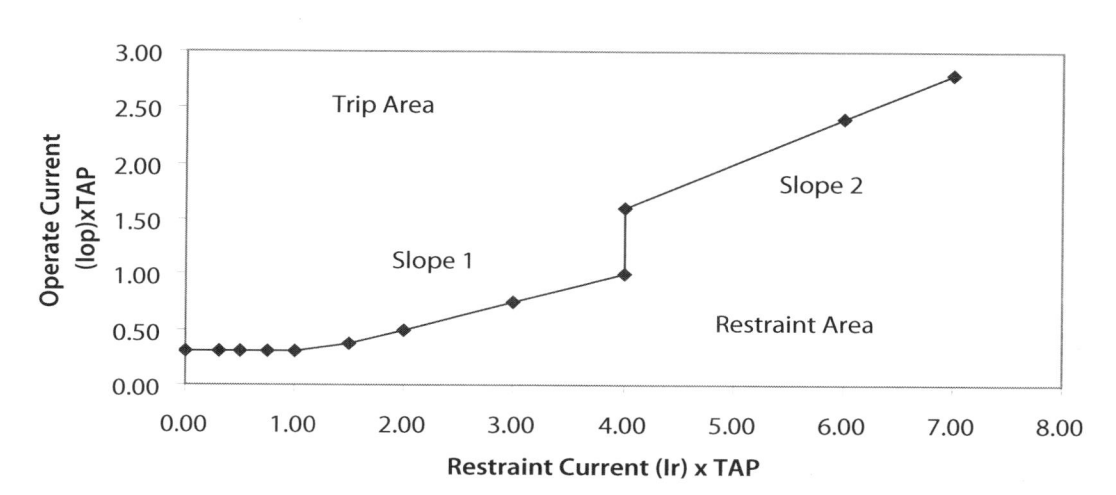

Figure 12-15: Percentage Differential Protection Dual Slope Characteristic Curve

2. Settings

The most common settings used in 87-elements are explained below:

A) Enable Setting

Many relays allow the user to enable or disable settings. Make sure that the element is ON, or the relay may prevent you from entering settings. If the element is not used, the setting should be disabled or OFF to prevent confusion. Some relays will also have "Latched" or 'Unlatched" options. A Latched option indicates that the output contacts will remain closed after a trip until a reset is performed and acts as a lockout relay. Unlatched indicates that the relay output contacts will open when the trip conditions are no longer present.

B) Minimum Pickup (Restrained)

The Minimum Pickup setting is used to provide more stability to the differential element by requiring a minimum amount of current to flow before the differential element will operate. This minimum operate current is used to prevent nuisance trips due to noise or metering errors at low current levels. This setting should be set around 0.3 times the nominal or Tap setting of the protected device.

C) Tap

The Tap setting defines the normal operate current based on the rated load of the protected equipment, the primary voltage, and the CT ratio. This setting is used as the per-unit operate current of the protected device and most differential settings are based on the Tap setting. Verify the correct Tap setting using the following formula.

$$\text{Amps}_{SEC} = \frac{CT_{SEC} \times Power}{\text{P-P Volts} \times \sqrt{3} \times CT_{PRI}}$$

D) Slope-1

The Slope-1 setting sets the ratio of operate current to restraint current that must be exceeded before the 87-element will operate. The slope setting is typically set at 20-30%.

~ 40%

E) Slope-2

The Slope-2 setting sets the ratio of operate current to restraint current that must be exceeded before the 87-element will operate if the restraint current exceeds a pre-defined or user-defined break point between Slope-1 and Slope-2.

F) Breakpoint

This setting defines whether Slope-1 or Slope-2 will be used for the differential calculation. The Breakpoint is defined as a multiple of Tap and if the restraint current exceeds the Breakpoint setting, the 87-element will use Slope-2 for its calculation.

G) Time Delay

The Time Delay setting sets a time delay between an 87-element pickup and trip. This is typically set at the minimum possible setting, but can be set as high as 3 cycles for maximum reliability on some relays.

H) Block

The Block setting defines a condition that will prevent the differential protection from operating such as a status input from another device. This setting is rarely used. If enabled, make sure the condition is not true when testing. Always verify correct blocking operation by operating the end-device instead of a simulation to ensure the block has been correctly applied.

3. Restrained-Differential Pickup Testing

Performing simple differential pickup testing is very similar to the test procedure for overcurrent protection with a few extra phases to test. Apply a current in one phase, and raise the current until the element operates. Repeat for all affected phases.

As usual, you must determine what the expected result should be before performing any test. Record the Pickup and Tap settings. If the relay does not have a Tap setting, refer to the manufacturer's literature to determine if the relay uses some other nominal setting such as rated secondary current (5A usually). Multiply the Pickup and Tap settings to determine the Minimum Pickup in amps. The Tap setting from our previous example is 4.16A and the Pickup setting is 0.3. The expected pickup current for this example would be 1.248A.

$$(4.16A \times 0.3 = 1.248A)$$

We will use a GE Power Management 489 relay with the following Phase Differential settings for the rest of the tests in this chapter.

- Phase Differential Trip = Unlatched
- Assign Trip Relays = 1
- Differential Trip Minimum Pickup = 0.1 x CT
- Differential Trip Slope-1 = 20%
- Differential Trip Slope-2 = 80%
- Differential Trip Delay = 0 cycles

The actual Minimum Pickup setting for this relay is 0.1 x CT because this relay does not have a Tap setting. The CT designation indicates the nominal CT secondary current (5.0A in North America and 1.0A in Europe). The expected Minimum Pickup is 0.5A.

$$(0.1 \times CT = 0.1 \times 5.0 \ A = 0.5 \ A)$$

If the pickup test results are significantly higher or lower than the expected result, refer to the next chapter's instructions for pickup testing because the algorithm for the relay under test probably uses a complex equation for restraint current and correction factors will apply.

A) Test-Set Connections

The most basic test-set connections use only one phase of the test-set with a test lead change between every pickup test. After the Winding-1 A-phase pickup test is performed, move the test leads from Winding-1 A-phase amps to B-phase amps and perform the test again. Repeat until all enabled phases are tested on all enabled windings.

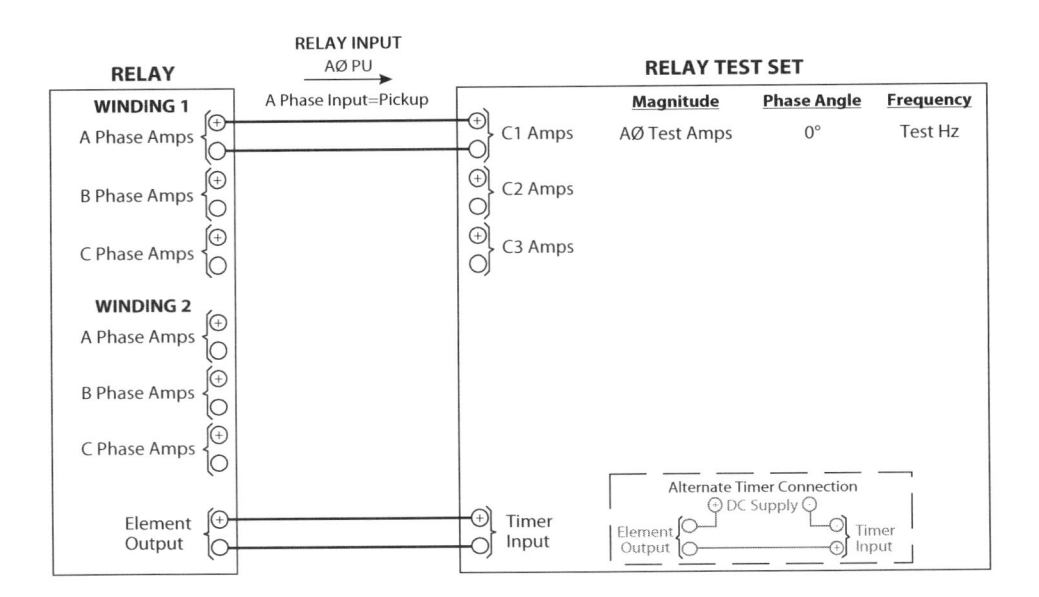

Figure 12-16: Simple 87-Element Test-Set Connections

You can also connect all three phases to one winding and change output channels instead of changing leads. After all Winding-1 tests are completed, move the test leads to Winding-2 and repeat.

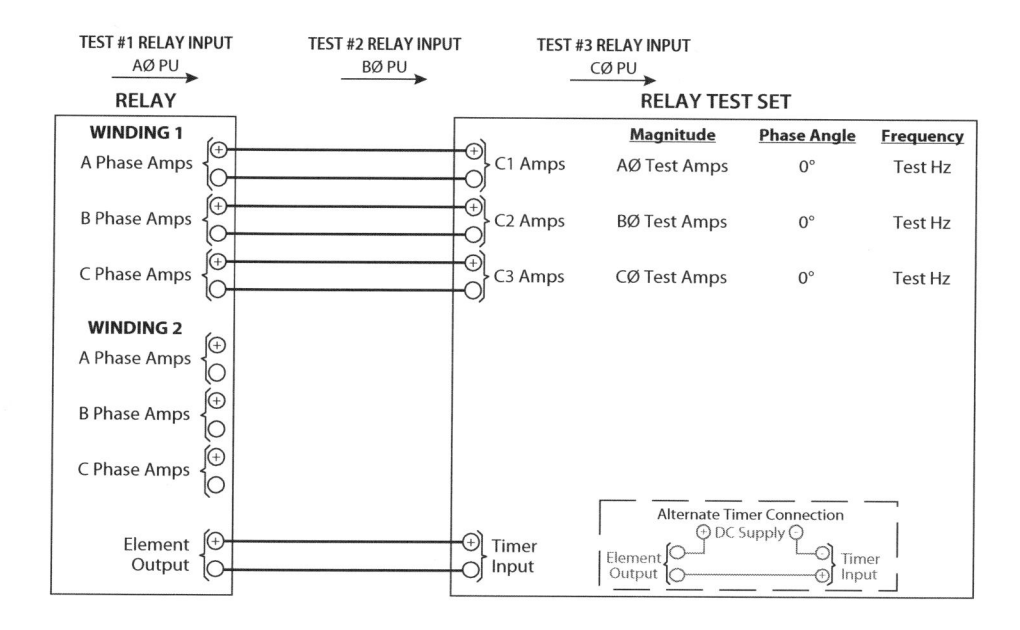

Figure 12-17: Simple 3-Phase 87-Element Test-Set Connections

If your test-set has six available current channels, you can use the following connection diagram and change the output channel for each test until all pickup values have been tested.

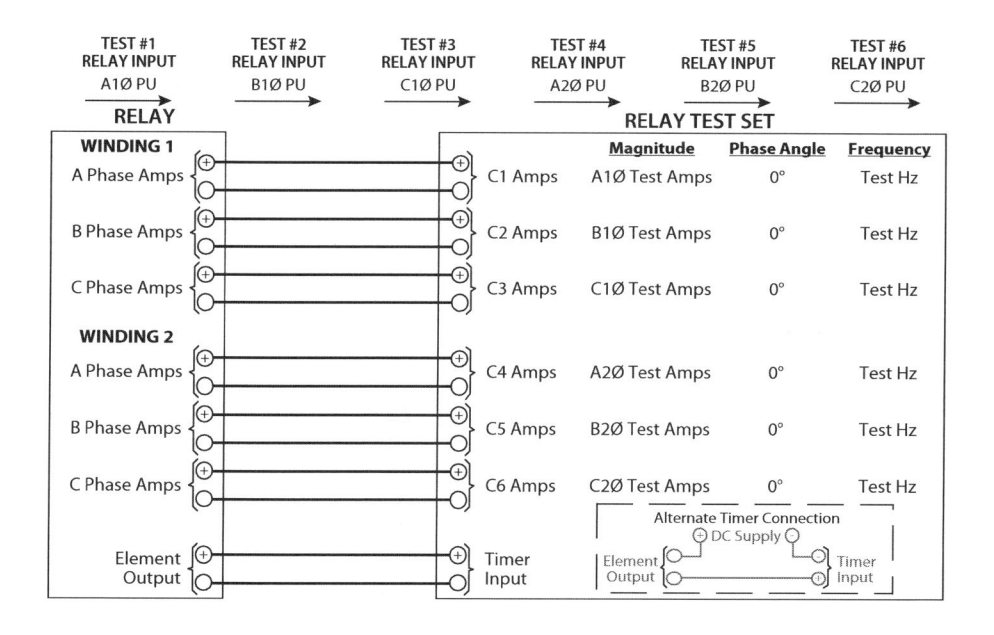

Figure 12-18: Simple 3-Phase 87-Element Test-Set Connections with Six Current Channels

B) Pickup Test Procedure

Use the following steps to determine pickup:

1. Determine how you will monitor pickup and set the relay accordingly, if required. (Pickup indication by LED, output contact, front panel display, etc. See the *Relay Test Procedures* section starting on page 109 for details.)

2. Set the fault current 5% higher than the pickup setting. For our example, set the fault current at 0.525A for an element with a 0.5A setpoint. Make sure pickup indication operates.

3. Slowly lower the current until the pickup indication is off. Slowly raise current until pickup indication is fully on. (Chattering contacts or LEDs are not considered pickup.) Record the pickup values on your test sheet. The following graph displays the pickup procedure.

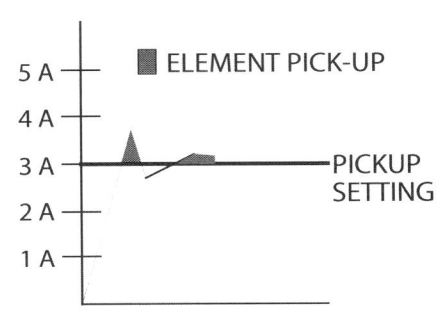

STEADY-STATE PICKUP TEST

Figure 12-19: Pickup Test Graph

4. Compare the pickup test result to the manufacturer's specifications and calculate the percent error as shown in the following example.

   The 87-element pickup setting is 0.5A and the measured pickup was 0.508A. Looking at the "OUTPUT AND NEUTRAL END CURRENT INPUTS" specification in Figure 12-20, we see that the acceptable metering error is 0.05A. ("Accuracy: @ < 2 x CT: ± 0.5% of 2 x CT" = 0.5% × 10A = 0.05A). The difference between the test result and setting is 0.008A (0.508A – 0.500A). We can immediately consider it acceptable as it falls within the metering accuracy specifications in Figure 12-20.

OUTPUT AND NEUTRAL END CURRENT INPUTS	
Accuracy:	@ < 2 x CT: ± 0.5% of 2 x CT
	@ > 2 x CT: ± 1% of 20 x CT
OUTPUT RELAYS	
Operate Time:	10ms
PHASE DIFFERENTIAL	
Pickup Accuracy:	as per Phase Current Inputs
Timing Accuracy:	+50ms @ 50/60 Hz or ± 0.5 % of total time

Figure 12-20: GE Power Management 489 Analog Input Specifications

Using these two sections, we can calculate the manufacturer's allowable percent error. Use "Accuracy: @ < 2 x CT: ± 0.5% of 2 x CT" from the specifications in Figure 12-20 because the applied current is less than 2x CT (20A), the allowable percent error is 10%.

$$\frac{\text{Maximum Accuracy Tolerance}}{\text{Setting}} \times 100 = \text{Allowable Percent Error}$$

$$\frac{0.5\% \times (2 \times CT)}{0.5A} \times 100 = \text{Allowable Percent Error}$$

$$\frac{0.005 \times (2 \times 5A)}{0.5A} \times 100 = \text{Allowable Percent Error}$$

$$\frac{0.005 \times 10A}{0.5A} \times 100 = \text{Allowable Percent Error}$$

$$\frac{0.05A}{0.5A} \times 100 = \text{Allowable Percent Error}$$

10% Allowable Percent Error

The measured percent error can be calculated using the percent error formula:

$$\frac{\text{Actual Value - Expected Value}}{\text{Expected Value}} \times 100 = \text{Percent Error}$$

$$\frac{0.508A - 0.500A}{0.500A} \times 100 = \text{Percent Error}$$

1.6% Error

5. Repeat the pickup test for all phase currents that are part of the differential scheme.

DIFFERENTIAL TEST RESULTS							
PICK UP	0.1	TIME DELAY	0				
SLOPE 1	10%						
SLOPE 2	20%						
MINIMUM PICK UP TESTS (Amps)							
	A PHASE	B PHASE	C PHASE	MFG	% ERROR		
W1 PICKUP	0.508	0.508	0.505	0.500	1.60	1.60	1.00
W2 PICKUP	0.500	0.506	0.508	0.500	0.00	1.20	1.60
W3 PICKUP	NA	NA	NA	0.500			

4. Restrained-Differential Timing Test Procedure

The timing test procedure is very straightforward. Apply a single-phase current 10% greater than the Minimum Pickup setting into each input related to the percent differential element and measure the time between the applied current and the relay trip signal. As with all tests, we first must discover what an acceptable result is. Use Figure 12-21 to determine the acceptable tolerances from the manufacturer's specifications.

PHASE DIFFERENTIAL

Pick-up Level:	0.05-1.00 x CT in steps of 0.01
Curve Shape:	Dual Slope
Time Delay:	0 - 100 cycles in steps of 1
Pickup Accuracy:	as per Phase Current Inputs
Timing Accuracy:	+50ms @ 50/60 Hz or ± 0.5 % of total time
Elements:	Trip

OUTPUT RELAYS

Configuration:	6 Electo-Mechanical Form C
Contact Material:	silver alloy
Operate Time:	10ms
Max Ratings for 100000 operations	

Figure 12-21: GE Power Management 489 Differential and Output Relay Specifications

The time delay setting for our example is 0 cycles which means that there is no *intentional* delay. However, there are software and hardware delays built into the relay that must be accounted for as shown in Figure 12-22 that state that the maximum allowable time is 60ms or 3.6 cycles.

REASON	DELAY
Phase Differential Timing Accuracy (Software)	50ms
Output Relay Operate Time (Hardware)	10ms
Total Time	60ms or 3.6 cycles

Figure 12-22: GE Power Management 489 Minimum Trip Time

Perform a timing test using the following steps:

1. Connect the test-set input(s) to the relay output(s) that are programmed to operate when the restrained-differential relay operates.

2. Configure your test-set to start a timer when current is applied and stop the timer and output channels when the appropriate input(s) operate.

3. Choose a connection diagram from Figures 16-18 on pages 14-15. Set a single-phase current at least 10% higher than the Minimum Pickup setting.
   Minimum Pickup × 110% = 0.5A × 1.1 = 0.55A

4. Run the test plan on the first phase related to restrained-differential. Record the test results.

5. Repeat on all phases related to the restrained-differential.

RESTRAINED DIFFERENTIAL TIMING TESTS (cycles)							
WINDING	TEST	A PHASE (cy)	B PHASE (cy)	C PHASE (cy)	MFG (cycles)	% ERROR		
W1	0.55A	2.410	2.490	2.500	3.600	OK	OK	OK
W2	0.55A	2.45	2.44	2.48	3.600	OK	OK	OK
W3	NA	NA	NA	NA	NA			

5. Restrained-Differential Slope Testing

Differential slope testing is one of the most complex relay tests that can be performed and requires careful planning, a good understanding of the differential relay's operating characteristics, and information from the manufacturer regarding the relay's characteristics. The instructions for this section use single-phase test techniques that will only work when applied to simple differential relays with identical CT ratios and no phase-angle shift between windings. The following test plan also assumes a simple slope calculation as described previously, but different relays may use different methods to determine slope. If your test results do not fall into an acceptable range with this test procedure, refer to the manufacturer's literature for the relay's slope calculation and apply the manufacturer's formula to the results. If the results still do not match the expected results, refer to the Restrained-Differential Slope Testing section of the next chapter for more details.

A simple differential slope test is performed by choosing a phase (AØ for example) and applying current into that phase for two windings that are part of the differential element. Let's look at the manufacturer's recommended connection diagram shown in Figure 12-23 for the GE Power Management 489 relay in our example.

Follow the AØ primary buss through the neutral CTs (Phase a) then follow the secondary CT to Terminal H3. This is where we will connect the first current from our test-set. Connect the neutral of the test-set current channel to G3 by following the other side of the CT to its relay terminal. Keep following the primary buss through the generator to the other CT and then follow the secondary to Terminal G6. This is the neutral of the CTs so we will connect the test-set's second current-channel-neutral to G6. Follow the other side of the CT to Terminal H6 which is where we will connect the second current-channel from the test-set. Follow the other phases to determine these connections when testing B or C-phases.

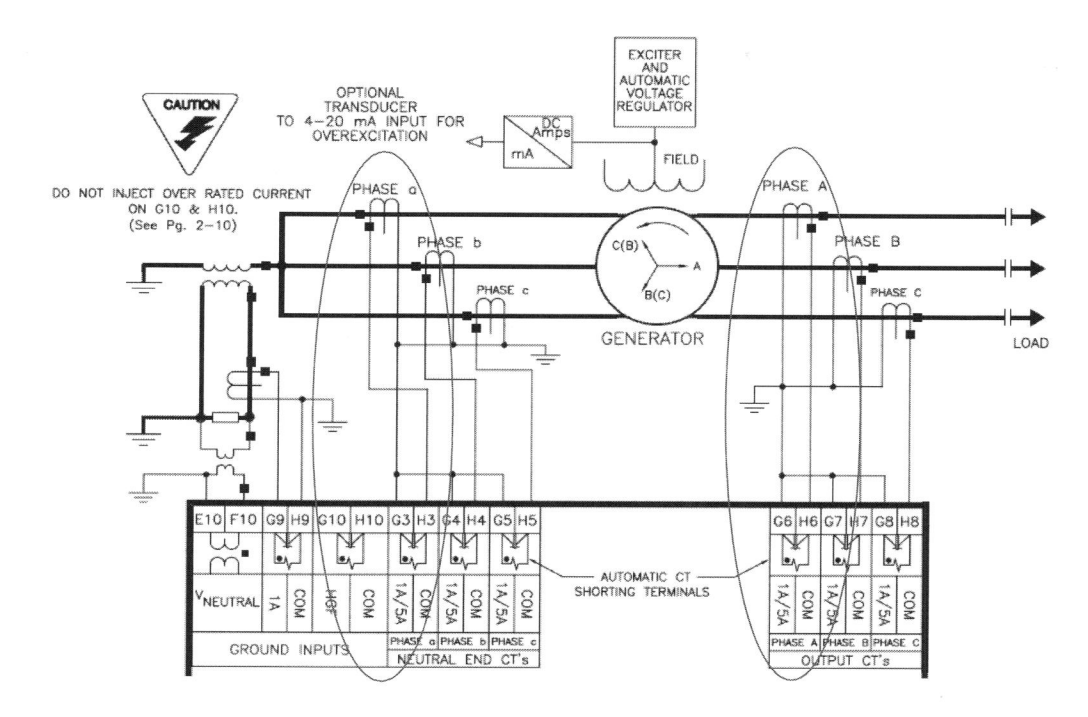

Figure 12-23: GE Power Management 489 Specifications

TEST-SET	A-PHASE	B-PHASE	C-PHASE
Channel 1 +	489 - H3	489 - H4	489 - H5
Channel 1 -	489 - G3	489 - G4	489 - G5
Channel 2 +	489 - H6	489 - H7	489 - H8
Channel 2 -	489 - G6	489 - G7	489 - G8

Figure 12-24: GE Power Management 489 Slope Test Connection Table

Notice that the polarity marks of the CT's are in opposite directions. If we were to apply both currents into the relay at the same phase-angle, all of the applied current would be differential or operate current. We would not be able to perform a slope test because the slope will always be 200% as shown in Figure 12-25.

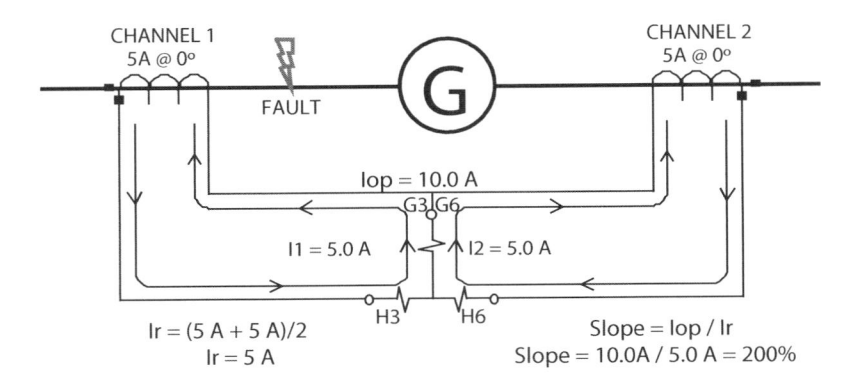

Figure 12-25: GE Power Management 489 Slope Test Connections Example #1

If this test was performed on a bench, the easy solution would be to switch the H3-G3 or H6-G6 connections. This quick fix isn't always possible when testing installed relays so we change the Channel 2 phase-angle to 180° to achieve the same result. If you look at Figure 12-26 carefully, you'll see that this is the normal running configuration for the GE 489 relay.

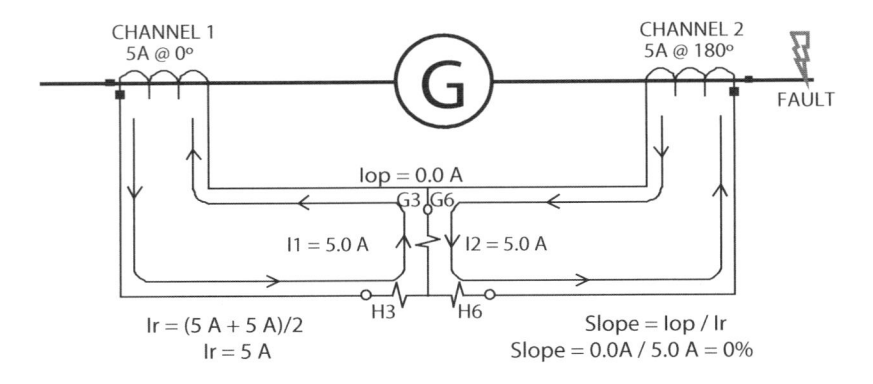

Figure 12-26: GE Power Management 489 Slope Test Connections Example #2

A) Test-Set Connections

The test-set connections for a simple slope test are as follows. Some tests require large amounts of current which will require the connection in Figure 12-28.

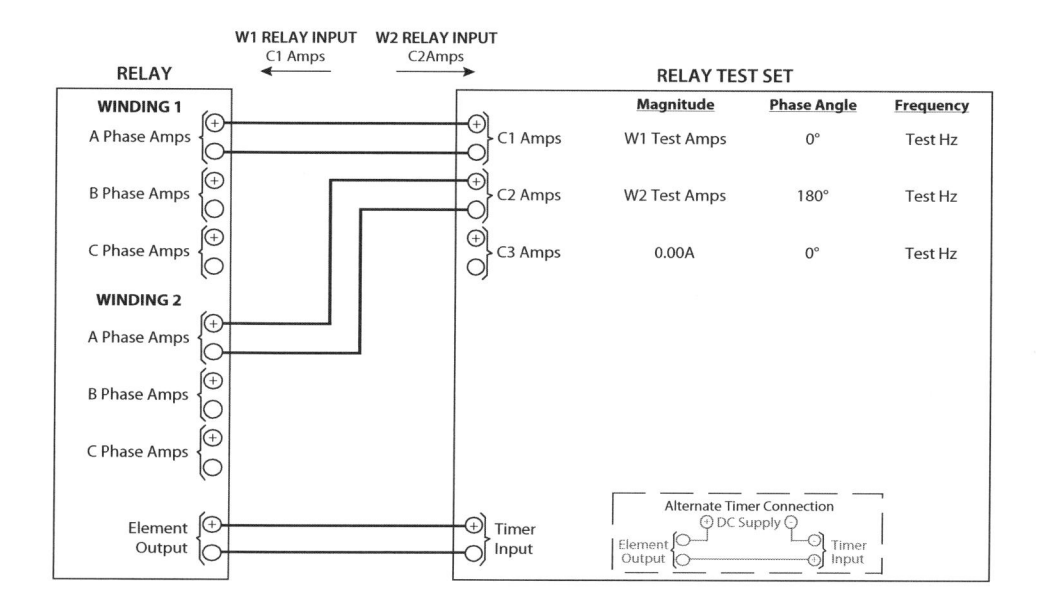

Figure 12-27: Simple 87-Element Slope Test-Set Connections

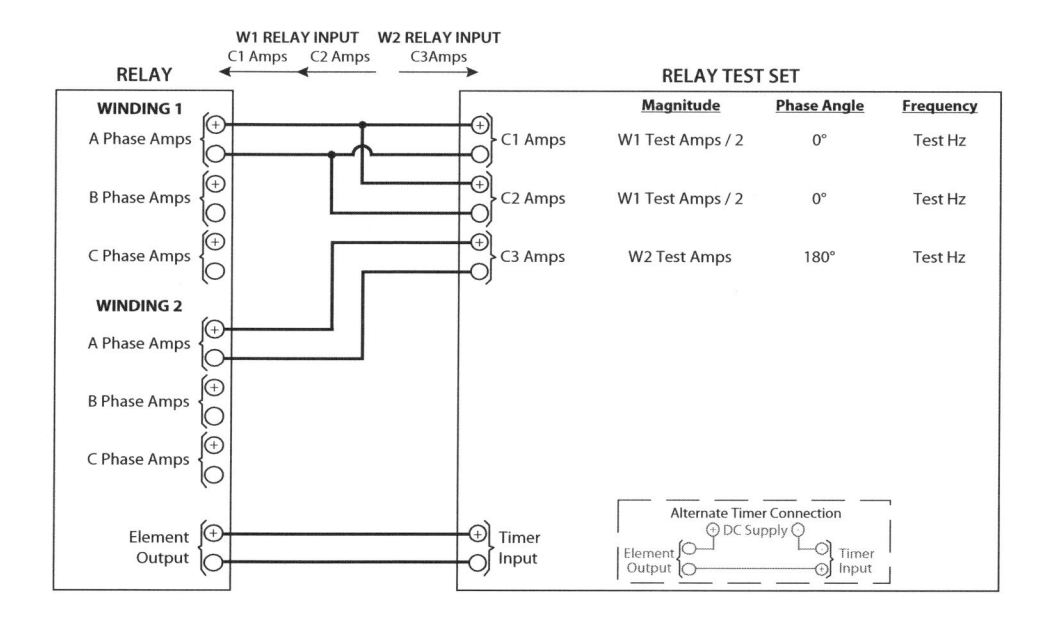

Figure 12-28: Simple 87-Element High Current Slope Test-Set Connections

B) Slope Test Procedure

The test procedure seems straightforward. Apply equal current into Winding-1 and Winding-2 with a 180° phase shift between windings, and raise one current until the relay operates. The difficult part of this procedure is determining what starting current to apply and what the expected pickup should be. We can start by re-plotting the characteristic curve shown in Figure 12-29 using the following relay settings.

- Phase Differential Trip = Unlatched
- Assign Trip Relays = 1
- Differential Trip Minimum Pickup = 0.1 x CT
- Differential Trip Slope-1 = 20%
- Differential Trip Slope-2 = 80%
- Differential Trip Delay = 0 cycles

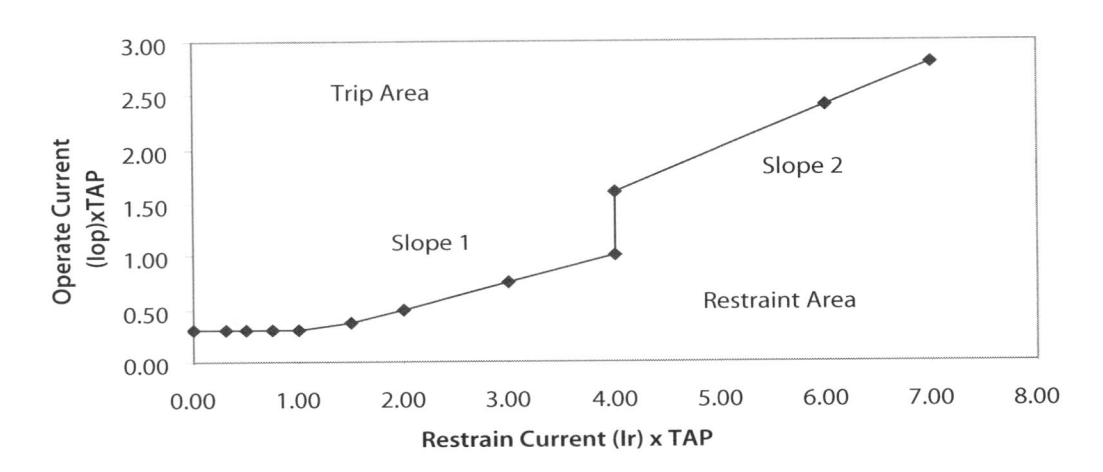

Differential Protection 20/80% Characteristic Curve

Figure 12-29: Percentage Differential Protection Dual Slope Characteristic Curve

We first need to change the restraint and operate current values to amps instead of multiples of Tap. Then we can calculate the transition point between the Minimum Pickup and Slope-1 operation. The Minimum Pickup is 0.50A ($0.1 \times$ CT) as we calculated earlier in this chapter. Slope-1 will be enabled when the operate current is greater than the minimum pickup. Different relays have different slope calculations and we can refer to the GE 489 relay instruction manual to determine the relay's differential calculation as shown in Figure 12-30.

FUNCTION:
The 489 percentage differential element has dual slope characteristic. This allows for very sensitive settings when fault current is low and less sensitive settings when fault current is high, more than 2 x CT, and CT performance may produce erroneous operate signals. The minimum pickup value sets an absolute minimum pickup in terms of operate current. The delay can be fine tuned to an application such that it still responds very fast, but rides through normal operational disturbances.

The Differential element for phase A will operate when:
$$I_{Operate} > k \times I_{Restraint}$$

$$I_{Operate} = \overline{I_A} - \overline{I_a}$$

$$I_{Restraint} = \frac{\left| I_A \right| - \left| I_a \right|}{2}$$

where: $I_{Operate}$ = operate current

$I_{Restraint}$ = restraint current

k = characteristic slope of differential element in percent
(Slope 1 if I_R < 2 X CT, Slope 2 if I_R > = 2 X CT)

I_A = phase current measured at the output CT

I_a = phase current measured at the neutral end CT

Differential elements for phase B and phase C operate in the same manner.

Figure 12-30: GE 489 Differential Formulas

Notice that the Ir current ($I_{Restraint}$) is defined as $\frac{\left| I_A \right| + \left| I_a \right|}{2}$ which can be translated into $\frac{\left| I_{W1} \right| + \left| I_{W2} \right|}{2}$ for our terminology. The bars on either side of each current indicate that we use absolute values when calculating the restraint current. We can drop the bars and re-define Ir as $\frac{I_{W1} + I_{W2}}{2}$. The Iop current ($I_{Operate}$) is defined as $\overline{I_A} - \overline{I_a}$ which can be translated into $\overline{I_{W1}} - \overline{I_{W2}}$ for our calculations. The bars above each current indicate a vector sum calculation and we can redefine the Iop calculation as $I_{W1} - I_{W2}$ because our vectors will always be 180° apart.

Use the first $I_{Operate}$ formula to calculate the transition between Minimum Pickup and Slope-1 where Slope-1 begins when Iop exceeds the Minimum Pickup setting (0.5A) defined previously. Notice that our generic formula on the right also works for this relay. Based on these calculations, our restraint current must be greater than 2.505A to test Slope-1.

$Iop > k \times Ir$	$20\% = 100 \times \dfrac{0.501}{Ir}$
$0.501 = 20\% \times Ir$	$Ir = \dfrac{100 \times 0.501}{20\%} = \dfrac{50.1}{20} = 2.505$
$Ir = \dfrac{0.501}{0.2} = 2.505$	$Ir = 2.505\ A$

Figure 12-31 displays the revised characteristic curve so far.

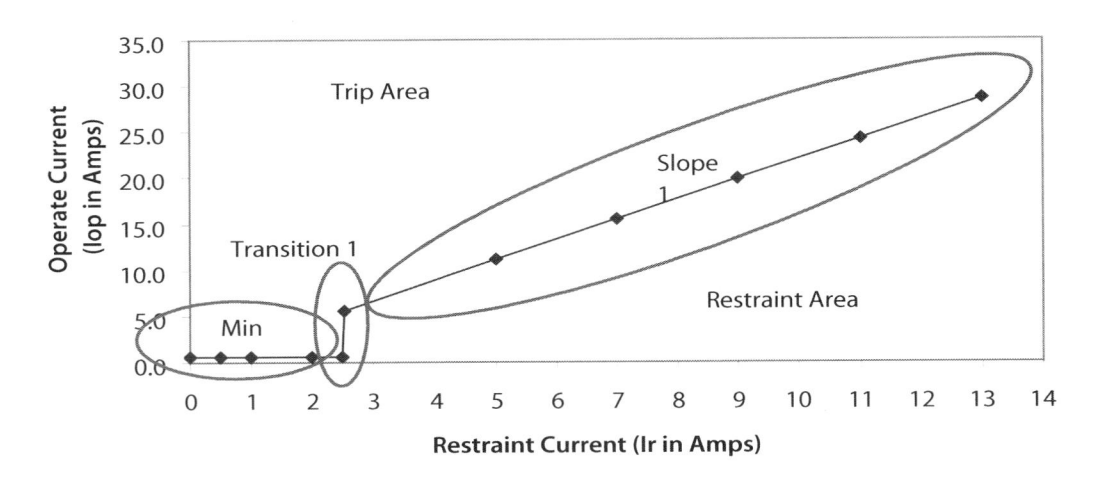

Figure 12-31: Example Characteristic Curve in Amps with 1ˢᵗ Transition Defined

We also need to determine the maximum restraint current before we accidentally start testing Slope-2 when testing Slope-1. We can use the slope formula ($Slope(\%) = 100 \times \dfrac{Iop}{Ir}$)

to calculate the restraint current (Ir) at the transition point between Slope-1 and Slope-2 but we first need to define the transition point. This is usually the Breakpoint setting, but this setting is not available in the GE 489 relay. A quick review of the manufacturer's operate description (Figure 12-32) of the manual informs us that the breakpoint is pre-defined as 2 times CT.

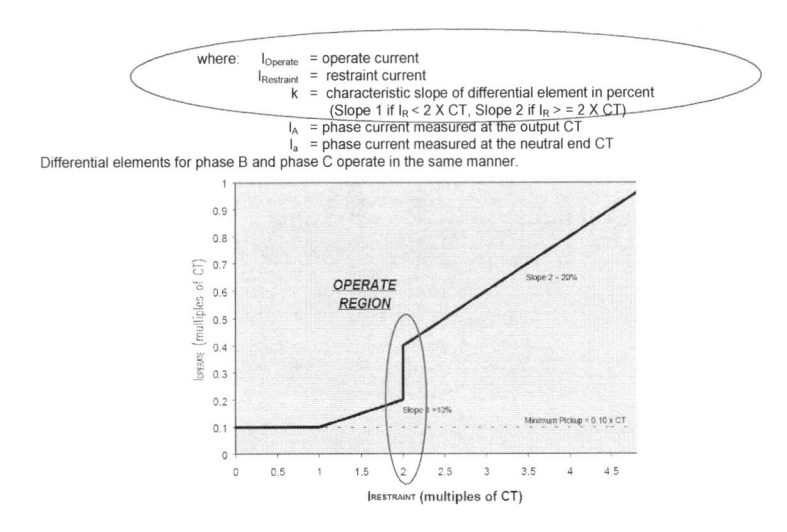

Figure 12-32: Percentage Differential Protection Dual Slope Characteristic Curve

You can use the same Minimum Pickup formulas to calculate the transition from Slope-1 to Slope-2. This time we're going to expand the formula to determine the actual W1 and W2 currents instead of operate and restraint currents.

$$Iop > k \times Ir$$

$$\left[I_{w1} - I_{w2}\right] = 0.2 \times \left[\frac{I_{w1} + I_{w2}}{2}\right]$$

$$\left[I_{w1} - I_{w2}\right] = \frac{0.2I_{w1} + 0.2I_{w2}}{2}$$

$$2 \times \left[I_{w1} - I_{w2}\right] = 0.2I_{w1} + 0.2I_{w2}$$

$$2I_{w1} - 2I_{w2} = 0.2I_{w1} + 0.2I_{w2}$$

$$2I_{w1} - 0.2I_{w1} = 0.2I_{w2} + 2I_{w2}$$

$$1.8I_{w1} = 2.2I_{w2}$$

$$I_{w1} = \frac{2.2I_{w2}}{1.8}$$

$$I_{w1} = 1.222I_{w2}$$

The maximum amount of Ir is defined as Ir<10A

$$Ir = \frac{I_{w1} + I_{w2}}{2}$$

$$9.99 = \frac{1.222I_{w2} + I_{w2}}{2}$$

$$19.98 = 2.222I_{w2}$$

$$8.99 = I_{w2}$$

Slope 1 expected test currents can be defined by:

$$I_{w1} = 1.222_{w2}$$

$$I_{w2} = \frac{I_{w1}}{1.222}$$

$$Slope(\%) = 100 \times \frac{Iop}{Ir} = 100 \times \frac{\left[I_{w1} - I_{w2}\right]}{\left[\frac{I_{w1} + I_{w2}}{2}\right]}$$

$$Slope(\%) = 100 \times \frac{2 \times \left[I_{w1} - I_{w2}\right]}{\left[I_{w1} + I_{w2}\right]}$$

$$Slope(\%) = 100 \times \frac{\left[2I_{w1} - 2I_{w2}\right]}{\left[I_{w1} + I_{w2}\right]}$$

$$Slope(\%) \times \left[I_{w1} + I_{w2}\right] = 100 \times \left[2I_{w1} - 2I_{w2}\right]$$

$$20 \times \left[I_{w1} + I_{w2}\right] = 100 \times \left[2I_{w1} - 2I_{w2}\right]$$

$$\left[20I_{w1}\right] + \left[20I_{w2}\right] = \left[200I_{w1} - 200I_{w2}\right]$$

$$20I_{w1} + 20I_{w2} = 200I_{w1} - 200I_{w2}$$

$$20I_{w1} + 20I_{w2} - 200I_{w1} = -200I_{w2}$$

$$20I_{w1} - 200I_{w1} = -200I_{w2} - 20I_{w2}$$

$$-180I_{w1} = -220I_{w2}$$

$$I_{w1} = \frac{220I_{w2}}{180}$$

$$I_{w1} = 1.222I_{w2}$$

The maximum amount of Ir is defined as Ir<10A

$$Ir = \frac{I_{w1} + I_{w2}}{2}$$

$$9.99 = \frac{1.222I_{w2} + I_{w2}}{2}$$

$$19.98 = 2.222I_{w2}$$

$$8.99 = I_{w2}$$

The new characteristic curve will look like Figure 12-33.

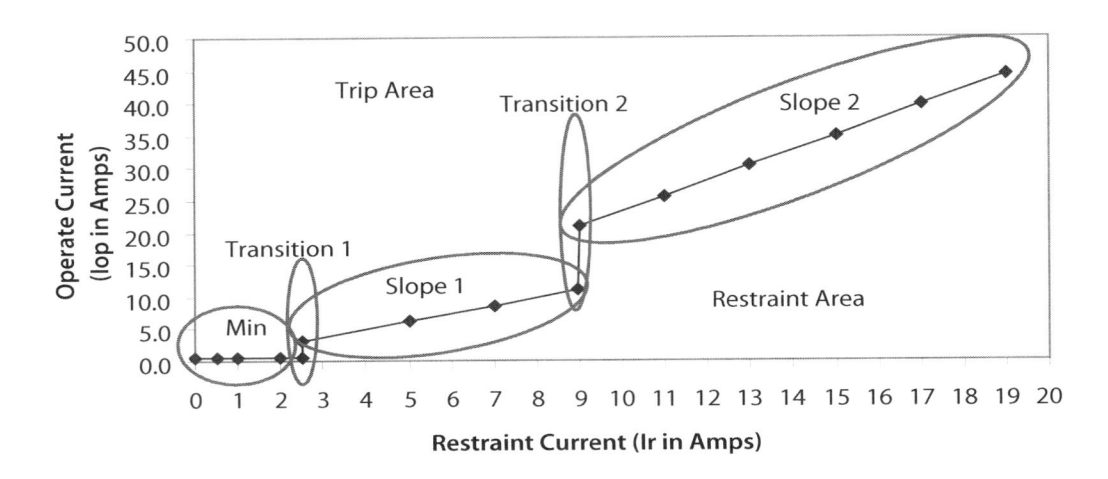

Differential Protection 20/80% Characteristic Curve

Figure 12-33: Percentage Differential Protection Dual Slope Characteristic Curve

Now we can define the expected test currents when Slope-2 is required.

$$Iop > k \times Ir$$

$$\left[I_{w1} - I_{w2}\right] = 0.8 \times \left[\frac{I_{w1} + I_{w2}}{2}\right]$$

$$\left[I_{w1} - I_{w2}\right] = \frac{0.8I_{w1} + 0.8I_{w2}}{2}$$

$$2 \times \left[I_{w1} - I_{w2}\right] = 0.8I_{w1} + 0.8I_{w2}$$

$$2I_{w1} - 2I_{w2} = 0.8I_{w1} + 0.8I_{w2}$$

$$2I_{w1} - 0.8I_{w1} = 0.8I_{w2} + 2I_{w2}$$

$$1.2I_{w1} = 2.8I_{w2}$$

$$I_{w1} = \frac{2.8I_{w2}}{1.2}$$

$$I_{w1} = 2.333I_{w2}$$

Slope 2 expected test currents can be defined by:

$$I_{w1} = 2.333I_{w2}$$

$$I_{w2} = \frac{I_{w1}}{2.333}$$

$$Slope(\%) = 100 \times \frac{Iop}{Ir} = 100 \times \frac{\left[I_{w1} - I_{w2}\right]}{\left[\frac{I_{w1} + I_{w2}}{2}\right]}$$

$$Slope(\%) = 100 \times \frac{2 \times \left[I_{w1} - I_{w2}\right]}{\left[I_{w1} + I_{w2}\right]}$$

$$Slope(\%) = 100 \times \frac{\left[2I_{w1} - 2I_{w2}\right]}{\left[I_{w1} + I_{w2}\right]}$$

$$Slope(\%) \times \left[I_{w1} + I_{w2}\right] = 100 \times \left[2I_{w1} - 2I_{w2}\right]$$

$$80 \times \left[I_{w1} + I_{w2}\right] = 100 \times \left[2I_{w1} - 2I_{w2}\right]$$

$$\left[80I_{w1}\right] + \left[80I_{w2}\right] = \left[200I_{w1} - 200I_{w2}\right]$$

$$80I_{w1} + 80I_{w2} = 200I_{w1} - 200I_{w2}$$

$$80I_{w1} + 80I_{w2} - 200I_{w1} = -200I_{w2}$$

$$80I_{w1} - 200I_{w1} = -200I_{w2} - 80I_{w2}$$

$$-120I_{w1} = -280I_{w2}$$

$$I_{w1} = \frac{280I_{w2}}{120}$$

$$I_{w1} = 2.333I_{w2}$$

The actual characteristic curve of this 87-element using Winding-1 and Winding-2 currents can be plotted where W2 (Restraint Current) is a group of arbitrary numbers and W1 (Operate Current) uses the following formulas:

- $If(I_{W2} < 8.99A \text{ and } Iop < 0.5A) \text{ then } I_{W1} = 0.5 + I_{W2}$
- $If(I_{W2} < 8.99A \text{ and } Iop > 0.5A) \text{ then } I_{W1} = 1.222I_{W2}$
- $If(I_{W2} > 10.0A) \text{ then } I_{W1} = 2.333I_{W2}$
- The spreadsheet calculation for these equations could be:
  $$I_{W1} = IF(I_{W2} < 8.99, IF((I_{W2} * 1.222) - I_{W2} < 0.5, 0.5 + I_{W2}, I_{W2} * 1.222), I_{W2} * 2.333)$$

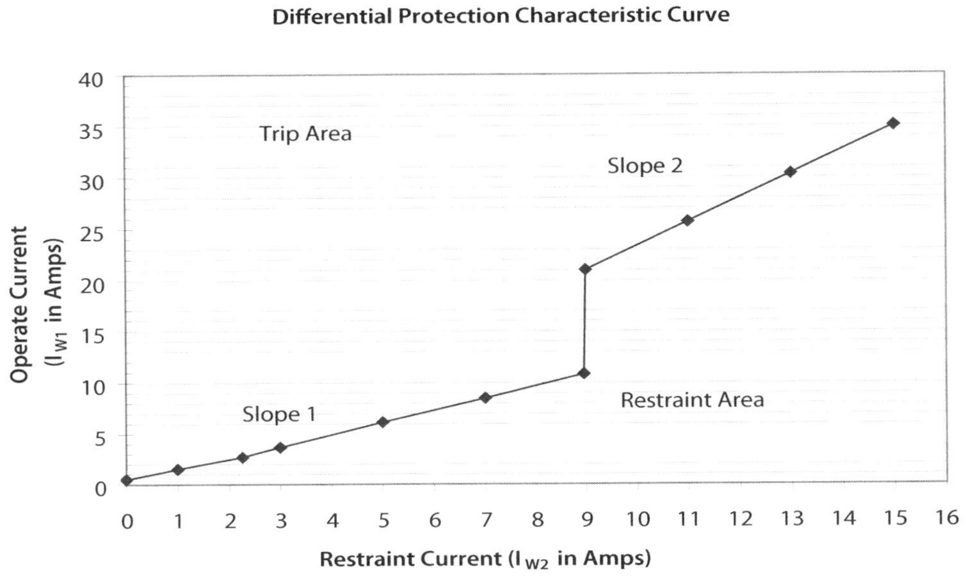

Differential Protection Characteristic Curve

Figure 12-34: Percentage Differential Protection Dual Slope Characteristic Curve in Amps

You could use this graph to determine the expected pickup current for a Slope-1 test by choosing a W2 current greater than the Minimum Pickup and less than Slope-2. Find that current on the x-axis, follow the current up until it crosses the line, then follow the crossover point back to the other axis to determine the expected pickup. For example, we could test Slope-1 by applying 5.0A to both windings and increase the W1 current until the relay operates. The relay should operate at approximately 6.0A as shown in Figure 12-35. Test Slope-2 by choosing a current below the Slope-2 portion of the graph (12.0A for example), follow 12.0A from the x-axis to the characteristic curve, and follow it to the y-axis. The expected pickup current for the Slope-2 test is 28A as per Figure 12-35.

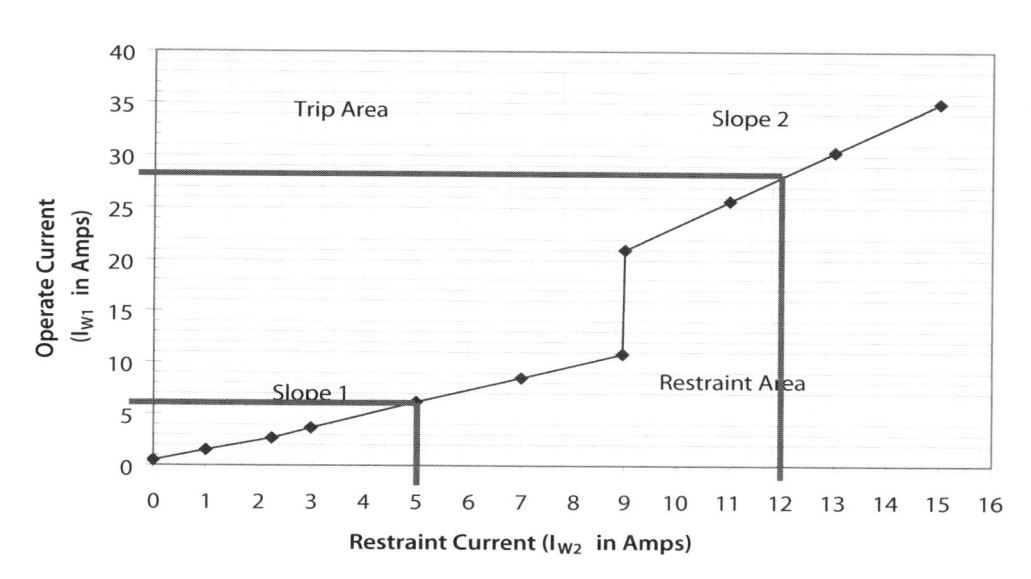

Differential Protection Characteristic Curve

Figure 12-35: Using Graphs to Determine Pickup Settings

It can be more convenient (and accurate) to use the formulas to determine the expected pickup current as shown below.

SLOPE-1	SLOPE-2
$I_{W1} = 1.222I_{W2}$	$I_{W1} = 2.333I_{W2}$
$I_{W1} = 1.222(5.0A)$	$I_{W1} = 2.333(12.0A)$
$I_{W1} = 6.11A$	$I_{W1} = 27.996A$

Follow these steps to test the Differential Slope settings:

1. Determine how you will monitor pickup and set the relay accordingly, if required. (Pickup indication by LED, output contact, front panel display, etc. See the *Relay Test Procedures* section starting on page 109 for details.)

2. Determine the Slope-1 restraint current by selecting a value between the Minimum-Pickup and Slope-2 transition points. Apply the restraint current through W2. The pickup indication should be ON because we have applied a 200% slope as per the calculations earlier in this chapter. The W2 current for our example will be 5.0A.

3. Apply identical current 180° from W2 into W1. The pickup indication should turn OFF because there is no operate current as the two applied currents cancel each other out.

4. Raise the W1 current until the element operates. This is the Slope-1 pickup. The measured pickup for our example is 6.175A.

5. Some organizations want more than one test point to determine slope. Repeat Steps 2-4 with another restraint current between the Minimum-Pickup and Slope-2 transition points until the required number of tests are completed.

6. Test Slope-2 by repeating Steps 2-4 with a restraint current greater than the Slope-1 to Slope-2 transition level. Remember that any test current greater than 10A can potentially damage the relay and should only be applied for the minimum time possible. You can switch to the pulse method to reduce the possibility of equipment damage. (Review the *Pickup Test Procedure if Pickup is Greater Than 10 Amps* section starting on page 204 for details.)

7. Compare the pickup test result to the manufacturer's specifications and calculate the percent error as shown in the following example.

Slope-1

In our example, we calculated that the 87-element should operate when W1 = 6.11A using the Slope-1 formula when W2 = 5.0A. We measured the pickup to be 6.175A. Looking at the "OUTPUT AND NEUTRAL END CURRENT INPUTS" specification in Figure 12-36, we see that the acceptable metering error is 0.05A per applied current or 0.10A including all applied currents ("Accuracy: @ < 2 x CT: ± 0.5% of 2 x CT" = $0.5\% \times 10$ A = 0.05 A). The difference between the test result and calculated pickup is 0.065A (6.175A – 6.11A = 0.065A) and the relay passes the test based on this criteria.

OUTPUT AND NEUTRAL END CURRENT INPUTS	
Accuracy:	@ < 2 x CT: ± 0.5% of 2 x CT
	@ > 2 x CT: ± 1% of 20 x CT
OUTPUT RELAYS	
Operate Time:	10ms
PHASE DIFFERENTIAL	
Pickup Accuracy:	as per Phase Current Inputs
Timing Accuracy:	+50ms @ 50/60 Hz or ± 0.5 % of total time

Figure 12-36: GE Power Management 489 Specifications

We can also calculate the manufacturer's allowable percent error. Use "Accuracy: @ < 2 x CT: ± 0.5% of 2 x CT" from the specifications in Figure 12-36 because the applied current is less than 2x CT (20A), the allowable percent error is 1.6%.

$$\frac{\text{Maximum Accuracy Tolerance}}{\text{Setting}} \times 100 = \text{Allowable Percent Error}$$

$$\frac{2 \times \left[0.5\% \times (2 \times CT)\right]}{6.11A} \times 100 = \text{Allowable Percent Error}$$

$$\frac{2 \times \left[0.005 \times (2 \times 5A)\right]}{6.11A} \times 100 = \text{Allowable Percent Error}$$

$$\frac{2 \times 0.005 \times 10A}{6.11A} \times 100 = \text{Allowable Percent Error}$$

$$\frac{0.10A}{6.11A} \times 100 = \text{Allowable Percent Error}$$

1.6% Allowable Percent Error

The measured percent error can be calculated using the percent error formula:

$$\frac{\text{Actual Value - Expected Value}}{\text{Expected Value}} \times 100 = \text{Percent Error}$$

$$\frac{6.175A - 6.11A}{6.11 \text{ A}} \times 100 = \text{Percent Error}$$

1.06% Error

Slope-2

In our example, we calculated that the 87-element should operate when W1 = 27.996A using the Slope-2 formula when W2 = 12.0A. We measured the pickup to be 28.712A. Looking at the "Output And Neutral End Current Inputs" specification in Figure 12-36, we see that the acceptable metering error is 1.00A per applied current or 2.00A including all applied currents ("Accuracy: @ > 2 x CT: ± 1.0% of 20x CT"=1% x 20 x 5A = 1.00A). The difference between the test result and calculated pickup is 0.716A (28.712A – 27.996A = 0.716A) and the relay passes the test based on this criteria.

We can also calculate the manufacturer's allowable percent error. Use "Accuracy: @ 2 x CT: ± 1.0% of 20 x CT" from the specifications in Figure 12-36 because the applied current is greater than 2 x CT (20A), the allowable percent error is 7.1%.

$$\frac{\text{Maximum Accuracy Tolerance}}{\text{Setting}} \times 100 = \text{Allowable Percent Error}$$

$$\frac{2 \times [1.0\% \times (20 \times CT)]}{27.996A} \times 100 = \text{Allowable Percent Error}$$

$$\frac{2 \times [0.01 \times (20 \times 5A)]}{27.996A} \times 100 = \text{Allowable Percent Error}$$

7.1% Allowable Percent Error

The measured percent error can be calculated using the percent error formula:

$$\frac{\text{Actual Value - Expected Value}}{\text{Expected Value}} \times 100 = \text{Percent Error}$$

$$\frac{28.712A - 27.996A}{27.996\ A} \times 100 = \text{Percent Error}$$

2.56% Error

Rule of Thumb

Remember that there are other factors that will affect the test result such as:

- Relay close time: The longer it takes the contact to close the higher the test result will be if ramping the pickup current.
- Test-set analog output error.
- Test-set sensing time.

Some of these error factors can be significant and a rule-of-thumb 5% error is usually applied to test results.

8. Repeat the pickup test for all phase currents that are part of the differential scheme. If more than two windings are used, change all connections and references to W2 to the next winding under test (W2 becomes W3 for W1-W3 tests).

SLOPE 1 TESTS (Amps)							
RESTRAINT	TEST	A PHASE (A)	B PHASE (A)	C PHASE (A)	MFG (A)	% ERROR	
W1 5 A W2		6.175	6.175	6.175	6.110	1.06	1.06	1.06
W1 8 A W2		9.896	9.899	9.902	9.776	1.23	1.26	1.29
W1 NA W3		NA	NA	NA	NA			

SLOPE 2 TESTS (Amps)							
RESTRAINT	TEST	A PHASE (A)	B PHASE (A)	C PHASE (A)	MFG (A)	% ERROR	
W1 12 A W2		28.712	28.712	28.712	27.996	2.56	2.56	2.56
W1 15 A W2		36.082	36.083	36.085	34.995	3.11	3.11	3.11
W1 NA W3		NA	NA	NA	NA			

Alternate Slope Calculation

There is another way to test slope that does not require as much preparation. Follow the same steps as the previous test and test 2 points on each slope and record the results.

Determine slope using the rise-over-run graphical method as shown in Figure 12-37 and the following formulas. Remember to use the $I_{OPERATE}$ ($I_{W1} - I_{W2}$) and $I_{RESTRAINT}$ ($\frac{I_{W1} + I_{W2}}{2}$) formulas.

Differential Protection Characteristic Curve

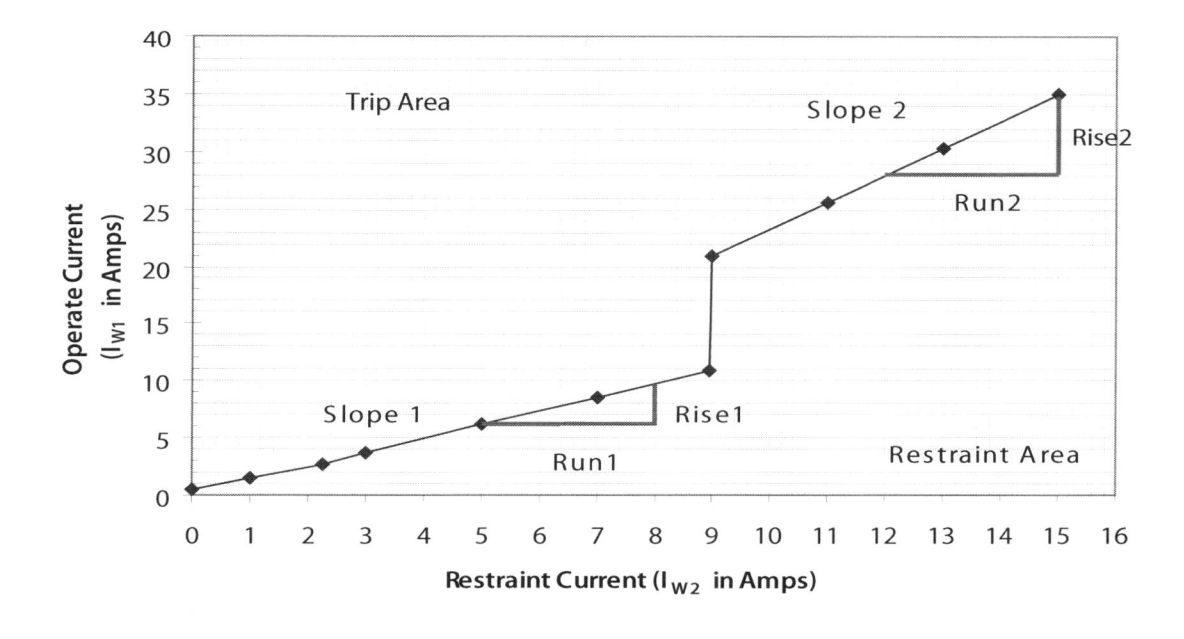

Figure 12-37: Determine Slope by Rise/Run Calculation

$$\%Slope1 = 100 \times \frac{Rise}{Run}$$

$$\%Slope1 = 100 \times \frac{I_{OPERATE}2 - I_{OPERATE}1}{I_{RESTRAINT}2 - I_{RESTRAINT}1}$$

$$\%Slope1 = 100 \times \frac{(I_{W1}2 - I_{W2}2) - (I_{W1}1 - I_{W2}1)}{\left(\frac{I_{W1}2 + I_{W2}2}{2}\right) - \left(\frac{I_{W1}1 + I_{W2}1}{2}\right)}$$

$$\%Slope1 = 100 \times \frac{(9.90A - 8A) - (6.175A - 5A)}{\left(\frac{9.90A + 8A}{2}\right) - \left(\frac{6.175A + 5A}{2}\right)}$$

$$\%Slope1 = 100 \times \frac{1.896A - 1.175A}{8.948A - 5.5875A}$$

$$\%Slope1 = 100 \times \frac{0.721}{3.3605}$$

$$Slope1 = 21.5\%$$

$$\%Slope2 = 100 \times \frac{Rise}{Run}$$

$$\%Slope2 = 100 \times \frac{I_{OPERATE}2 - I_{OPERATE}1}{I_{RESTRAINT}2 - I_{RESTRAINT}1}$$

$$\%Slope2 = 100 \times \frac{(I_{W1}2 - I_{W2}2) - (I_{W1}1 - I_{W2}1)}{\left(\frac{I_{W1}2 + I_{W2}2}{2}\right) - \left(\frac{I_{W1}1 + I_{W2}1}{2}\right)}$$

$$\%Slope2 = 100 \times \frac{(36.08A - 15A) - (28.71A - 12A)}{\left(\frac{36.08A + 15A}{2}\right) - \left(\frac{28.71A + 12A}{2}\right)}$$

$$\%Slope2 = 100 \times \frac{21.082A - 16.712A}{25.541A - 20.356A}$$

$$\%Slope2 = 100 \times \frac{4.37}{5.185}$$

$$\%Slope2 = 100 \times 0.843$$

$$Slope2 = 84.3\%$$

SLOPE 1 TESTS (Amps)										
RESTRAINT TEST		A PHASE (A)	B PHASE (A)	C PHASE (A)	SLOPE (%)			% ERROR		
W1 5.0 W2		6.175	6.175	6.175						
W1 8.0 W2		9.896	9.899	9.902	21.5	21.5	21.6	7.28	7.67	8.07
W1 NA W3		NA	NA	NA						
SLOPE 2 TESTS (Amps)										
RESTRAINT TEST		A PHASE (A)	B PHASE (A)	C PHASE (A)	SLOPE (%)			% ERROR		
W1 12.0 W2		28.712	28.712	28.712						
W1 15.0 W2		36.082	36.083	36.085	84.3	84.3	84.3	5.35	5.37	5.39
W1 NA W3		NA	NA	NA						

6. Tips and Tricks to Overcome Common Obstacles

The following tips or tricks may help you overcome the most common obstacles.

- Before you start, apply current at a lower value and review the relay's measured values to make sure your test-set is actually producing an output and your connections are correct.
- 180° apart?
- If the pickup tests are off by a $\sqrt{3}$, 1.5, or 0.577 multiple; the relay probably is applying correction factors and the test procedures in the next chapter should be used.
- Are you applying the currents into the same phase relationships?
- Did you calculate the correct Tap?
- Don't forget to use the entire operate and restrained calculations.

Chapter 13

Percent Differential (87) Element Testing

The previous chapter discussed differential protection on a single-phase basis, which is fine when discussing the simplest differential protective elements typically used for generator or motor protection. There are several new problems to be resolved when differential protection is applied to transformers which add complexity to our testing procedure:

- The simple definition of differential protection (current in equals current out) no longer applies (now power in equals power out) because the primary and secondary sides of the transformer have different voltage levels. Different voltage levels create different CT ratios and we now have to compensate for different Tap values on either side of the protected system.
- The situation becomes more complex with Delta-Wye or Wye-Delta transformers. The phase-angle on the primary and secondaries of the transformer are no longer 180 degrees apart. Electromechanical relays solved this problem by changing the CT connections to compensate for the phase shift, but most digital relays use Wye-connected CTs and the phase shift is compensated by relay settings.
- Transformer differential relays are designed to ignore the large inrush current that occur when a transformer is initially energized.
- The relay should trip if we close into a fault. Additional protection is applied to operate quickly to clear the fault.

1. Application

We will start with a description of zones of protection typically used in protection schemes.

A) Zones of Protection

Zones of protection are defined by strategic CT placement in an electrical system and there are often overlapping zones to provide maximum redundancy. The differential protection should operate when a fault occurs inside the zone and ignore faults outside the zone. The following Figure 13-1 depicts a typical electrical system with multiple zones of protection.

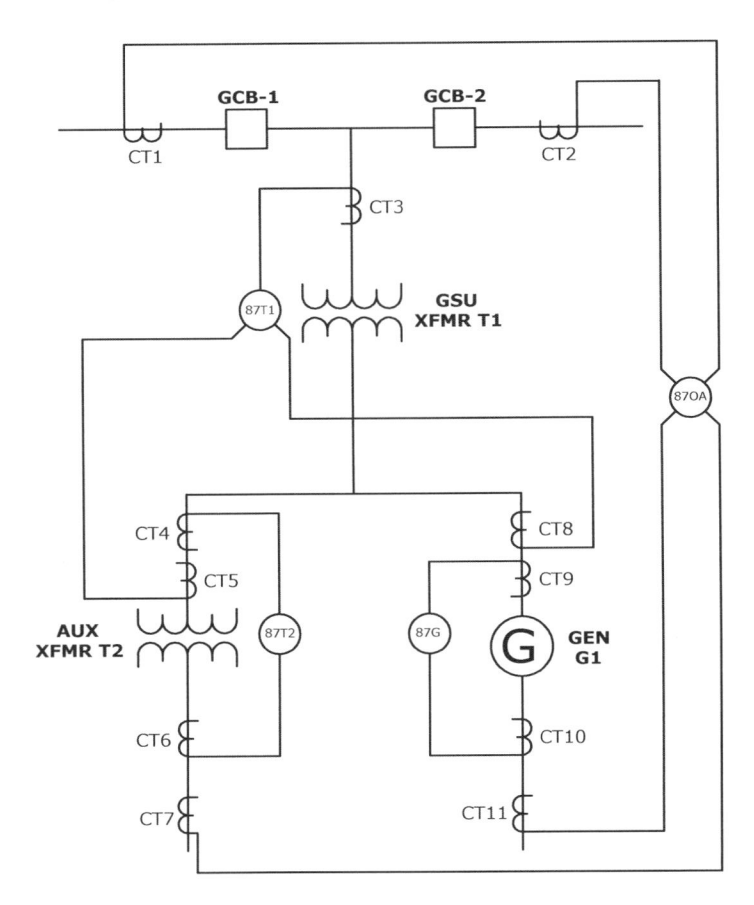

Figure 13-1: Zones of Protection Example

There are four differential relays in Figure 13-1 that protect each major piece of equipment and the overall plant.

- Relay 87OA is an overall differential relay with 4 inputs that will operate for any fault within the plant as defined by the locations of its input CTs; CT1, CT2, CT7, and CT11.

- GSU XFMR T1 is protected by differential relay 87T1 that will operate if a fault is detected between CT3, CT5, and CT8.
- 87T2 protects AUX XFMR T2 between CT4 and CT6. The overlap between CT4 and CT5 provides 100% protection between the two transformers. Conversely, the gap between CT8 and CT9 means that a fault in that gap will only trip the 87OA relay and will be difficult to locate.

B) Tap

i) Simple Differential Protection

The previous chapter demonstrated differential protection on a single-phase basis which should be applied to all three phases as shown in Figure 13-2. This is the same protection but displayed on a 3 phase basis. Because there is no voltage difference or phase shift between windings, we were able to isolate one phase at a time to perform our testing. The CT ratios on both sides were identical and we could ignore the Tap setting as well.

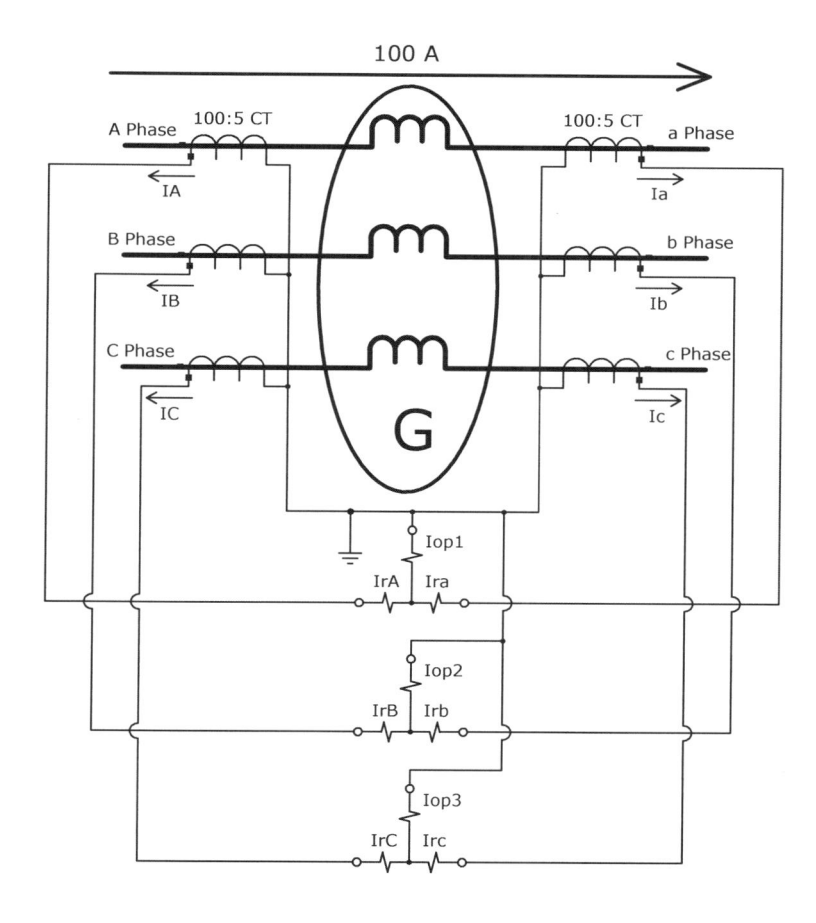

Figure 13-2: 3 Phase Generator Differential Protection

Tap settings are very important when more complicated differential schemes are applied such as the 115,000V to 34,500V; 30 MVA transformer depicted in Figure 13-3.

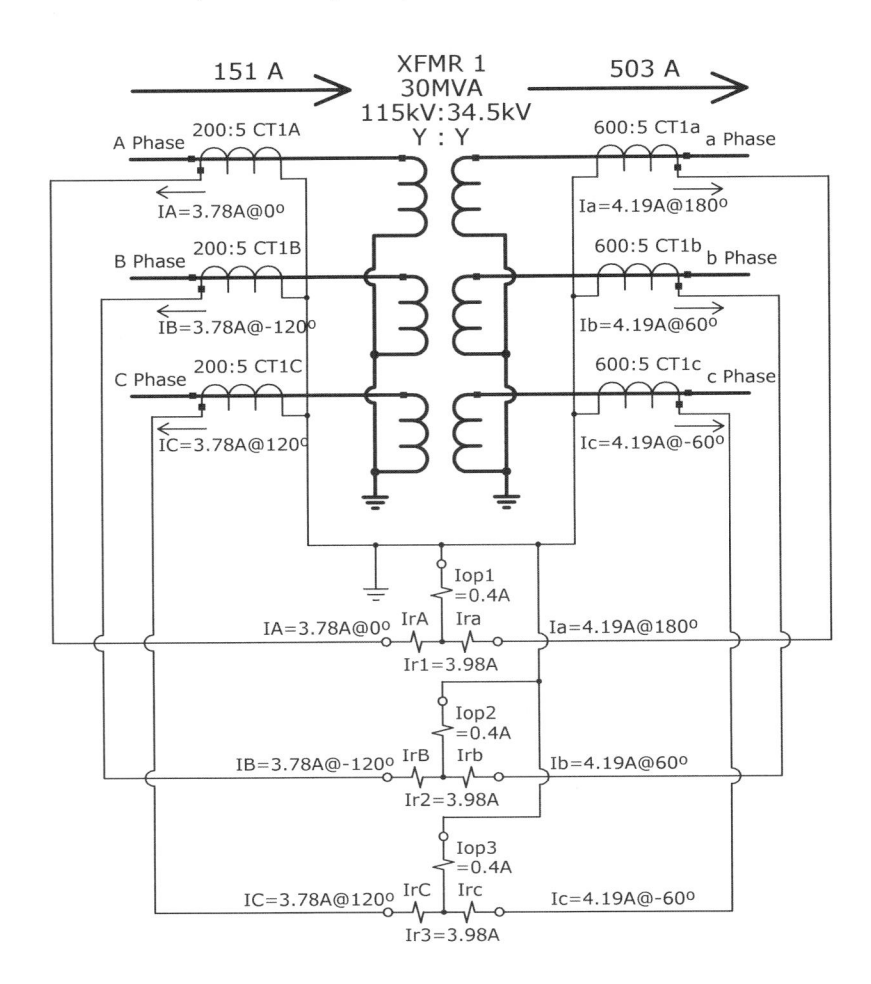

Figure 13-3: 3 Phase Transformer Differential Protection

A quick review of Figure 13-3 (that does not apply Tap settings) shows that the Iop current under normal conditions in each phase is 0.4A. If we apply the percent-slope formula from the previous chapter, we discover that there is approximately a 10% slope under normal operating conditions.

$$\%\text{Slope} = 100 \times \frac{I_{OP}}{I_{Restraint}} = 100 \times \frac{0.4}{3.98} = 10.05\%$$

Remember that these are ideal conditions and there is an additional 10% allowable CT error under symmetrical conditions and 20% error under asymmetrical conditions. We can apply Winding Taps to compensate for the different nominal currents.

Figure 13-4 displays the same protection with a 3.80A Tap setting on Winding-1 and 4.2A on Winding-2. Different Tap settings are applied, so it is now easier to use % of Tap or per-unit for all of our differential calculations.

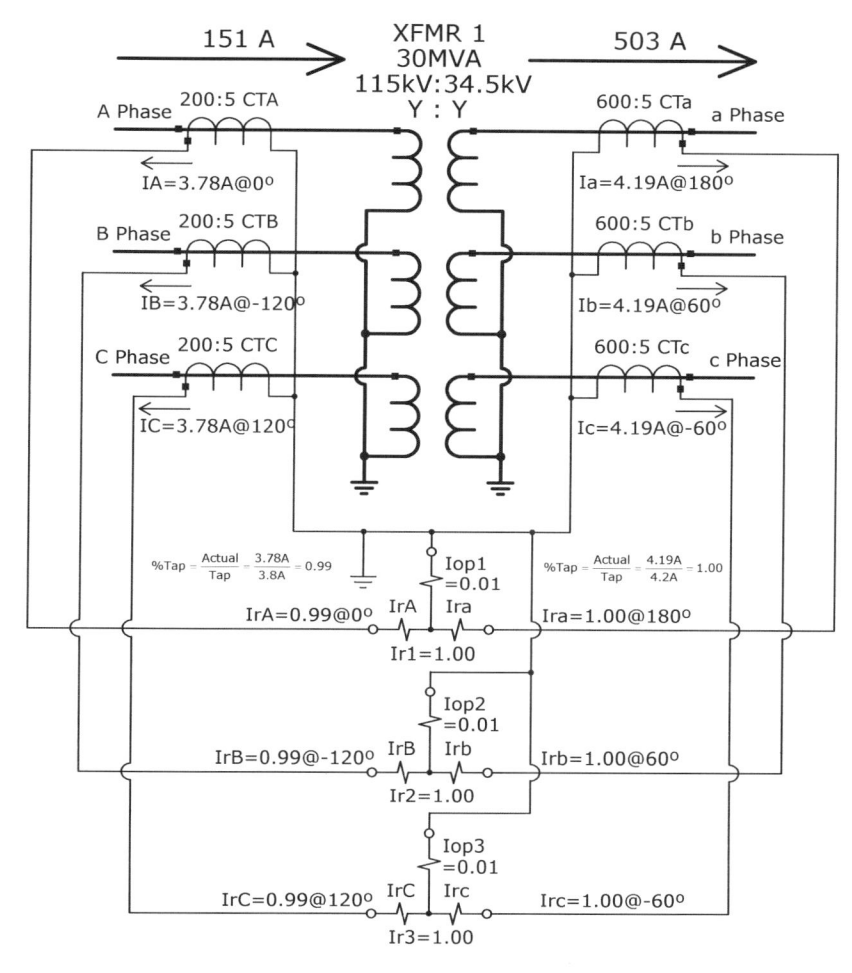

Figure 13-4: 3 Phase Transformer Differential Protection Using Tap Settings

Now the measured slope is less than 1% under normal conditions in an ideal situation.

$$\%\text{Slope} = 100 \times \frac{I_{OP}}{I_{Restraint}} = 100 \times \frac{0.01}{1.00} = 0.01\%$$

Notice that all of the numbers in the slope calculation do not include A or amps because these numbers are now related in percent of Tap or per-unit (PU). Understanding the Tap settings and correctly applying them to your differential calculations is vital when testing more complicated differential protection.

ii) Transformer Differential Protection with E-M Relays

Unfortunately, most transformers do not have Wye-Wye connections and many have Delta-Wye or Wye-Delta Connections. The Wye-Delta transformer in Figure 13-5 has essentially the same characteristics (Voltage, MVA, etc) but the transformer secondary has been reconfigured for a delta connection which completely changes the differential protection scheme. We will go through all of the changes in Figure 13-5 on a step-by-step basis to help you understand all of the challenges and it does get quite involved. Remember that we will discuss much simpler methods that require less analysis later in the chapter.

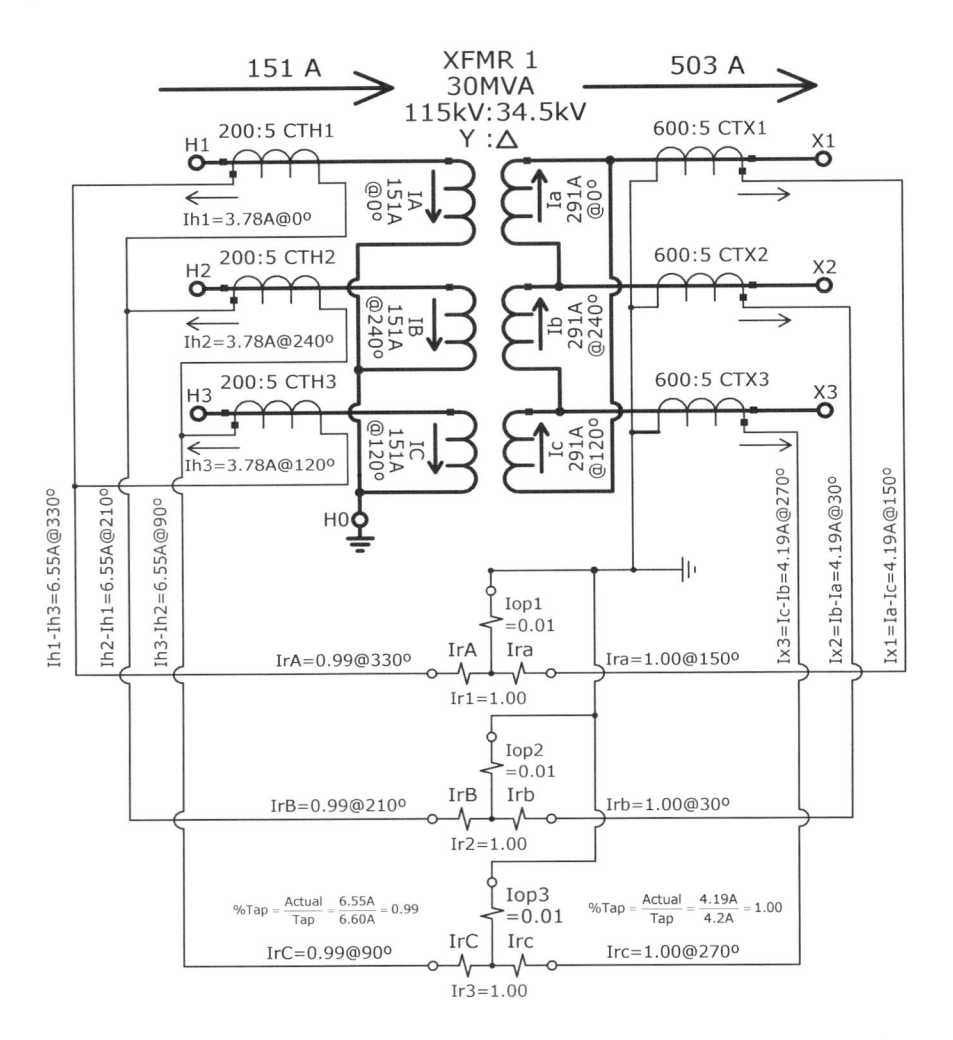

Figure 13-5: Wye-Delta Transformer Differential Protection Using CT Connections

Let's start with the primary connections. The High and Low side currents (and corresponding CT ratios) entering and leaving the transformer have not changed because the MVA and voltage ratings haven't changed. However, we had to relabel many of the measured currents because of the new connections. The current entering H1, H2, and H3 is now labeled with the capital H designation to indicate current measured at the primary bushings. The H-winding is connected in Wye and the primary winding currents equal the bushing current. (CTH1 = IA, CTH2 = IB, CTH3= IC).

The CT secondary currents have been relabeled with a lower case Ih1, Ih2, and Ih3 to indicate that they are the secondary currents of the bushing CTs. These CTs have been connected in Delta to compensate for the Delta-Wye shift of the transformer.

The current entering the differential relay's transformer primary restraint coils is labeled IrA, IrB, and IrC and is NOT the CT secondary current. The restraint current is the resultant of the delta connection. If you follow the circuit back from IrA, you will discover that IrA is connected to the polarity of Ih1 and the non-polarity of Ih3. IrB is connected to the polarity of Ih2 and the non-polarity of Ih1. IrC is connected to the polarity of Ih3 and the non-polarity of Ih2. We must add these currents together vectorially to determine the restraint currents. Add the currents connected to the polarity marks to the currents connected to the non-polarity marks rotated by 180° as shown below. Remember, if you add two vectors that have equal magnitudes and are 60° apart, you can determine the resultant vector by multiplying the magnitude by $\sqrt{3}$ and split the difference in angles

Ih1 + −Ih3 3.78 @ 0° + −(3.78 @ 120°) 3.78 @ 0° + (3.78 @ 120° + 180°) 3.78 @ 0° + 3.78 @ 300° 6.55 @ 330°	Ih3 = 3.78 Amp Ih1 = 3.78Amp
Ih2 + −Ih1 3.78 @ 240° + −(3.78 @ 0°) 3.78 @ 240° + (3.78 @ 0° + 180°) 3.78 @ 240° + 3.78 @ 180° 6.55 @ 210°	Ih1 = 3.78Amp
Ih3 + −Ih2 3.78 @ 120° + −(3.78 @ 240°) 3.78 @ 120° + (3.78 @ 240° + 180°) 3.78 @ 120° + 3.78 @ 420° 3.78 @ 120° + 3.78 @ 60° (420° − 360°) 6.55 @ 90°	Ih1 = 3.78Amp −Ih2 = 3.78 Amp IrA = 6.55 Amp

The primary winding Tap setting must change because the measured current at the relay is $\sqrt{3}$ larger due to the delta-connected CTs. The new Tap for the primary winding is 6.6A.

The transformer secondary winding is connected in Delta and, therefore, the CTs are connected in Wye to compensate for the phase shift between the primary and secondary windings. The transformer ratio is the same as the Wye-Wye example which means that the current magnitude flowing through the X-Bushings have not changed. The phase-angles and winding currents will change due to the Delta connection.

The Ia current magnitude can be calculated one of two ways. We can use logic and the Delta-Wye rule of thumb that states the current inside the delta is $\sqrt{3}$ smaller than the current leaving the delta. $Ia = \dfrac{IX1}{\sqrt{3}} = \dfrac{503A}{\sqrt{3}} = 290.4A$. Or we can calculate the Ia, I,b, and Ic

currents using the transformer ratio. (Remember that transformer ratios are listed as phase-to-phase values.)

$\dfrac{Ia}{IA} = \dfrac{VAN}{Vab}$	$\dfrac{Ib}{IB} = \dfrac{VBN}{Vbc}$	$\dfrac{Ic}{IC} = \dfrac{VCN}{Vca}$
$Ia = \dfrac{IA \times VAN}{Vab}$	$Ib = \dfrac{IB \times VBN}{Vbc}$	$Ic = \dfrac{IC \times VCN}{Vca}$
$Ia = \dfrac{151A@0° \times \left(VAB/\sqrt{3}\right)}{34.5kV}$	$Ib = \dfrac{151A@240° \times \left(VBC/\sqrt{3}\right)}{34.5kV}$	$Ic = \dfrac{151A@120° \times \left(VCA/\sqrt{3}\right)}{34.5kV}$
$Ia = \dfrac{151A@0° \times \left(115kV/\sqrt{3}\right)}{34.5kV}$	$Ib = \dfrac{151A@240° \times \left(115kV/\sqrt{3}\right)}{34.5kV}$	$Ic = \dfrac{151A@120° \times \left(115kV/\sqrt{3}\right)}{34.5kV}$
$Ia = 291A@0°$	$Ib = 291A@240°$	$Ic = 291A@120°$

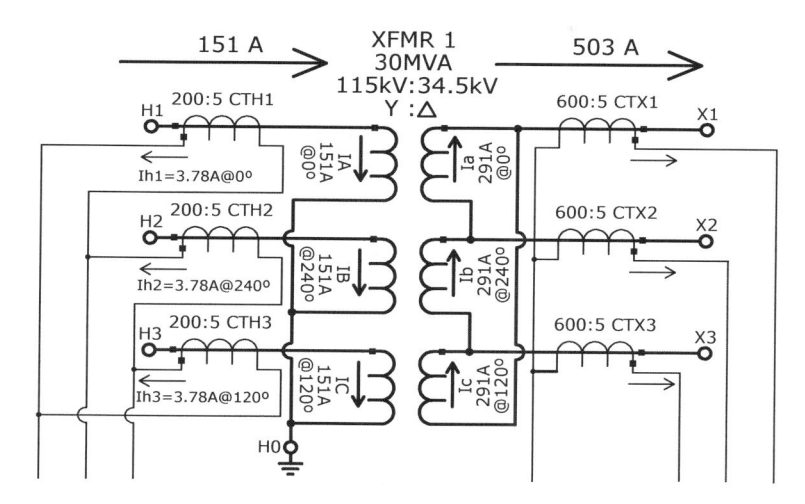

We can calculate the X-Bushing currents(Ia, Ib, and Ic) using the same calculations from the Delta CTs on the high side. IX1 is connected to Ia and –Ic, IX2 is connected to Ib and –Ia, and IX3 is connected to Ic and –Ib. The current is entering the non-polarity mark of the CTs and the CT secondary current (1x) will be the primary current divided by the CT ratio and rotated by 180°.

TRANSFORMER CURRENT	CT OUTPUT
$IX1 = Ia + -Ic$ $IX1 = 291@0° + -(291@120°)$ $IX1 = 291@0° + (291@120° + 180°)$ $IX1 = 291@0° + 291@300°$ $IX1 = 503@330°$	$Ix1 = \dfrac{1X1}{CT\ Ratio} + 180°$ $Ix1 = \dfrac{503A@330°}{(600/5)} + 180°$ $Ix1 = \dfrac{503A@330°}{120} + 180°$ $Ix1 = 4.19A@330° + 180°$ $Ix1 = 4.19A@510°$ $Ix1 = 4.19A@150°\ (510° - 360°)$
$IX2 = Ib + -Ia$ $IX2 = 291@240° + -(291@0°)$ $IX2 = 291@240° + (291@0° + 180°)$ $IX2 = 291@240° + 291@180°$ $IX2 = 503@210°$	$Ix2 = \dfrac{1X2}{CT\ Ratio} + 180°$ $Ix2 = \dfrac{503A@210°}{120} + 180°$ $Ix2 = 4.19A@210° + 180°$ $Ix2 = 4.19A@30°$
$IX3 = Ic + -Ib$ $IX3 = 291@120° + -(291@240°)$ $IX3 = 291@120° + (291@240° + 180°)$ $IX3 = 291@120° + 291@60°$ $IX3 = 503@90°$	$Ix3 = \dfrac{1X3}{CT\ Ratio} + 180°$ $Ix3 = \dfrac{503A@90°}{120} + 180°$ $Ix3 = 4.19A@90° + 180°$ $Ix2 = 4.19A@270°$

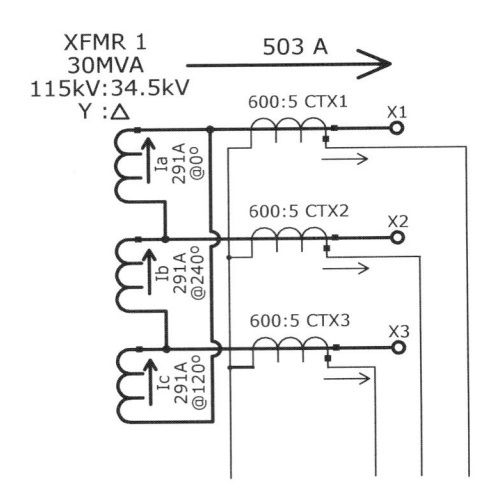

We can now calculate the restraint currents for each winding and phase and the operate current for each phase. Remember that all calculations are in percent of Tap or per-unit.

$IrA = \dfrac{Ih1 - Ih3}{Tap1}$	$IrB = \dfrac{Ih2 - Ih1}{Tap1}$	$IrC = \dfrac{Ih3 - Ih2}{Tap1}$
$IrA = \dfrac{6.55A @ 330°}{6.60A}$	$IrB = \dfrac{6.55A @ 210°}{6.60A}$	$IrC = \dfrac{6.55A @ 90°}{6.60A}$
$IrA = 0.99 @ 330°$	$IrB = 0.99 @ 210°$	$IrC = 0.99 @ 90°$
$Ira = \dfrac{Ix1}{Tap2}$	$Irb = \dfrac{Ix2}{Tap2}$	$Irc = \dfrac{Ix3}{Tap2}$
$Ira = \dfrac{4.19A @ 150°}{4.2A}$	$Irb = \dfrac{4.19A @ 30°}{4.2A}$	$Irc = \dfrac{4.19A @ 270°}{4.2A}$
$Ira = 1.00 @ 150°$	$Irb = 1.00 @ 30°$	$Irc = 1.00 @ 270°$
$Ir1 = \left\lvert \dfrac{IrA + Ira}{2} \right\rvert$	$Ir2 = \left\lvert \dfrac{IrB + Irb}{2} \right\rvert$	$Ir3 = \left\lvert \dfrac{IrC + Irc}{2} \right\rvert$
$Ir1 = \left\lvert \dfrac{0.99 + 1.00}{2} \right\rvert$	$Ir2 = \left\lvert \dfrac{0.99 + 1.00}{2} \right\rvert$	$Ir3 = \left\lvert \dfrac{0.99 + 1.00}{2} \right\rvert$
$Ir1 = \left\lvert \dfrac{1.99}{2} \right\rvert$	$Ir2 = \left\lvert \dfrac{1.99}{2} \right\rvert$	$Ir3 = \left\lvert \dfrac{1.99}{2} \right\rvert$
$Ir1 = 1.00$	$Ir2 = 1.00$	$Ir3 = 1.00$
$Iop1 = \vec{IrA} + \vec{Ira}$	$Iop2 = \vec{IrB} + \vec{Irb}$	$Iop3 = \vec{IrC} + \vec{Irc}$
$Iop1 = 0.99@330° + 1.00@150°$	$Iop2 = 0.99@210° + 1.00@30°$	$Iop3 = 0.99@90° + 1.00@270°$
$Iop1 = 0.01$	$Iop2 = 0.01$	$Iop3 = 0.01$

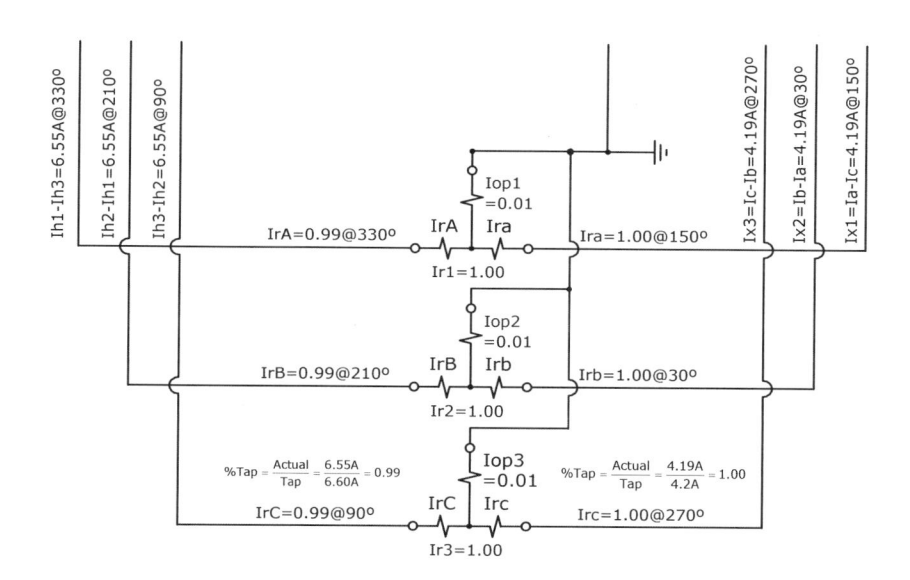

There is no reason to perform all three calculations when dealing with a balanced, 3-phase condition as described previously. As you can see, all the phases have the same operate and restraint currents. Most calculations assume a balanced, 3-phase condition and are usually only performed on A-phase.

iii) Transformer Differential Protection with Digital Relays

Most digital relays have correction factors and other algorithms to compensate for the phase shift between Delta-Wye transformers, and the Tap setting is relatively simple. The CTs are almost always connected in Wye and the Tap setting calculation is very similar to the simple differential protection described previously. Figure 13-6 displays the same transformer described in Figure 13-5 with a digital relay.

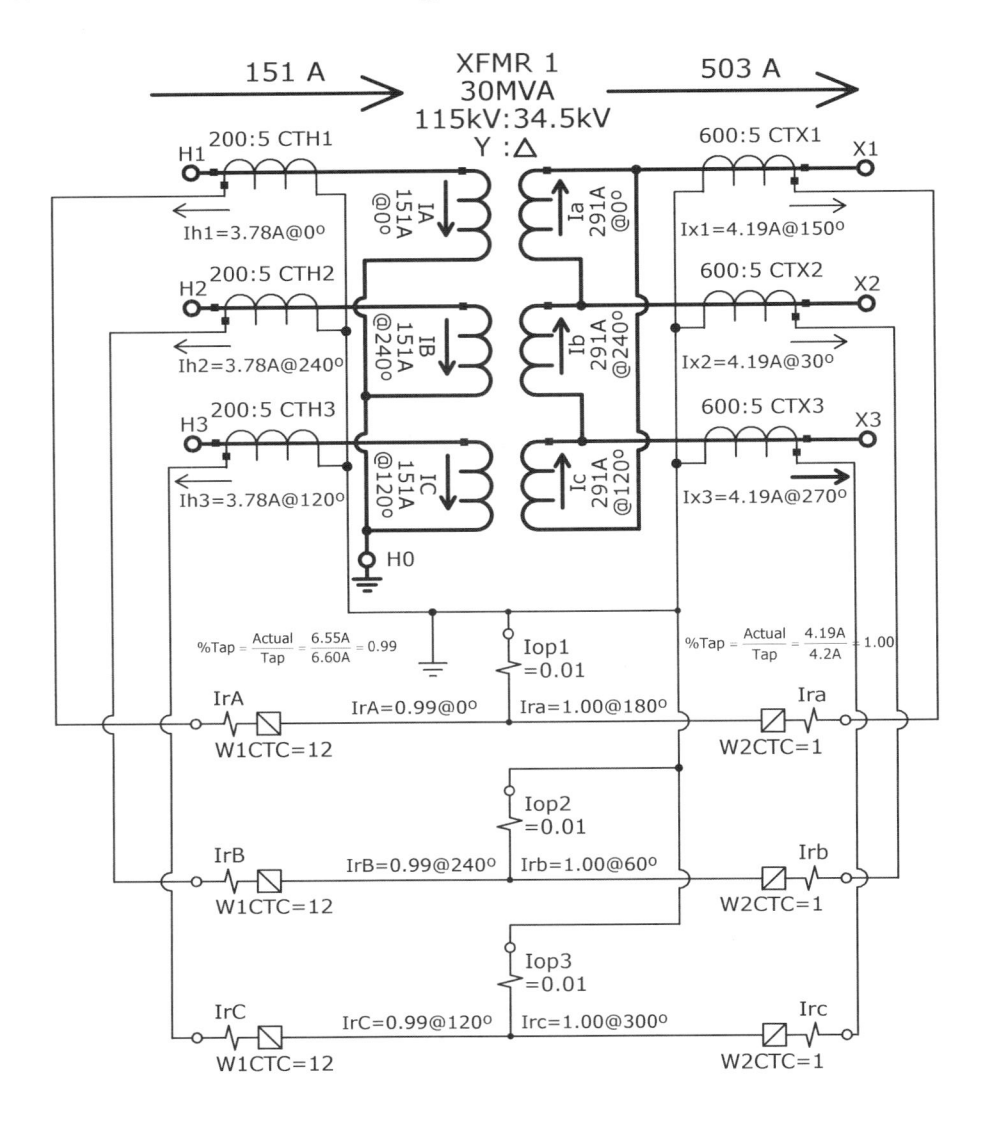

Figure 13-6: Wye-Delta Transformer Differential Protection Using CT Connections

Protecting this transformer with a digital relay has simplified the connections substantially. Both winding CTs are connected in Wye and are connected to the restrained coils (remember all of the coils have been replaced with digital algorithms now) and we use the same actual-current vs. Tap-current to determine the per-unit or percent-of-Tap current. The ◪ symbol represents the relay's phase compensation settings which are "W1CTC=12" and "W2CTC=1" for this application. (Don't worry, these settings will be explained in the next section.) The phase compensation angle replaces the different CT configurations required by electromechanical relays. The end result is the same with operate currents of 0.01A.

C) Phase-Angle Compensation

Modern relays use formulas to compensate for the phase shifts between windings on transformers and can even compensate for unwanted zero-sequence currents. All CTs are typically Wye-connected and the relay's settings are changed to match the application. Unfortunately there is no standard between relays or relay manufacturers and most problems in transformer differential applications occur trying to apply the correct compensation settings.

All phase-angle compensation settings are based on the transformer windings and their phase relationship using different codes depending on the manufacturer or relay model. The winding is always designated with a "D" or "d" for Delta-connected windings; and "Y" or "y" for Wye-connected windings.

The third designation is the phase shift between windings and can be designated by a connection description, phase-angle, or clock position. Beckwith Electric and older SEL relays use the connection description as shown in the following figures:

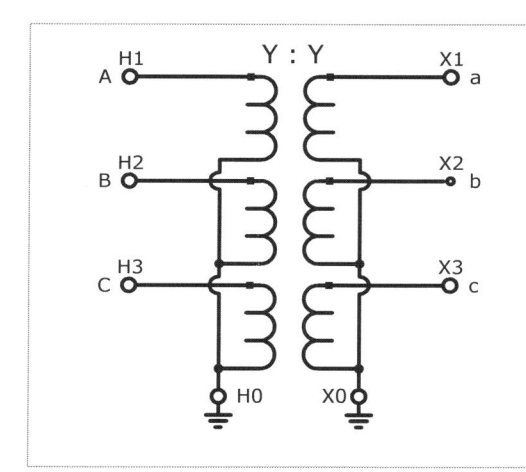

PHASE-ANGLE COMPENSATION

- "Y" is the primary designation for the Wye-connected winding.
- "Y" is the secondary designation for the Wye-connected winding.
- This is a "YY" transformer.

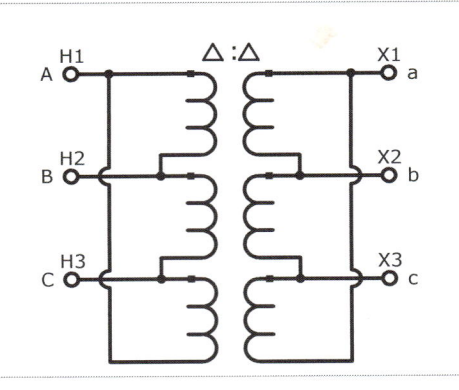

PHASE-ANGLE COMPENSATION

- "D" is the primary designation for the Delta-connected winding.
- "D" is the secondary designation for the Delta-connected winding.
- Both windings are "AC" because the H1 and X1 bushing is connected to A and C-phases.
- This is a "DACDAC" transformer.

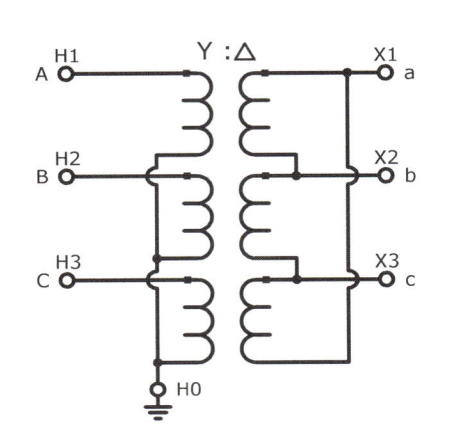

PHASE-ANGLE COMPENSATION IF H IS WINDING-1

- "Y" is the W1 designation.
- "D" is the W2 designation.
- "AC" is the third designation because the X1 bushing is connected to a and C-phases.
- This is a "YDAC" transformer.

PHASE=ANGLE COMPENSATION IF X IS W1

- "D" is the W1 designation.
- "AC" is the second designation because the W1 winding is Delta and the X1 bushing is connected to A and C-phases.
- "Y" is the W2 designation.
- This is a "DACY" transformer.

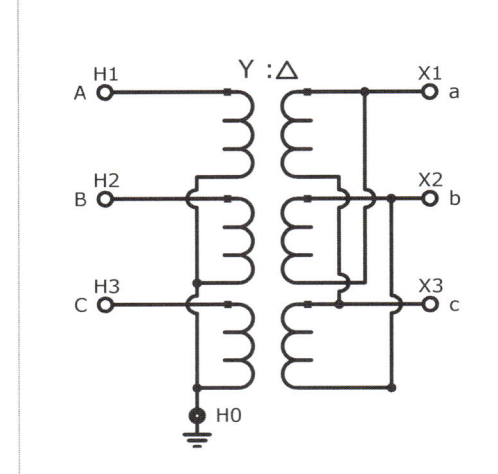

PHASE-ANGLE COMPENSATION IF H IS WINDING-1

- "Y" is the W1 designation.
- "D" is the W2 designation.
- "AB" is the third designation because the X1 bushing is connected to a and b phases.
- This is a "YDAB" transformer.

PHASE-ANGLE COMPENSATION IF X IS WINDING-1

- "D" is the W1 designation.
- "AB" is the second designation because the W1 winding is Delta and the X1 bushing is connected to a and b phases.
- "Y" is the W2 designation.
- This is a "DABY" transformer.

Some transformer differential relays define the phase relationship between windings in their settings using the actual phase displacement or clock positions as a simple reference. 0° is shifted to a vertical line to make the clock reference work, and Figures 44 and 45 display the phase-angle/clock references. It is extremely important to understand what reference the relay uses for phase-angles. Some relays, such as SEL, use a leading phase-angle reference as shown in Figure 13-7 where the angles increase in the counter-clockwise direction. Other relays, such as GE Multilin, use a lagging reference and the angles increase in the clockwise direction.

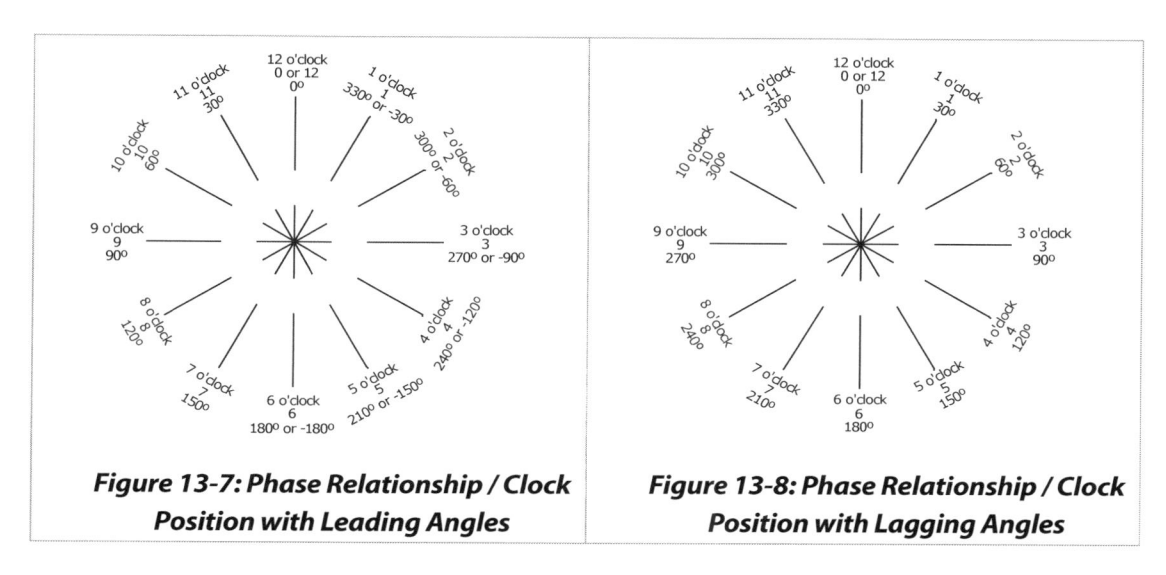

Figure 13-7: Phase Relationship / Clock Position with Leading Angles

Figure 13-8: Phase Relationship / Clock Position with Lagging Angles

Many people assume that the high voltage winding is the primary winding but this is not true. The primary winding could be the winding that is first energized [e.g. the grid (high voltage) side of a generator step-up transformer], the winding that is connected to the primary source [e.g. the generator (low voltage) side of a generator set-up transformer], or arbitrarily chosen by the design engineer. The primary winding in transformer differential protection is determined by the CT connections to the relay. The CT's connected to the following terminals in the most common relays are considered to be the primary windings to the relay:

- GE/Multilin SR-745—Terminals H1, H2, H3, G1, G2, G3
- GE T-60—f1a, f2a, f3a, f1b, f2b, f3b
- Beckwith Electric Co. Inc. M-3310—51, 49, 47, 50, 48, 46
- Schweitzer Engineering Laboratories SEL-587—101, 103, 105, 102, 104, 106
- Schweitzer Engineering Laboratories SEL-387—Z01, Z03, Z05, Z02, Z04, Z06

There is an easy and hard way to determine transformer compensation settings that can be gleaned from the transformer nameplate or the single and three line drawings. I always recommend looking at the nameplate instead of drawings to ensure you have the most accurate information.

1) The Hard Way

The hard way requires a three line drawing of the transformers and a little vector addition. We'll review the 6 most common transformer connections and determine the vector relationship between windings.

The first example is a Wye-Wye transformer with Wye-connected CTs as shown in the following example.

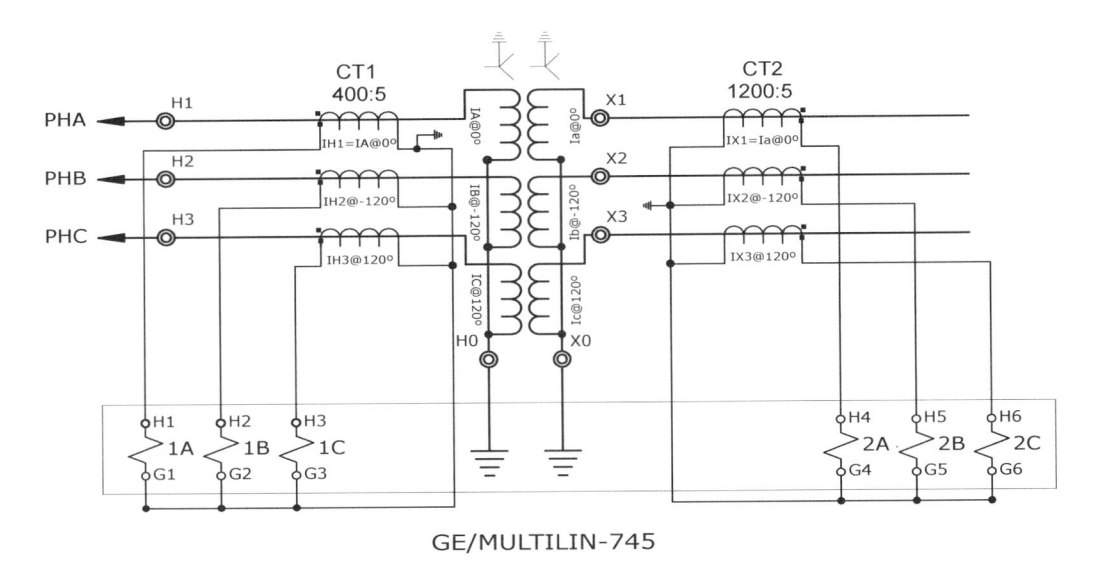

GE/MULTILIN-745

Figure 13-9: Wye-Wye Transformer

You should always start with the Wye-connected winding when determining phase relationships and the primary winding current phasors are plotted in the first vector diagram. Follow the current flowing through the H1 terminal into the IA winding. The currents are the same and all three phase currents (primary) are plotted in the first phasor diagram of Figure 13-9.

As you follow the current through the transformer's AØ primary winding, an opposite current flows out the aØ secondary winding and flows through the X1 bushing. The current flowing into the H1 bushing has the same phase relationship as the current flowing through the X1 Bushing. The secondary current is plotted in the second phasor diagram 180° from the actual current flowing into the relay to make the relationship between windings easier to understand.

Only the H1 and X1 currents are used when determining the vector relationships and are plotted in the third phasor diagram. Using this information, we can determine the transformer vector relationship is Y0y0; "Y" for the primary winding connection, "y" for the secondary connection, and "0" to show that both windings are at the 0 o'clock position. We usually drop the first 0 and the correct notation would be "Yy0".

It is important to note that if the phasor diagrams were displayed on the actual relay software, 1 o'clock would be 30° and not -30° as shown because this relay uses the phase relationship in Figure 13-10. All examples in this book use the Figure 13-10 characteristic to better illustrate the difference between connections.

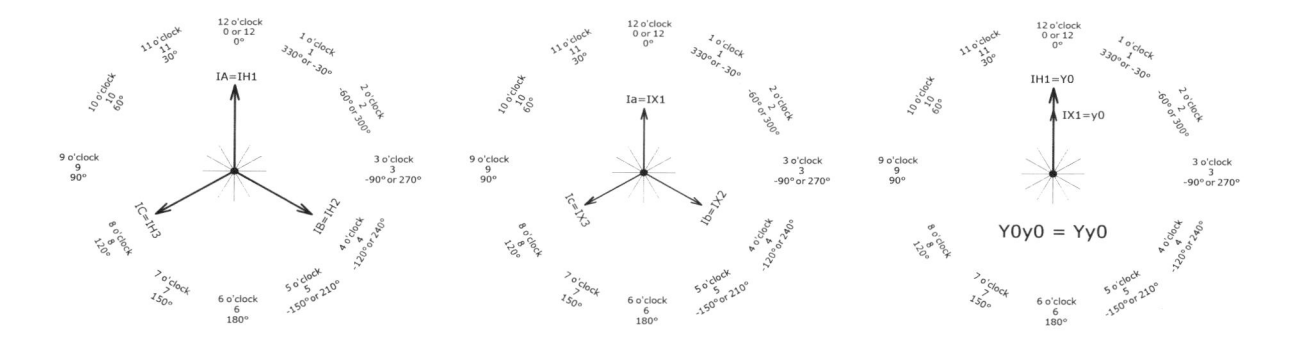

Figure 13-10: Wye-Wye Transformer Phasor Diagram

Our next example uses a Delta-Delta transformer. There is no Wye-connected winding in this transformer configuration and we will start with the primary winding as defined by the CTs connected to F1a, F2a, F3a, F1b, F2b, and F3b.

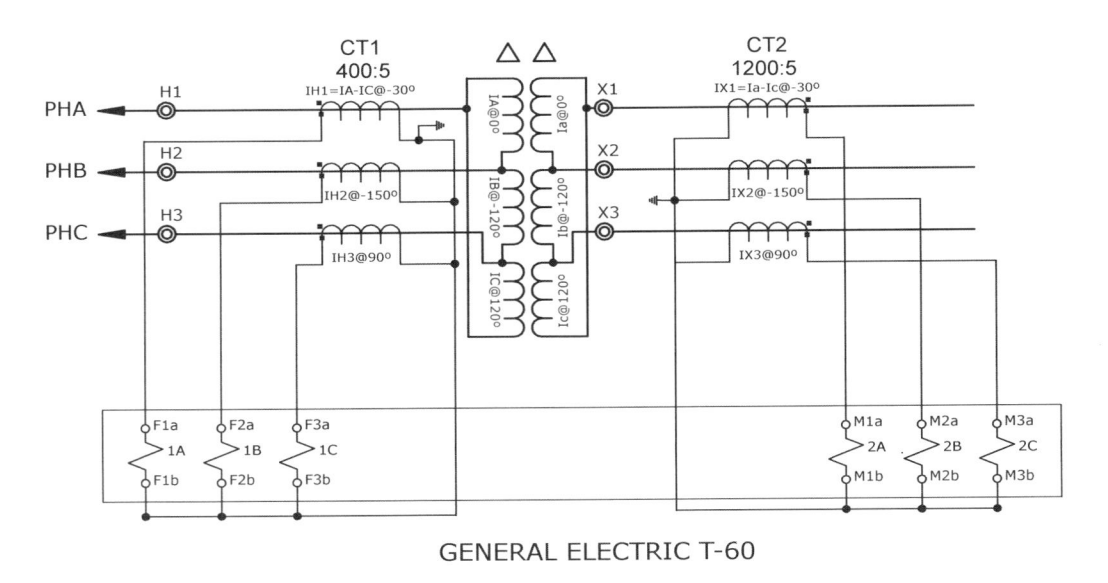

GENERAL ELECTRIC T-60

Figure 13-11: Delta-Delta Transformer Connections

The current flowing through H1 is the transformer "IA and -IC" currents because H1 is connected to the polarity of the AØ winding and the non-polarity of the CØ winding. The first phasor diagram demonstrates the IA and -IC addition and the final result "IH1" is found at the 1 o'clock position or -30°.

The current flowing through X1 is "Ia and -Ic" because X1 is connected to the polarity of the aØ winding and the non-polarity of the cØ winding. The secondary current is plotted in the second phasor diagram 180° from the actual current flowing into the relay to make the relationship between windings easier to understand. The second phasor diagram demonstrates the Ia and -Ic addition and the final result "IX1" is found at the 1 o'clock position or -30°.

Only the H1 and X1 currents are used when determining the vector relationships and are plotted in the third phasor diagram. Using this information, we can determine the transformer vector relationship is D1d1; "D" for the primary winding connection, "d" for the secondary connection, and "1" to show that both windings are at the 1 o'clock position. It is important to note that if the phasor diagrams were displayed on the actual relay software, 1 o'clock would be 30° and not -30° as shown because this relay uses the phase relationship in Figure 13-12. All examples in this book use the Figure 13-11 characteristic to better illustrate the difference between connections.

The primary winding should always be at the 0 or 12 o'clock position and we can rotate both phasors as shown in the fourth phasor diagram to display a "D0d0" transformer. The first "0" is usually dropped from the designation and the Delta-Delta connection is displayed as "Dd0."

You could also change the reference instead of the phasors as shown in the fifth phasor diagram.

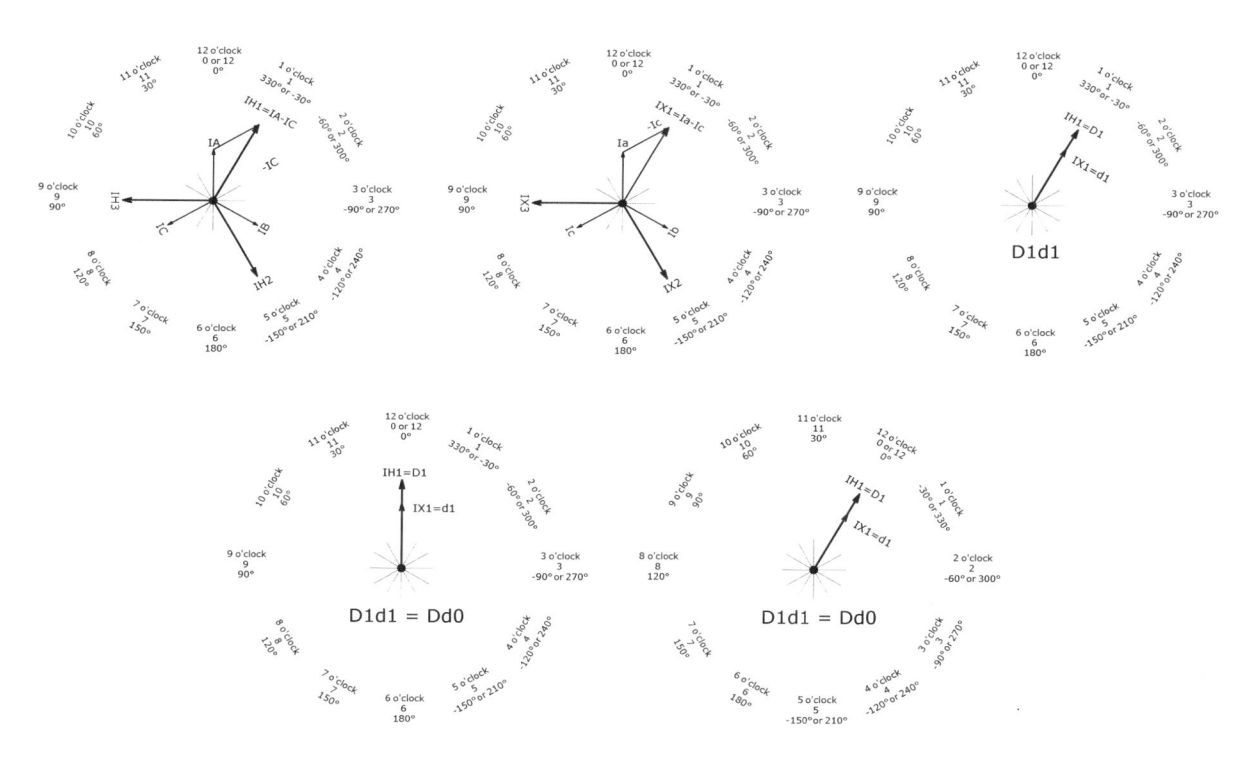

Figure 13-12: Delta-Delta Transformer Phasors

The next transformer is a Wye-Delta transformer with a Wye side primary because the Wye side, H-Bushing CTs are connected to 47, 49, 51, 46, 48, and 50.

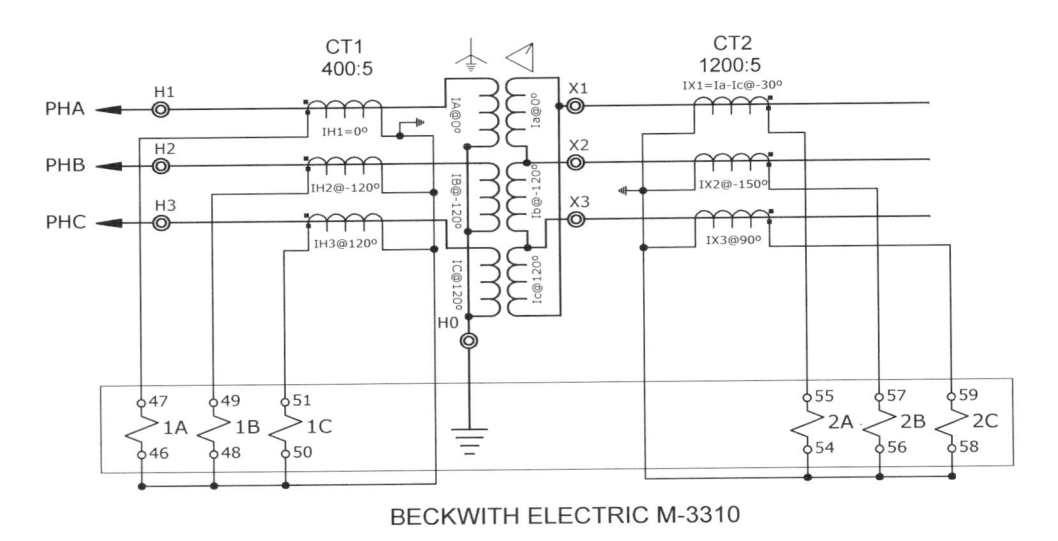

BECKWITH ELECTRIC M-3310

Figure 13-13: Wye-Delta Transformer Connections

Starting with the Wye currents, draw a phasor diagram of the current flowing through the primary bushings (IH1, IH2, & IH3) which are equal to the phase (IA, IB, & IC) currents as shown in the first phasor diagram. The current flowing through X1 is "Ia and -Ic" because X1 is connected to the polarity winding of the aØ winding and the non-polarity of the cØ winding. The secondary current is plotted in the second phasor diagram 180° from the actual current flowing into the relay to make the relationship between windings easier to understand. The second phasor diagram demonstrates the Ia and -Ic addition and the final result "IX1" which is found at the 1 o'clock position or -30°. The third phasor diagram displays the relationship between the primary and secondary windings which can be translated to "Y0d1" or "Yd1" using correct notation.

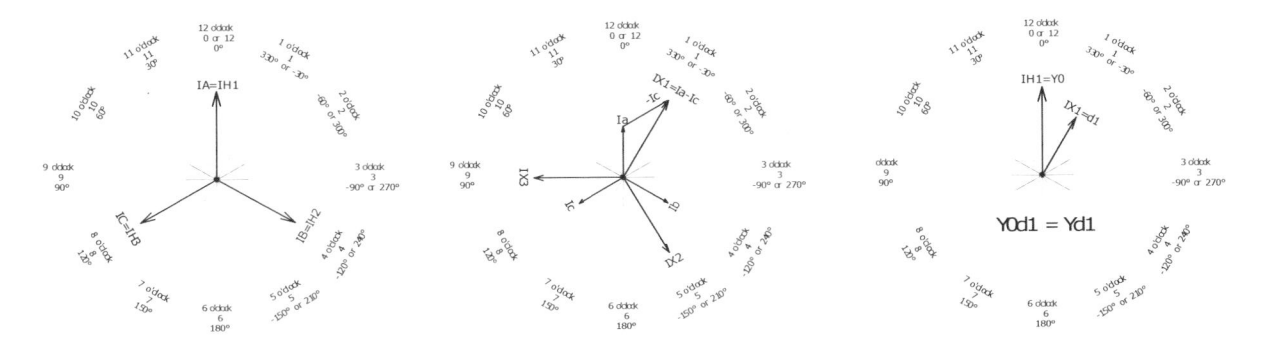

Figure 13-14: Wye-Delta Transformer Phasor Diagrams

The next example is a Wye-Delta transformer with a different delta configuration. The Wye or high voltage winding is the primary winding because the Wye side CTs are connected to the SEL-587 Relay Terminals Z01, Z03, Z05, Z02, Z04, and Z06.

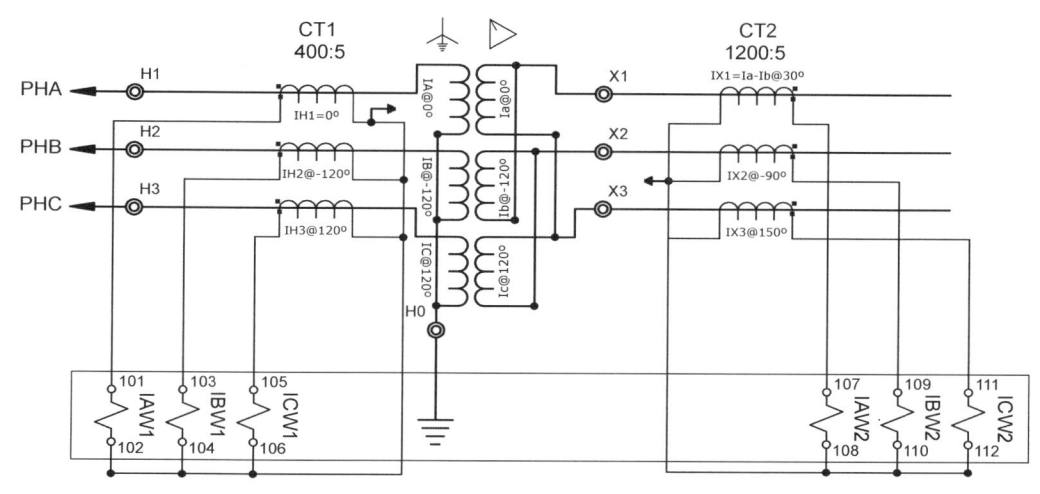

Figure 13-15: Wye-Delta Alternate Transformer Connections

Starting with the Wye currents, draw a phasor diagram of the current flowing through the primary bushings (IH1, IH2, & IH3) which are equal to the phase (IA, IB, & IC) currents as shown in the first phasor diagram.

The second phasor diagram displays the Delta-winding phasors with the phase currents in phase with the Wye-connected winding and the line current equal to "Ia-Ib" because the X1 bushing is connected to the aØ winding and the bØ non-polarity winding.

The third phasor shows the phase relationship between windings and can be described as Y0d11 or "Yd11".

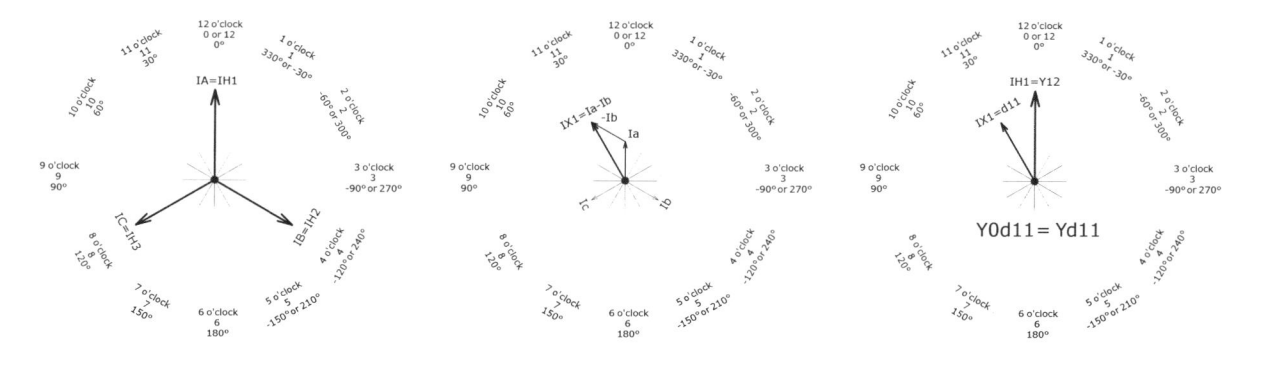

Figure 13-16: Wye-Delta Alternate Transformer Phasor Diagrams

The example in Figure 13-17 can easily become a Delta-Wye transformer by switching the protective relay connections as shown in the next example.

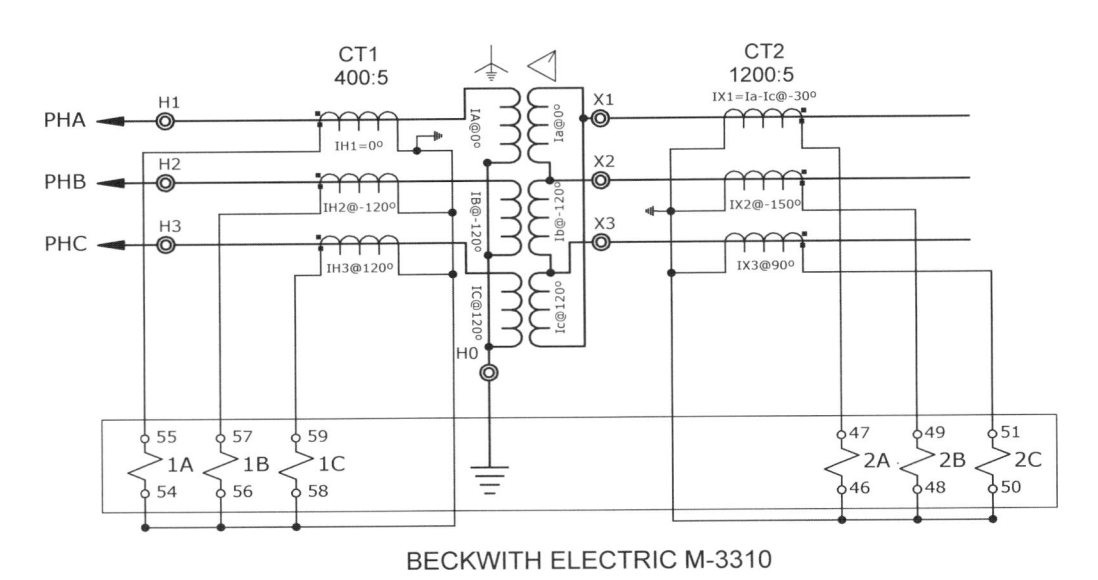

Figure 13-17: Delta-Wye Transformer Connections

You should always start with the Wye-winding and the phase and line currents are drawn in the second phasor diagram because the Wye-winding is the secondary winding.

The first diagram follows the steps from Figure 13-17 for the Delta-winding because it is now the primary winding. The third diagram displays the phasor relationship between windings and the transformer is described as D1y0 based on this diagram.

The primary winding must always be at 12 o'clock and we can rotate the phasors or the clock to obtain the correct description D0y11 or "Dy11" as shown in the fourth and fifth phasor diagrams.

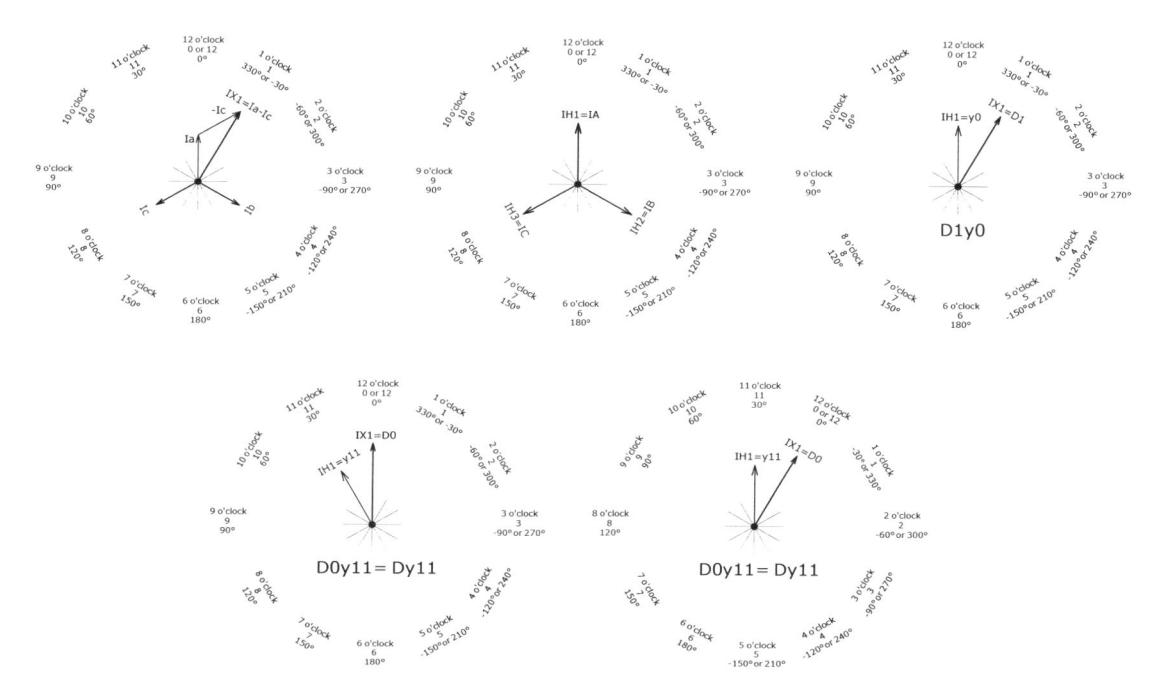

Figure 13-18: Delta-Wye Transformer Connections

The example in Figure 13-19 can easily become a Delta-Wye transformer by switching the protective relay connections as shown in the next example where the Delta-winding is connected to SEL-387 Terminals Z01, Z03, Z05, Z02, Z04, and Z06

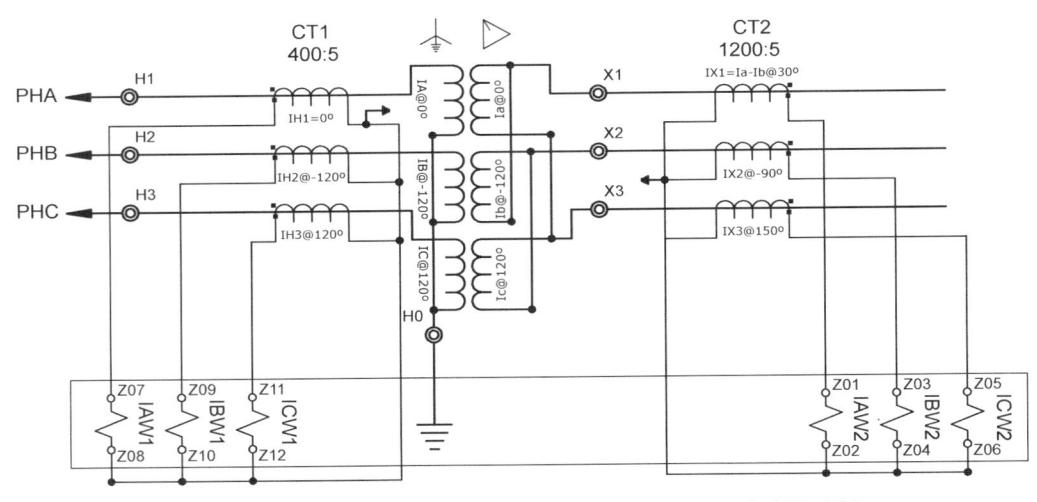

Figure 13-19: Delta-Wye Alternate Transformer Connections

Starting with the Wye currents, draw a phasor diagram of the current flowing through the primary bushings (IH1, IH2, & IH3) which are equal to the phase (IA, IB, & IC) currents as shown in the second phasor diagram.

The first phasor diagram displays the Delta-winding phasors with the phase currents in phase with the Wye-connected winding and the line current equal to "Ia-Ib" because the X1 bushing is connected to the aØ winding and the bØ non-polarity winding.

The third diagram displays the phasor relationship between windings and the transformer is described as D11y0 based on this diagram.

The primary winding must always be at 12 o'clock and we can rotate the phasors or the clock to obtain the correct description D0y1 or "Dy1" as shown in the fourth and fifth phasor diagrams.

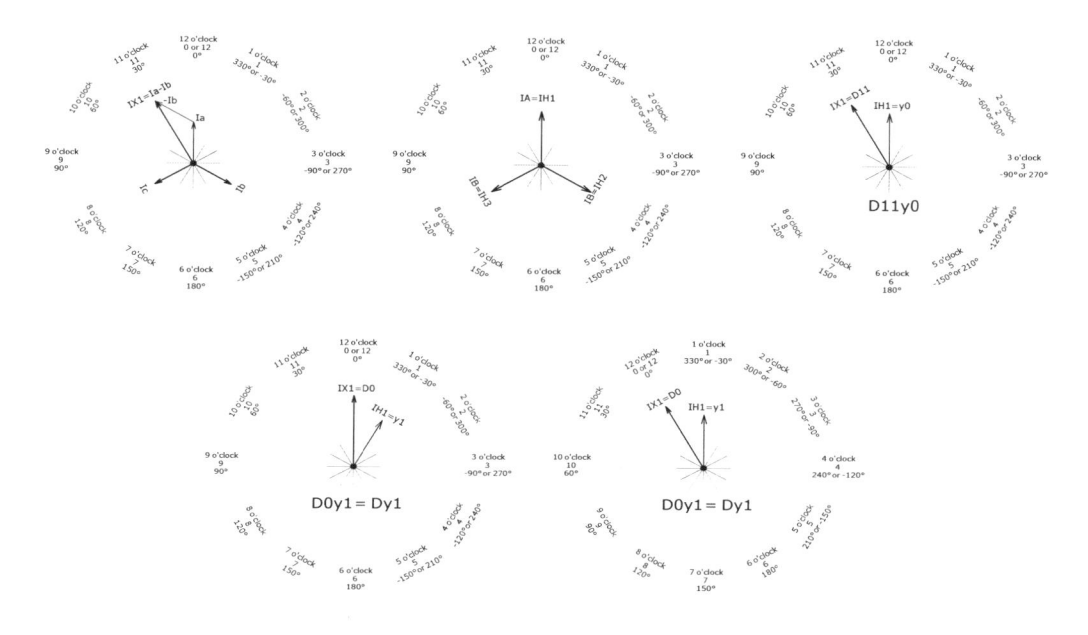

Figure 13-20: Delta-Wye Alternate Transformer Phasor Diagrams

2) The Easy Way

The easy way requires a single line drawing, three line drawings, or the transformer nameplate which should all have a set of symbols like ⊥ ◁ that describe the phase relationship between windings. This symbol ⊥ represents a Wye-connected winding and symbol ◁ represents a Delta-connected winding. These are the only two symbols you need to determine phase relationship after reviewing the CT connections to determine which winding the relay considers to be the primary winding.

After the primary winding is determined, redraw its connection symbol with AØ at the 12 o'clock position as shown in the next example. Redraw the secondary winding symbol to the right of the primary winding with the same phase relationship as shown on the nameplate. The secondary winding symbol's AØ position will be either in the 11, 12, or 1 o'clock position. You can use any phase at the 12 o'clock position to make it easier to translate as long as the same phase is used for both windings. The following table displays the most common transformer connections and their phase relationships.

PHASE SYMBOL	DESCRIPTION
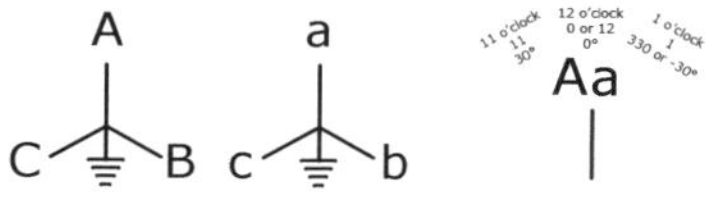	Yy0
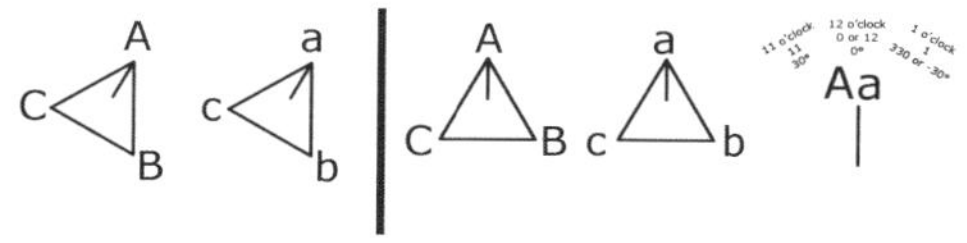	Dd0
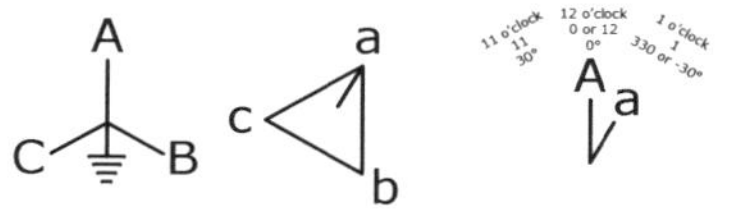	Yd1
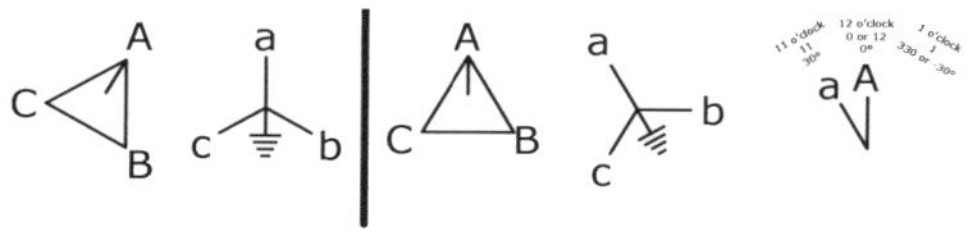	Dy11
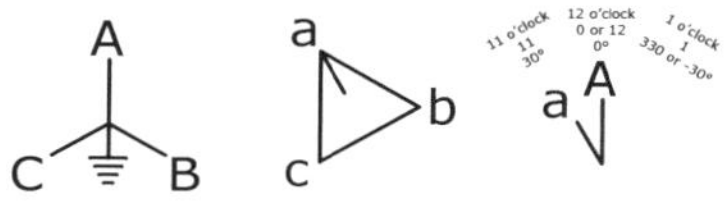	Yd11
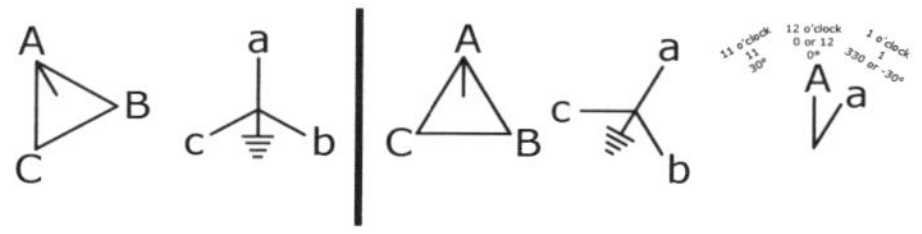	Dy1

Figure 13-21: Transformer Nameplate Phase Relationships

2. Settings

The following differential settings are described below to help understand what each setting means.

A) Enable Setting

Many relays allow the user to enable or disable settings. Make sure that the element is ON or the relay may prevent you from entering settings. If the element is not used, the setting should be disabled or OFF to prevent confusion. Some relays will also have "Latched" or 'Unlatched" options. A Latched option indicates that the output contacts will remain closed after a trip until a reset is performed and acts as a lockout relay. Unlatched indicates that the relay output contacts will open when the trip conditions are no longer present.

This setting can also be applied by enabling individual windings such as the SEL-387 "E87n" setting where "n" is the winding number. If all of the E87W settings are set to "N", differential protection is disabled.

B) Number of Windings

This setting defines the number of windings to be used for differential protection. At least two must be selected with a typical maximum of four windings. Determine the number of CT inputs used and ensure the correct number of windings is selected.

C) Phase-Angle Compensation

This setting defines the phase-angle compensation between windings as described in the previous section. Depending on the relay manufacturer, this setting could be defined in one step using a descriptor such as "TRCON=DABY" or each winding could be defined by "W1CTC=11" and "W2CTC=12". *This is the most important setting and MUST be correct.*

A table of the most common transformer connections and relays can be found in the following table.

CONNECTION		GE/MULTILIN SR-745	GE T-60, T-30
Yy0		Y/y0°	W1 Connection=Wye W1 Angle=0 W2 Connection=Wye W2 Angle=0
Dd0		D/d0°	W1 Connection=Delta W1 Angle=0° W2 Connection=Delta W2 Angle=0°
Yd1		Y/d30°	W1 Connection=Wye W1 Angle=0° W2 Connection=Delta W2 Angle=30°
Dy11		D/y330°	W1 Connection=Delta W1 Angle=0° W2 Connection=Wye W2 Angle=330°
Yd11		Y/d330°	W1 Connection=Wye W1 Angle=0° W2 Connection=Delta W2 Angle=330°
Dy1		D/y30°	W1 Connection=Delta W1 Angle=0° W2 Connection=Wye W2 Angle=30°

CONNECTION		BECKWITH ELECTRIC M-3310	SEL-387	SEL-587
Yy0	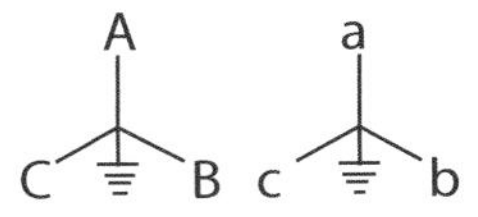	Transformer Connection 01. Yyyy	W1CTC=12 W2CTC=12	TRCON=YY CTCON=YY
Dd0	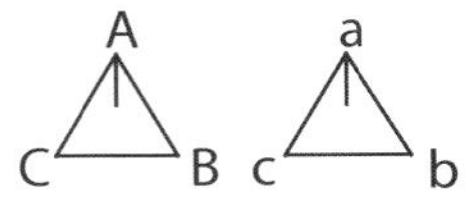	Transformer Connection 10. DACDACyy	W1CTC=0 W2CTC=0	TRCON=DACDAC CTCON=YY
Yd1	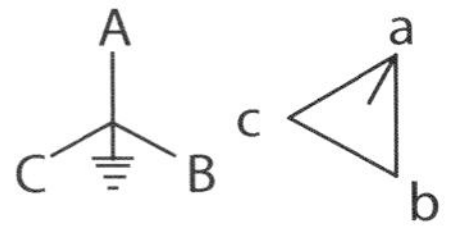	Transformer Connection 02. YDACyy	W1CTC=12 W2CTC=1	TRCON=YDAC CTCON=YY
Dy11	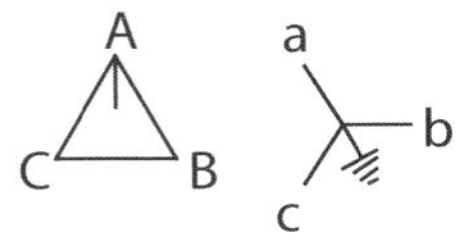	Transformer Connection 05. DACYyy	W1CTC=0 W2CTC=11	TRCON=DACY CTCON=YY
Yd11	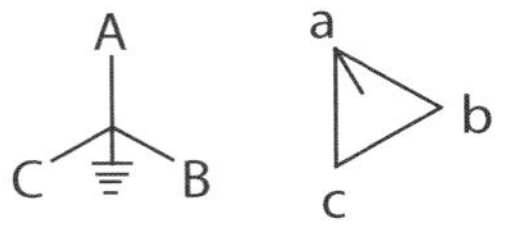	Transformer Connection 03. YDAByy	W1CTC=12 W2CTC=11	TRCON=YDAB CTCON=YY
Dy1	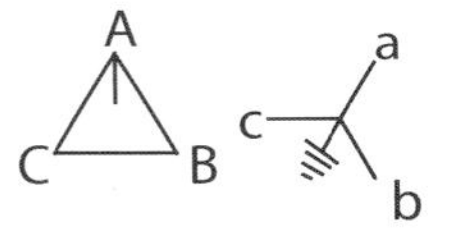	Transformer Connection 04. DABYyy	W1CTC=0 W2CTC=1	TRCON=DABY CTCON=YY

Figure 13-22: Common Phase-Angle Compensation Settings

D) MVA

This setting is used in conjunction with the Winding Voltage setting to automatically calculate the Tap values for each winding. Different engineers use different criteria for the MVA setting (minimum MVA, nominal MVA, maximum MVA) but the setting should NEVER exceed the maximum transformer MVA.

E) Winding Voltage

This setting is used in conjunction with the MVA setting to automatically calculate the Tap values for each winding.

F) Minimum Pickup (Restrained)

The Minimum Pickup setting is used to provide more stability to the differential element by requiring a minimum amount of current to flow before the differential element will operate. This minimum operate current is used to prevent nuisance trips due to noise or metering errors at low current levels. This setting should be set around 0.3 x the nominal or Tap current of the protected device.

G) Tap

The Tap setting defines the normal operate current based on the rated load of the protected equipment, the primary voltage, and the CT ratio. This setting is used as the per-unit operate current of the protected device and most differential settings are based on the Tap setting. Verify the correct Tap setting using the following formula.

$$TAP = \frac{CT_{SEC} \times Power}{\text{P-P Volts} \times \sqrt{3} \times CT_{PRI}}$$

H) Slope-1

The Slope-1 setting sets the ratio of operate current to restraint current that must be exceeded before the 87-element will operate as described previously. The slope setting is typically set at 20-30%.

I) Slope-2

The Slope-2 setting sets the ratio of operate current to restraint current that must be exceeded before the 87-element will operate if the restraint current exceeds a pre-defined or user-defined breakpoint between Slope-1 and Slope-2.

J) Breakpoint

This setting defines whether Slope-1 or Slope-2 will be used to determine if the relay should trip. The Breakpoint is defined as a multiple of Tap and if the restraint current exceeds the Breakpoint setting, the 87-element will use Slope-2 for its calculation.

K) Time Delay

The Time Delay setting sets a time delay (typically in cycles) between an 87-element pickup and trip. The Time Delay is typically set at the minimum possible setting but can be set as high as three cycles for maximum reliability on some relays.

L) Block

The block setting defines a condition that will prevent the differential protection from operating such as a status input from another device. This setting is rarely used. If enabled, make sure the condition is not true when testing. Always verify correct blocking operation by operating the end-device instead of a simulation to ensure the block has been correctly applied.

M) Harmonic Inhibit Parameters

This setting determines what harmonic the differential protection will use to detect transformer inrush or over-excitation which could cause a nuisance trip. This setting is typically set for 2nd, 4th, or 5th harmonics.

3. Current Transformer Connections

Zones of protection are defined by the CT locations which makes the CT connections the most important aspect of transformer differential protection. The CT ratio, polarity, and location must match the manufacturer's requirements and relay settings, or the differential protection may never work correctly.

CT connections are confusing for some people, but the connections can be correctly interpreted by following the flow of power. Use the following figures to determine the correct connections for our examples. Power flows from the generator to the system in this application so we can trace the flow of power from "To Generator" to "To System." Current flows into the CT polarity mark and out the CT secondary polarity mark (or H1 & X1). Follow the line from the polarity mark to Relay Terminal H1. Current flows through the relay coil and exits the relay from Terminal G1. Follow G1 back to the CT to ensure the CT secondary connections are a closed loop. Repeat this step for the other two phases and ground, if applicable. Using this procedure on Figure 13-23 we discover that the Generator AØ(1A) current flows into H1, BØ(1B) flows into H2, and CØ(1C) flows into H3.

The secondary current flows from "Winding No. 2" to "To System" and flows into the non-polarity mark of the CT. The secondary current flows out of the non-polarity mark, through the two neutral jumpers and into G4. The current flows through the relay and out H4 to the CT polarity mark. Using this procedure on Figure 13-23 we discover that the System AØ(2a) current is connected to H4, BØ(2b) to H5, and CØ(2c) to H6.

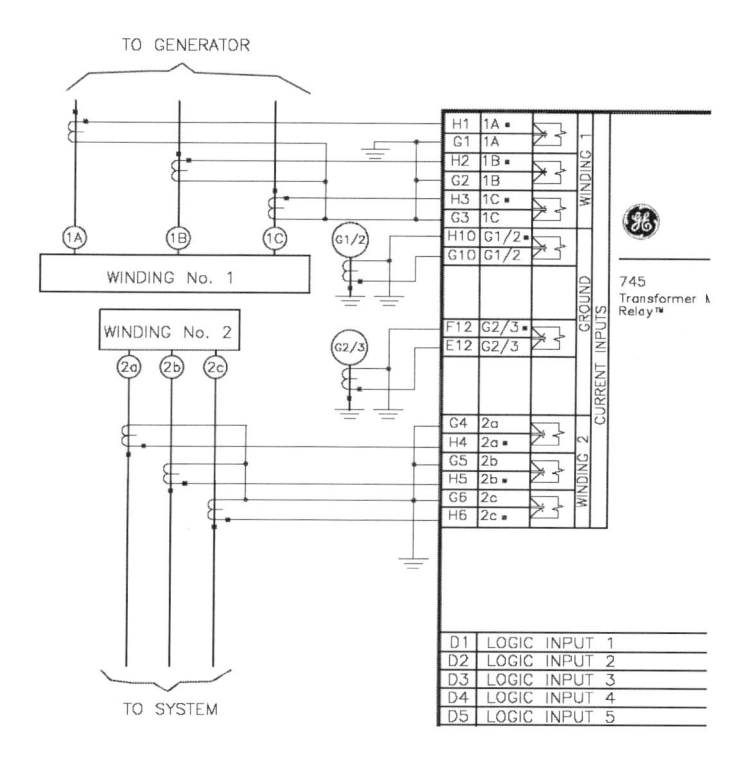

Figure 13-23: Zones of Protection Example

Another way to quickly determine correct CT polarity is to pick a reference point and compare the manufacturer's drawings to the application drawings. Choose the CT side away from the transformer in the examples and follow them to H1, H2, and H3 for each phase. Choose the CT side facing away from the transformer on the secondary side and follow the wiring to H4, H5, and H6. Always ensure that the CTs are a closed loop and that only one grounding point is connected for each isolated circuit. Many people get fixated on the polarity marks which, as shown in Figures 13-24 and 13-25, are not as important as the actual CT connections.

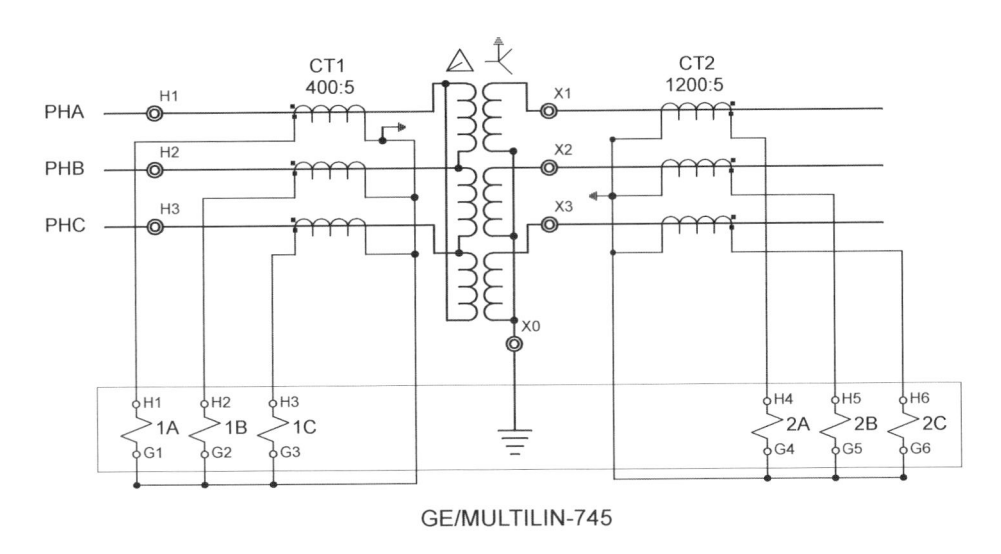

Figure 13-24: CT Connections Example #1

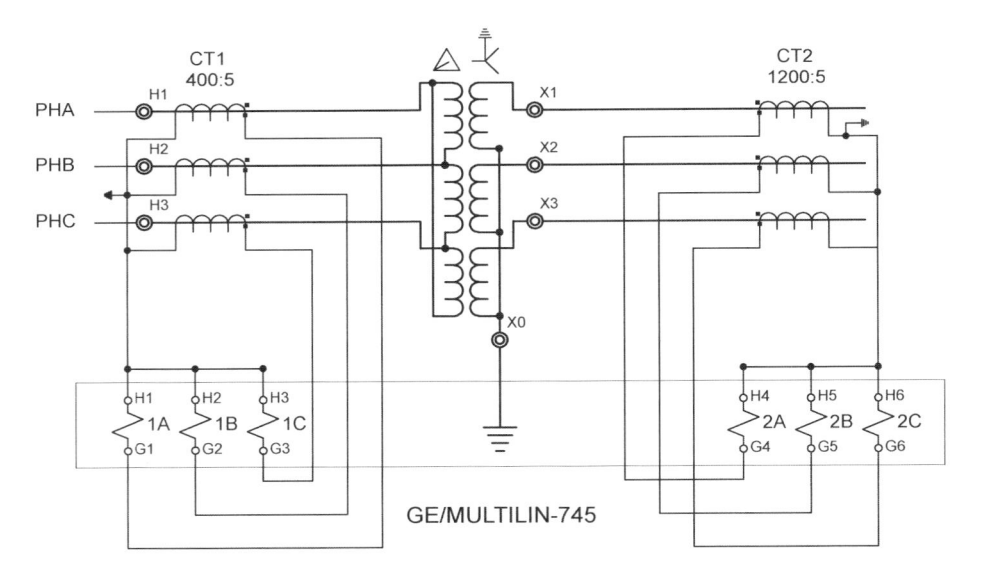

Figure 13-25: CT Connections Example #2

The following figures represent the connection drawings for the most popular differential relays applied today.

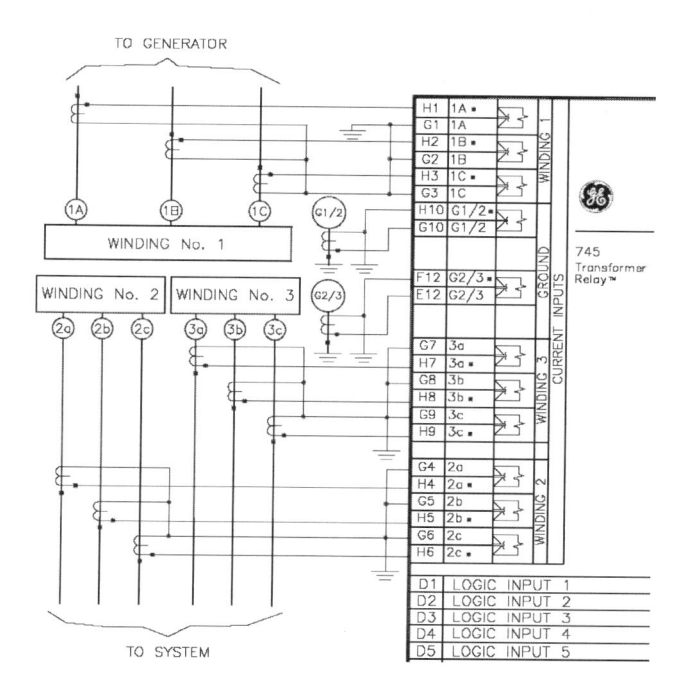

Figure 13-26: GE/Multilin SR-745 Transformer Protective Relay Connections

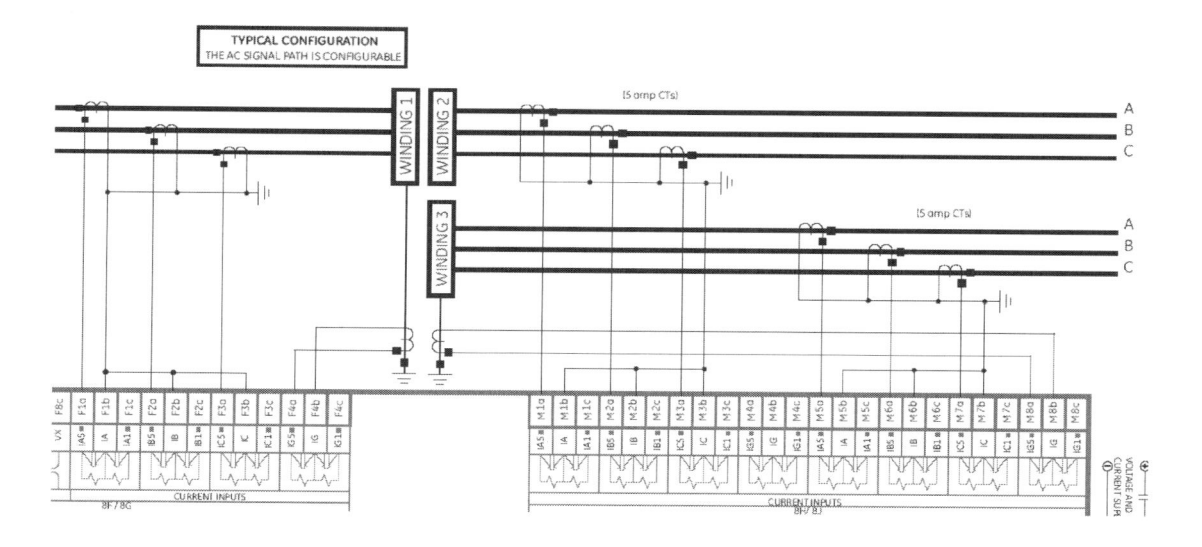

Figure 13-27: GE T-60 Transformer Protective Relay Connections

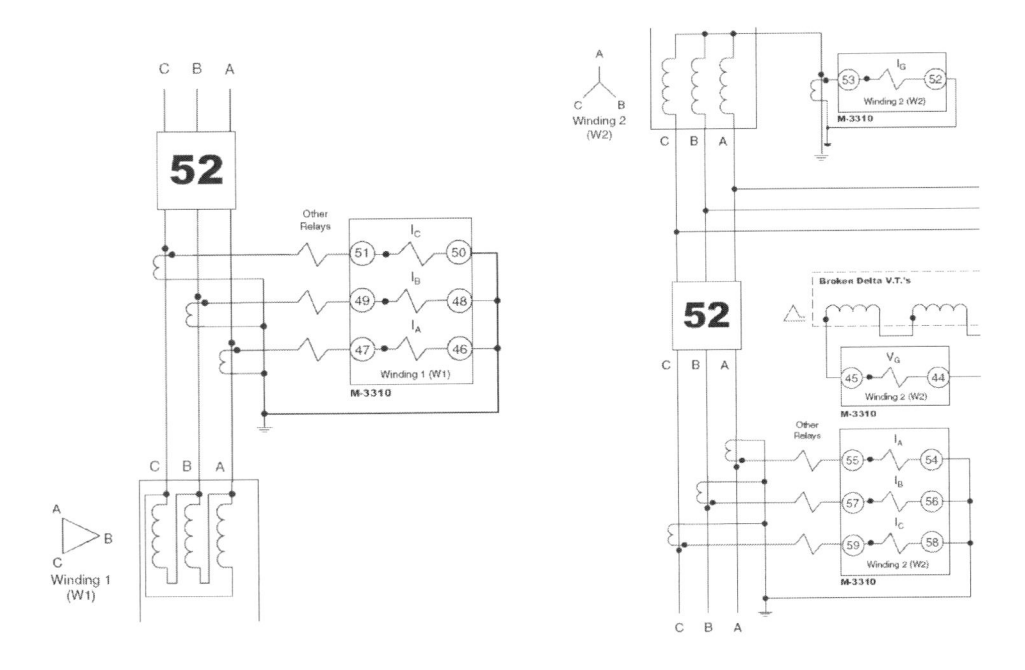

Figure 13-28: Beckwith Electric M-3310 Transformer Protective Relay Connections

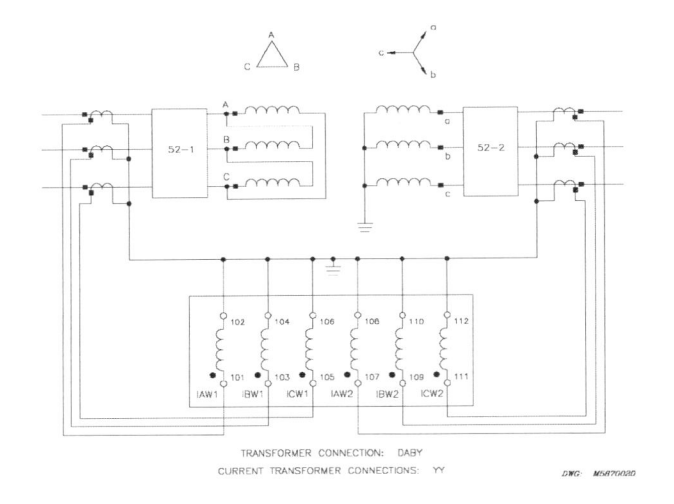

Figure 13-29: Schweitzer Electric SEL-587 Transformer Protective Relay Connections

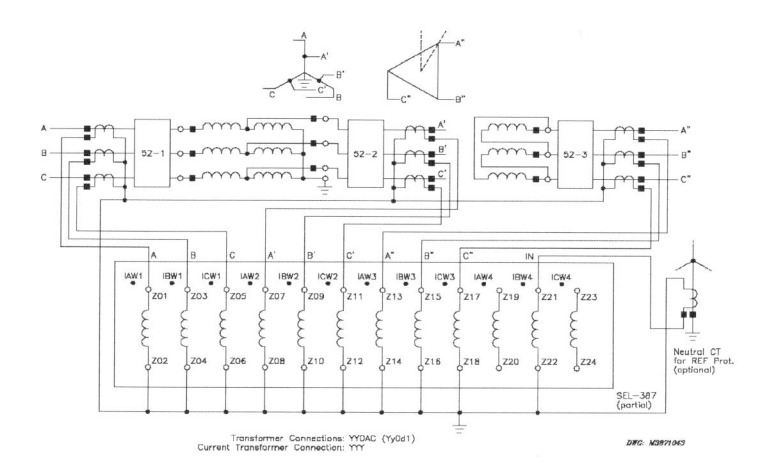

Figure 13-30: Schweitzer Electric SEL-387 Transformer Protective Relay Connections

4. 3-Phase Restrained-Differential Pickup Testing

Performing differential pickup testing is very similar to the test procedure for overcurrent protection with a few extra phases to test. Because 3-phase transformers typically have correction factors to compensate for transformer phase shifts, Minimum Pickup tests should be performed using 3-phase balanced currents. The test is performed by applying and raising current in one winding until the element operates. Repeat for all differential windings.

You must determine what the expected result should be before performing any test. Record the Pickup and Tap settings. If the relay does not have a Tap setting, refer to the manufacturer's literature to determine if the relay uses some other nominal setting such as rated secondary current (5A usually). Multiply the Pickup and Tap settings to determine the Minimum Pickup in amps.

We will use an SEL-387E relay with the following differential settings for the rest of the tests in this chapter.

	Winding-1	Winding-2
CTR_ (CT Ratio)	240	1600
MVA (XFMR MVA Ration)	230	
W_CTC (Winding Connection)	12	1
VWDG_ (Winding Voltage)	230	18
Tap_	2.41	4.61
O87P	0.30	
SLP1	20	
SLP2	60	
IRS1	3.0	

This relay has a Minimum Pickup setting (O87P) and we can determine the 3-phase Minimum Pickup using the formula $O87P \times TAP$. It's always a good idea to verify the Tap settings as described in the previous chapter using the transformer nameplate as shown below for this transformer.

WINDING-1	WINDING-2
$TAP1 = \dfrac{CT_{SEC} \times Power}{\text{P-P Volts} \times \sqrt{3} \times CT_{PRI}}$	$TAP2 = \dfrac{CT_{SEC} \times Power}{\text{P-P Volts} \times \sqrt{3} \times CT_{PRI}}$
$TAP1 = \dfrac{1 \times 230,000,000}{230,000 \times \sqrt{3} \times 240}$	$TAP2 = \dfrac{1 \times 230,000,000}{18,000 \times \sqrt{3} \times 1600}$
$TAP1 = 2.40569$	$TAP2 = 4.6109$

The Tap settings appear to be appropriate for the application and are very close to the SEL-387E generated Tap settings. We can determine the Minimum Pickup for each winding with the following formula.

WINDING-1	WINDING-2
Minimum Pickup = O87P × Tap	Minimum Pickup = O87P × Tap
Minimum Pickup = 0.3 × 2.41	Minimum Pickup = 0.3 × 4.61
Minimum Pickup = 0.723A	Minimum Pickup = 1.383A

A) 3-Phase Test-Set Connections

Connect all three phases to one winding. After all Winding-1 tests are completed, move the test leads to Winding-2 and repeat.

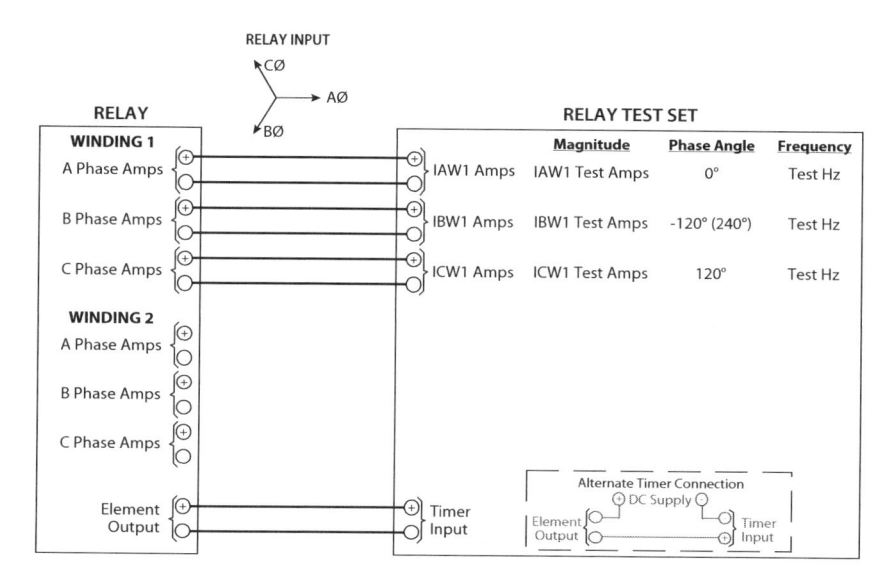

Figure 13-31: Simple 3-Phase 87-Element Test-Set Connections

If your test-set has six available current channels, you can use the following connection diagram, and change the output channel for each test until all pickup values have been tested.

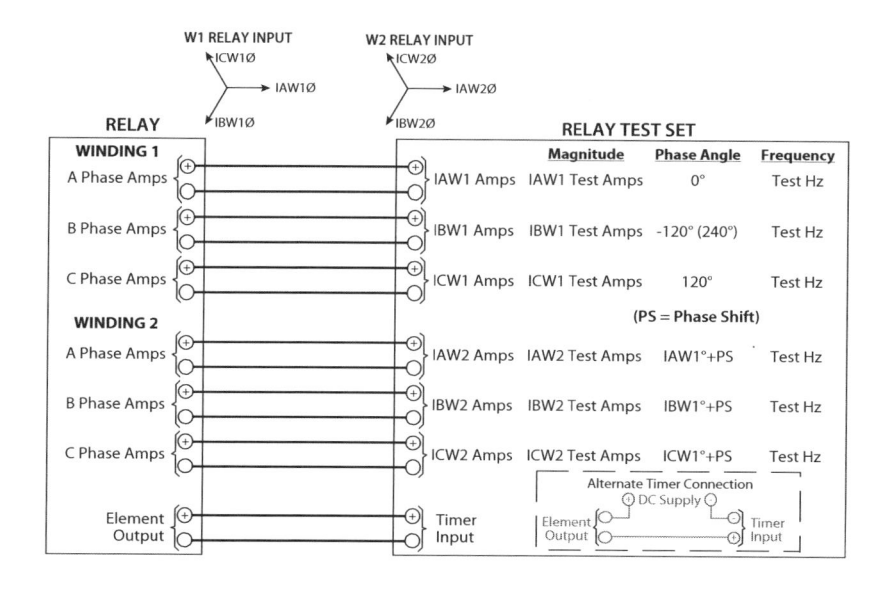

Figure 13-32: Simple 3-Phase 87-Element Test-Set Connections with Six Current Channels

B) 3-Phase Pickup Test Procedure

Use the following steps to determine pickup.

1. Determine how you will monitor pickup and set the relay accordingly, if required. (Pickup indication by LED, output contact, front panel display, etc. See the *Relay Test Procedures* section starting on page 109 for details.)

2. Group all three currents together so that you can change the magnitude of all three simultaneously while maintaining 120° between phases. Set the test current 5% higher than the pickup setting as shown in the following table.

	WINDING-1	WINDING-2
	Min Pickup × 105% = Test Amps	Min Pickup × 105% = Test Amps
	0.723A × 1.05 = Test Amps	1.383A × 1.05 = Test Amps
	0.75915A	1.452A
AØ Test Amps	0.759A@0°	1.452A@0°
BØ Test Amps	0.759A@240°	1.452A @240°
CØ Test Amps	0.759A@120°	1.452A @120°

3. Apply 3-phase test current and make sure pickup indication operates.

4. Slowly lower all three currents simultaneously until the pickup indication is off. Slowly raise the currents until pickup indication is fully on (Chattering contacts or LEDs are not considered pickup). Record the pickup values on your test sheet. The following graph displays the pickup procedure.

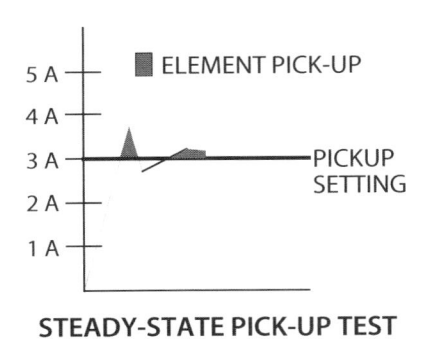

STEADY-STATE PICK-UP TEST

Figure 13-33: Pickup Test Graph

5. Compare the pickup test result to the manufacturer's specifications and calculate the percent error as shown in the following example.

    The 87-element pickup setting is 0.723A and the measured pickup was 0.731A. Looking at the "Differential Element" specification in Figure 13-34, we see that the acceptable metering error is ±5% ±0.10A.

Differential Element

Unrestrained Pickup Range:	1–20 in per unit of tap
Restrained Pickup Range:	0.1–1.0 in per unit of tap
Pickup Accuracy (A secondary)	
5 A Model:	±5% ±0.10 A
1 A Model:	±5% ±0.02 A
Unrestrained Element Pickup Time (Min/Typ/Max):	0.8/1.0/1.9 cycles
Restrained Element (with harmonic Blocking) Pickup Time (Min/Typ/Max):	1.5/1.6/2.2 cycles
Restrained Element (with harmonic Restraint) Pickup Time (Min/Typ/Max):	2.62/2.72/2.86 cycles

Figure 13-34: SEL-387E Specifications

We can calculate the manufacturer's allowable percent error for Winding-1.

$$100 \times \frac{(5\% \times \text{Setting}) + 0.1A}{\text{Setting}} = \text{Allowable Percent Error}$$

$$100 \times \frac{(5\% \times 0.723) + 0.1A}{0.723} = \text{Allowable Percent Error}$$

18.831% Allowable Percent Error

The measured percent error can be calculated using the percent error formula:

$$\frac{\text{Actual Value - Expected Value}}{\text{Expected Value}} \times 100 = \text{Percent Error}$$

$$\frac{0.731A - 0.723A}{0.723A} \times 100 = \text{Percent Error}$$

1.1% Error

The Winding-1 test is within the manufacturer's tolerance and passed the test.

6. Repeat the pickup test for all windings that are part of the differential scheme.

DIFFERENTIAL TEST RESULTS					
PICK UP	0.3	TIME DELAY	0		
SLOPE 1	20%	TAP1	2.41		
SLOPE 2	60%	TAP2	4.61		
MINIMUM PICK UP TESTS (Amps)					
		3 PHASE		MFG	% ERROR
W1 PICKUP		0.731		0.723	1.11
W2 PICKUP		1.398		1.383	1.08
W3 PICKUP		NA		NA	

C) 1-Phase Test-Set Connections

The most basic test-set connection uses only one phase of the test-set with a test lead change between every pickup test. After the Winding-1 A-phase pickup test is performed, move the test leads from Winding-1 A-phase amps to B-phase amps and perform the test again. Repeat until all enabled phases are tested on all windings.

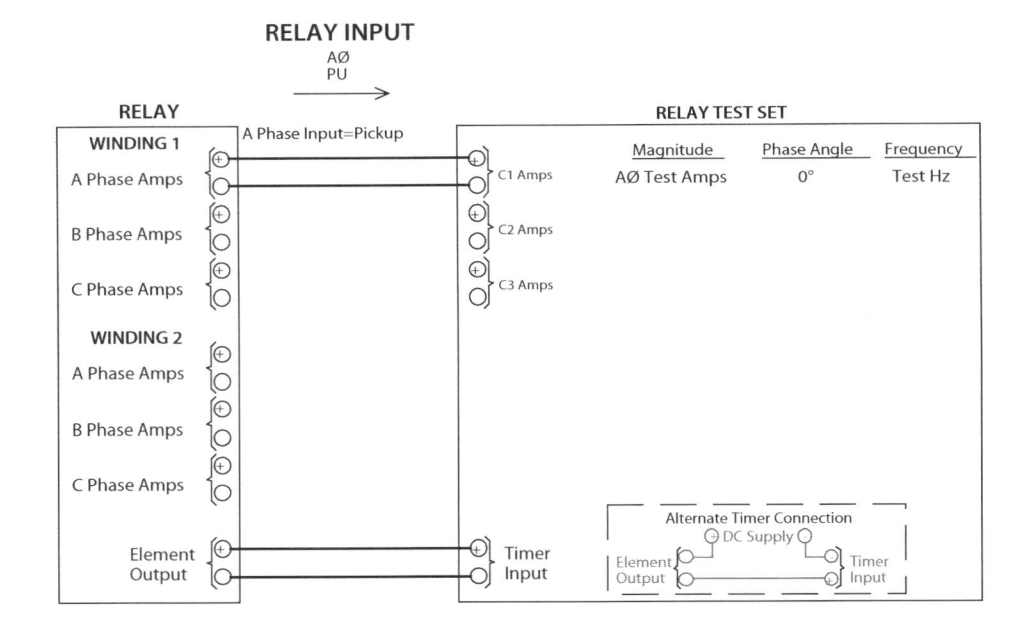

Figure 13-35: Simple 87-Element Test-Set Connections

You can also connect all three phases to one winding and change output channels instead of changing leads. After all Winding-1 tests are completed, move the test leads to Winding-2 and repeat.

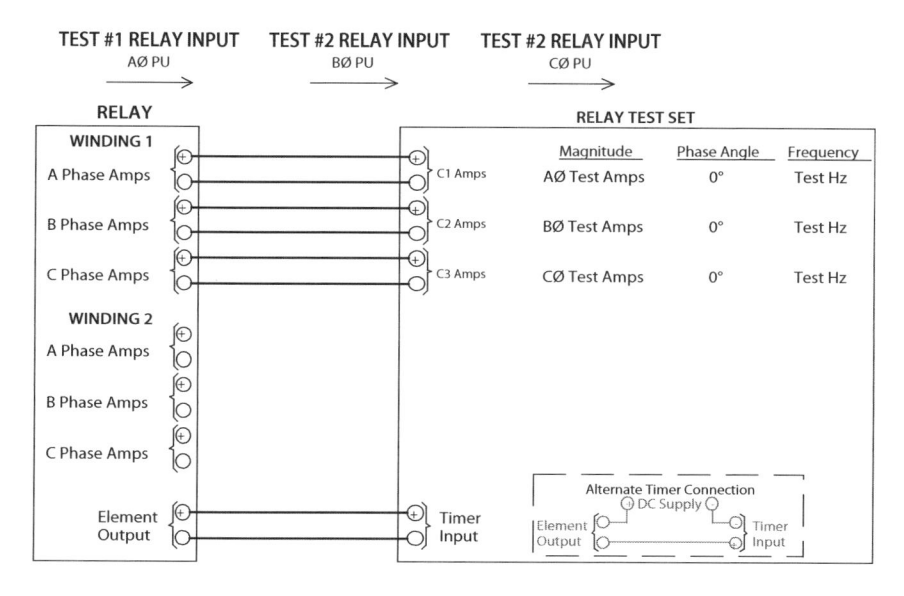

Figure 13-36: Simple 3-Phase 87-Element Test-Set Connections

If your test-set has six available current channels, you can use the following connection diagram and change the output channel for each test until all pickup values have been tested.

Figure 13-37: Simple 3-Phase 87-Element Test-Set Connections with Six Current Channels

D) 1-Phase Pickup Test Procedure

A single-phase test procedure is slightly more complicated because correction factors may apply. Digital transformer relays use algorithms to compensate for the different transformer phase shifts as described earlier and apply compensation factors to compare windings with different configurations. The compensation factors for the example SEL-387 can be found in the Testing and Troubleshooting section of the instruction manual as shown in the following table.

WnCTC SETTING	A
0	1
Odd: 1, 3, 5, 7, 9, 11	$\sqrt{3}$
Even: 2, 4, 6, 8, 10, 12	1.5

However, the table does not seem to be correct when the WnCTC equals 12. We have found through experimentation that the correction factor is 1 when WnCTC equals 12.

Use the following steps to determine pickup.

1. Determine how you will monitor pickup and set the relay accordingly, if required. (Pickup indication by LED, output contact, front panel display, etc. See the *Relay Test Procedures* section starting on page 109 for details.)

2. Determine the expected pickup:

WINDING-1	WINDING-2
$MinimumPickup = O87P \times TAP1 \times A$	$MinimumPickup = O87P \times TAP2 \times A$
$MinimumPickup = 0.3 \times 2.41 \times (W1CTC(12))1$	$MinimumPickup = 0.3 \times 4.61 \times (W1CTC(1))\sqrt{3}$
$MinimumPickup = 0.723$	$MinimumPickup = 2.395$

3. Set the fault current 10% higher than the pickup setting. For example, set the fault current at 0.795A ($0.723 \times 1.1 = 0.795A$) for Winding-1 or 2.634A ($2.395 \times 1.1 = 2.634A$) for Winding-2. Make sure pickup indication operates.

4. Slowly lower the current until the pickup indication is off. Slowly raise current until pickup indication is fully on (Chattering contacts or LEDs are not considered pickup). Record the pickup values on your test sheet. The following graph displays the pickup procedure.

STEADY-STATE PICK-UP TEST

Figure 13-38: Pickup Test Graph

5. Compare the pickup test result to the manufacturer's specifications and calculate the percent error as shown in the following example.

The 87-element pickup Winding-2 pickup setting is 2.395A and the measured pickup was 2.421A. Looking at the "Differential" specification in Figure 13-39, we see that the acceptable metering error is ±5% ±0.10A.

Differential Element

Unrestrained Pickup Range:	1–20 in per unit of tap
Restrained Pickup Range:	0.1–1.0 in per unit of tap
Pickup Accuracy (A secondary)	
5 A Model:	±5% ±0.10 A
1 A Model:	±5% ±0.02 A
Unrestrained Element Pickup Time (Min/Typ/Max):	0.8/1.0/1.9 cycles
Restrained Element (with harmonic blocking) Pickup Time (Min/Typ/Max):	1.5/1.6/2.2 cycles
Restrained Element (with harmonic restraint) Pickup Time (Min/Typ/Max):	2.62/2.72/2.86 cycles

Figure 13-39: SEL-387E Specifications

We can calculate the manufacturer's allowable percent error for Winding-2.

$$100 \times \frac{(5\% \times \text{Setting}) + 0.1A}{\text{Setting}} = \text{Allowable Percent Error}$$

$$100 \times \frac{(5\% \times 2.395) + 0.1A}{2.395} = \text{Allowable Percent Error}$$

$$100 \times \frac{0.21975}{2.395} = \text{Allowable Percent Error}$$

9.18% Allowable Percent Error

The measured percent error can be calculated using the percent error formula:

$$\frac{\text{Actual Value - Expected Value}}{\text{Expected Value}} \times 100 = \text{Percent Error}$$

$$\frac{2.421A - 2.395A}{2.395A} \times 100 = \text{Percent Error}$$

1.1% Error

The Winding-2 test is within the manufacturer's tolerance of 9.18% and passed the test.

6. Repeat the pickup test for all phase currents that are part of the differential scheme.

DIFFERENTIAL TEST RESULTS						
PICK UP	0.3	TIME DELAY	0			
SLOPE 1	20%	TAP1	2.41	W1CTC	12	
SLOPE 2	60%	TAP2	4.61	W2CTC	1	
MINIMUM PICK UP TESTS (Amps)						
	A PHASE	B PHASE	C PHASE	MFG	% ERROR	
W1 PICKUP	0.731	0.732	0.730	0.723	1.11	1.24	0.97
W2 PICKUP	2.421	2.422	2.425	2.395	1.07	1.11	1.24
W3 PICKUP	NA	NA	NA	NA			

5. Restrained-Differential Timing Test Procedure

It is very important to ensure that all phases are assigned to operate the output contact because some relays, such as the SEL-387, allow the designer to assign differential trips for each phase individually. For example, the SEL-387 word bit "87R1" is an A-phase differential trip and "87R" is a differential trip on any phase. Assigning the incorrect word bit to the final output is an easy mistake that has occurred in the field. I strongly recommend performing single-phase timing tests on every current-input connected to the differential element using the final trip output to ensure that the relay trips on all phases and all windings.

The timing test procedure is very straightforward. Apply a single-phase current 10% greater than the Minimum Pickup setting into each input related to the percent differential element and measure the time between the applied current and the relay trip signal. As with all tests, we first must discover what an acceptable result is. Use Figure 13-40 to determine the acceptable tolerances from the manufacturer's specifications.

Differential Element

Unrestrained Pickup Range:	1–20 in per unit of tap
Restrained Pickup Range:	0.1–1.0 in per unit of tap
Pickup Accuracy (A secondary)	
5 A Model:	±5% ±0.10 A
1 A Model:	±5% ±0.02 A
Unrestrained Element Pickup Time (Min/Typ/Max):	0.8/1.0/1.9 cycles
Restrained Element (with harmonic Blocking) Pickup Time (Min/Typ/Max):	1.5/1.6/2.2 cycles
Restrained Element (with harmonic Restraint) Pickup Time (Min/Typ/Max):	2.62/2.72/2.86 cycles

Output Contacts: Standard:

Make: 30 A; Carry: 6 A continuous carry at 70°C, 4 A continuous at 85°C;
1 s Rating: 50 A: MOV protected: 270 Vac, 360 Vdc, 40 J;
Pickup time: Less than 5 ms; Dropout time: Less than 5 ms typical.
Breaking Capacity (10000 operations):

Figure 13-40: SEL-387 Differential and Output Relay Specifications

The time delay setting for our example is 0 cycles which means that there is no *intentional* delay. However, there are software and hardware delays built into the relay that must be accounted for. Figure 13-41 indicates that the maximum allowable time is 60ms or 3.6 cycles.

REASON	DELAY
Restrained Element (with harmonic restraint) Pickup Time (Max) [Software]	2.86 cycles (47.7 ms)
Output Contacts Operate Time [Hardware]	5 ms (0.3 cycles)
TOTAL TIME	**52.7ms OR 3.16 CYCLES**

Figure 13-41: SEL-387 Differential Minimum Trip Time

Perform a timing test using the following steps:

1. Connect the test-set input(s) to the relay output(s) that are programmed to operate when the restrained-differential relay operates.

2. Configure your test-set to start a timer when current is applied and stop the timer and output channels when the appropriate input(s) operate(s).

3. Choose a connection diagram from Figures 13-35 to 13-37. Set a single-phase current at least 10% higher than the Minimum Pickup setting using the calculations in the "1-Phase Pickup Test Procedure" section of this chapter. (0.795A ($0.723 \times 1.1 = 0.795$A) for Winding-1 or 2.634A ($2.395 \times 1.1 = 2.634$A) for Winding-2)

4. Run the test plan on the first phase related to restrained-differential. Record the test results.

5. Repeat the test on all phases related to the restrained-differential.

RESTRAINED DIFFERENTIAL TIMING TESTS (cycles)								
WINDING	TEST	A PHASE (cy)	B PHASE (cy)	C PHASE (cy)	MFG (cycles)	% ERROR		
W1	0.723 A	2.410	2.490	2.500	3.16	OK	OK	OK
W2	2.634A	2.45	2.44	2.48	3.16	OK	OK	OK
W3	NA	NA	NA	NA	NA			

6. 3-Phase Restrained-Differential Slope Testing

Differential slope testing is one of the most complex relay tests that can be performed and requires careful planning; a good understanding of the differential relay's operating fundamentals; and information from the manufacturer regarding the relay's characteristics. This section will discuss 3-phase testing that will require at least six current channels to be performed correctly. Refer to the "1-Phase Restrained-Differential Slope Testing" section of this chapter if your test equipment has less than six-channels.

Six-channel restrained-differential testing is actually very similar to the simple testing discussed in the previous chapter after the phase-angles between windings have been applied correctly. Six-phase restrained-differential testing is achieved by applying 3-phase balanced currents in each winding, mimicking an ideal-world steady state scenario, and then increasing all three currents in one winding simultaneously until the relay operates. The math is also simplified when you apply balanced 3-phase conditions.

We discussed how to interpret the phase-angle-shift settings previously in this chapter which can be summarized by the following figure:

CONNECTION	BECKWITH ELECTRIC M-3310	SEL-387	SEL-587
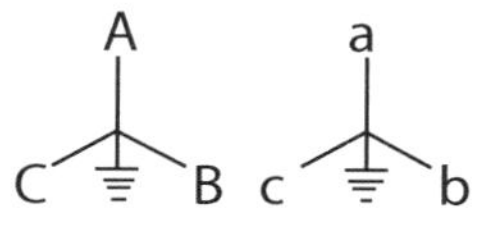	Transformer Connection 01. Yyyy	W1CTC=12 W2CTC=12	TRCON=YY CTCON=YY
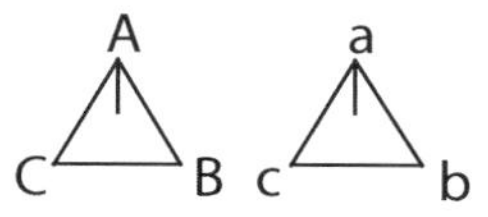	Transformer Connection 10. DACDACyy	W1CTC=0 W2CTC=0	TRCON=DACDAC CTCON=YY
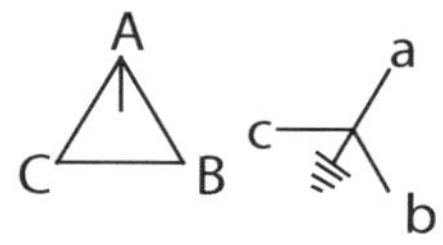	Transformer Connection 04. DABYyy	W1CTC=0 W2CTC=1	TRCON=DABY CTCON=YY
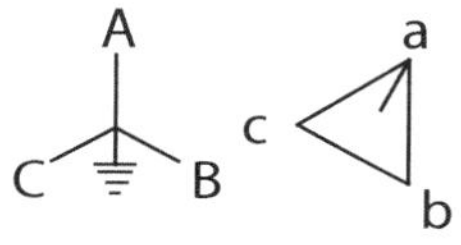	Transformer Connection 02. YDACyy	W1CTC=12 W2CTC=1	TRCON=YDAC CTCON=YY
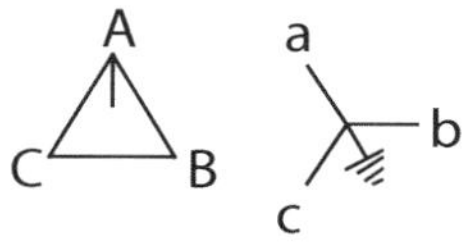	Transformer Connection 05. DACYyy	W1CTC=0 W2CTC=11	TRCON=DACY CTCON=YY
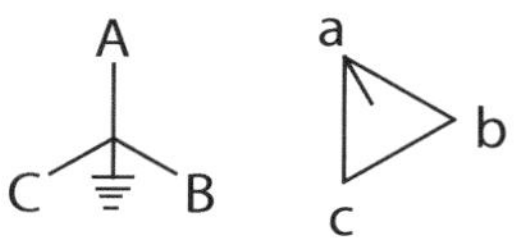	Transformer Connection 03. YDAByy	W1CTC=12 W2CTC=11	TRCON=YDAB CTCON=YY

Figure 13-42: Common Phase-Angle Compensation Settings

Let's figure out the test-settings for balanced full-load conditions for each case. The full load condition for Winding-1 is the Tap1 setting. All three phases for Winding-1 should start at Tap1. All three phases for Winding-2 should start at Tap2.

The first case is SEL-387 when both W1CTC and W2CTC=12. Both settings are the same so there is no phase-angle shift for this connection. The Capital "A" (Winding-1) and lower case "a" (Winding-2) are both at 0° so our A-phase test conditions will be at 0°. The other two phases will be 120° apart to create a 3-phase balanced condition. A-phase for Winding-2 will start at 180° to be opposite Winding-1 and the other two phases will be 120° apart as shown in the following table.

CONNECTION	SEL-387	W1AØ	W1BØ	W1CØ	W2AØ	W2BØ	W2CØ
A a C ⏚ B c ⏚ b	W1CTC=12 W2CTC=12	Tap1@ 0°	Tap1@ -120°	Tap1@ 120°	Tap2@ 180°	Tap2@ 60°	Tap2@ -60°

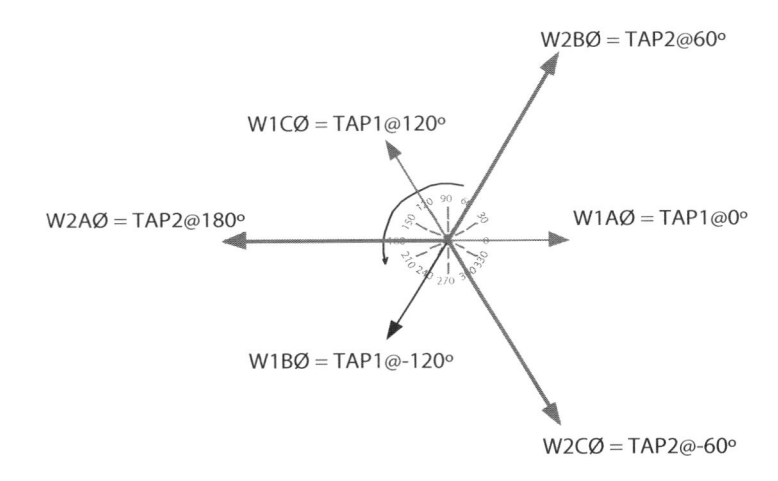

Figure 13-43: Yy12 or Yy0 3-Phase Differential Restraint Test Connections

The next case is SEL-387 when both W1CTC and W2CTC=0. Both settings are the same so there is no phase-angle shift for this connection. The Capital "A" (Winding-1) and lower case "a" (Winding-2) are both at 0° so our A-phase test conditions will be at 0°. The other two phases will be 120° apart to create a 3-phase balanced condition. A-phase for Winding-2 will start at 180° to be opposite Winding-1 and the other two phases will be 120° apart as shown in the following table.

CONNECTION	SEL-387	W1AØ	W1BØ	W1CØ	W2AØ	W2BØ	W2CØ
A a C B c b	W1CTC=0 W2CTC=0	Tap1@ 0°	Tap1@ -120°	Tap1@ 120°	Tap2@ 180°	Tap2@ 60°	Tap2@ -60°

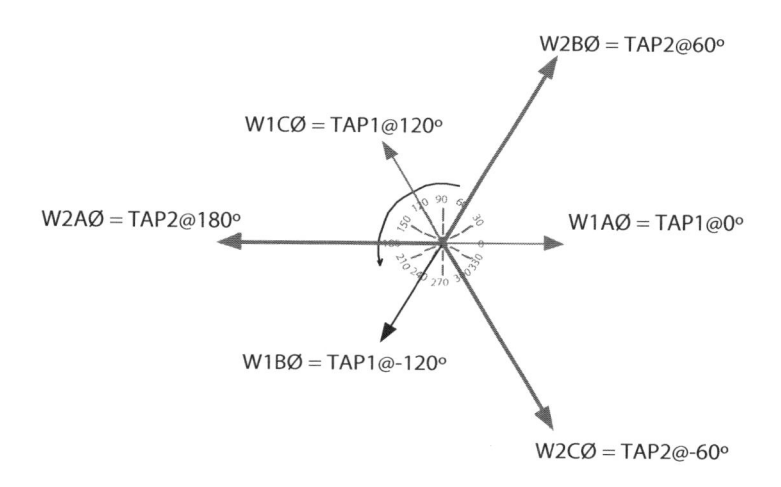

Figure 13-44: Dd0 3-Phase Differential Restraint Test Connections

The next case is SEL-387 with W1CTC=0 and W2CTC =1. The Capital "A" (Winding-1) is at $0°$ so our A-phase test conditions will be at $0°$. The other two phases will be $120°$ apart to create a 3-phase, balanced condition. The lower case "a" is at -30° so there is a phase shift between windings. Remember that we apply Winding-2 current $180°$ out-of-phase so the A-phase for Winding-2 will start at $150°$ $(-30° + 180°)$ to be opposite Winding-1. The other two phases will be $120°$ apart as shown in the following table.

CONNECTION	SEL-387	W1AØ	W1BØ	W1CØ	W2AØ	W2BØ	W2CØ
	W1CTC=0 W2CTC=1	Tap1@ 0°	Tap1@ -120°	Tap1@ 120°	Tap2@ 150°	Tap2@ 30°	Tap2@ -90°

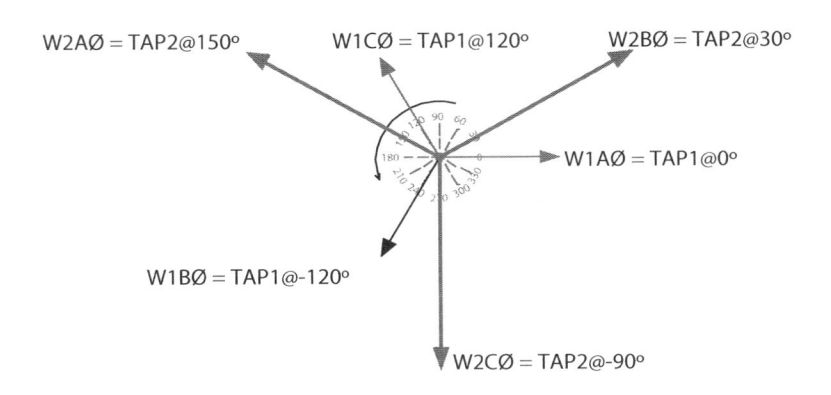

Figure 13-45: Dy1 3-Phase Differential Restraint Test Connections

The next case is SEL-387 with W1CTC=12 and W2CTC =1. The Capital "A" (Winding-1) is at $0°$ so our A-phase test conditions will be at $0°$. The other two phases will be $120°$ apart to create a 3-phase, balanced condition. The lower case "a" is at -30° so there is a phase shift between windings. Remember that we apply Winding-2 current $180°$ out-of-phase so the A-phase for Winding-2 will start at $150°$ $(-30° + 180°)$ to be opposite Winding-1. The other two phases will be $120°$ apart as shown in the following table.

CONNECTION	SEL-387	W1AØ	W1BØ	W1CØ	W2AØ	W2BØ	W2CØ
	W1CTC=12 W2CTC=1	Tap1@ 0°	Tap1@ -120°	Tap1@ 120°	Tap2@ 150°	Tap2@ 30°	Tap2@ -90°

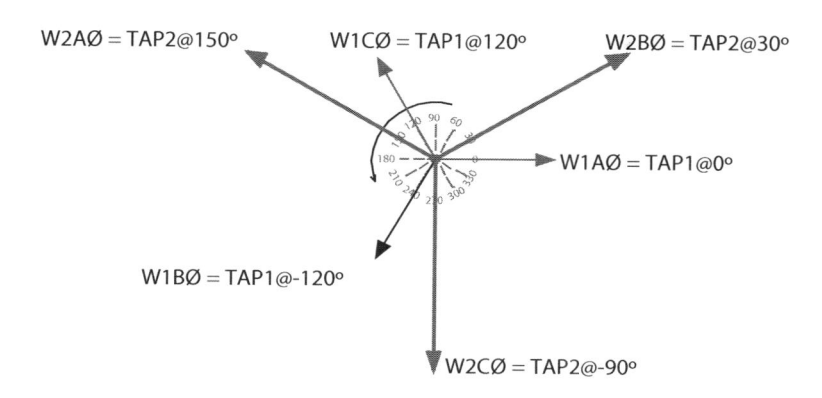

Figure 13-46: Yd1 3-Phase Differential Restraint Test Connections

The next case is SEL-387 with W1CTC=0 and W2CTC =11 The Capital "A" (Winding-1) is at 0° so our A-phase test conditions will be at 0°. The other two phases will be 120° apart to create a 3-phase balanced condition. The lower case "a" is at 30° so there is a phase shift between windings. Remember that we apply Winding-2 current 180° out-of-phase so the A-phase for Winding-2 will start at 210° (-150°) (30° + 180°) to be opposite Winding-1. The other two phases will be 120° apart as shown in the following table.

CONNECTION	SEL-387	W1AØ	W1BØ	W1CØ	W2AØ	W2BØ	W2CØ
	W1CTC=0 W2CTC=11	Tap1@ 0°	Tap1@ -120°	Tap1@ 120°	Tap2@ -150°	Tap2@ 90°	Tap2@ -30°

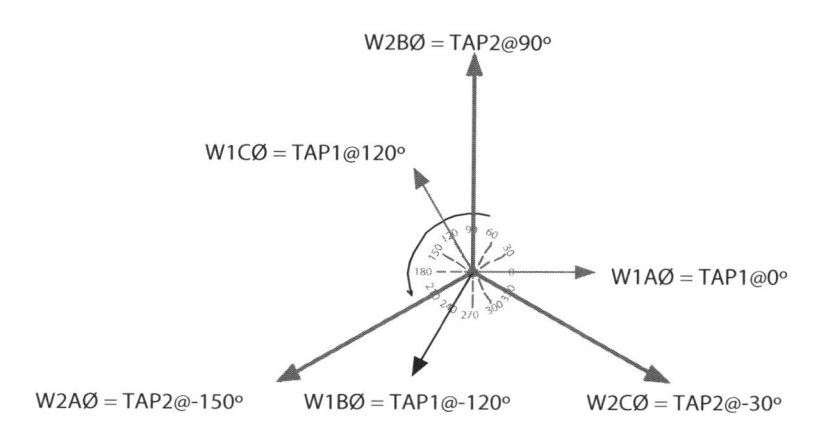

Figure 13-47: Dy11 3-Phase Differential Restraint Test Connections

The next case is SEL-387 with W1CTC=12 and W2CTC =11 The Capital "A" (Winding-1) is at 0° so our A-phase test conditions will be at 0°. The other two phases will be 120° apart to create a 3-phase, balanced condition. The lower case "a" is at 30° so there is a phase shift between windings. Remember that we apply Winding-2 current 180° out-of-phase so the A-phase for Winding-2 will start at 210° (-150°) (30° + 180°) to be opposite Winding-1. The other two phases will be 120° apart as shown in the following table.

CONNECTION	SEL-387	W1AØ	W1BØ	W1CØ	W2AØ	W2BØ	W2CØ
A a C B b c	W1CTC=12 W2CTC=11	Tap1@ 0°	Tap1@ -120°	Tap1@ 120°	Tap2@ -150°	Tap2@ 90°	Tap2@ -30°

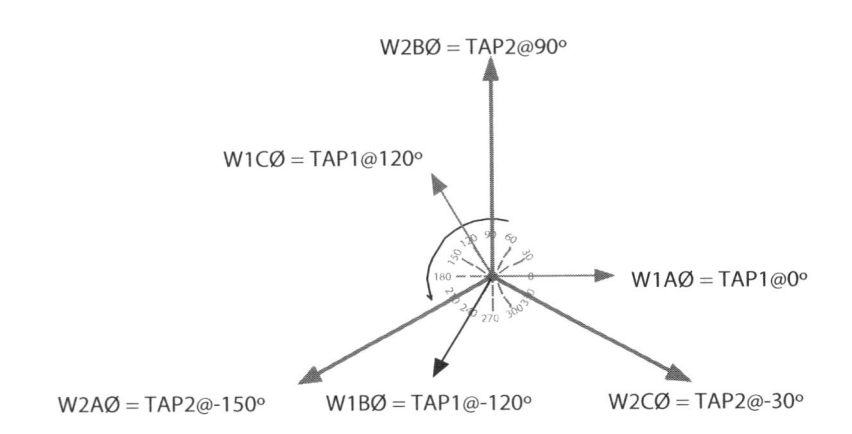

Figure 13-48: Yd11 3-Phase Differential Restraint Test Connections

A) Test-Set Connections

The connection diagram for the SEL-387 is as follows:

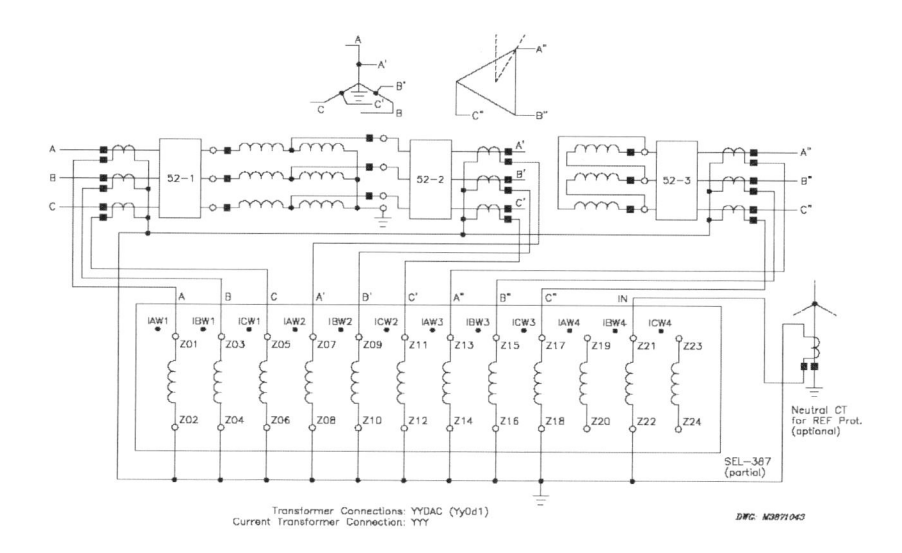

Figure 13-49: Schweitzer Electric SEL-387 Transformer Protective Relay Connections

Follow the AØ primary buss through the Phase A CTs then follow the CT secondary to Terminal Z01(IAW1). This is where we will connect the first current from our test-set. Connect the neutral of the test-set current channel to Z02 by following the other side of the CT to its relay terminal. Keep following the primary buss through the transformer to the Phase A' CT and then follow the secondary to Terminal Z08. This is the neutral of the CTs so we will connect the test-set's second current-channel-neutral-terminal to Z08. Follow the other side of the CT to Terminal Z07 (IAW2) which is where we will connect the second channel current from the test-set. Follow the other phases to determine the following connections when testing B or C-phases.

The test-set connections for a 3-Phase Restrained-Differential Slope test are displayed in the next figure.

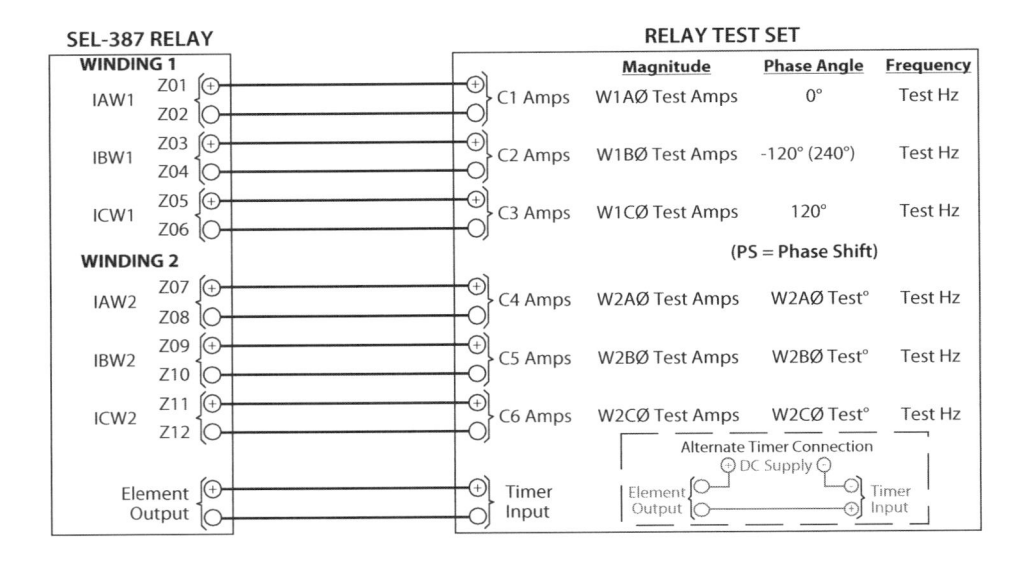

Figure 13-50: 3-Phase Restrained-Differential Slope Test-Set Connections

B) Pre-Test Calculations

There are several different methods to test the restrained-differential slope. The first method involves a significant amount of calculations to determine exactly how the element is supposed to operate, what test points will work best, and what results to expect. This section will discuss this method in detail. Remember that there is an easier way that will be discussed later in this chapter.

The following information details the example test settings and expected characteristic curve.

	WINDING-1	WINDING-2
W_CTC	12	1
Tap_	2.41	4.61
O87P	0.30	
SLP1	20	
SLP2	60	
IRS1	3.0	

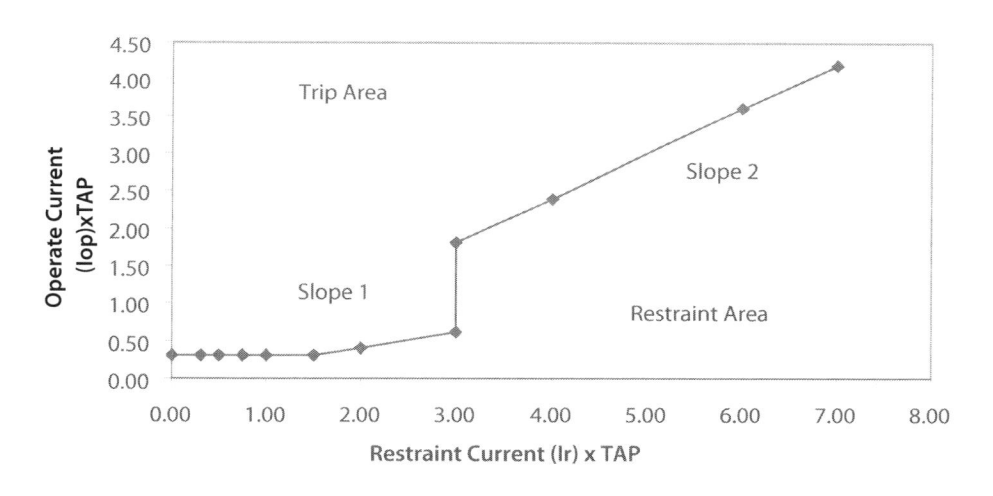

Figure 13-51: Percentage Differential Protection Dual Slope Characteristic Curve

We first need to change the restraint and operate current values to amps instead of multiples of Tap. Then we can calculate the transition point between the Minimum Pickup and Slope-1 operation. The Winding-1 Minimum Pickup is 0.723A $(0.3 \times \text{TAP1})$ as we calculated earlier in this chapter. Slope-1 will start operating when the operate current is greater than the minimum pickup. Different relays have different slope calculations and we can refer to the SEL-387 relay instruction manual to determine the relay's differential calculation as shown in Figures 89 and 90.

Method: Decide where you want to cross the differential characteristic by picking a restraint value IRT, which is a vertical line on the graph. Because this test is for the SLP1 threshold, select a point above the O87P intersection point and below IRS1. If SLP2 = OFF, IRS1 and SLP2 are not functional.

$$\text{O87P} \bullet \frac{100}{\text{SLP1}} < \text{IRT} < \text{IRS1}$$

The value of IOP corresponding to the selected IRT equals the following:

$$\text{IOP} = \frac{\text{SLP1}}{100} \bullet \text{IRT}$$

Both IRT and IOP are in multiples of tap.

Figure 13-52: SEL-387 Slope-1 Differential Formulas

The resulting A-phase, B-phase, and C-phase currents from each winding then go to the differential elements 87R-1, -2, and -3, respectively. In each element a phasor addition sums the winding currents. The magnitude of this result is IOP. The magnitudes of the winding currents are then summed in a simple scalar addition and divided by two. This result is IRT. For example, for a balanced through-load current of 4 amps, these calculations produce ideal results of IOP = 0 and IRT = 4.

Figure 13-53: SEL-387 Definition of IOP and IRT

Notice that the IRT current is defined as "The magnitudes of the winding currents are then summed in a simple scalar addition and divided by 2." This can be translated into $\frac{I_{W1}+I_{W2}}{2}$ because we will use balanced 3-phase currents at the theoretical normal operating angles. The IOP current is defined as "In each element a phasor addition sums the winding currents." The magnitude of this result is IOP which can be translated into $I_{W1}-I_{W2}$ for our calculations because we will use balanced 3-phase currents at the theoretical normal operating angles. Remember that both of these formulas are still in per-unit and not actual Amp values.

Use the first IOP formula to calculate the transition between Minimum Pickup and Slope-1.	We should calculate the expected Winding-1 current for a Winding-2 current less than 6.915A.
$O87P \times \dfrac{100}{SLP1} < IRT$	$IOP = O87P = I_{W1} - I_{W2}$
$0.3 \times \dfrac{100}{20} < IRT$	$I_{W1} = O87P + I_{W2}$
$0.3 \times 5 < IRT$	$\dfrac{I_{AW1}}{TAP1} = O87P + \dfrac{I_{AW2}}{TAP2}$
$1.5 < IRT$	$\dfrac{I_{AW1}}{2.41} = 0.3 + \dfrac{I_{AW2}}{4.61}$
$1.5 < \dfrac{I_{W1}+I_{W2}}{2}$	$I_{AW1} = \left(0.3 + \dfrac{I_{AW2}}{4.61}\right) \times 2.41$
Min IRT for Slope1 $= 1.5 \times TAP2$	
Min IRT for Slope1 $= 1.5 \times 4.61A$	$I_{AW1} = (0.3 \times 2.41) + \left(\dfrac{I_{AW2}}{4.61} \times 2.41\right)$
Min IRT for Slope1 $= 6.915A$	$I_{AW1} = 0.723 + 0.523 I_{AW2}$

Now we can calculate the Slope-1 expected results:	Remember that these values are in per-unit. Change to amps with the following formulas:
$$IOP = \frac{SLP1}{100} \times IRT$$ $$IOP = \frac{20}{100} \times IRT$$ $$IOP = 0.20 \times IRT$$ $$\left[I_{W1} - I_{W2} \right] = 0.20 \times \frac{\left[I_{W1} + I_{W2} \right]}{2}$$ $$2 \times \left[I_{W1} - I_{W2} \right] = 0.20 \times \left[I_{W1} + I_{W2} \right]$$ $$\frac{2 \times \left[I_{W1} - I_{W2} \right]}{0.20} = \left[I_{W1} + I_{W2} \right]$$ $$10 \times \left[I_{W1} - I_{W2} \right] = I_{W1} + I_{W2}$$ $$10 I_{W1} - 10 I_{W2} = I_{W1} + I_{W2}$$ $$10 I_{W1} - I_{W1} = I_{W2} + 10 I_{W2}$$ $$9 I_{W1} = 11 I_{W2}$$ $$I_{W1} = \frac{11}{9} I_{W2}$$ $$I_{W1} = 1.222 I_{W2}$$ $$9 I_{W1} = 11 I_{W2}$$ $$\frac{9}{11} I_{W1} = I_{W2}$$ $$I_{W2} = 0.818 I_{W1}$$	$$I_{W1} = 1.222 I_{W2}$$ $$\frac{I_{AW1}}{TAP1} = 1.222 \frac{I_{AW2}}{TAP2}$$ $$\frac{I_{AW1}}{2.41} = 1.222 \frac{I_{AW2}}{4.61}$$ $$4.61 I_{AW1} = 2.41 \times 1.222 \times I_{AW2}$$ $$4.61 I_{AW1} = 2.945 I_{AW2}$$ $$I_{AW1} = \frac{2.945}{4.61} I_{AW2}$$ $$I_{AW1} = 0.639 I_{AW2}$$ $$\frac{1}{0.6388} I_{AW1} = I_{AW2}$$ $$I_{AW2} = 1.565 I_{AW1}$$

We also need to determine the maximum restraint current before we accidentally start testing Slope-2 when testing Slope-1.

We can use the following formula to calculate the restraint current (Ir) at the transition point between Slope-1 and Slope-2 by setting the restraint slightly below the IRS1 setting.	Don't forget that the calculation so far has been in per-unit. The maximum W2 current in amps is calculated as follows:
$IRT < IRS1$	
$IRT < 3$	
$IRT = 2.99$	
$\dfrac{\left[I_{W1} + I_{W2}\right]}{2} = 2.99$	$\dfrac{I_{AW2}}{TAP2} = 2.69$
$\dfrac{\left[1.222I_{W2} + I_{W2}\right]}{2} = 2.99$	$\dfrac{I_{AW2}}{4.61} = 2.69$
$1.222I_{W2} + I_{W2} = 2 \times 2.99$	$I_{AW2} = 2.69 \times 4.61$
$2.222I_{W2} = 5.98$	$I_{AW2} = 12.406A$
$I_{W2} = \dfrac{5.98}{2.222}$	
$I_{W2} = 2.69$	

Now we can define the expected test currents when Slope-2 is required. The following information is found in the test section of the SEL-387 instruction manual:

Method: Decide where you want to cross the differential characteristic by picking a restraint value IRT, which is a vertical line on the graph. ISince this test is for the SLP2 threshold, select a point above the IRS1 setting.

$$IRT > IRS1$$

The value of IOP that corresponds to the selected IRT is as follows:

$$IOP = \frac{SLP2}{100} \bullet IRT + IRS1 \bullet \left(\frac{SLP1 - SLP2}{100} \right)$$

Both IRT and IOP are in multiples of tap.

Figure 13-54: SEL-387 Slope-2 Differential Formulas

$$IOP = \left[\frac{SLP2}{100} \times IRT \right] + \left[IRS1 \times \left(\frac{SLP1 - SLP2}{100} \right) \right]$$

$$IOP = \left[\frac{60}{100} \times IRT \right] + \left[3.0 \times \left(\frac{20 - 60}{100} \right) \right]$$

$$IOP = 0.6IRT + \left[3.0 \times \left(\frac{-40}{100} \right) \right]$$

$$IOP = 0.6IRT + \left[3.0 \times -0.40 \right]$$

$$IOP = 0.6IRT + -1.2$$

$$\left[I_{W1} - I_{W2} \right] = 0.6 \times \frac{\left[I_{W1} + I_{W2} \right]}{2} - 1.2$$

$$I_{W1} - I_{W2} + 1.2 = 0.6 \times \frac{\left[I_{W1} + I_{W2} \right]}{2}$$

$$I_{W1} - I_{W2} + 1.2 = 0.3 \times \left[I_{W1} + I_{W2} \right]$$

$$I_{W1} - I_{W2} + 1.2 = 0.3I_{W1} + 0.3I_{W2}$$

$$I_{W1} - 0.3I_{W1} + 1.2 = 0.3I_{W2} + I_{W2}$$

$$0.7I_{W1} + 1.2 = 1.3I_{W2}$$

$$0.7I_{W1} + 1.2 = 1.3I_{W2}$$

$$0.7I_{W1} = 1.3I_{W2} - 1.2$$

$$I_{W1} = \frac{1.3I_{W2} - 1.2}{0.7}$$

$$I_{W1} = \frac{1.3I_{W2}}{0.7} - \frac{1.2}{0.7}$$

$$I_{W1} = 1.857I_{W2} - 1.714$$

$$0.7I_{W1} + 1.2 = 1.3I_{W2}$$

$$I_{W2} = \frac{0.7I_{W1} + 1.2}{1.3}$$

$$I_{W2} = \frac{0.7I_{W1}}{1.3} + \frac{1.2}{1.3}$$

$$I_{W2} = 0.538I_{W1} + 0.923$$

Remember that the calculations so far are in per-unit. The following calculations change the per-unit to actual current applied.

$$I_{W1} = 1.857 I_{W2} - 1.714$$

$$\frac{I_{AW1}}{TAP1} = 1.857 \times \frac{I_{AW2}}{TAP2} - 1.714$$

$$\frac{I_{AW1}}{2.41} = 1.857 \times \frac{I_{AW2}}{4.61} - 1.714$$

$$\frac{I_{AW1}}{2.41} = 0.403 I_{AW2} - 1.714$$

$$I_{AW1} = \left(0.403 I_{AW2} - 1.714\right) \times 2.41$$

$$I_{AW1} = \left(0.403 I_{AW2} \times 2.41\right) - \left(1.714 \times 2.41\right)$$

$$I_{AW1} = 0.971 I_{AW2} - 4.131$$

$$W2 = 0.538 W1 + 0.923$$

$$\frac{I_{AW2}}{TAP2} = 0.538 \frac{I_{AW1}}{TAP1} + 0.923$$

$$\frac{I_{AW2}}{4.61} = 0.538 \frac{I_{AW1}}{2.41} + 0.923$$

$$\frac{I_{AW2}}{4.61} = 0.223 I_{AW1} + 0.923$$

$$I_{AW2} = \left(0.223 I_{AW1} + 0.923\right) \times 4.61$$

$$I_{AW2} = \left(0.223 I_{AW1} \times 4.61\right) + \left(0.923 \times 4.61\right)$$

$$I_{AW2} = 1.029 I_{AW1} + 4.255$$

The actual characteristic curve of this 87-element using Winding-1 and Winding-2 currents can be plotted where W2 (Restraint Current) is a group of arbitrary numbers and W1 (Operate Current) uses the following formulas:

- $If(I_{AW2} < 6.915A$ then $I_{AW1} = 0.723 + 0.523 I_{AW2}$
- $If(I_{AW2} > 6.915A$ and $<12.406A)$ then $I_{AW1} = 0.639 I_{AW2}$
- $If(I_{AW2} > 12.406A)$ then $I_{AW1} = 0.971 I_{AW2} - 4.131$
- The spreadsheet calculation for these equations could be:

$$\text{EXPECTED} = IF(I_{AW2} < 6.915, 0.723 + (0.523 * I_{AW2}), IF(I_{AW2} < 12.406, 0.639 * I_{AW2}, (0.971 * I_{AW2}) - 4.131))$$

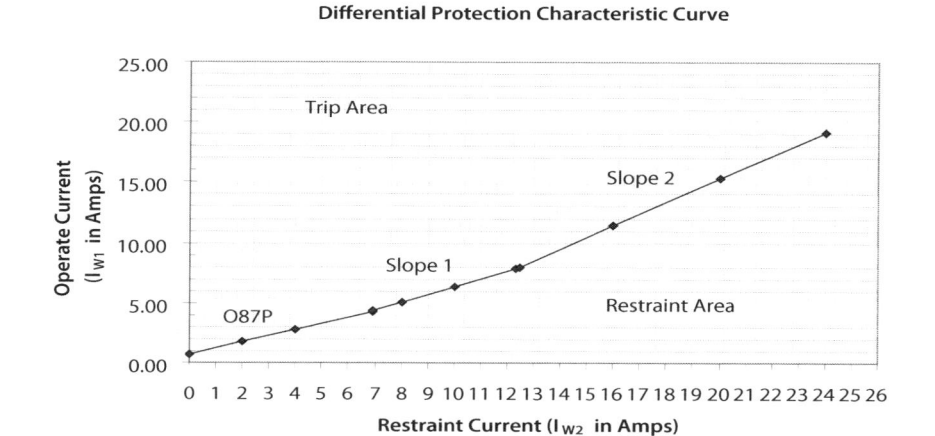

Differential Protection Characteristic Curve

Figure 13-55: Percentage Differential Protection Dual Slope Characteristic Curve in Amps

You could use this graph to determine the expected pickup current for a Slope-1 test by choosing a W2 current after the Minimum Pickup and less than Slope-2. Find that current on the x-axis, follow the current up until it crosses the line, then follow the crossover point back to the other axis to determine the expected pickup. For example, we could test Slope-1 by applying 2xTap to both windings and increase the W1 current until the relay operates. The relay should operate at approximately 5.9A as shown in Figure 13-56. Test Slope-2 by choosing a current below the Slope-2 portion of the graph (4xTap=18.44A), follow 18.44A from the x-axis to the characteristic curve, and follow it to the y-axis. The expected pickup current for the Slope-2 test is 13.7A as per Figure 13-56.

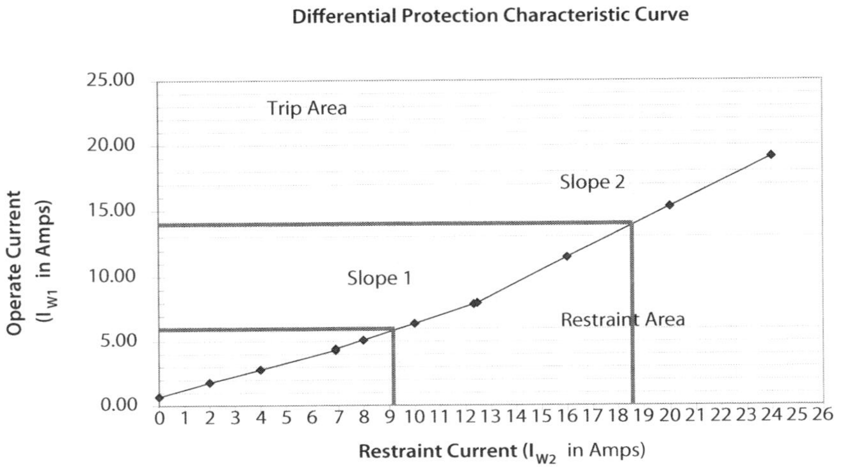

Differential Protection Characteristic Curve

Figure 13-56: Using Graphs to Determine Pickup Settings

It is more convenient to use formulas to determine the expected pickup current as shown below.

SLOPE-1	SLOPE-2
$I_{AW1} = 0.639 I_{AW2}$ $I_{AW1} = 0.639 (9.22A)$ $I_{AW1} = 5.892A$	$I_{AW1} = 0.971 I_{AW2} - 4.131$ $I_{AW1} = 0.971 (18.44A) - 4.131$ $I_{AW1} = 17.905A - 4.131$ $I_{AW1} = 13.774A$

The test sheet for this test could be similar to the test sheet below with the "EXPECTED" calculation:

$$EXPECTED = IF(I_{AW2} < 6.915, 0.723 + (0.523 * I_{AW2}), IF(I_{AW2} < 12.406, 0.639 * I_{AW2}, (0.971 * I_{AW2}) - 4.131))$$

DIFFERENTIAL TEST RESULTS							
PICK UP		0.3	TIME DELAY		0		
SLOPE 1		20%	TAP1		2.41	W1CTC	12
SLOPE 2		60%	TAP2		4.61	W2CTC	1
SLOPE TESTS (Amps)							
TEST	IAW2 (A)		I_W1 PHASE (A)		EXPECTED (A)	% ERROR	
1	10.0	A	6.401		6.390	0.17	
2	12.0	A	7.69		7.668	0.29	
3	16.0	A	11.421		11.405	0.14	
4	18.0	A	13.368		13.347	0.16	

C) Post-Test Calculations

It is often easier to pick random points and perform a calculation to determine if the test result is correct rather than perform all of the calculations above. We're using the same information for this test as summarized below:

	WINDING-1	WINDING-2
W_CTC	12	1
Tap_	2.41	4.61
O87P	0.30	
SLP1	20	
SLP2	60	
IRS1	3.0	

IRT is defined as "The magnitudes of the winding currents are then summed in a simple scalar addition and divided by 2." This can be translated into $\frac{I_{W1} + I_{W2}}{2}$ because we will use balanced 3-phase currents at the theoretical normal operating angles. IOP is defined as "In each element a phasor addition sums the winding currents. The magnitude of this result is IOP which can be translated into $I_{W1} - I_{W2}$ for our calculations because we will use balanced 3-phase currents at the theoretical normal operating angles. Remember that both of these formulas are still in per-unit and not actual Amp values.

O87P	SLP1	SLP2
$O87P \times \dfrac{100}{SLP1} < IRT$ $0.3 \times \dfrac{100}{20} < IRT$ $0.3 \times 5 < IRT$ $1.5 < IRT$	$IOP = \dfrac{SLP1}{100} \times IRT$ $IOP = \dfrac{20}{100} \times IRT$ $IOP = 0.20 \times IRT$	$IOP = \left[\dfrac{SLP2}{100} \times IRT\right] + \left[IRS1 \times \left(\dfrac{SLP1 - SLP2}{100}\right)\right]$ $IOP = \left[\dfrac{60}{100} \times IRT\right] + \left[3.0 \times \left(\dfrac{20-60}{100}\right)\right]$ $IOP = 0.6IRT + \left[3.0 \times \left(\dfrac{-40}{100}\right)\right]$ $IOP = 0.6IRT + \left[3.0 \times -0.40\right]$ $IOP = 0.6IRT - 1.2$

The following test results were recorded:

DIFFERENTIAL TEST RESULTS					
PICK UP	0.3	TIME DELAY	0		
SLOPE 1	20%	TAP1	2.41	W1CTC	12
SLOPE 2	60%	TAP2	4.61	W2CTC	1
PICKUP TESTS							
TEST	IAW2 (A)		IAW1 (A)	IRT	IOP	EXPECTED	ACTUAL	ERROR (%)
---	---	---	---	---	---	---	---	---
1	0.0	A	0.726					
2	2.0	A	1.779					
3	4.0	A	2.825					
4	6.0	A	3.878					
5	8.0	A	5.128					
6	10.0	A	6.401					
7	12.0	A	7.69					
8	14.0	A	9.47					
9	16.0	A	11.421					
10	18.0	A	13.368					
11	20.0	A	15.309					
12	22.0	A	17.25					
13	24.0	A	19.197					
14	26.0	A	21.143					
15	28.0	A	23.091					

First we apply the formula for IRT	Then we apply the formula for IOP
$$IRT = \frac{I_{W1} + I_{W2}}{2}$$ $$IRT = \frac{\dfrac{I_{AW1}}{TAP1} + \dfrac{I_{AW2}}{TAP2}}{2}$$	$$IOP = I_{W1} - I_{W2}$$ $$IOP = \frac{I_{AW1}}{TAP1} - \frac{I_{AW2}}{TAP2}$$
$$TEST4\ IRT = \frac{\dfrac{3.878}{2.41} + \dfrac{6.0}{4.61}}{2}$$ $$TEST4\ IRT = 1.455$$	$$TEST4\ IOP = \frac{3.878}{2.41} - \frac{6.0}{4.61}$$ $$TEST4\ IOP = 0.308$$
$$TEST6\ IRT = \frac{\dfrac{6.401}{2.41} + \dfrac{10.0}{4.61}}{2}$$ $$TEST6\ IRT = 2.413$$	$$TEST6\ IOP = \frac{6.401}{2.41} - \frac{10.0}{4.61}$$ $$TEST6\ IOP = 0.487$$
$$TEST10\ IRT = \frac{\dfrac{13.368}{2.41} + \dfrac{18.0}{4.61}}{2}$$ $$TEST10\ IRT = 4.726$$	$$TEST10\ IOP = \frac{13.368}{2.41} - \frac{18.0}{4.61}$$ $$TEST10\ IOP = 1.642$$

Now our test sheet is shown below:

DIFFERENTIAL TEST RESULTS							
PICK UP	0.3	TIME DELAY	0				
SLOPE 1	20%	TAP1	2.41	W1CTC		12	
SLOPE 2	60%	TAP2	4.61	W2CTC		1	
PICKUP TESTS							
TEST	IAW2 (A)	IAW1 (A)	IRT	IOP	EXPECTED	ACTUAL	ERROR (%)
1	0.0 A	0.726	0.151	0.301			
2	2.0 A	1.779	0.586	0.304			
3	4.0 A	2.825	1.020	0.305			
4	6.0 A	3.878	1.455	0.308			
5	8.0 A	5.128	1.932	0.392			
6	10.0 A	6.401	2.413	0.487			
7	12.0 A	7.69	2.897	0.588			
8	14.0 A	9.47	3.483	0.893			
9	16.0 A	11.421	4.105	1.268			
10	18.0 A	13.368	4.726	1.642			
11	20.0 A	15.309	5.345	2.014			
12	22.0 A	17.25	5.965	2.385			
13	24.0 A	19.197	6.586	2.759			
14	26.0 A	21.143	7.206	3.133			
15	28.0 A	23.091	7.828	3.508			

With IOP and IRT, the expected result formulas are fairly simple.

If IRT <1.5, IOP = 0.3	1.5 < IRT < 3 = Slope-1	IRT > 3 = Slope-2
	$IOP = \dfrac{SLP1}{100} \times IRT$ $\dfrac{IOP}{IRT} = \dfrac{SLP1}{100}$ $SLP1 = 100 \times \dfrac{IOP}{IRT}$	$IOP = \left[\dfrac{SLP2}{100} \times IRT\right] + \left[IRS1 \times \left(\dfrac{SLP1 - SLP2}{100}\right)\right]$ $IOP = \left[\dfrac{SLP2}{100} \times IRT\right] - 1.2$ $IOP + 1.2 = \left[\dfrac{SLP2}{100} \times IRT\right]$
	$Test6\ SLP1 = 100 \times \dfrac{IOP}{IRT}$ $Test6\ SLP1 = 100 \times \dfrac{0.487}{2.413}$ $Test6\ SLP1 = 20.18\%$	$\dfrac{IOP + 1.2}{IRT} = \dfrac{SLP2}{100}$ $SLP2 = 100 \times \dfrac{IOP + 1.2}{IRT}$
	$Test7\ SLP1 = 100 \times \dfrac{0.588}{2.897}$ $Test7\ SLP1 = 20.30\%$	$Test10\ SLP2 = 100 \times \dfrac{IOP + 1.2}{IRT}$ $Test10\ SLP2 = 100 \times \dfrac{1.642 + 1.2}{4.726}$ $Test10\ SLP2 = 60.13\%$

The final test sheet could look like the following:

DIFFERENTIAL TEST RESULTS							
PICK UP		0.3	TIME DELAY	0			
SLOPE 1		20%	TAP1	2.41	W1CTC		12
SLOPE 2		60%	TAP2	4.61	W2CTC		1
PICKUP TESTS							
TEST	IAW2 (A)	IAW1 (A)	IRT	IOP	EXPECTED	ACTUAL	ERROR (%)
1	0.0 A	0.726	0.151	0.301	0.30	0.30	0.41
2	2.0 A	1.779	0.586	0.304	0.30	0.30	1.44
3	4.0 A	2.825	1.020	0.305	0.30	0.30	1.51
4	6.0 A	3.878	1.455	0.308	0.30	0.31	2.54
5	8.0 A	5.128	1.932	0.392	20.00	20.32	1.59
6	10.0 A	6.401	2.413	0.487	20.00	20.18	0.89
7	12.0 A	7.69	2.897	0.588	20.00	20.29	1.46
8	14.0 A	9.47	3.483	0.893	60.00	60.08	0.13
9	16.0 A	11.421	4.105	1.268	60.00	60.13	0.22
10	18.0 A	13.368	4.726	1.642	60.00	60.15	0.24
11	20.0 A	15.309	5.345	2.014	60.00	60.13	0.21
12	22.0 A	17.25	5.965	2.385	60.00	60.11	0.18
13	24.0 A	19.197	6.586	2.759	60.00	60.12	0.20
14	26.0 A	21.143	7.206	3.133	60.00	60.13	0.21
15	28.0 A	23.091	7.828	3.508	60.00	60.14	0.24

The formulas used for this test report are:

- IRT=((IAW2 / TAP2) + (IAW1 / TAP1)) / 2
- IOP=(IAW1/TAP1)-(IAW2/TAP2)
- EXPECTED=IF(IRT<(O87P × 100 / SLP1),O87P,IF(IRT<3,SLP1,SLP2))
- ACTUAL=IF(IRT<(O87P × 100 / SLP1),IOP,IF(IRT<3,100*(IOP/IRT),100*((IOP+1.2)/IRT)))

D) Alternate Slope Calculation

There is another way to test slope that does not require as much preparation or calculation. Follow the Post-Test Calculation in the previous section to the point that IRT and IOP are calculated and pick two test points from each slope. You can tell which test will be Slope-1 if the IRT is between 1.5 and 3. Slope-2 will occur if the IRT is greater than 3.

DIFFERENTIAL TEST RESULTS						
PICK UP	0.3	TIME DELAY	0			
SLOPE 1	20%	TAP1	2.41	W1CTC		12
SLOPE 2	60%	TAP2	4.61	W2CTC		1
PICKUP TESTS						
TEST	IAW2 (A)	IAW1 (A)	IRT	IOP	EXPECTED	ACTUAL	ERROR (%)
6	10.0 A	6.401	2.413	0.487			
7	12.0 A	7.69	2.897	0.588			
10	18.0 A	13.368	4.726	1.642			
11	20.0 A	15.309	5.345	2.014			

You can use the graphical Rise over Run formula to determine the slope if you know the correct IOP and IRT for any two test results for a given slope as shown in the following example.

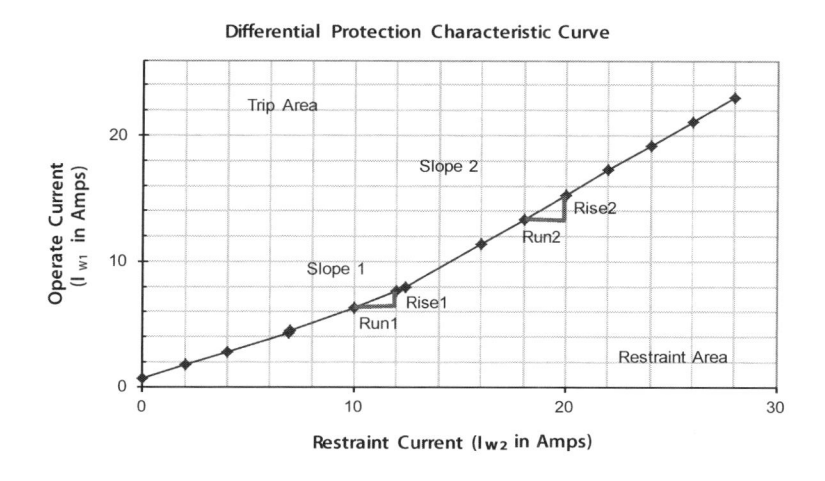

Figure 13-57: Determine Slope by Rise/Run Calculation

Determine slope using the rise-over-run graphical method as shown in Figure 13-57 and the following formulas:

$$\%Slope1 = 100 \times \frac{Rise}{Run}$$

$$\%Slope1 = 100 \times \frac{IOP7 - IOP6}{IRT7 - IRT6}$$

$$\%Slope1 = 100 \times \frac{0.588 - 0.487}{2.897 - 2.413}$$

$$\%Slope1 = 100 \times \frac{0.101}{0.484}$$

$$\%Slope1 = 20.86\%$$

$$\%Slope2 = 100 \times \frac{Rise}{Run}$$

$$\%Slope2 = 100 \times \frac{IOP11 - IOP10}{IRT11 - IRT10}$$

$$\%Slope2 = 100 \times \frac{2.014 - 1.642}{5.345 - 4.726}$$

$$\%Slope2 = 100 \times \frac{0.372}{0.619}$$

$$\%Slope2 = 60.09\%$$

The final test sheet could look like the following:

DIFFERENTIAL TEST RESULTS							
PICK UP		0.3	TIME DELAY	0			
SLOPE 1		20%	TAP1	2.41	W1CTC		12
SLOPE 2		60%	TAP2	4.61	W2CTC		1
PICKUP TESTS							
TEST	IAW2 (A)	IAW1 (A)	IRT	IOP	EXPECTED	ACTUAL	ERROR (%)
6	10.0 A	6.401	2.413	0.487			
7	12.0 A	7.69	2.897	0.588	20.00	20.86	4.28
10	18.0 A	13.368	4.726	1.642			
11	20.0 A	15.309	5.345	2.014	60.00	59.97	-0.06

E) 3-Phase Differential Slope Test Procedure

The slope test procedure seems straightforward. Apply 3-phase balanced current into Winding-1 and Winding-2 using the phase shift described in the previous section of this chapter, and raise all three currents in one winding until the relay operates. The difficult part of this procedure is determining what starting current to apply and what the expected pickup should be. You can use one of the following test procedures to test the relay slope settings:

- **Pre-Test Calculations**—Use the calculations to determine the actual restraint and expected operate currents and insert these values into the procedure.
- **Post-Test Calculations**—Start at 0.00A restraint current and determine pickup. Increase the restraint current and determine pickup and repeat until the O87P, Slope-1, and Slope-2 settings are tested. Calculate IRT after each test to determine whether O87P, Slope-1, or Slope-2.

Follow these steps to test the Differential Slope settings using six current channels:

1. Determine how you will monitor pickup and set the relay accordingly, if required. (Pickup indication by LED, output contact, front panel display, etc. See the *Relay Test Procedures* section starting on page 109 for details.)

2. Determine the Slope-1 restraint current by selecting a value between the Minimum-Pickup and Slope-2 transition points. Apply balanced 3-phase restraint current through W2 at the phase-angle described in the "3-Phase Restrained-Differential Slope Testing" chapter of this chapter. The pickup indication should be ON because we have applied a 200% slope as per the calculations earlier in this document. The applied W2 current for our example will be:

CONNECTION	SEL-387	W1AØ	W1BØ	W1CØ	W2AØ	W2BØ	W2CØ
	W1CTC=12 W2CTC=1	0A 0°	0A -120°	0A 120°	10.0A@ 150°	10.0A@ 30°	10.0A@ -90°

3. We are applying current greater than 5.0A into the relay so this test should be completed as quickly as possible. You may want to apply the previous step for a moment to ensure the 87-element operates and setup the next step offline.

4. Apply an equal, but opposite (accounting for phase shift) 3-phase current in W1. You can use the Tap ratios to determine what the current should be. For example:

$$W1 = W2 \times \frac{TAP1}{TAP2}$$

$$W1 = 10 \times \frac{2.41}{4.61}$$

$$W1 = 10 \times 0.523$$

$$W1 = 5.23$$

CONNECTION	SEL-387	W1AØ	W1BØ	W1CØ	W2AØ	W2BØ	W2CØ
	W1CTC=12 W2CTC=1	5.23A 0°	5.23A -120°	5.23A 120°	10.0A@ 150°	10.0A@ 30°	10.0A@ -90°

The relay should not operate.

5. Raise the W1 current until the element operates. Because the applied current is higher than 10.0A, you could also use the pulse or jog method to minimize the amount of current applied to the relay. (Review the *Pickup Test Procedure if Pickup is Greater Than 10 Amps* section starting on page 204 for details.) The measured pickup for our example is 6.401A.

6. Some organizations want more than one test point to determine slope. Repeat Steps 2-5 with another restraint current between the Minimum-Pickup and Slope-2 transition points until the required number of tests are completed.

7. Test Slope-2 by repeating Steps 2-6 with a restraint current greater than the Slope-1 to Slope-2 transition level.

8. Compare the pickup test result to the manufacturer's specifications and calculate the percent error as shown in the following example.

DIFFERENTIAL TEST RESULTS							
PICK UP		0.3	TIME DELAY	0			
SLOPE 1		20%	TAP1	2.41	W1CTC		12
SLOPE 2		60%	TAP2	4.61	W2CTC		1
SLOPE TESTS (Amps)							
TEST	IAW2 (A)		I_W1 PHASE (A)		EXPECTED (A)	% ERROR	
1	10.0	A	6.401		6.390	0.17	
2	12.0	A	7.69		7.668	0.29	
3	16.0	A	11.421		11.405	0.14	
4	18.0	A	13.368		13.347	0.16	

Slope-1

In the TEST 1 example, we calculated that the 87-element should operate when IAW1 = 6.390A using the Slope-1 formula when IAW2 = 10.0A. We measured the pickup to be 6.175A. Looking at the "Differential Element" specification below, we see that the acceptable metering error is $\pm 0.5\%$ $\pm 0.1A$. All of the results in the test sheet are less than 0.5% so we can consider them acceptable.

Differential Element

Unrestrained Pickup Range:	1–20 in per unit of tap
Restrained Pickup Range:	0.1–1.0 in per unit of tap
Pickup Accuracy (A secondary)	
5 A Model:	±5% ±0.10 A
1 A Model:	±5% ±0.02 A

$$\frac{\text{IAW1-EXPECTED}}{\text{EXPECTED}} \times 100 = \text{Percent Error}$$

$$\frac{6.401\text{-}6.390}{6.390} \times 100 = \text{Percent Error}$$

0.17% Allowable Percent Error

We can calculate the manufacturer's allowable percent error. Using "$\pm 0.5\%$ $\pm 0.1A$" from the specifications above, the allowable percent error is 1.6%.

$$\frac{\text{Maximum Accuracy Tolerance}}{\text{Expected}} \times 100 = \text{Allowable Percent Error}$$

$$\frac{(0.5\% \times \text{EXPECTED}) + 0.1A}{\text{Expected}} \times 100 = \text{Allowable Percent Error}$$

$$\frac{(0.005 \times 6.39A) + 0.1A}{6.39A} \times 100 = \text{Allowable Percent Error}$$

$$\frac{(0.032) + 0.1A}{6.39A} \times 100 = \text{Allowable Percent Error}$$

$$\frac{0.132}{6.39A} \times 100 = \text{Allowable Percent Error}$$

2.1% Allowable Percent Error

Rule of Thumb

Remember that there are other factors that will affect the test result such as:

- Relay close time. (The longer it takes the contact to close the larger the test result will be if ramping the pickup current.)
- Test-set analog output error.
- Test-set sensing time.

Some of these error factors can be significant and a rule-of-thumb 5% error is usually applied to test results.

9. Repeat the pickup test for all windings that are part of the differential scheme. If more than two windings are used, change all connections and references to W2 to the next winding under test. (W2 becomes W3 for W1-to-W3 tests)

7. 1-Phase Restrained-Differential Slope Testing

3-Phase differential slope testing is relatively straight-forward as you simulate balanced 3-phase conditions that the relay would expect to see in an ideal world to make all of the calculations easier. The 1-Phase testing procedure described in this section uses the same calculations and test procedure described in the previous "3-Phase Differential Slope Testing" section but the connections are completely different. Schweitzer Engineering Laboratories does have a test procedure and calculation for single-phase testing but I have found that the procedure described here is easier to apply and calculate.

Our example will use the same settings used earlier in this chapter for the SEL-387 which are summarized to:

	WINDING-1	WINDING-2
E87W_	Y	Y
W_CT	Y	Y
W_CTC	12	1
VWDG_	230	18
Tap_	2.41	4.61
O87P	0.30	
SLP1	20	
SLP2	60	
IRS1	3.0	

A) Understanding the Test-Set Connections

Wye-Wye and Delta-Delta connections are not a problem as there is no phase shift and the current in any phase-winding is reflected in the same phases on the other windings. However in a Delta-Wye or Wye-Delta transformer, the current in one Wye phase CT is reflected into two phases of the Delta CT as shown in Figure 13-58.

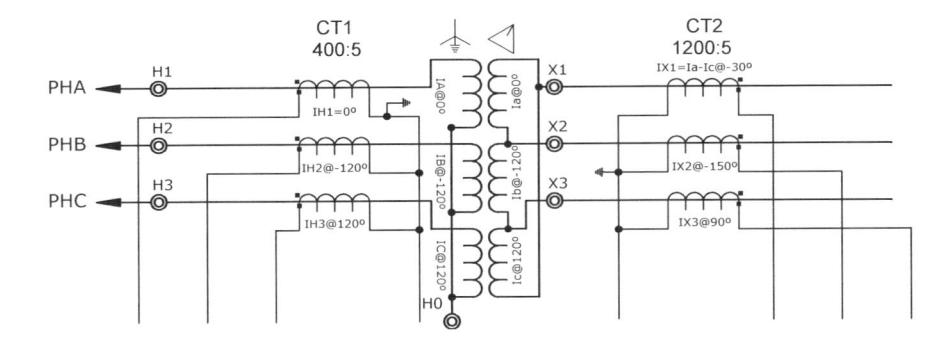

Figure 13-58: YDac Transformer Connection

When testing transformer differential elements on a single-phase basis, we must apply the W1 current to the phases W2 usually measures. Then apply the W2 current to the phases

W1 normally measures. For the example in Figure 13-58, connect the W1 test current to IAW1 and ICW1 in series and the W2 test current is applied to IAW2 for an A-phase test. Connect the B-phase W1 test current to IBW1 and IAW1 and W2 test current to IBW2. Connect the C-phase W1 test current to ICW1 and IBW1 and W2 test current to ICW2.

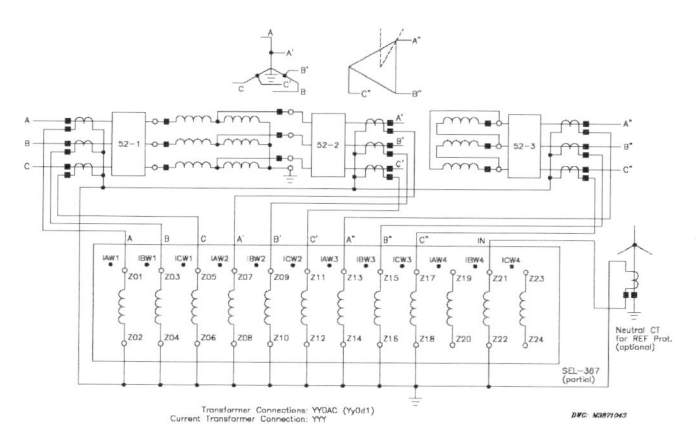

Another problem with 1-Phase testing is the unbalance or zero sequence current that is applied to the relay. Zero-sequence is a problem in transformer differential protection because a phase-to-ground fault outside the transformer on the power system can cause zero-sequence current to flow through one winding of the transformer and not appear in the other winding due to a delta or tertiary winding. Most digital relays apply zero-sequence filtering to prevent a transformer differential trip for ground faults on the power system. Zero-sequence filtering is applied on the SEL-387 relay when the W_CTC setting equals 12. A phase-to-phase connection is required to compensate for the zero-sequence compensation. When both W_CTC settings are equal, both windings will use a phase-to-phase connection.

Use the table in Figure 13-59 to determine the correct 1-Phase test connections.

GE T-60, T-30	SR-745	Beckwith	SEL-587	SEL-387		A Phase		B Phase		C-Phase	
				W1	W2	W1	W2	W1	W2	W1	W2
W1=0, W2=0	Y/y0°	Yyyy	YY	12	12	A-B	A-B	B-C	B-C	C-A	C-A
W1=0, W2=30°	Y/d30°		YDAC	12	1	A-C	A-N	B-A	B-N	C-B	C-N
W1=0, W2=330°	Y/d330°		DACY	12	11	A-B	A-N	B-C	B-N	C-A	C-N
				12	0	A-B	A-B	B-C	B-C	C-A	C-A
D/d0°	D/d0°		DACDAC	0	0	A-N	A-N	B-N	B-N	C-N	C-N
				0	12	A-B	A-B	B-C	B-C	C-A	C-A
W1=0, W2=30°	D/y30°		DABY	0	1	A-C	A-N	B-A	B-N	C-B	C-N
W1=0, W2=330°	D/y330°		DACY	0	11	A-B	A-N	B-C	B-N	C-A	C-N
				1	0	A-N	A-C	B-N	B-A	C-N	C-B
				1	12	A-N	A-C	B-N	B-A	C-N	C-B
				1	1	A-N	A-N	B-N	B-N	C-N	C-N
				11	0	A-N	A-B	B-N	B-C	C-N	C-A
				11	12	A-N	A-B	B-N	B-C	C-N	C-A
				11	11	A-N	A-N	B-N	B-N	C-N	C-N

Figure 13-59: Transformer Relay Connections for Single-Phase Differential Testing

B) Test-Set Connections

The connection diagram for the SEL-387 is as follows:

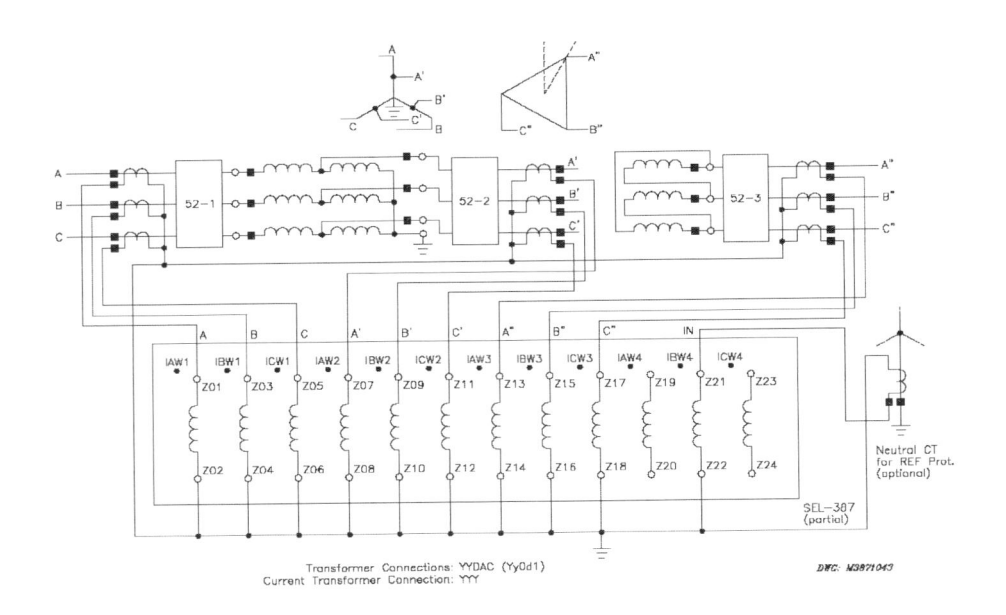

Figure 13-60: Schweitzer Electric SEL-387 Transformer Protective Relay Connections

The winding settings for this transformer are W1CTC=12 and W2CTC=1. Using the table in Figure 13-61, the following test-set connections will be required for each phase:

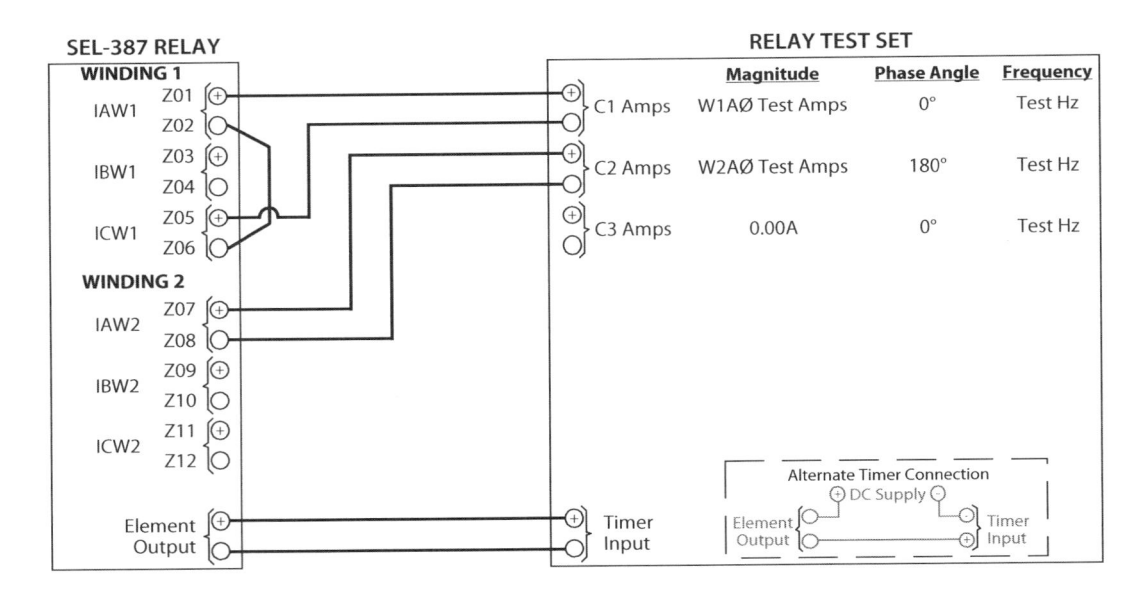

Figure 13-61: 1-Phase Restrained-Differential Slope Test-Set AØ Yd1 Connections

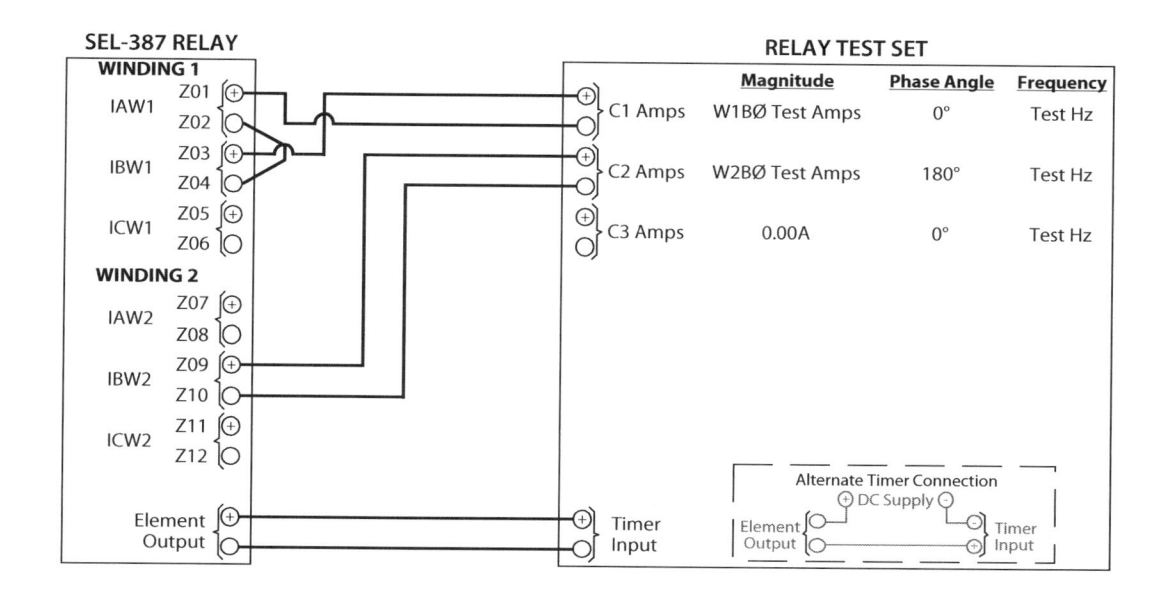

Figure 13-62: 1-Phase Restrained-Differential Slope Test-Set BØ Yd1 Connections

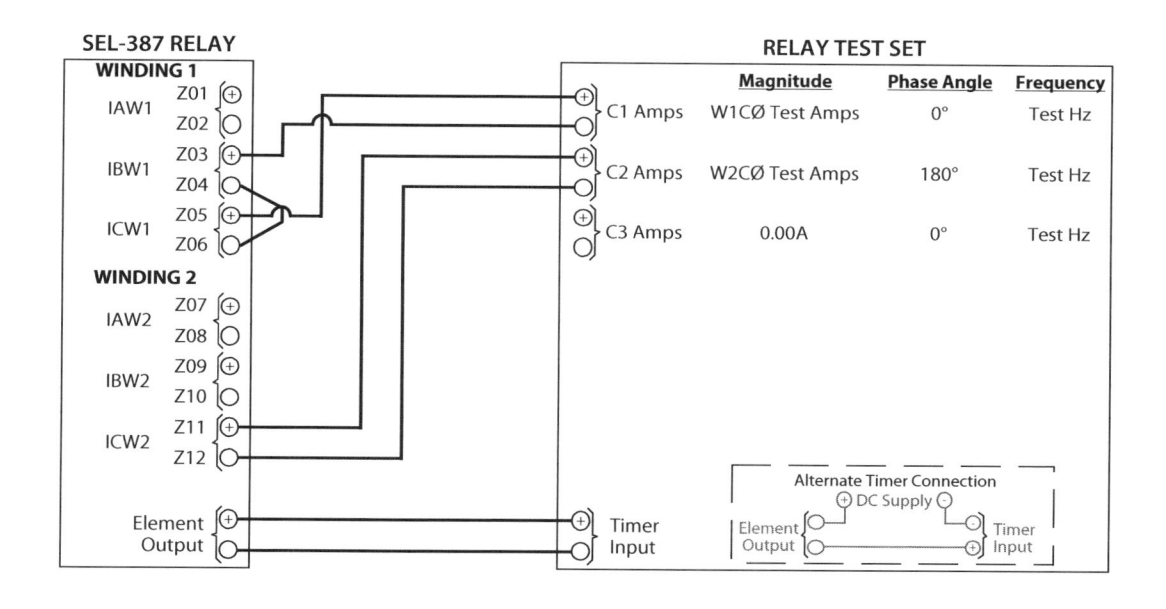

Figure 13-63: 1-Phase Restrained-Differential Slope Test-Set CØ Yd1 Connections

C) 1-Phase Differential Slope Test Procedure

Follow these steps to test the Differential Slope settings using the Post-Test Calculation method from the previous "3-Phase Differential Slope Testing" Section:

1. Determine how you will monitor pickup and set the relay accordingly, if required. (Pickup indication by LED, output contact, front panel display, etc. See the *Relay Test Procedures* section starting on page 109 for details.)

2. Apply a restraint current into the first test phase of Winding-2. The pickup indication should be on.

3. Apply an equal, but opposite current in W1 using the Tap ratios to determine what the current magnitude should be. For example:

$$W1 = W2 \times \frac{TAP1 \times A1}{TAP2 \times A2}$$

$$W1 = 5 \times \frac{2.41 \times 1}{4.61 \times 1.732}$$

$$W1 = 5 \times 0.302$$

$$W1 = 1.510$$

4. The pickup indication should now be off. Raise the Winding-1 current until the relay operates. Record the value on your test sheet. (Remember that you can also use the pulse or jog method to minimize the amount of current applied to the relay. Review the *Pickup Test Procedure if Pickup is Greater Than 10 Amps* section starting on page 204 for details.)

5. Calculate the IRT, IOP, and Slope using the formulas in the Post-Test Calculation section of the "3-Phase Differential Slope Testing" in this document. Use a 1.732 correction factor for windings with odd W_CTC settings and use the IRT calculation to determine if the test was O87P, Slope-1, or Slope-2. Raise the W2 current and repeat 2-5 until all elements of the differential protection have been tested.

- $IRT = \left[(IAW2 / (TAP2 \times A2)) + (IAW1 / (TAP1 \times A1)) \right] / 2$

- $IOP = \left[IAW1 / (TAP1 \times A1) \right] - \left[IAW2 / (TAP2 \times A2) \right]$

- EXPECTED=IF(IRT<(O87P $\times$ 100 / SLP1),O87P,IF(IRT<3,SLP1,SLP2))

- ACTUAL=IF(IRT<(O87P $\times$ 100 / SLP1),IOP,IF(IRT<3,100*(IOP/IRT),100*((IOP+1.2)/IRT)))

6. Compare the pickup test result to the manufacturer's specifications and calculate the percent error as shown in the following example.

DIFFERENTIAL TEST RESULTS								
PICK UP		0.3	TIME DELAY		0			CORR
SLOPE 1		20%	TAP1		2.41	W1CTC	12	1
SLOPE 2		60%	TAP2		4.61	W2CTC	1	1.732
PICKUP TESTS								
TEST	IAW2 (A)		IAW1 - ICW1 (A)	IRT	IOP	EXPECTED	ACTUAL	ERROR (%)
1	0.0	A	0.736	0.153	0.305	0.30	0.31	1.80
2	5.0	A	2.239	0.778	0.303	0.30	0.30	0.94
3	15.0	A	5.54	2.089	0.420	20.00	20.11	0.57
4	20.0	A	7.394	2.786	0.563	20.00	20.21	1.06
5	25.0	A	9.923	3.624	0.986	60.00	60.33	0.54
6	30.0	A	12.726	4.519	1.523	60.00	60.26	0.44
TEST	IBW2 (A)		IBW1 - IAW1 (A)	IRT	IOP	EXPECTED	ACTUAL	ERROR (%)
7	0.0	A	0.732	0.152	0.304	0.30	0.30	1.24
8	5.0	A	2.24	0.778	0.303	0.30	0.30	1.08
9	15.0	A	5.544	2.090	0.422	20.00	20.19	0.93
10	20.0	A	7.382	2.784	0.558	20.00	20.05	0.26
11	25.0	A	9.902	3.620	0.978	60.00	60.16	0.26
12	30.0	A	12.711	4.516	1.517	60.00	60.17	0.28
TEST	ICW2 (A)		ICW1 - IBW1(A)	IRT	IOP	EXPECTED	ACTUAL	ERROR (%)
13	0.0	A	0.732	0.152	0.304	0.30	0.30	1.24
14	6.0	A	2.544	0.904	0.304	0.30	0.30	1.38
15	15.0	A	5.535	2.088	0.418	20.00	20.02	0.12
16	21.0	A	7.753	2.924	0.587	20.00	20.08	0.38
17	24.0	A	9.342	3.441	0.871	60.00	60.17	0.28
18	27.0	A	11.035	3.980	1.197	60.00	60.23	0.38

7. Repeat Steps 2-6 for all three phases.

8. Repeat the pickup test for all windings that are part of the differential scheme. If more than two windings are used, change all connections and references to W2 to the next winding under test. (W2 becomes W3 for W1 to W3 tests)

8. Harmonic Restraint Testing

Transformers are inductive machines that require a magnetic field to operate. Under normal operating conditions, there are very small losses inside the transformer that are typically less than 3%. A large inrush of current is required in the first few cycles after energization to create the magnetic field necessary for transformer operation. This current only occurs in the first-energized winding and can be up to 10x the transformer's nominal current that, in any other circumstance, would be the definition of transformer differential and the differential relay should operate.

We do not want the transformer differential to operate every time a transformer is energized, and there are several options available to prevent transformer inrush from tripping the differential relay. It is possible to block the 87-element for a preset time after a circuit breaker is closed, but an internal transformer fault would not be isolated until the time delay had passed that will cause additional damage for the extended time caused by this workaround. You could increase the pickup setting, but this would prevent the relay from operating during high-impedance faults.

The most common technique used to prevent differential operation during inrush conditions is called Harmonic Restraint or Harmonic Blocking. Figure 13-64 displays a typical transformer inrush waveform. As you can see, this is not your typical sine wave. There is a significant DC offset in the first few cycles where the center point between the positive and negative peaks is not the x-axis. Also the waveform is extremely distorted and doesn't display the nice round peaks we normally see.

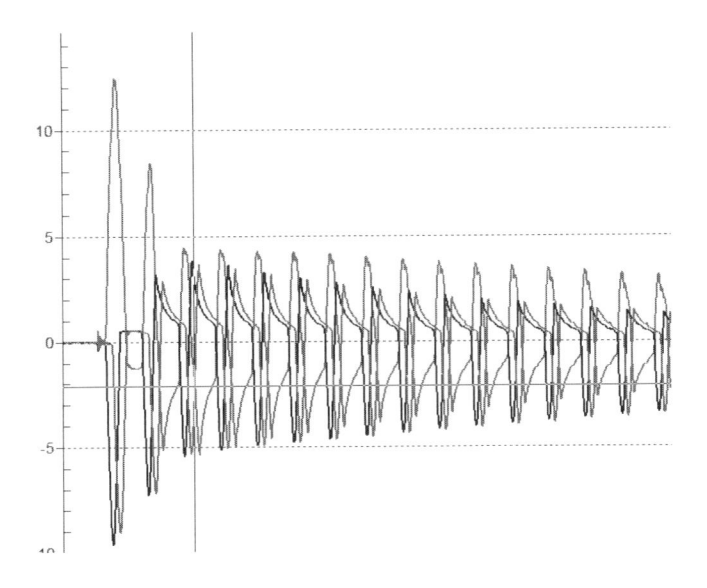

Figure 13-64: Transformer Inrush Waveform

Protective relay designers analyzed these waveforms and realized that the distortion was caused by even harmonic content that typically only occurred during transformer energization and not during transformer faults. Harmonic detectors were built into transformer differential relays which measure the percent of 2nd, 4th and/or 5th harmonics in a waveform. These harmonic detectors will prevent differential operation if the harmonics exceed a pre-defined setpoint.

A) Harmonic Restraint Test Connections

There are two possible connections for testing Harmonic Restraint depending on your test equipment. If your test equipment allows you to directly adjust the applied harmonic for an output current channel, you can use the standard timing test connections diagram shown in Figure 13-65.

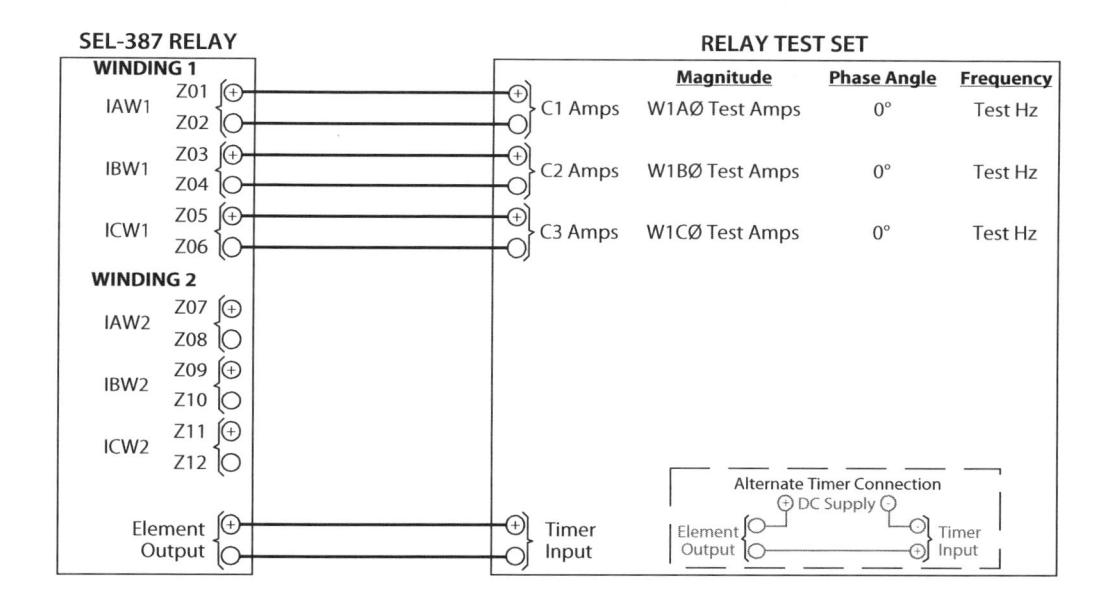

Figure 13-65: Simple 3-Phase 87-Element Test-Set Connections

Otherwise you will need two channels to test the Harmonic Restraint connected in parallel as shown in Figure 13-66. This test configuration includes one channel at the fundamental frequency and the second channel set at the harmonic frequency. The harmonic percentage is the ratio of harmonic current to fundamental current.

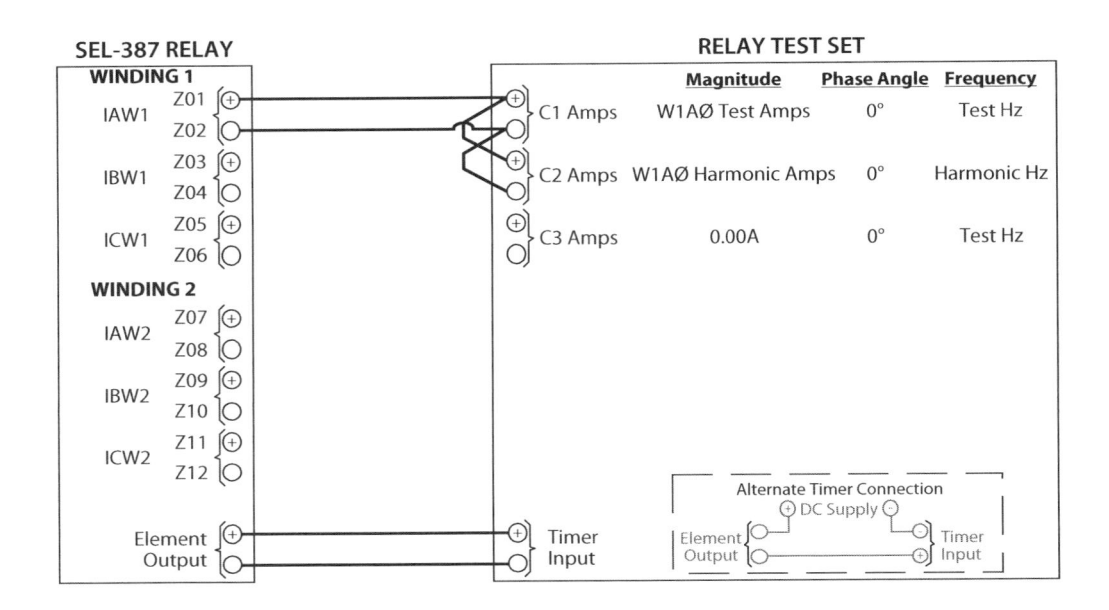

Figure 13-66: Simple, Higher Current 3-Phase 87-Element Test-Set Connections

B) Harmonic Restraint Test Procedure with Harmonics

Harmonic Restraint application can be very different depending on the relay manufacturer and model. Some relays will only restrain when the current is pulsed to look like the sudden inrush of a transformer. Other relays may require the harmonic to start below the setting and ramp up. Some other relays may require the harmonic to be above the setpoint and ramp down so you actually measure the dropout instead of the pickup. We will modify our timing test procedure to apply the pulse method which should work on most relays. Use the following procedure to test the Harmonic Restraint when a harmonic setting is available on your test-set.

1. Turn the test-set channel harmonic setting on and choose the correct harmonic. (Typically the 2nd, 4th, or 5th Harmonic.)

2. Apply the connection diagram from Figure 13-65. Connect the test-set input(s) to the relay output(s) that are programmed to operate when the restrained-differential relay operates.

3. Set a single-phase current at least 200% higher than the Minimum Pickup setting using the calculations in the "1-Phase Pickup Test Procedure" section of this chapter. (1.446A $(0.723 \times 2.0 = 1.446A)$ for Winding-1 or 4.79A $(2.395 \times 2.0 = 4.79A)$ for Winding-2.)

4. Apply the current and make sure the relay differential element operates.

5. Set the harmonic percentage 5% higher than the pickup setting. (25% for a 20% setting)

6. Apply the current and make sure the relay differential element does not operate.

7. Reduce the harmonic percentage to 5% below the pickup setting. (15% for a 20% setting)

8. Apply the test current and ensure the relay operates.

9. Increase the harmonic in 1% increments and apply until the relay does not operate.

10. Return to the last test harmonic when the element operated and increase the harmonic in 0.1% increments and apply until the relay does not operate. This is the Harmonic Restraint pickup result. Record on your test sheet.

11. Repeat Steps 3-10 for all phases.

DIFFERENTIAL TEST RESULTS								
PICK UP	0.3	TIME DELAY	0	PCT2	15			
SLOPE 1	20%	TAP1	2.41	PCT4	15			
SLOPE 2	60%	TAP2	4.61	PCT5	35			
HARMONIC RESTRAINT TESTS (Amps)								
TEST	HARMONIC	IAW1 (%)	IBW1 (%)	ICW1 (%)	EXPECTED (%)	% ERROR		
1	2nd	14.8	15.2	15.1	15.0	-1.3	1.3	0.7
2	4th	15.2	15.3	15.6	15.0	1.3	2.0	4.0
3	5th	34.8	35.6	35.9	35.0	-0.6	1.7	2.6

C) Harmonic Restraint Test Procedure with Current

Testing without a harmonic setting on your test-set requires two current channels connected in parallel with one current channel set at the fundamental frequency (60 Hz in our example) and the second current at the harmonic frequency. Harmonics are described in multiples of the fundamental frequency so a 2nd harmonic would be 120 Hz in our example or 100 Hz in a system with a 50Hz nominal frequency.

As described in the previous section, different relays have different operating characteristics when applying harmonic restraint and we'll use the pulse method with the following procedure.

1. Apply the connection diagram from Figure 13-66. Connect the test-set input(s) to the relay output(s) that are programmed to operate when the restrained-differential relay operates.

2. Set the current Channel 1 at the fundamental frequency (60Hz) with a magnitude at least 200% higher than the Minimum Pickup setting using the calculations in the "1-Phase Pickup Test Procedure" section of this chapter. ($1.446A$ $(0.723 \times 2.0 = 1.446A)$ for Winding-1 or $4.79A$ $(2.395 \times 2.0 = 4.79A)$ for Winding-2) Apply the current and make sure the relay differential element operates.

3. Set the current 2 Channel frequency to the full harmonic frequency. (2nd = 120 Hz, 4th = 240Hz, 5th = 300Hz) We'll call this "harmonic current" from now on.

4. Multiply the harmonic pickup setting (PCT2=15%) by the applied current from Step 2 $(1.446 \times 0.15 = 0.217A)$. This is your expected current.

5. Set your test-set's harmonic current output 10% higher than the expected harmonic current $(0.217 \times 1.10 = 0.239A)$. Apply the fundamental and harmonic currents simultaneously and the relay should not operate.

6. Reduce the harmonic percentage to 5% below the expected current (0.217 x 0.95 = 0.206A) and apply the fundamental and harmonic currents simultaneously. The relay should operate.

7. Increase the harmonic current in 1% increments and apply until the relay does not operate.

8. Return to the last test harmonic when the element operated, and increase the harmonic in 0.1% increments and apply until the relay does not operate. This is the Harmonic Restraint pickup result. Record on your test sheet.

9. Repeat Steps 2-8 for all phases.

DIFFERENTIAL TEST RESULTS							
PICK UP	0.3	TIME DELAY	0	PCT2	15		
SLOPE 1	20%	TAP1	2.41	PCT4	15		
SLOPE 2	60%	TAP2	4.61	PCT5	35		
HARMONIC RESTRAINT TESTS (Amps)							
HARM	FUND	IAW1 (A)	IBW1 (A)	ICW1 (A)	EXPECTED (A)	% ERROR		
PCT2	1.446 A	0.214	0.220	0.218	0.217	-1.4	1.3	0.6
PCT4	1.446 A	0.220	0.221	0.226	0.217	1.3	2.0	4.0
PCT5	1.446 A	0.503	0.515	0.519	0.506	-0.6	1.7	2.6

D) Evaluate Results

Use the harmonic specifications below to determine if the test results are acceptable. All of the example test results are less than 5% so we do not need to proceed further.

Harmonic Element

Pickup Accuracy (A secondary)
5 A Model: ±5% ±0.10 A
1 A Model: ±5% ±0.02 A
Time Delay Accuracy: ±0.1% ±0.25 cycle

We could calculate the exact allowable percent error with the following calculation which is a huge number because of the very low expected current:

$$\frac{\text{Maximum Accuracy Tolerance}}{\text{EXPECTED}} \times 100 = \text{Allowable Percent Error}$$

$$\frac{(5.0\% \times \text{EXPECTED}) + 0.1A}{\text{EXPECTED}} \times 100 = \text{Allowable Percent Error}$$

$$\frac{(0.05 \times 0.217) + 0.1A}{0.217A} \times 100 = \text{Allowable Percent Error}$$

$$\frac{(0.011) + 0.1A}{0.217A} \times 100 = \text{Allowable Percent Error}$$

$$\frac{0.111}{0.217A} \times 100 = \text{Allowable Percent Error}$$

51% Allowable Percent Error

9. Tips and Tricks to Overcome Common Obstacles

The following tips or tricks may help you overcome the most common obstacles:

- Before you start, apply current at a lower value and review the relay's measured values to make sure your test-set is actually producing an output and your connections are correct.
- Are the 2 winding currents $180°$ apart?
- If the pickup tests are off by a $\sqrt{3}$, 1.5, or 0.577 multiple; the relay probably is applying correction factors, and you should review the manufacturer's literature.
- Are you applying the currents into the same phase relationships?
- Did you calculate the correct Tap?
- Don't forget to use the complete operate and restrained calculations.
- Always use a Wye-connected winding's phase A as the reference when setting angles. If a Delta-winding is set as the reference, the phase shift will look backwards.
- Slope-2 tests near the break point can yield seemingly erratic results. Try a test point further from the break point.

Chapter 14

Unrestrained-Differential Testing

Electrical engineers are always concerned about close-onto-fault conditions, especially with transformer protection. What would happen if a transformer circuit breaker closed onto a fault in the first-energized winding? It is conceivable that the inrush condition could mask the fault for several cycles and cause additional damage or system disturbances. Additional protection can be applied which is set higher than the expected inrush current to detect faults on the first-energized winding of the transformer.

The unrestrained-differential element operation is very much like the early differential relays described in the "Application" section of this chapter as shown in Figure 14-1. There are no restraint coils to prevent mis-operations at high current levels and, in fact, the element is typically set at 8-10x Tap.

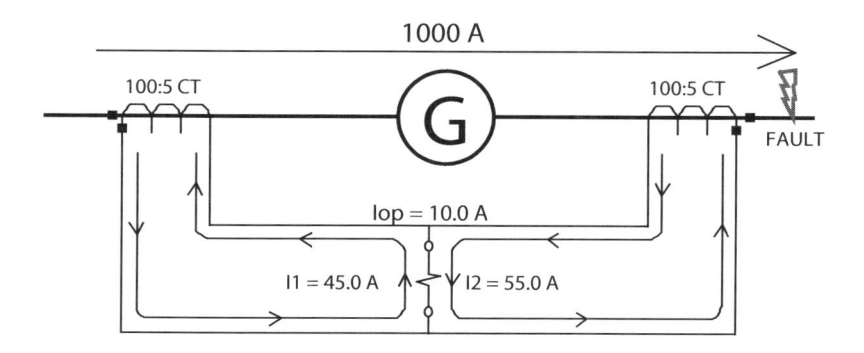

Figure 14-1: Simple Differential Protection with Worst Case CT Error and External Fault

With pickup settings in the 8-10x Tap range, unrestrained-differential protection becomes nearly identical to the test procedure described in the *Pickup Test Procedure if Pickup is Greater Than 10 Amps* section starting on page 204.

1. Settings

A) Enable Setting

Many relays allow the user to enable or disable settings. Make sure that the element is ON or the relay may prevent you from entering settings. If the element is not used, the setting should be disabled or OFF to prevent confusion. Some relays will also have "Latched" or 'Unlatched" options. A Latched option indicates that the output contacts will remain closed after a trip until a reset is performed and acts as a lockout relay. Unlatched indicates that the relay output contacts will open when the trip conditions are no longer present.

B) Minimum Pickup

The Minimum Pickup setting is usually set in multiples of Tap and is the minimum amount of current required for the relay to operate.

C) Tap

The Tap setting defines the normal operate current based on the rated load of the protected equipment, the primary voltage, and the CT ratio. This setting is used as the per-unit operate current of the protected device and most differential settings are based on the Tap setting. Verify the correct Tap setting using the following formula:

$$TAP = \frac{CT_{SEC} \times Power}{\text{P-P Volts} \times \sqrt{3} \times CT_{DDT}}$$

D) Time Delay

The Time Delay setting sets a time delay between an 87-element pickup and trip. This setting is typically set at the minimum possible setting but can be set as high as 3 cycles for maximum reliability on some relays.

E) Block

The Block setting defines a condition that will prevent the differential protection from operating such as a status input from another device. This setting is rarely used. If enabled, make sure the condition is not true when testing. Always verify correct blocking operation by operating the end-device instead of a simulation to ensure the block has been correctly applied.

2. Test-Set Connections

The most basic test-set connection uses only one phase of the test-set with a test lead change between every pickup test. After the Winding-1 A-phase pickup test is performed, move the test leads from Winding-1 A-phase amps to B-phase amps and perform the test again. Repeat until all enabled phases are tested.

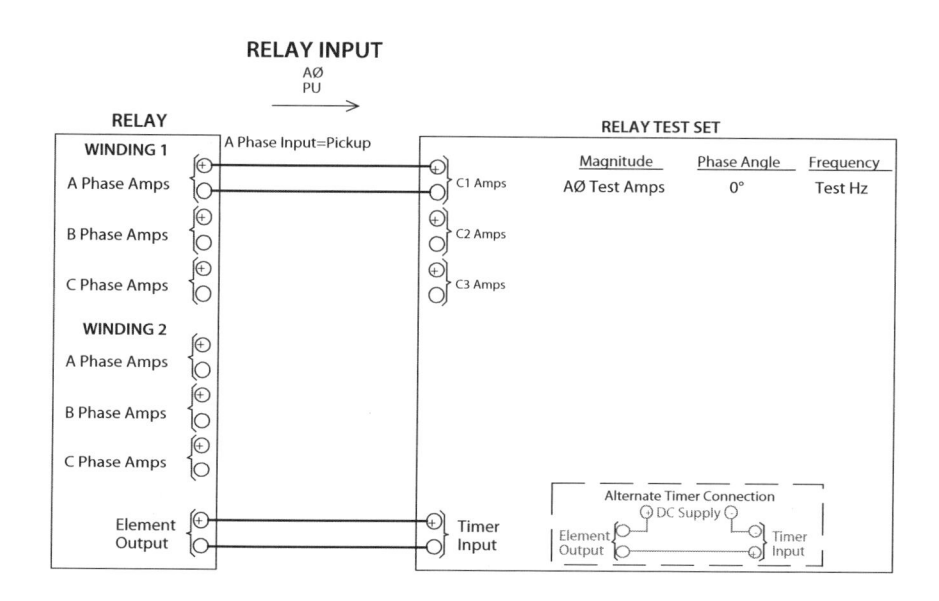

Figure 14-2: Simple 87U-Element Test-Set Connections

This element will require a lot of current so the connection in Figure 14-3 will probably be required if you treat the element as a 50-element.

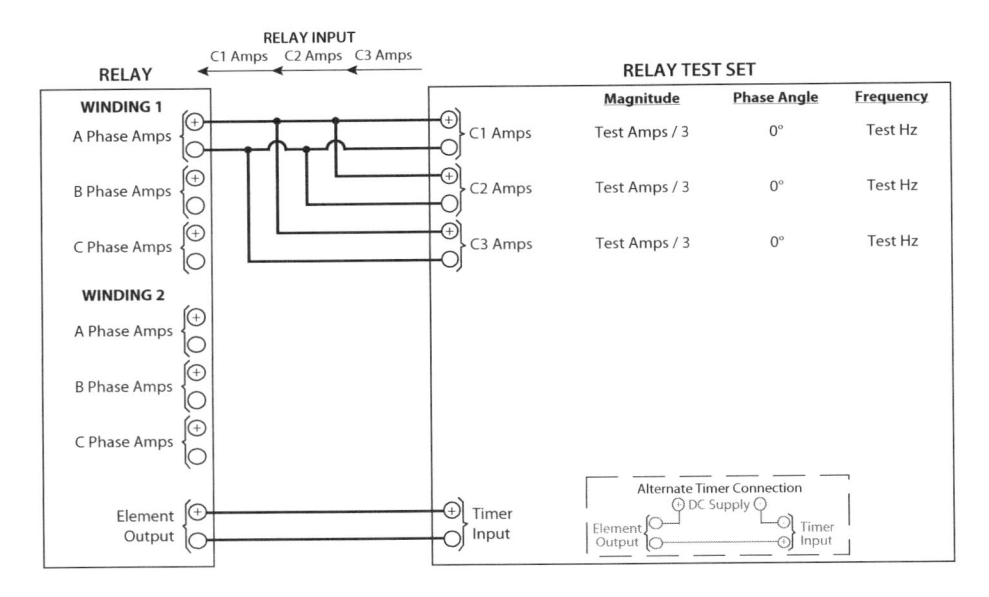

Figure 14-3: Parallel 87U-Element Test-Set Connections

If we realize that two different relay inputs are used to create the operate current, we can minimize the impact on any relay input by splitting the test current into the analog inputs that operate the unrestrained-differential element as shown in Figure 14-4. This connection also checks two windings simultaneously and reduces test times.

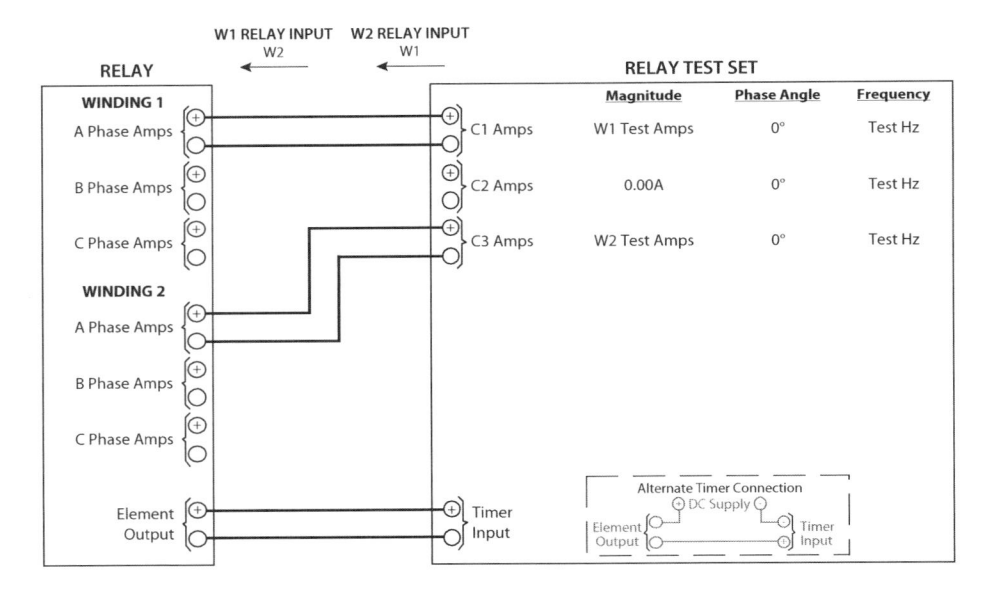

Figure 14-4: Parallel 87U-Element Test-Set Connections with Equal W_CTC Settings

The connection diagram in Figure 14-5 only works correctly if both W_CTC settings are identical. If the W_CTC settings are not identical, we need to apply the chart in Figure 14-6 to apply a connection diagram such as shown in the following connection drawings:

GE T-60, T-30	SR-745	Beckwith	SEL-587	SEL-387		A Phase		B Phase		C-Phase	
				W1	W2	W1	W2	W1	W2	W1	W2
W1=0, W2=0	Y/y0°	Yyyy	YY	12	12	A-B	A-B	B-C	B-C	C-A	C-A
W1=0, W2=30°	Y/d30°	YDACyy	YDAC	12	1	A-C	A-N	B-A	B-N	C-B	C-N
W1=0, W2=330°	Y/d330°	YDAByy	DACY	12	11	A-B	A-N	B-C	B-N	C-A	C-N
				12	0	A-B	A-B	B-C	B-C	C-A	C-A
D/d0°	D/d0°	DACDACyy	DACDAC	0	0	A-N	A-N	B-N	B-N	C-N	C-N
				0	12	A-B	A-B	B-C	B-C	C-A	C-A
W1=0, W2=30°	D/y30°	DAByyy	DABY	0	1	A-C	A-N	B-A	B-N	C-B	C-N
W1=0, W2=330°	D/y330°	DACyy	DACY	0	11	A-B	A-N	B-C	B-N	C-A	C-N
				1	0	A-N	A-C	B-N	B-A	C-N	C-B
				1	12	A-N	A-C	B-N	B-A	C-N	C-B
				1	1	A-N	A-N	B-N	B-N	C-N	C-N
				11	0	A-N	A-B	B-N	B-C	C-N	C-A
				11	12	A-N	A-B	B-N	B-C	C-N	C-A
				11	11	A-N	A-N	B-N	B-N	C-N	C-N

Figure 14-5: Transformer Relay Connections for Single-Phase Differential Testing

The winding settings for our example transformer are W1CTC=12 and W2CTC=1. Using the table in Figure 14-6, the following test-set connections will be required for A-phase. Use the chart to determine connections for B-phase and C-phase testing.

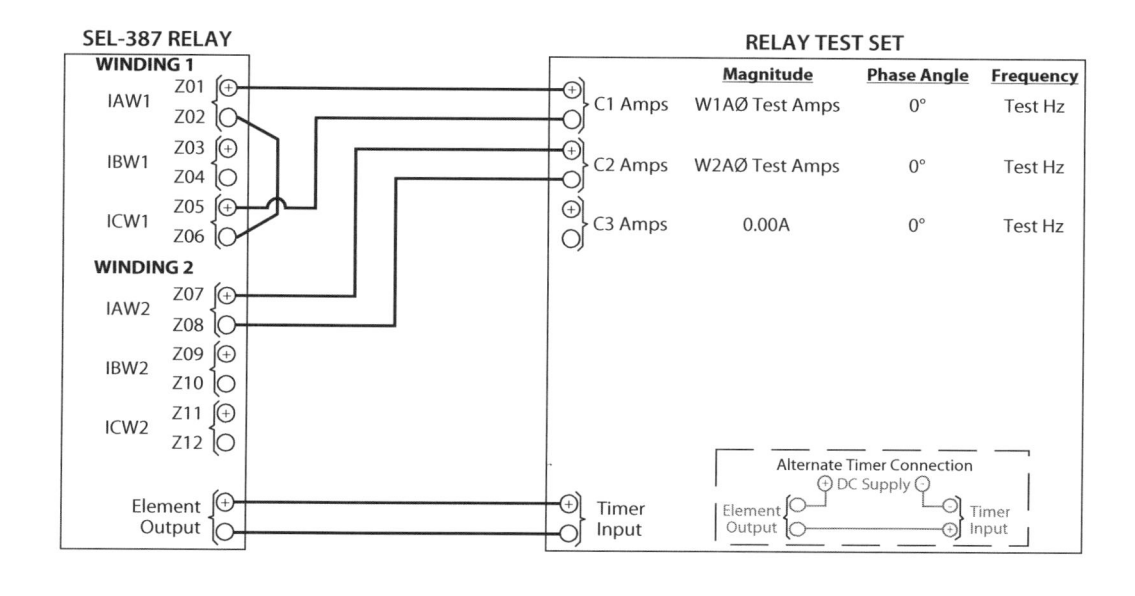

Figure 14-6: 1-Phase Differential Test-Set AØ Yd1 Connections

3. Simple Pickup Test Procedure

The simple unrestrained-differential pickup test procedure is identical to an Instantaneous Overcurrent test procedure;

- Apply a current into one-phase higher than the pickup setting and make sure pickup indication operates.
- Apply a current lower than pickup setting and make sure pickup indication does NOT operate.
- Raise current until pickup indication operates.
- Repeat for all phases.

A single-phase test procedure is slightly more complicated because correction factors may apply. Digital transformer relays use algorithms to compensate for the different transformer phase shifts as described earlier and apply compensation factors to compare windings with different configurations. The compensation factors for the example SEL-387 can be found in the Testing and Troubleshooting section of the instruction manual as shown in the following table:

WNCTC SETTING	A
0	1
Odd: 1, 3, 5, 7, 9, 11	$\sqrt{3}$
Even: 2, 4, 6, 8, 10, 12	1.5

However, the table does not seem to be correct when the WnCTC equals 12. We have found through experimentation that the correction factor is actually 1.

The following settings will be used for our SEL-387 example.

- Tap1 = 2.41
- Tap2 = 4.61
- U87P = 10 (xTap)

Use the following steps to determine pickup:

1. Determine how you will monitor pickup and set the relay accordingly, if required. (Pickup indication by LED, output contact, front panel display, etc. See the *Relay Test Procedures* section starting on page 109 for details.)

2. Determine the expected pickup.

WINDING-1	WINDING-2
Minimum Pickup = U87P × TAP1 × A	Minimum Pickup = U87P × TAP2 × A
Minimum Pickup = $10 \times 2.41 \times (12 = 1)1$	Minimum Pickup = $10 \times 4.61 \times (1 = \sqrt{3})\sqrt{3}$
Minimum Pickup = 24.1A	Minimum Pickup = 79.845A

3. Set the fault current 5% higher than the pickup setting. For example, set the fault current at 25.305A ($24.1 \times 1.05 = 25.305$A) for Winding-1 or 83.837A ($79.845 \times 1.05 = 83.837$A) for Winding-2. Make sure pickup indication operates.

4. Reduce the fault current 5% below the expected current and apply the fault current momentarily. The relay should not operate.

5. Increase the fault current in 1% increments and apply until the relay does operate.

6. Return to the last test current when the element did not operate and increase the fault current in 0.1% increments and apply until the relay operates. This is the pickup result. Record on your test sheet.

7. Repeat Steps 2-6 for all phases.

8. Compare the pickup test result to the manufacturer's specifications and calculate the percent error as shown in the following example.

The 87U-element pickup Winding-1 pickup setting is 24.1A and the measured pickup was 24.21A. Looking at the "Differential" specification in Figure 14-7 we see that the acceptable metering error is ±5% ±0.10A.

Differential Element

Unrestrained Pickup Range:	1–20 in per unit of tap
Restrained Pickup Range:	0.1–1.0 in per unit of tap
Pickup Accuracy (A secondary)	
5 A Model:	±5% ±0.10 A
1 A Model:	±5% ±0.02 A
Unrestrained Element Pickup Time (Min/Typ/Max):	0.8/1.0/1.9 cycles
Restrained Element (with harmonic blocking) Pickup Time (Min/Typ/Max):	1.5/1.6/2.2 cycles
Restrained Element (with harmonic restraint) Pickup Time (Min/Typ/Max):	2.62/2.72/2.86 cycles

Figure 14-7: SEL-387E Specifications

We can calculate the manufacturer's allowable percent error for Winding-1.

$$100 \times \frac{(5\% \times \text{Setting}) + 0.1A}{\text{Setting}} = \text{Allowable Percent Error}$$

$$100 \times \frac{(5\% \times 24.1) + 0.1A}{24.1} = \text{Allowable Percent Error}$$

$$100 \times \frac{1.305}{24.1} = \text{Allowable Percent Error}$$

5.41% Allowable Percent Error

The measured percent error can be calculated using the percent error formula below:

$$\frac{\text{Actual Value - Expected Value}}{\text{Expected Value}} \times 100 = \text{Percent Error}$$

$$\frac{24.21A - 24.1A}{24.1A} \times 100 = \text{Percent Error}$$

0.46% Error

The Winding-1 test is within the manufacturer's tolerance of 5.41% and passed the test.

9. Repeat the pickup test for all phase currents that are part of the differential scheme.

DIFFERENTIAL TEST RESULTS							
87U PICK UP	10	TIME DELAY	0				
TAP1	2.41	TAP2	4.61				
UNRESTRAINED PICKUP TESTS (Amps)							
		IAW_ (A)	IBW_ (A)	ICW_ (A)	EXPECTED (A)	% ERROR		
W1		24.210	24.150	24.130	24.100	0.5	0.2	0.1
W2		79.850	78.860	79.840	79.845	0.0	-1.2	0.0

4. Alternate Pickup Test Procedure

Unrestrained-differential elements are, by definition, high-current applications. Adding a 1.73 correction factor for single-phase testing can often put the expected pickup above the maximum output of your test-set. We can use the testing philosophy from the "1-Phase Restrained-Differential Slope Testing" section of the previous chapter to apply equal currents in two windings simultaneously while in-phase to minimize the required current to test the Unrestrained-Differential element.

The following settings will be used for our SEL-387 example.

- W1CTC = 12
- W1CTC = 1
- Tap1 = 2.41
- Tap2 = 4.61
- U87P = 10 (xTap)

1. Connect the test-set as per Figures 14-2 to 14-4 and the descriptions in the "1-Phase Restrained-Differential Slope Testing" section of the previous chapter.

2. Determine how you will monitor pickup and set the relay accordingly, if required. (Pickup indication by LED, output contact, front panel display, etc. See the *Relay Test Procedures* section starting on page 109 for details.)

3. Use the following formula to calculate equal currents for any multiple of Tap using the following table for correction factors.

WNCTC SETTING	A
0	1
Odd: 1, 3, 5, 7, 9, 11	$\sqrt{3}$
Even: 2, 4, 6, 8, 10, 12	1.5

However, the table does not seem to be correct when the WnCTC equals 12. We have found through experimentation that the correction factor is, in fact, 1.

$$\frac{\text{Test Amps}}{\text{TAP1} \times \text{A1}} + \frac{\text{Test Amps}}{\text{TAP2} \times \text{A2}} = \text{Test@Tap}$$

$$\text{TAP1} \times \text{A1} \times \left(\frac{\text{Test Amps}}{\text{TAP1} \times \text{A1}} + \frac{\text{Test Amps}}{\text{TAP2} \times \text{A2}} \right) = \text{Test@Tap} \times \text{TAP1} \times \text{A1}$$

$$\left(\text{Test Amps} + \frac{\text{TAP1} \times \text{A1} \times \text{Test Amps}}{\text{TAP2}} \right) = \text{Test@Tap} \times \text{TAP1} \times \text{A1}$$

$$\text{TAP2} \times \text{A2} \times \left(\text{Test Amps} + \frac{\text{TAP1} \times \text{Test Amps}}{\text{TAP2}} \right) = \text{Test@Tap} \times \text{TAP1} \times \text{A1} \times \text{TAP2} \times \text{A2}$$

$$\left(\text{TAP2} \times \text{A2} \times \text{Test Amps} \right) + \left(\text{TAP1} \times \text{A1} \times \text{Test Amps} \right) = \text{Test@Tap} \times \text{TAP1} \times \text{A1} \times \text{TAP2} \times \text{A2}$$

$$\left(\text{TAP2} \times \text{A2} + \text{TAP1} \times \text{A1} \right) \text{Test Amps} = \text{Test@Tap} \times \text{TAP1} \times \text{A1} \times \text{TAP2} \times \text{A2}$$

$$\text{Test Amps} = \frac{\text{Test@Tap} \times \text{TAP1} \times \text{A1} \times \text{TAP2} \times \text{A2}}{\left[\left(\text{TAP1} \times \text{A1} \right) + \left(\text{TAP2} \times \text{A2} \right) \right]}$$

$$\text{Test Amps} = \frac{\text{Test@Tap} \times \text{TAP1} \times \text{A1} \times \text{TAP2} \times \text{A2}}{\left[\left(\text{TAP1} \times \text{A1} \right) + \left(\text{TAP2} \times \text{A2} \right) \right]}$$

$$\text{Test Amps} = \frac{10 \times 2.41 \times 1 \times 4.61 \times 1.732}{\left[\left(2.41 \times 1 \right) + \left(4.61 \times 1.732 \right) \right]}$$

$$\text{Test Amps} = \frac{192.43}{10.39}$$

$$\text{Test Amps} = 18.52$$

4. Momentarily apply 10% more current ($\text{Pickup} \times 1.1 = 18.52\text{A} \times 1.1 = 20.732\text{A}$) from all applied test-set current channels and ensure the relay operates.

5. Lower the currents to 95% of the pickup setting ($\text{Pickup} \times 0.95 = 18.52\text{A} \times 0.95 = 17.594\text{A}$) and momentarily apply. The pickup indication should remain OFF.

6. Raise the currents in 1% increments and momentarily apply until the pickup indication turns ON.

7. Subtract 1% and increase in 0.1% steps until the pickup operates. This is the 87U pickup. Use the following formula to determine the pickup in per-unit:

$$\text{Per Unit Pickup} = \frac{\text{Test Amps}}{\text{TAP1} \times \text{A1}} + \frac{\text{Test Amps}}{\text{TAP2} \times \text{A2}}$$

8. Change all connections for B-phase and repeat Steps 4-7.

9. Change all connections for C-phase and repeat Steps 4-7.

10. Evaluate the results using "Pickup Accuracy (A Secondary): 5 A Model" from the SEL-387 specification in Figure 14-8.

Differential Element

Unrestrained Pickup Range:	1–20 in per unit of tap
Restrained Pickup Range:	0.1–1.0 in per unit of tap
Pickup Accuracy (A secondary)	
5 A Model:	±5% ±0.10 A
1 A Model:	±5% ±0.02 A
Unrestrained Element Pickup Time (Min/Typ/Max):	0.8/1.0/1.9 cycles

Figure 14-8: SEL-387 Differential Element Specifications

DIFFERENTIAL TEST RESULTS						
87U PICK UP	10	TIME DELAY	0			
TAP1	2.41	TAP2	4.61			
UNRESTRAINED PICKUP TESTS (Amps)						
	87U1	87U2	87U3	EXPECTED (A)	% ERROR	
	18.550	18.530	18.540	18.520	0.2	0.1	0.1
	IAW1,IAW2&ICW2	IBW1,IBW2&IAW2	ICW1,ICW2&IBW2			

5. Timing Test Procedure

1. Connect the relay output contact to the test-set input.

2. Apply the connections from the Simple or Alternate Pickup Test Procedure from the previous sections in this chapter.

3. Set the test current 110% of the expected current using the same test procedure in Step 2.

4. Configure a test that applies the test current and starts the fault timer simultaneously. The timer and current output should stop when the relay output operates.

5. Apply the test and record the results.

6. Change the connections for the next phase and repeat the test.

7. Change the connections for the next phase and repeat.

8. Evaluate the results using "Unrestrained Element Pickup Time" from the SEL-387 specification in Figure 14-7 and the Output Contact "Pickup Time" specifications in Figure 14-8.

Output Contacts: Standard:

> Make: 30 A; Carry: 6 A continuous carry at 70°C, 4 A continuous at 85°C;
> 1 s Rating: 50 A; MOV protected: 270 Vac, 360 Vdc, 40 J;
> Pickup time: Less than 5 ms; Dropout time: Less than 5 ms typical.

Figure 14-9: Preferred SEL-387 Output Contact Specifications

Maximum Operating Time = Unrestrained Element Pickup Time(Max) + Output Contact Pickup Time

Maximum Operating Time = 1.9 cycles + < 5ms

$$\text{Maximum Operating Time} = \frac{1.9 \text{ cycles}}{60 \text{ cycles}} + < 5\text{ms}$$

Maximum Operating Time = 31.67ms + < 5ms

Maximum Operating Time = 36.67ms or 2.2 cycles

6. Tips and Tricks to Overcome Common Obstacles

The following tips or tricks may help you overcome the most common obstacles.

- Before you start, apply current at a lower value and review the relay's measured values to make sure your test-set is actually producing an output and your connections are correct.
- Are the 2 winding currents $0°$ apart?
- If the pickup tests are off by a $\sqrt{3}$, 1.5, or 0.577 multiple; the relay probably is applying correction factors and you should review the manufacturer's literature.
- Are you applying the currents into the same phase relationships?
- Did you calculate the correct Tap?

Chapter 15

Line Distance (21) Element Testing

Line distance (21) protection is primarily used to protect transmission lines or other electrical equipment where the equipment's impedance characteristic can be calculated or modeled with software applications. The protective relay calculates the ratio of measured voltage and current and uses Ohm's Law to continually monitor the measured impedance while energized. The measured impedance is compared to the equipment impedance characteristic created by the relay settings and the relay will trip if the measured impedance falls within the setting characteristic. This method of protection creates a Zone of Protection which allows the relay to selectively trip when internal faults occur and ignore external faults. The relay impedance characteristic can also be applied with a larger characteristic and a time delay to provide backup protection for external equipment.

Understanding impedance protection can be difficult so we will start by reviewing the application as shown in Figure 1. Imagine that you are trying to protect the following transmission system using overcurrent protection and a fault occurs on the line between W and X (Line WX).

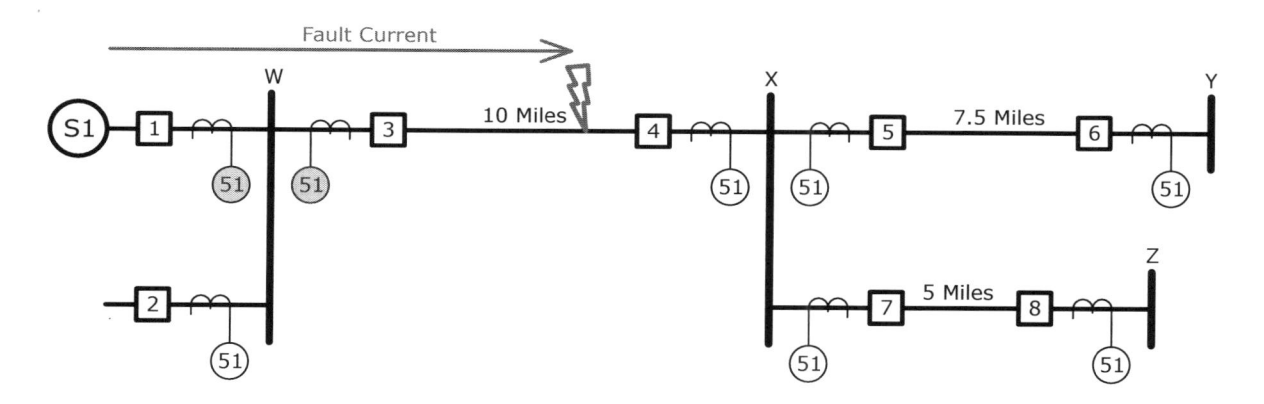

Figure 15-1: Radial Transmission System with Overcurrent Protection

Either of the simple overcurrent relays (51-1, 51-3) connected to circuit breakers 1 (CB1) and (CB3) could operate. If the 51-1 relay operates first, all of the loads connected to CB2 will be de-energized unnecessarily and cause complaints and lost revenue. We could mitigate this scenario by adding directional protection as shown in the following diagram and described in *Chapter 11: Directional Overcurrent (67) Element Testing*. With directional overcurrent (67) protection applied, the only relay that will operate is 67-3 which will isolate Line WX and the load connected to CB2 will remain in service.

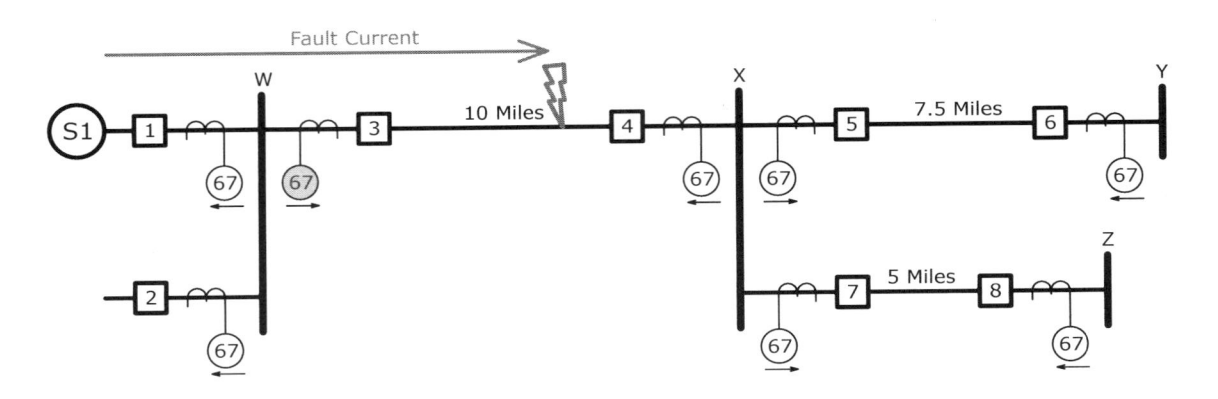

Figure 15-2: Typical Radial Transmission System with Directional Overcurrent Protection

A system connected with one source is called a radial feed which is not a common scenario in an electrical grid. Most transmission and distribution systems have multiple sources. Let's see what happens when an additional source is connected to Bus Z in Figure 15-3. Now we have the same problem with relay 67-8. If it operates first, Buses X and Y will be de-energized for no legitimate reason.

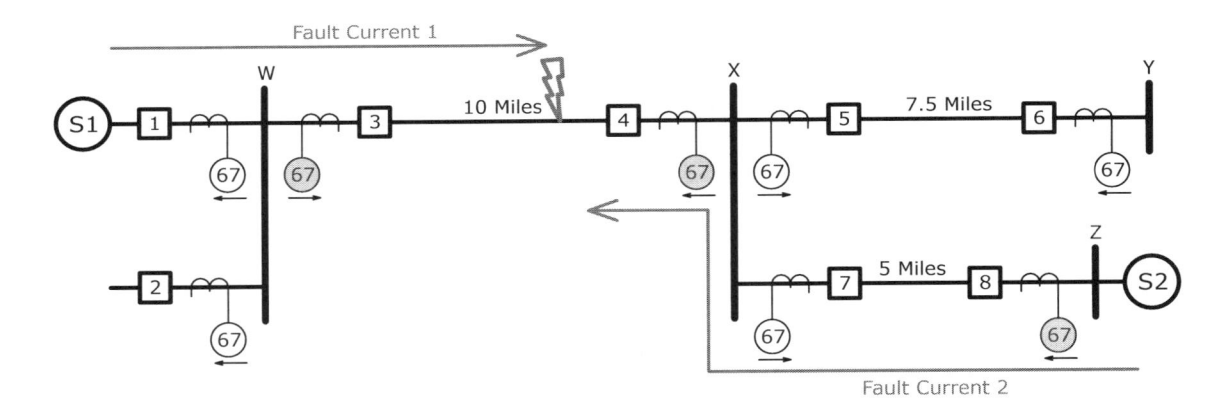

Figure 15-3: Typical Transmission System with Directional Overcurrent Protection

In order to provide the best protection, we need to create a zone of protection for each transmission line so that the relays will only operate for faults on their transmission lines. Differential protection as described in previous publications of *Chapter 12: Simple Percent Differential (87) Element Testing* would be the perfect solution for this application. We can measure the current entering and leaving each transmission line, and the relay will operate if it detects a significant difference. The following figure demonstrates this protection scheme where both CB3 and CB4 will open when the 87 relay operates with minimal disruption to the surrounding electrical system.

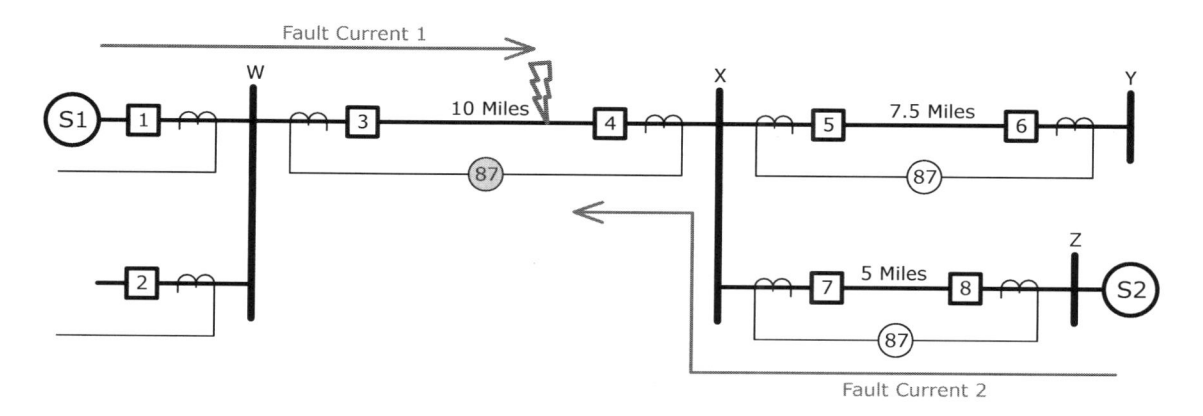

Figure 15-4: Typical Transmission System with Line-Differential Protection

This Line-Differential scheme works very well with modern technology but there are additional costs associated because both sides of the transmission line must communicate the measured current in real time with no delays. This technology was not available in the early 20th century so a new protection scheme was created called Impedance Protection.

Impedance protection looks at the transmission system in our example and asks, "What are the known characteristics of a transmission line?" We know the approximate length of the line, the nominal voltage, and the size and material of the current carrying conductor.

We could measure the actual resistance of the transmission line, but that would require specialized equipment and manpower. It is more cost effective to measure the impedance of a small length of the conductor material and then multiply that resistance by the length of the line. It is important to remember that we are discussing an AC circuit and a transmission line will have resistive and reactive component as shown in the following figure. Notice that the actual conductor by itself can be calculated with the formula $Z_{Ohms/Mile} = R_a + j(X_a)$. When three conductors are placed into a 3-phase system, an additional reactance must be applied which will vary according to the line length and conductor spacing. The new formula for transmission line impedance is $Z_{Ohms/Mile} = R_a + j(X_a + X_d)$.

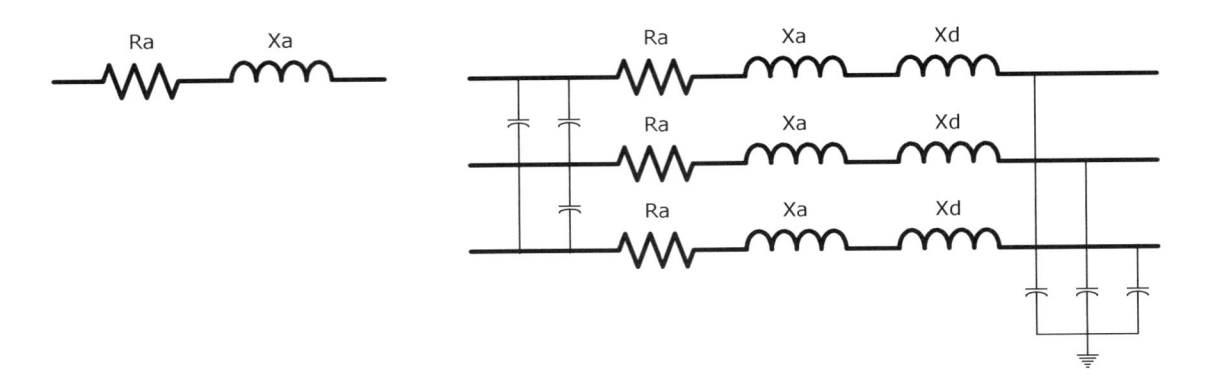

Figure 15-5: Equivalent Transmission Line Impedance

The reactance (j) or X component is always significantly larger than the resistance of the transmission line and the line angle will grow closer to 90° as the nominal voltage increases. For example, a transmission line connected to a 34.5kV system will likely have a fault angle in the neighborhood of 60° lag. A relatively identical fault on a 230kV system, however, will have a nearly 90° fault angle due to the higher system voltage because the reactance component is several multiples greater than the resistive component at the higher voltage.

Let's assume that the transmission lines in our example are rated for 1 Ohm/mile at the nominal voltage. Line WX has a 10 Ohm impedance, Line XY has a 7.5 Ohm impedance, and Line XZ has a 5 Ohm impedance. For simplicity's sake, let's generalize and state that the line angle for all lines is 75°. Because we are using impedance relays to protect these lines, we can use ohms on our drawing to better visualize the system.

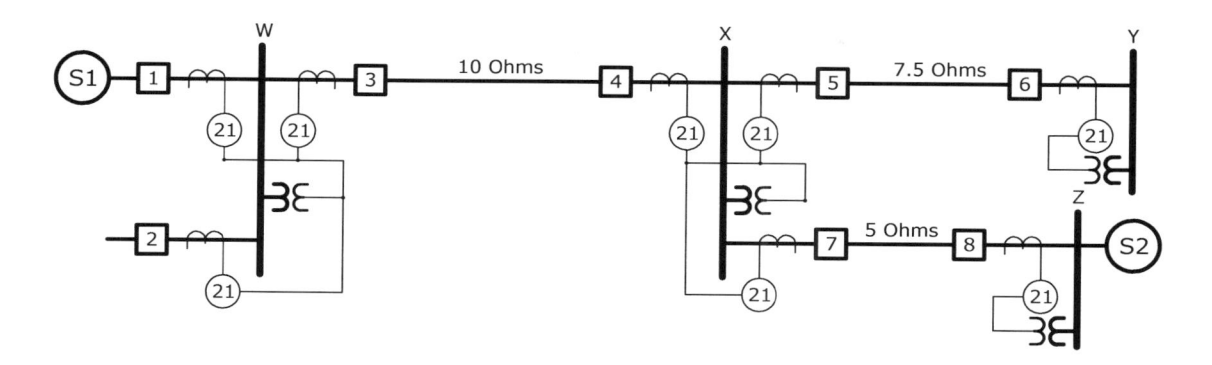

Figure 15-6: Typical Transmission System with Early Line Distance Protection (Primary Ohms)

Most relays measure secondary ohms so we should re-draw the system using secondary ohms. Let's assume that the PT ratio is 1200:1 for all PTs and the CT ratio is 2000:5 for all CTs. The formula for calculating secondary ohms is derived in the following calculation:

$$\left[\text{if } Z = \frac{V}{I}\right] \text{ and } \left[V_{Sec} = \frac{V_{Pri}}{PT_{RATIO}}\right] \text{ and } \left[I_{Sec} = \frac{I_{Pri}}{CT_{RATIO}}\right]$$

$$Z_{Sec} = \frac{V_{Sec}}{I_{Sec}} = \frac{\left(\dfrac{V_{Pri}}{PT_{RATIO}}\right)}{\left(\dfrac{I_{Pri}}{CT_{RATIO}}\right)} = \frac{V_{Pri} \times CT_{RATIO}}{I_{Pri} \times PT_{RATIO}} = \frac{V_{Pri}}{I_{Pri}} \times \frac{CT_{RATIO}}{PT_{RATIO}} = Z_{Pri} \times \frac{CT_{RATIO}}{PT_{RATIO}}$$

Figure 15-7: Primary to Secondary Impedance Calculation

The conversion for our example will be

$$Z_{Sec} = Z_{Pri}\frac{CT_{RATIO}}{PT_{RATIO}} = Z_{Pri}\frac{\left(\dfrac{2000}{5}\right)}{1200} = Z_{Pri}\frac{400}{1200} = Z_{Pri} \times \frac{1}{3}$$

The line calculations will be

$$\text{LineWX } Z_{Sec} = Z_{Pri} \times \frac{1}{3} = 10 \times \frac{1}{3} = 3.333\Omega$$

$$\text{LineXY } Z_{Sec} = Z_{Pri} \times \frac{1}{3} = 7.5 \times \frac{1}{3} = 2.5\Omega$$

$$\text{LineXZ } Z_{Sec} = Z_{Pri} \times \frac{1}{3} = 5 \times \frac{1}{3} = 1.667\Omega$$

Our system will look like the following figure from the relay's perspective.

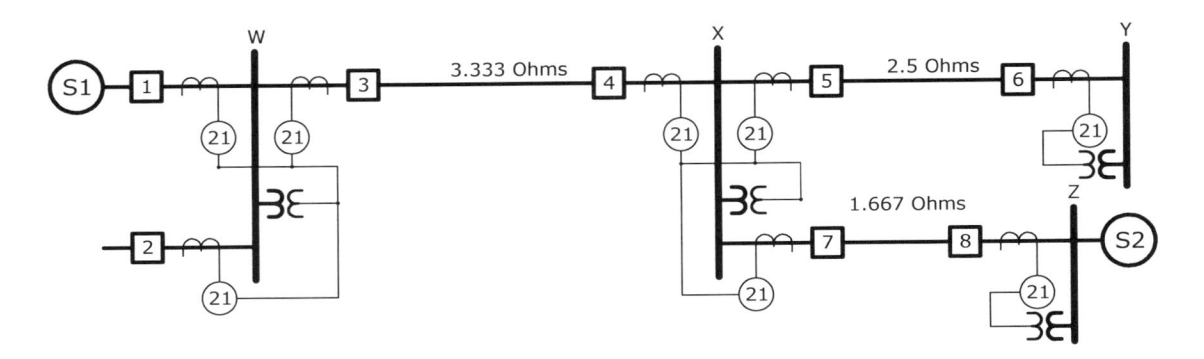

Figure 15-8: Typical Transmission System with Early Line Distance Protection (Secondary Ohms)

Now that we have calculated the line impedance and angle for the transmission lines, we can build a relay that will monitor the voltage *and* current to calculate the measured impedance. The 21-element relays connected to Line WX could be set for 3.33Ω to protect the entire line. A 3.33Ω setting could cause problems because our line impedance is based on a calculation that makes quite a few assumptions such as; exact line length, # of splices, and exact conductor spacing. There are many real-world factors that could make the actual line impedance larger or smaller than the calculated value. Most design engineers set the 21-element at 70-85% of the line to ensure they do not over-reach, so the Line WX relays will be set to 2.667Ω, Line XY relays will be set at 2.0Ω, and Line XZ relays will be set at 1.33Ω.

1. Impedance Relays

The first impedance relays were constructed with a current coil, a voltage coil, and a balance beam contact to apply Ohm's Law $Z = \dfrac{V}{I}$ without a microprocessor as shown in the next figure. The voltage coil is designed to pull the contacts apart and the current coil will attempt to pull the contacts together. If enough current flows through the current coil, the contacts will close and the relay will operate. The coil strength is adjusted by taps and resistors to allow different pickup settings. A normal impedance (Z) is very large because of the low current under normal conditions. The fault current is very high and the fault voltage drops during a fault which causes the impedance (Z) to decrease. Unlike most protective elements, the relay will operate when the measured value shrinks.

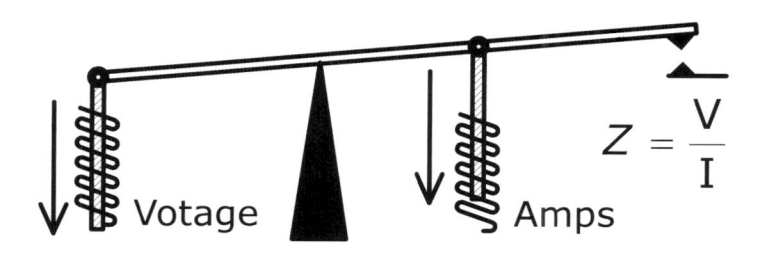

Figure 15-9: Equivalent Transmission Line Impedance

Our standard phasor diagrams and inverse-time diagram are not going to be very helpful if we want to easily understand this new protective element, so a new diagram was created which plots impedance. The X-axis is resistance and the Y-axis is reactance and the diagram is considered to have four quadrants. The following figure displays a phasor diagram and corresponding impedance diagram. Most current vectors and impedances are found in Quadrant I which represents a load with a lagging power factor or inductive component. Current vectors and impedances found in Quadrant IV indicate capacitive loads. Quadrant III is the opposite of Quadrant I and indicates inductive (normal) current in the opposite direction which can be considered behind the relay. Quadrant II is the opposite of Quadrant IV and indicates capacitive current in the opposite direction which can be considered to be behind the relay.

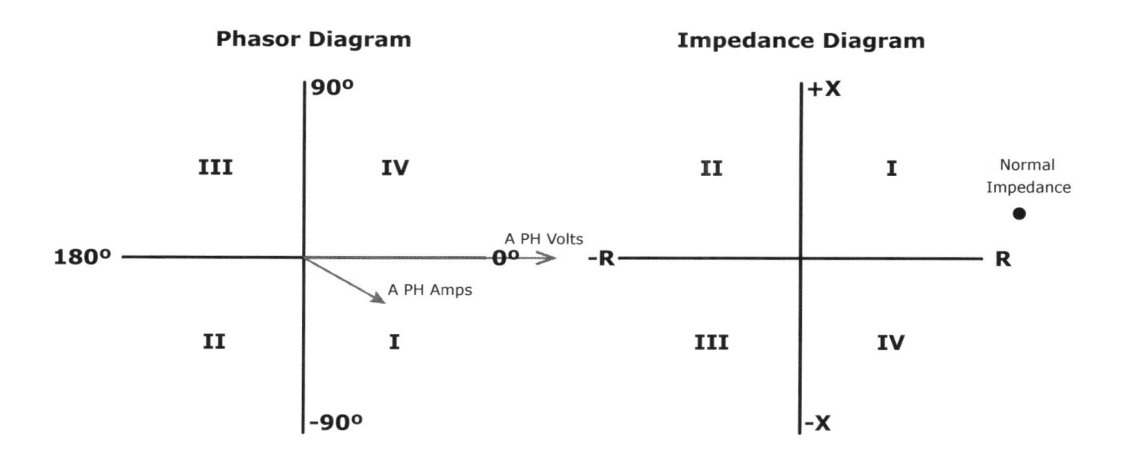

Figure 15-10: Phasor Diagram vs. Impedance Diagram Under Normal Conditions

One very important concept to remember with both diagrams is that the relay location is located at the origin of the X and Y axis. It is also important to remember that the relay location is actually defined by the location of its source PTs. The CT location determines the direction of the fault. Most impedance calculations for relay testing assume identical faults for all phases, and therefore only one phase is usually plotted and the test magnitudes are the same for each tested phase with appropriately rotated angles.

Notice that there are some differences between the Phasor Diagram and Impedance Diagram in the previous figure. The current and impedance are both located in Quadrant I, but Quadrant I is located in different areas on the different diagrams. The calculation to convert a Phasor Diagram to an Impedance Diagram is Ohm's Law $Z = \dfrac{V\angle\phi}{I\angle\theta}$ using complex numbers that can be modified to $Z = \dfrac{V}{I}\angle\phi - \theta$. The calculation for our example is $Z = \dfrac{V}{I}\angle\phi - \theta = \dfrac{69.28}{1}\angle(0° - -30°) = 69.28\Omega\angle30°$. The -30° current angle changed to a +30° which explains why the quadrants are different in each diagram.

Now let's look at a 3-phase fault on Line WX. The secondary fault voltage is 20V, the secondary fault current is 10A, and the current lags the voltage by 80°. Notice that the impedance is much closer to the origin which indicates that the fault impedance is closer to the relay than the normal impedance.

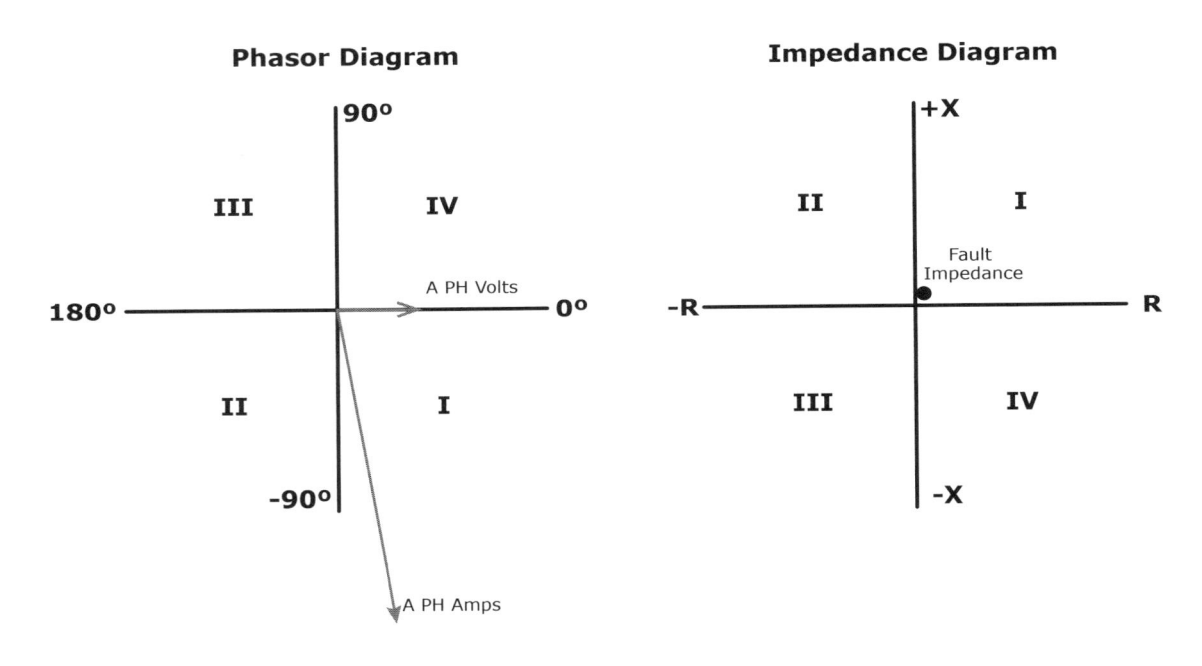

Phasor Diagram   **Impedance Diagram**

Figure 15-11: Phasor Diagram vs. Impedance Diagram Under Fault Conditions

The first impedance relays were set at approximately 80% of the line as described earlier so the following relay settings were applied; Line WX = 2.667Ω, Line XY = 2.0Ω, and Line XZ = 1.333Ω. The following figure plots the Line WX relay settings on the Impedance Diagram as depicted by the circle. If the measured impedance falls inside the circle, the relay will trip. The fault impedance falls inside the circle so the relay will operate if this fault were to occur. The normal impedance was so large it could not be plotted to scale and the relay will not trip.

Impedance Diagram

Figure 15-12: Impedance Relay Operating Characteristics

While this relay is more selective than an overcurrent relay, we still have a big problem. The relay characteristic operates in all quadrants and will trip for faults in all directions as shown in the following figure. The circles in this diagram indicate the operating characteristics of relays 21-3 and 21-4. Notice that the centers of the circles are located at the appropriate buss and not at the circuit breaker. Remember that relays calculate impedance based on CT and PT inputs and the PT source is located at the bus. Also notice that these relays will trip on faults just about anywhere in the system. This system is not much better than the original overcurrent protection.

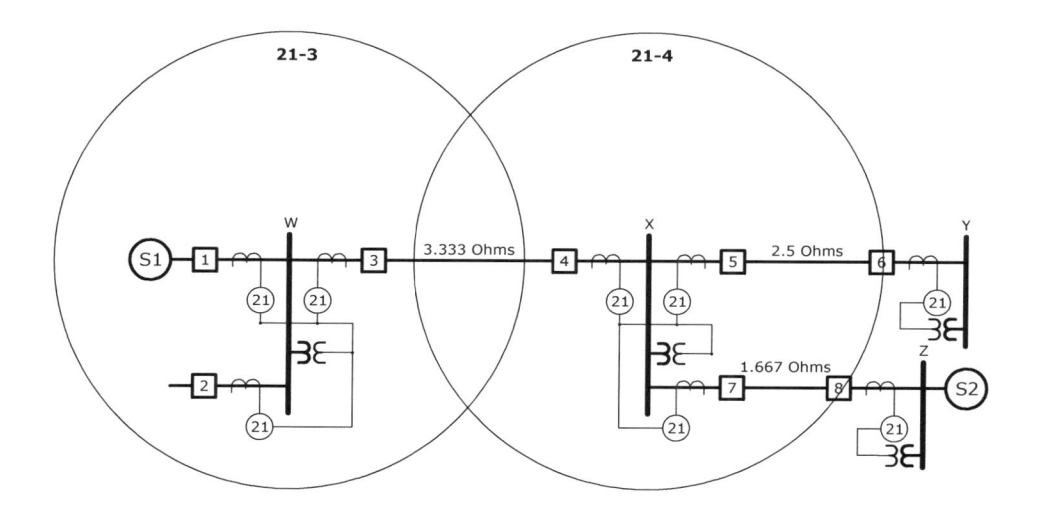

Figure 15-13: Impedance Relay Zone of Protection

A blocking element was added to these relays which prevented them from operating in the reverse direction as shown in the following figures. While this is an improvement, a significant portion of the new characteristic falls on the X-axis. This could cause nuisance trips under normal conditions such as a long line that is heavily loaded. The characteristic on the Impedance Diagram is offset by the Line Angle.

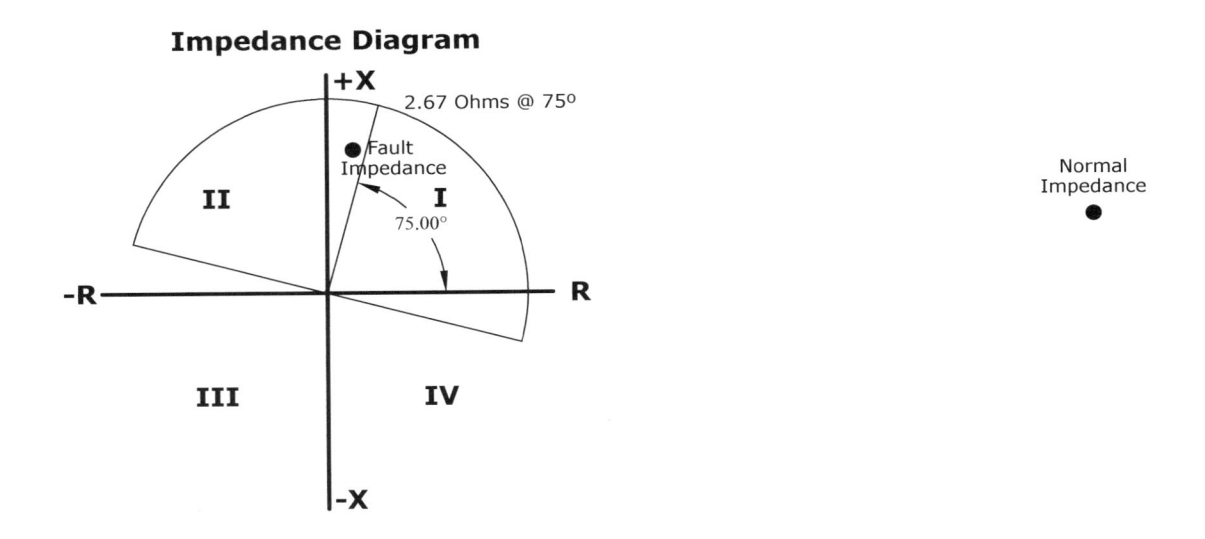

Figure 15-14: Directional Impedance Relay Operating Characteristics

This system drawing displays the new characteristic.

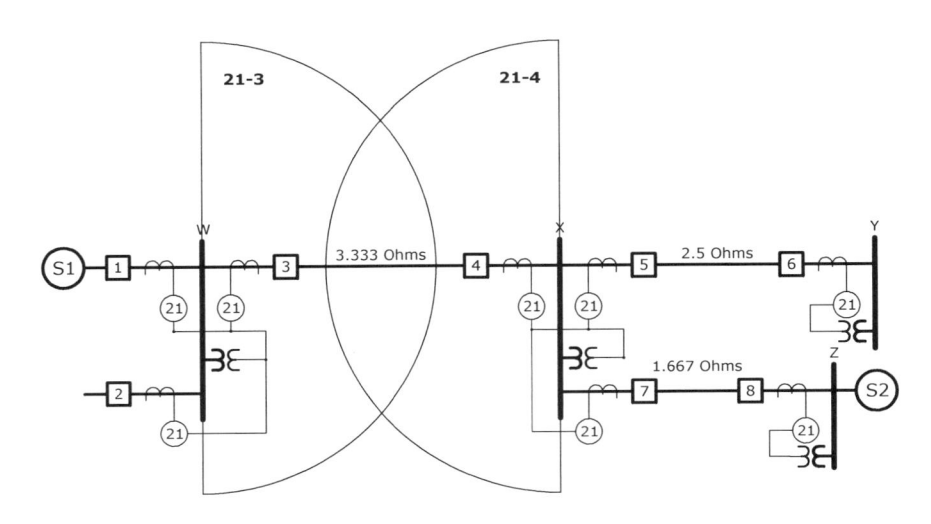

Figure 15-15: Directional Impedance Relay Zone of Protection

A) The MHO Impedance Characteristic

The next generation of relay kept the circle characteristic but shifted the entire circle by the line angle as shown in the following figures. This is called the MHO characteristic and is the most popular impedance element in service today.

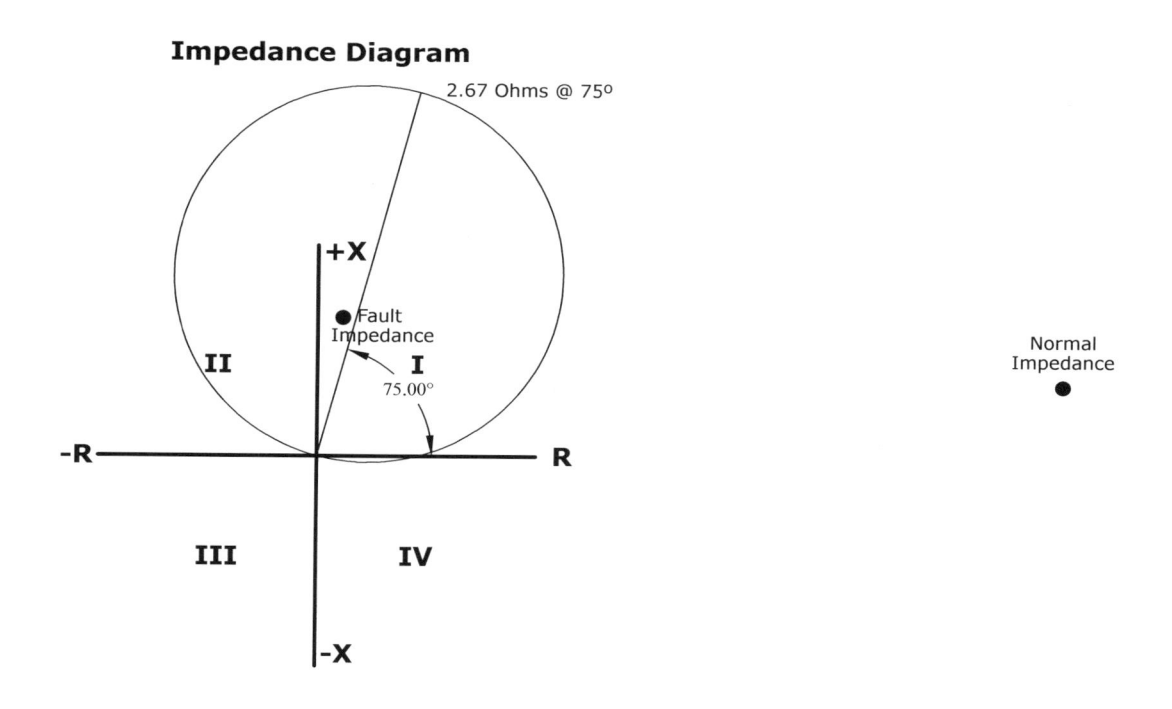

Figure 15-16: MHO Impedance Relay Operating Characteristics

The MHO characteristic was created using the technology of the time, but it is still in use on most modern relays. Digital relays can apply impedance protection in many different shapes including a lens, tomato, quad, or MHO with blinders to meet specific line parameters as shown in the following diagrams. You might ask yourself why we need all of these different shapes if we know the line impedance and angle. These impedance characteristics are applied because we do not know the fault impedance of the arc or foreign material that caused the fault and an impedance window is created to account for the unknowns. The MHO characteristic may be the simplest, but it has proven to be reliable for several decades.

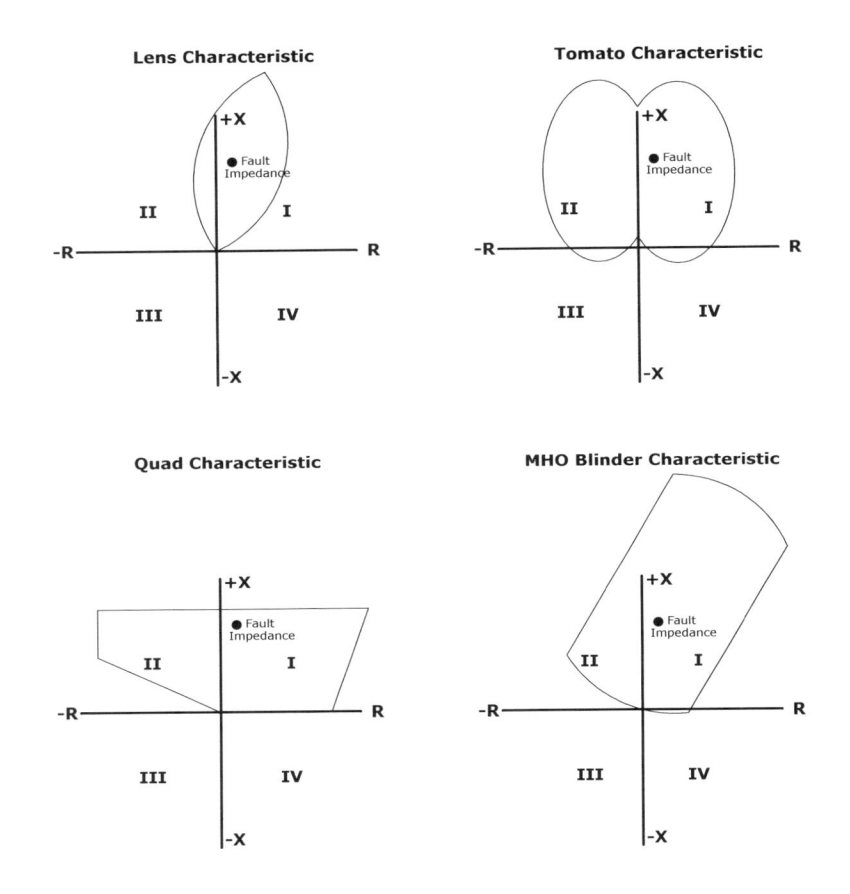

Figure 15-17: Alternate Impedance Relay Operating Characteristics

B) 2 Zone Protection Schemes

You can now see how the MHO characteristic will protect a line. Any fault that occurs in the overlapping area (which covers 60% of the line) causes both line relays to trip instantaneously. But what happens when a fault occurs on the 40% when only one relay trips? The other circuit breaker will continue to expose the system to the fault until it gets worse or some other device operates which will likely be a backup device that will isolate more equipment than necessary.

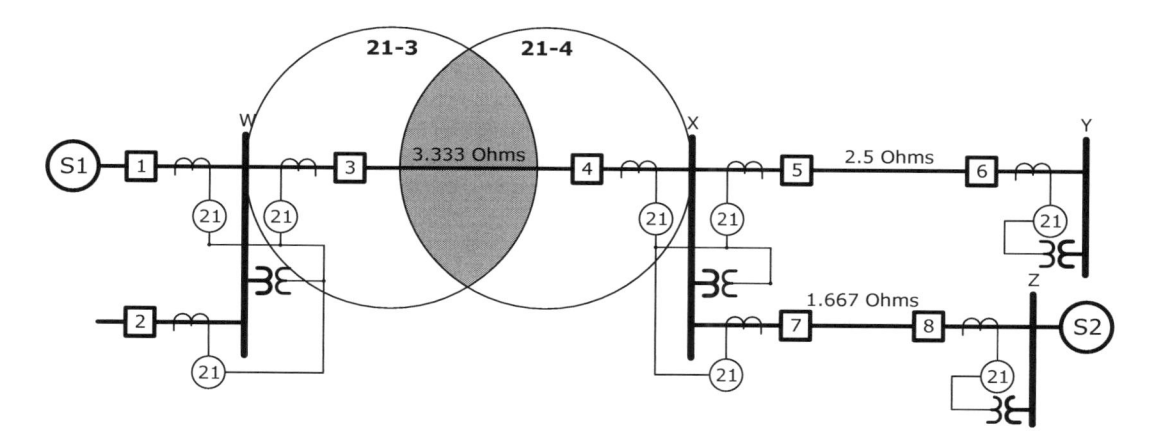

Figure 15-18: MHO Impedance Relay Zone of Protection

We don't want to expand the circles to cover the entire line as described earlier so we will add another relay or element that will cover 120% of the line and over-reach into the surrounding equipment. This new zone of protection is usually referred to as Zone 2 and is set with a time delay larger than any breaker-failure relays on the system which is typically in the 16-20 cycle range. Zone 2 provides two important functions: 100% protection of the protected line and backup protection for surrounding lines. Now when a fault occurs in the shaded region, the second relay will operate 18 cycles later and isolate the fault from the system unless another relay operates first.

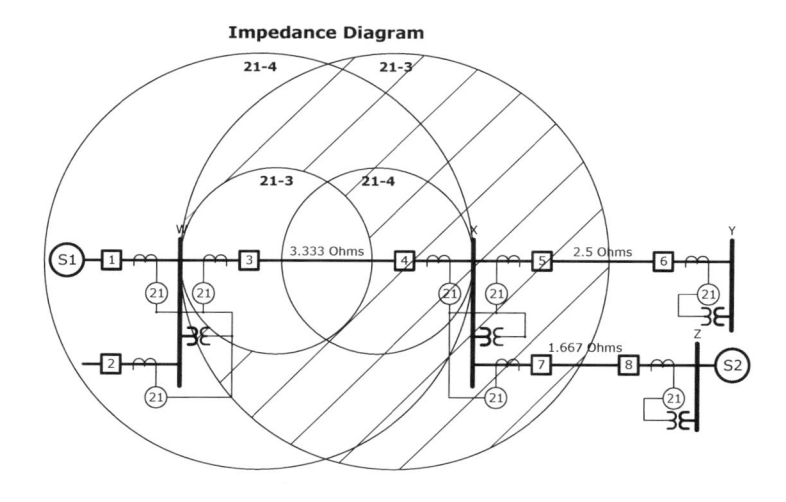

Figure 15-19: 2 Zone MHO Impedance Relay Zone of Protection

The dual-zone impedance characteristic is depicted in the following figure:

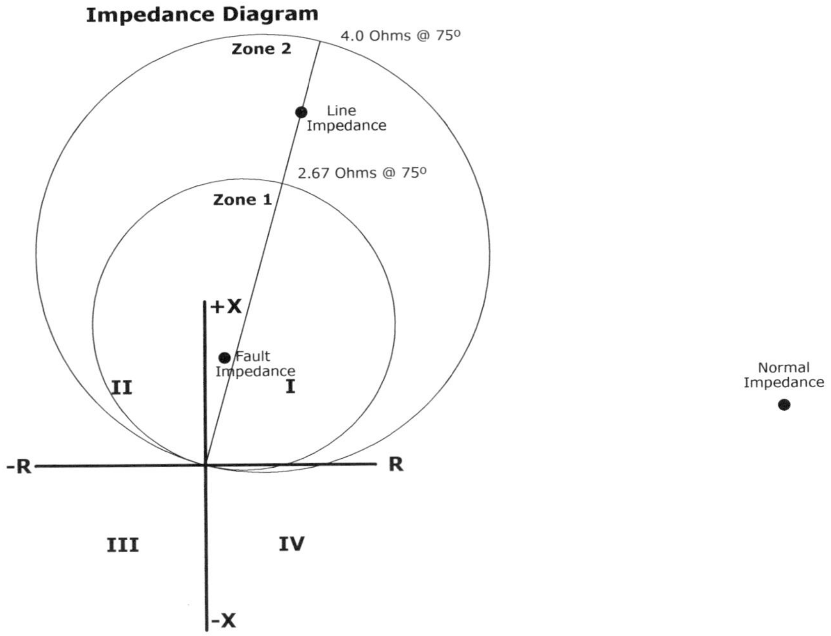

Figure 15-20: 2 Zone MHO Impedance Diagram

C) 3 Zone Protection Schemes

A third zone of protection was also added in the past with a bigger characteristic and a longer time delay, typically 60 cycles. Protection schemes with three forward-looking zones of protection was largely abandoned by the industry after investigations of some major system disruptions in the United States determined that incorrectly applied Zone 3 elements caused or contributed to the disturbance. The following figures depict a protection scheme with three forward-looking zones.

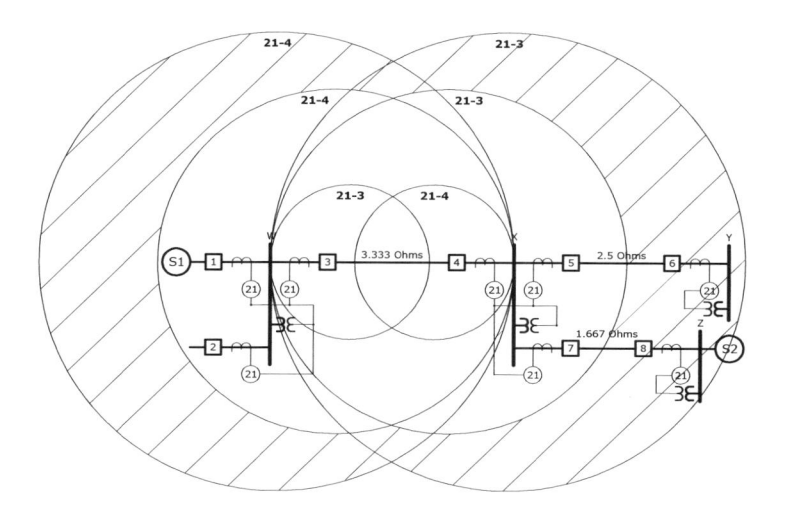

Figure 15-21: 3 Forward-Zone MHO Impedance Relay Zone of Protection

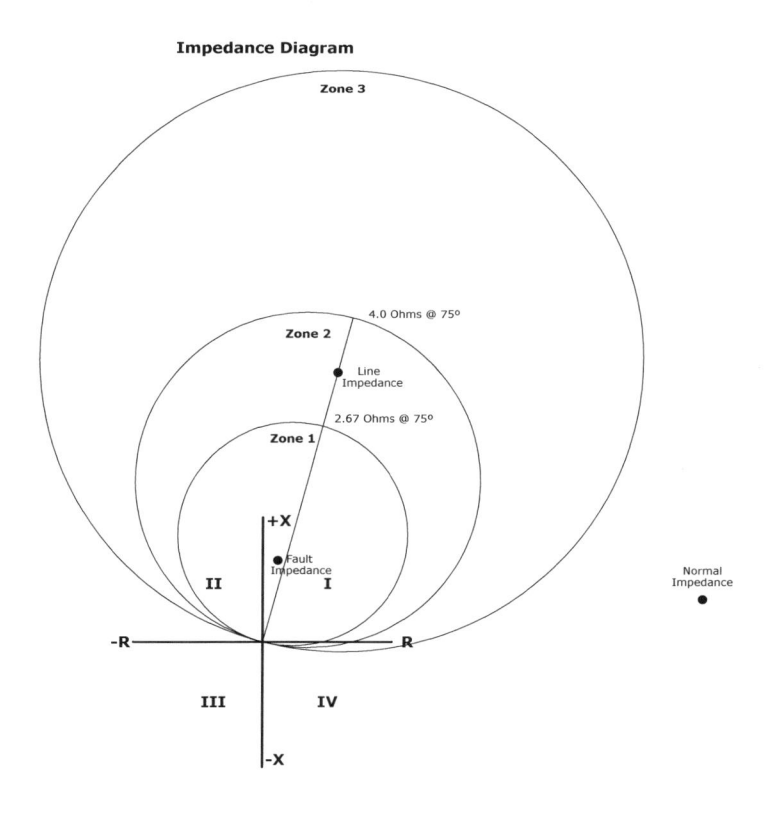

Figure 15-22: 3-Forward-Zone MHO Impedance Diagram

Zone 3 is typically applied in the reverse direction to provide backup protection or to communicate that a fault is behind the local relay to a remote line relay. When two line relays communicate via output contacts or direct communication, they can provide 100% protection of the transmission line. The Zone 1 protection will not change but additional logic is applied to Zone 2 protection for better selectivity. If a Zone 2 fault is detected by relay 21-3 and relay 21-4 does NOT signal a reverse Zone 3 fault, the fault must be on the line, and 21-3 can trip quickly isolating the line after accounting for communication delays. If a Zone 2 fault is detected by relay 21-3 and relay 21-4 signals a reverse Zone 3 fault, the fault must NOT be on the line and 21-3 will trip after the normal delay for backup protection or not trip at all if backup protection is not desired. The following figures depict a 2-Forward-1-Reverse 3 Zone protective scheme.

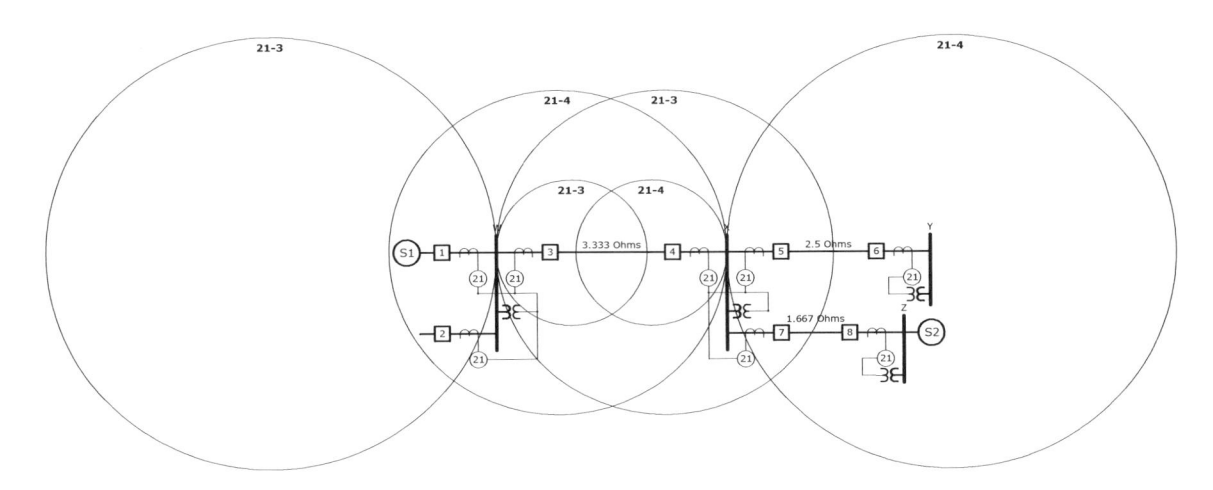

Figure 15-23: -Forward-1-Reverse 3 Zone MHO Impedance Relay Zone of Protection

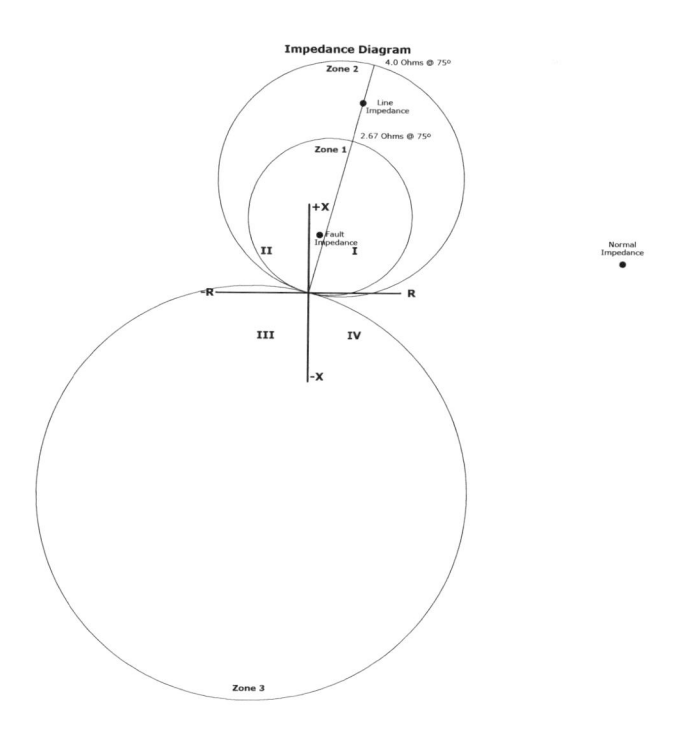

Figure 15-24: 2-Forward-1-Reverse 3 Zone MHO Impedance Diagram

2. Settings

The most common settings used in 21-elements are explained below:

A) Enable Setting

Many relays allow the user to enable or disable settings. Make sure that the element is ON or the relay may prevent you from entering settings. If the element is not used, the setting should be disabled or OFF to prevent confusion. Some relays will also have "Latched" or 'Unlatched" options. A Latched option indicates that the output contacts will remain closed after a trip until a reset is performed and acts as a lockout relay. Unlatched indicates that the relay output contacts will open when the trip conditions are no longer present.

B) Positive-Sequence Line Impedance Magnitude

The Positive-Sequence Line Impedance Magnitude setting is a calculated value from system modeling to help the relay determine the protected line's characteristics when used in conjunction with the positive-sequence angle and zero-sequence settings. This value is typically used by the relay to determine the fault location only and is not part of the 21-element characteristics.

C) Positive-Sequence Line Impedance Angle

The Positive-Sequence Line Impedance Angle setting is a calculated value from system modeling to help the relay determine the protected line's characteristics when used in conjunction with the positive-sequence magnitude and zero-sequence settings. This value is typically used by the relay to determine the fault location and is used in the 21-element characteristics. This is one of the most important settings in 21-element testing and should be recorded.

D) Zero-Sequence Line Impedance Magnitude

The Zero-Sequence Line Impedance Magnitude setting is a calculated value from system modeling to help the relay determine the protected line's characteristics when used in conjunction with the zero-sequence angle and positive-sequence settings. This value is typically used by the relay to determine the fault location only and not part of the 21-element characteristics.

E) Zero-Sequence Line Impedance Angle

The Zero-Sequence Line Impedance Angle setting is a calculated value from system modeling to help the relay determine the protected line's characteristics when used in conjunction with the zero-sequence magnitude and positive-sequence settings. This value is typically used by the relay to determine the fault location only and not part of the 21-element characteristics.

F) Line Length

The line length setting is used in the fault locating feature of some relays to assist operators when the relay operates. If a 21-element operates, the relay will look at the fault characteristics and compare them to the positive and zero sequence settings to determine where the fault happened on the protected line. The predicted fault location will then be displayed on the front panel and event record. This setting could be a percentage of the line (100) or the actual line length in km or miles.

G) Zero Sequence Factor Magnitude

The Zero Sequence Factor Magnitude is also referred to as K-factor and is used by the relay to compensate for the ground return path when a fault includes a phase-to-ground component. This setting is very important when performing phase-to-ground testing as all expected and actual values will be affected by this setting in conjunction with the Zero Sequence Factor Angle.

H) Zero Sequence Factor Angle

The Zero Sequence Factor Angle is also referred to as the K-factor angle and is used by the relay to compensate for the ground return path when a fault includes a phase-to-ground component. This setting is very important when performing phase-to-ground testing as all expected and actual values will be affected by this setting in conjunction with the Zero Sequence Factor Magnitude.

I) Zone Shape

Most 21-elements use the MHO shape that is basically a circle characteristic with an offset and Maximum Torque Angle. Modern relays can allow the design engineer to choose other shapes to provide protection better suited to the application. For example, a very long line that is heavily loaded could cause nuisance trips under certain conditions with an MHO circle and a lens or tomato shaped curve could provide better protection.

J) Zone Blinders

Zone Blinders are used to disable or block a 21-element operation in certain areas of the characteristic curve. Blinders are typically straight lines drawn through a characteristic and faults that happen on the wrong side of the blinder line will not cause a 21-element operation. Blinders are often used in out-of-step (78) protection and could be used in the scenario described in the Zone Shape description.

K) Zone Supervision / Zone Fault Detector

The Zone Supervision setting sets a minimum amount of current that must flow before the 21-element is enabled to prevent nuisance trips under normal conditions. The minimum current settings ensure that a fault is occurring instead of a normal condition.

L) Zone Voltage Level

21-elements measure the voltage, current, and the phase angles between voltage and current to determine the fault magnitude and direction. If the voltage is below the metering accuracy of the relay, the measured phase angle may not be correct and cause nuisance trips for faults in the wrong direction. The Zone Voltage Level setting sets a minimum voltage that must be present to ensure correct 21-element detection.

M) MHO Phase Distance Zones

The MHO Phase Distance Zones setting determines how many overlapping zones of protection will be used for phase-to-phase and 3-phase faults.

N) MHO Ground Distance Zones

The MHO Ground Distance Zones setting determines how many overlapping zones of protection will be used for faults with a phase-to-ground component.

O) Quadrilateral Ground Distance Zones

The Quadrilateral Ground Distance Zones setting determines how many overlapping zones of protection will be used for faults with a phase-to-ground component using a quadrilateral characteristic instead of the more-common MHO circle.

P) Zone Pickup

The Zone Pickup sets the diameter of the circle (or reach) for each 21-element zone that is enabled.

Q) Zone Offset/Zone Reverse Reach and Angle

The Zone Offset setting applies an offset to the 21-element that moves the circle off of the origin of an impedance diagram to meet a specific application criteria. This element is rarely used for 21-elements but is a common setting in other impedance protective applications such as Loss-of-Field (40) and Out-of-Step (78) protection.

R) Zone Time Delay

The Zone Time Delay setting sets a time delay between a 21-element pickup and trip for each zone. Time delays typically increase as the size of the characteristic increases.

S) Zone Direction

21-elements are designed to operate for faults in the forward direction for faults on the protected line or in the reverse direction to provide backup protection or blocking signals for communication schemes. This setting determines the direction for each zone.

T) Block

The Block setting defines a condition that will prevent the 21-element from operating such as a status input from another device. This setting is rarely used. If enabled, make sure the condition is not true when testing. Always verify correct blocking operation by operating the end-device instead of a simulation to ensure the block has been correctly applied.

3. Preventing Interference in Digital Relays

Protective elements in digital relays often interact with other elements inside a digital relay to provide the most comprehensive protection possible and prevent nuisance trips. 21-elements are often linked with the Loss-of-Potential or Fuse-Fail protection (60-element). The 60-element typically monitors the negative-sequence voltage (V_2) and negative-sequence current (I_2) to determine if an event is a true fault on the system. When a fault occurs on a system, the faulted voltage will decrease and the faulted current will increase causing a V_2 increase due to the unbalanced voltages with a simultaneous increase in I_2. If a potential transformer (PT) fuse operates, one voltage will suddenly decrease and cause a V_2 rise. The PT fuse does not affect the generated current in any way so there will be no corresponding rise in I_2 and the 60-element will operate.

The 21-element uses voltage *and* current to detect a fault. Therefore, the impedance calculations will be affected when a PT fuse operates which could cause the 21-element to operate. We never want a relay to operate unless there is a real fault on the system, so many 21-elements are blocked when a 60-element is detected. In many cases, applying a test case with no prefault condition will trigger the 60-element that prevents the 21-element from operating. Many test procedures instruct the user to disable the 60-element which will allow you to test the 21-element without interference. This solution requires setting changes which carries some risk along with the fact that you are no longer testing the in-service condition when the changes are made. In modern test-sets, this problem is easily overcome by applying nominal prefault voltages before every test. Never apply prefault current unless you have an accurate model of the system under test because the relay will calculate the source impedance based on the prefault conditions and modify the 21-element characteristics to compensate. The applied test vectors must also be correct for this technique to work and they will be discussed in the test descriptions that follow.

Another element that can interfere with 21-element tests is called Switch-On-To-Fault (SOTF). The SOTF element is applied to immediately trip if the breaker is closed onto a fault. This element typically operates using the Zone 2, 21-element; or a ground instantaneous overcurrent (50) element. If the connected circuit breaker transitions from open to closed and a Zone 2 21-element or 50G element is detected within a short time delay, the SOTF element will operate to isolate the system from the fault as quickly as possible. The SOTF can make Zone 2,3,4, and/or 5 timing tests difficult and can trick a relay tester into thinking that a Zone 1, 21-element operated when, in fact, the SOTF element operated instead. Many test procedures disable the SOTF logic when testing but there are inherent risks whenever a setting is changed and you are not testing in-service parameters which, depending on your reasons for testing, can negate your results. The key to preventing SOTF interference is understanding what the SOTF logic is using to detect the breaker status. If the SOTF uses:

- **Voltage**—simply applying nominal prefault voltage before every test will prevent SOTF from operating.
- **Current detection**—apply a prefault current slightly higher than the breaker status pickup setting. The actual pickup test may be affected by this technique but it may not be noticeable with small prefault currents.
- **Contact status**—connect an output contact from your test-set to the relay input that creates a closed condition when generating currents and/or voltages; and an open condition when the test-set is off. You could also place a jumper on the relay input to apply a closed breaker condition at all times, but this technique can cause the breaker-fail protection to operate or cause every test to seal-in which will require a manual reset before every test depending on the relay settings.

4. 3-Phase Line Distance Protection Testing

We will use a GE D-60 for the examples in this section. The following settings are the key settings for testing the Zone 1 21-element protection.

Z1 Function = Enabled	Is Zone 1 enabled?
Z1 Direction = F	Is the zone protection looking in the forward or reverse direction?
Z1 Shape = Mho	What is the shape of the characteristic? MHO is a standard circle characteristic. Review Figure 15-17 for examples of other characteristics.
Z1 Reach = 1.23	This is the diameter of the circle and the pickup setting at the maximum torque angle. (MTA)
Z1 RCA = 87 deg	This is the maximum torque angle (MTA). The MHO circle is created by starting the Z1 Reach magnitude at the Z1 RCA as the circle diameter.
Z1 Comp Limit = 90 deg	This setting changes the shape of the MHO characteristic if the value is any number other than 90º.
Z1 DIR RCA = 87 deg	This setting applies a reference for blinders to the MHO element to allow the protective zone to be customized.
Z1 DIR Comp Limit = 87 deg	This setting determines where the blinders will be positioned.
Z1 Supervision = 0.200 pu	This is the minimum amount of phase current that must flow to enable the 21-element. This element can cause problems with very large impedance settings (>8Ω).
Z1 Volt level = 0.000 pu	This is the minimum amount of voltage that must be present to provide a reliable directional reference for the Z1 element.
Z1 Delay = 0.000 s	The time delay between element pickup and operation.
Z1 Block = SRC1 VT FUSE FAIL OP	Any blocking condition that will prevent the element from operating. In this case, the 21-element is blocked if a fuse failure is detected.
Z1 Target = Latched	If "latched", the element will stay ON until a reset condition is initiated. If "unlatched", the element will turn OFF when the pickup conditions are removed.
Z1 Events = Enabled	Will the Z1 events be recorded in the Relay's Sequence of Events record?

The following diagrams from the GE D-60 manual illustrate how the Z1 element settings affect the Z1 characteristic.

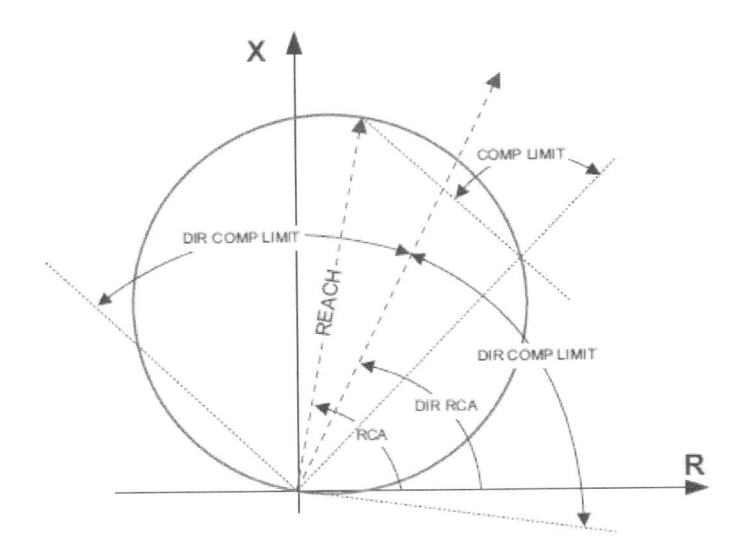

Figure 15-25: Directional MHO Distance Characteristic

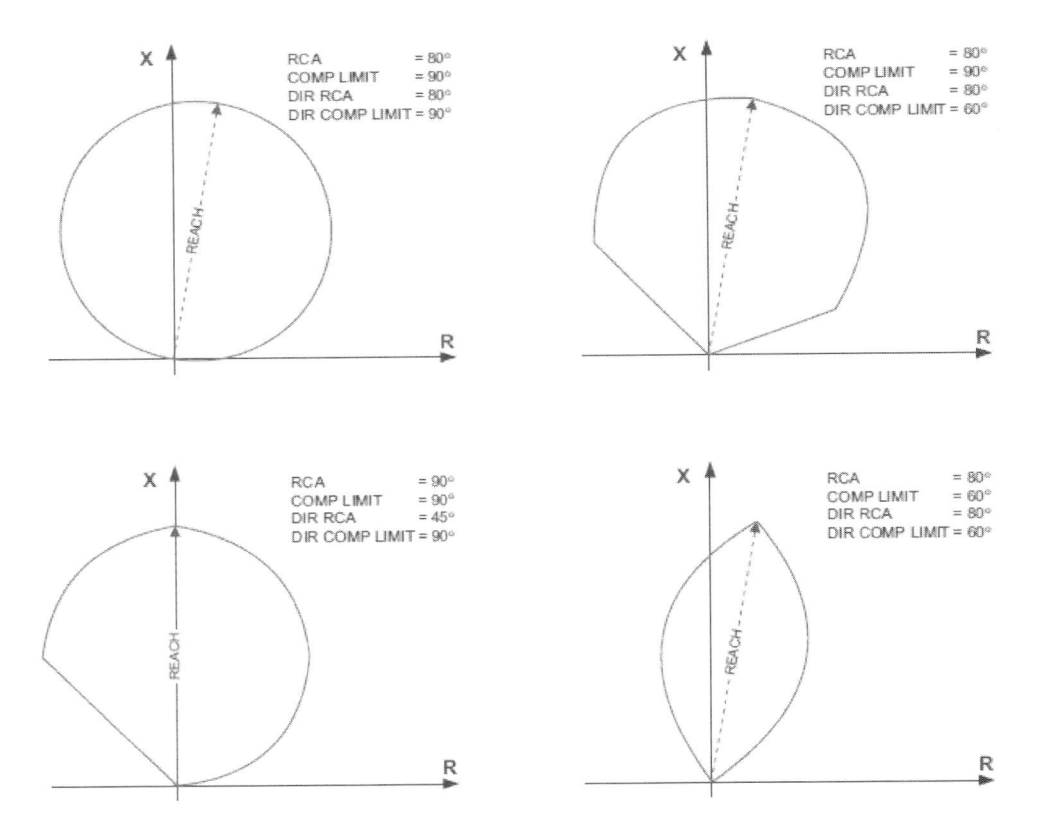

Figure 15-26: MHO Distance Characteristic Sample Shapes

The following figure plots the example characteristic.

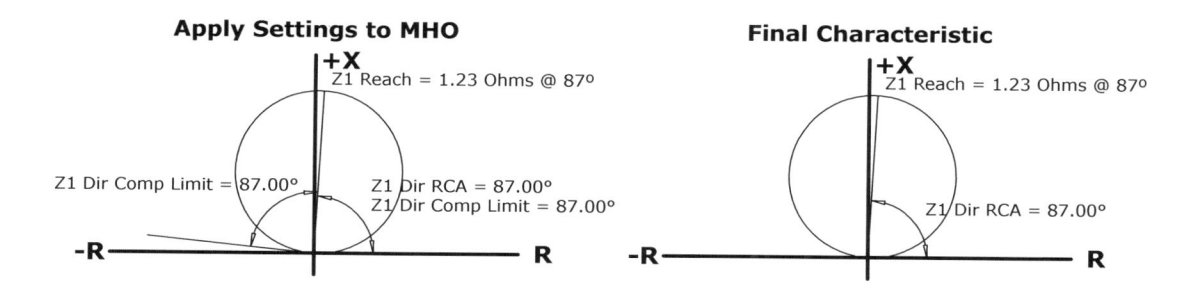

Figure 15-27: 3-Phase Example MHO Distance Characteristic

A) Test-Set Connections

Apply 3-phase balanced currents and voltages when testing the 3-phase, 21-element.

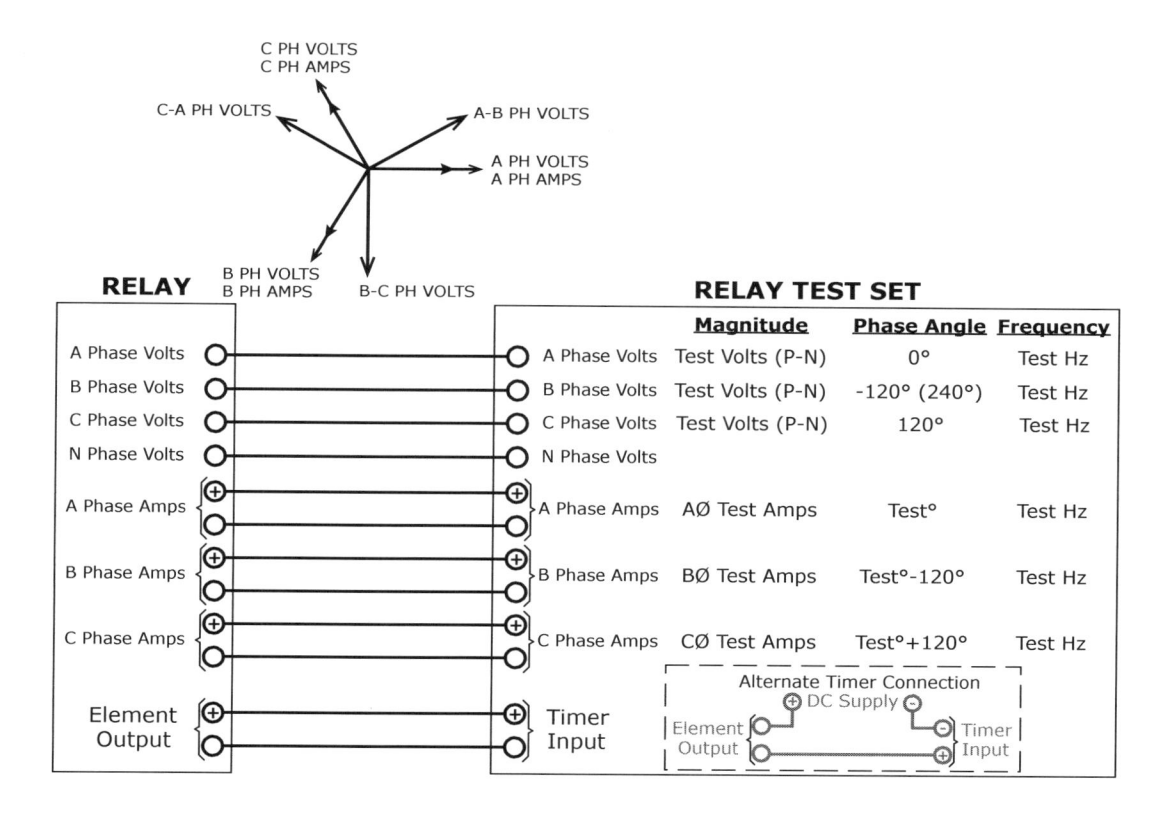

Figure 15-28: 3-Phase Test-Set Connections

B) MTA Test Procedure

The first test you can perform is the Maximum Torque Angle (MTA) test which verifies the RCA setting inside the relay. This test is performed by setting a fixed impedance lower than the reach setting and then varying the phase angle. Record the pickup angle on either side of the circle and calculate the average of the two results. The entire procedure is shown in the following figure. Remember that angles in an impedance diagram are reversed so you must multiply your test result by -1 to get the MTA in impedance terms.

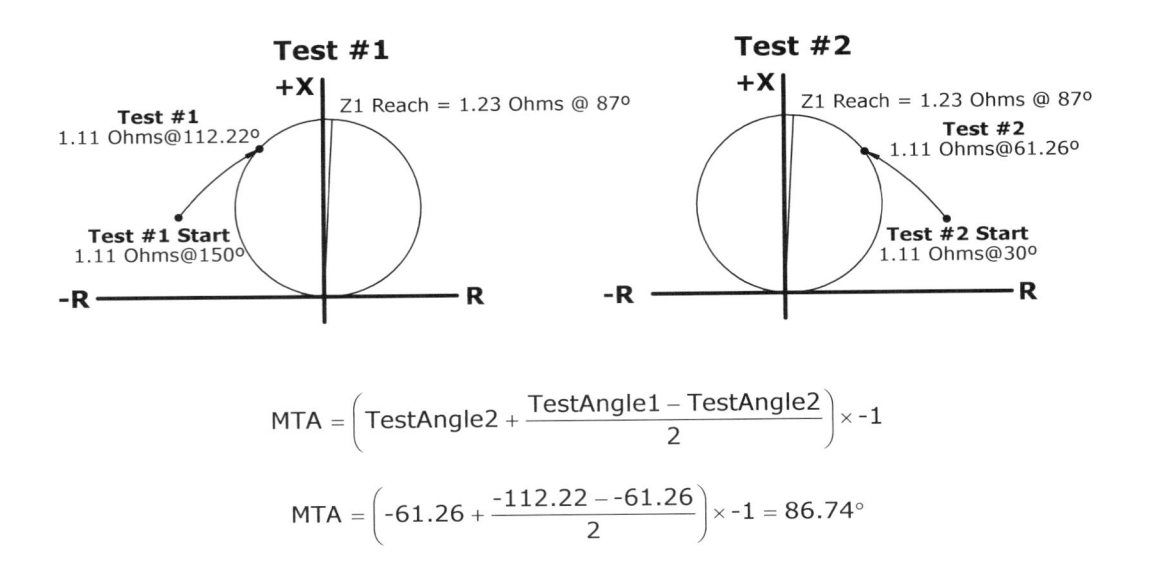

$$MTA = \left(TestAngle2 + \frac{TestAngle1 - TestAngle2}{2} \right) \times -1$$

$$MTA = \left(-61.26 + \frac{-112.22 - -61.26}{2} \right) \times -1 = 86.74°$$

Figure 15-29: 3-Phase Example MHO Distance MTA Test

The MTA test could be performed as part of the Reach Test Procedure described later in the chapter.

Follow the steps below to perform an MTA test.

1. Determine how you will monitor pickup and set the relay accordingly, if required. (Pickup indication by LED, output contact, front panel display, etc. See the *Relay Test Procedures* section starting on page 109 for details.)

2. Calculate the test configuration using one of these test methods:

   a. If the Z1 Reach setting is greater than 0.66 Ohms, arbitrarily determine the maximum amount of current to be applied to the relay. 10.000A is a good rule of thumb for maximum continuous current when testing most relays. Calculate the 3-phase balanced voltage required to apply 90% of the Z1 Reach setting using the following formula based on Ohm's Law.

$$V = I \times R$$

$$FaultVolts = 0.9 \times FaultAmps \times Z1Reach$$

$$FaultVolts = 0.9 \times 9.0 \times 1.23$$

$$FaultVolts = 9.96V$$

b. If the Z1 Reach setting is less than 0.66Ω, use the traditional formula described in this step with a 5.0V voltage setting. If more than 10.0A is required, the test should be performed as quickly as possible to prevent equipment damage.

Traditional test techniques reverse the formula in the previous step because electromechanical relays required a minimum voltage of approximately 20V to function reliably. Digital relays typically only require 3.0V to properly determine direction when a prefault voltage is applied. The traditional formula can also be used to calculate the test parameters in the first step, but you may need to make several calculations to keep the test current at reasonable levels.

$I = \dfrac{V}{R}$	$I = \dfrac{V}{R}$
$FaultAmps = \dfrac{FaultVolts}{0.9 \times Z1Reach}$	$FaultAmps = \dfrac{FaultVolts}{0.9 \times Z1Reach}$
$FaultAmps = \dfrac{20.0V}{0.9 \times 1.23}$	$FaultAmps = \dfrac{9.96V}{0.9 \times 1.23}$
$FaultAmps = 18.067A$	$FaultAmps = 9.00A$

3. Configure a test with two states (prefault and fault) as shown in the following chart:

PREFAULT	FAULT
V1 = Nominal Volts @ 0°	V1 = FaultVolts @ 0°
V2 = Nominal Volts @ -120°	V2 = FaultVolts @ -120°
V3 = Nominal Volts @ 120°	V3 = FaultVolts @ 120°
I1 = 0.00A	I1 = FaultAmps @ V1° - Z1 RCA
I2 = 0.00A	I2 = FaultAmps @ V2° - Z1 RCA
I3 = 0.00A	I3 = FaultAmps @ V3° - Z1 RCA

Figure 15-30: 3-Phase Example MHO Distance MTA Test Configuration

The test configuration for the example settings are shown in the following chart:

PREFAULT	FAULT
V1 = 69.28V @ 0°	V1 = 9.96V @ 0°
V2 = 69.28V @ -120°	V2 = 9.96V @ -120°
V3 = 69.28V @ 120°	V3 = 9.96V @ 120°
I1 = 0.00A	I1 = 9.000A @ -87° (0° - 87°)
I2 = 0.00A	I2 = 9.000A @ 153°(-120° - 87°)
I3 = 0.00A	I3 = 9.000A @ 33°(120° - 87°)

4. Configure your test-set to change all three current angles equally and simultaneously.

5. Apply the prefault condition to the relay and verify the metering is correct, if required.

6. Apply the fault condition and ensure the 21-element operates.

7. Increase all three phase-current-angles equally until the 21-element drops out.

8. Decrease all three phase-current-angles until the 21-element picks up. Record the AØ angle at pickup. (-112.22° in our example)

9. Continue decreasing all three phase-current-angles until the 21-element drops out.

10. Increase all three phase-current-angles until the 21-element picks up. Record the AØ angle at pickup. (-61.26° in our example).

11. Calculate the average of the two angles and add the result to the smaller angle. Record the calculated value on your test sheet. Steps 6-11 are shown in the following figures:

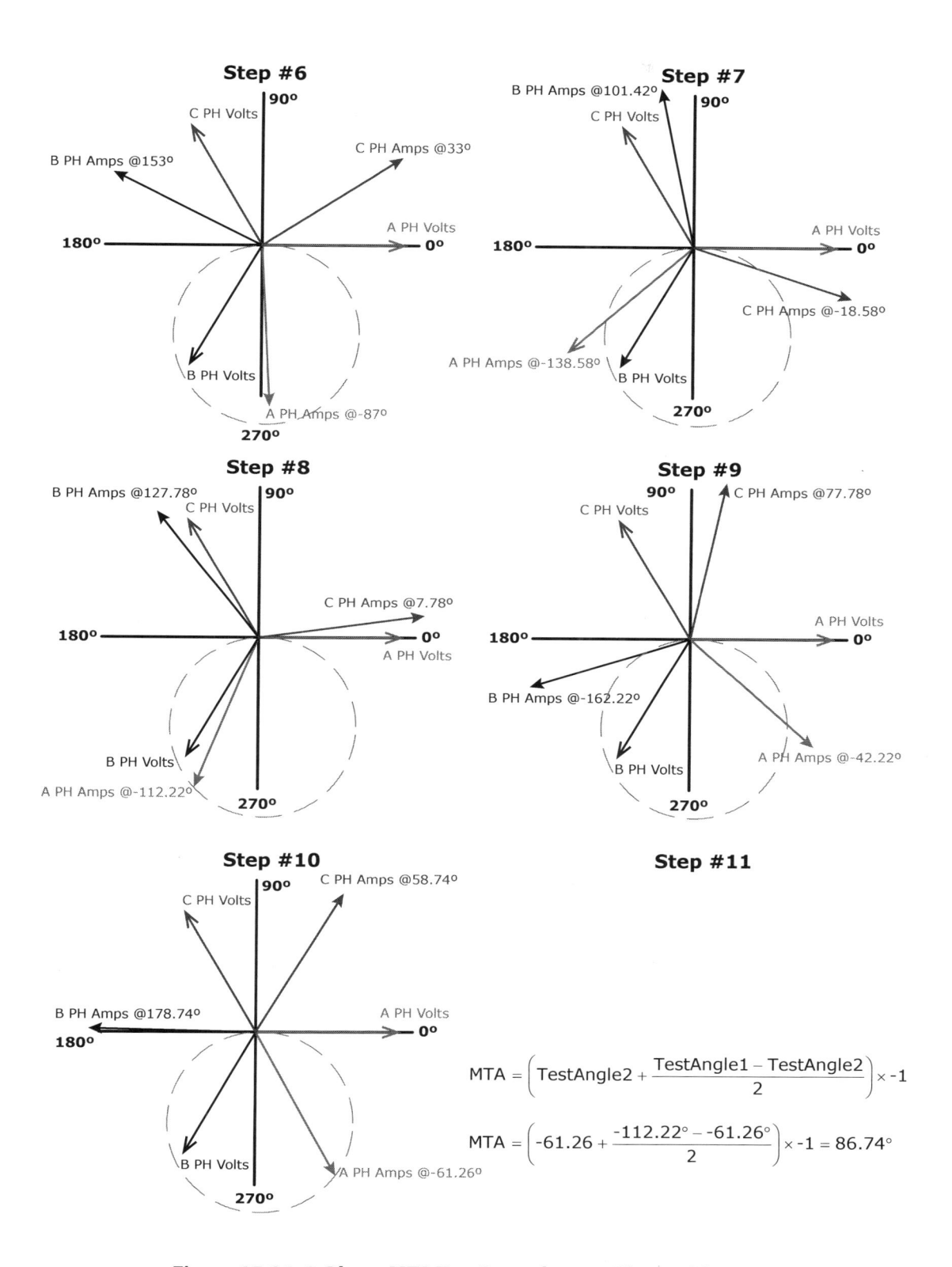

Figure 15-31: 3-Phase MTA Test Procedure on Phasor Diagram

Step #6, Step #7, Step #8, Step #9, Step #10, Step #11

$$MTA = \left(TestAngle2 + \frac{TestAngle1 - TestAngle2}{2} \right) \times \text{-}1$$

$$MTA = \left(\text{-}61.26 + \frac{\text{-}112.22° - \text{-}61.26°}{2} \right) \times \text{-}1 = 86.74°$$

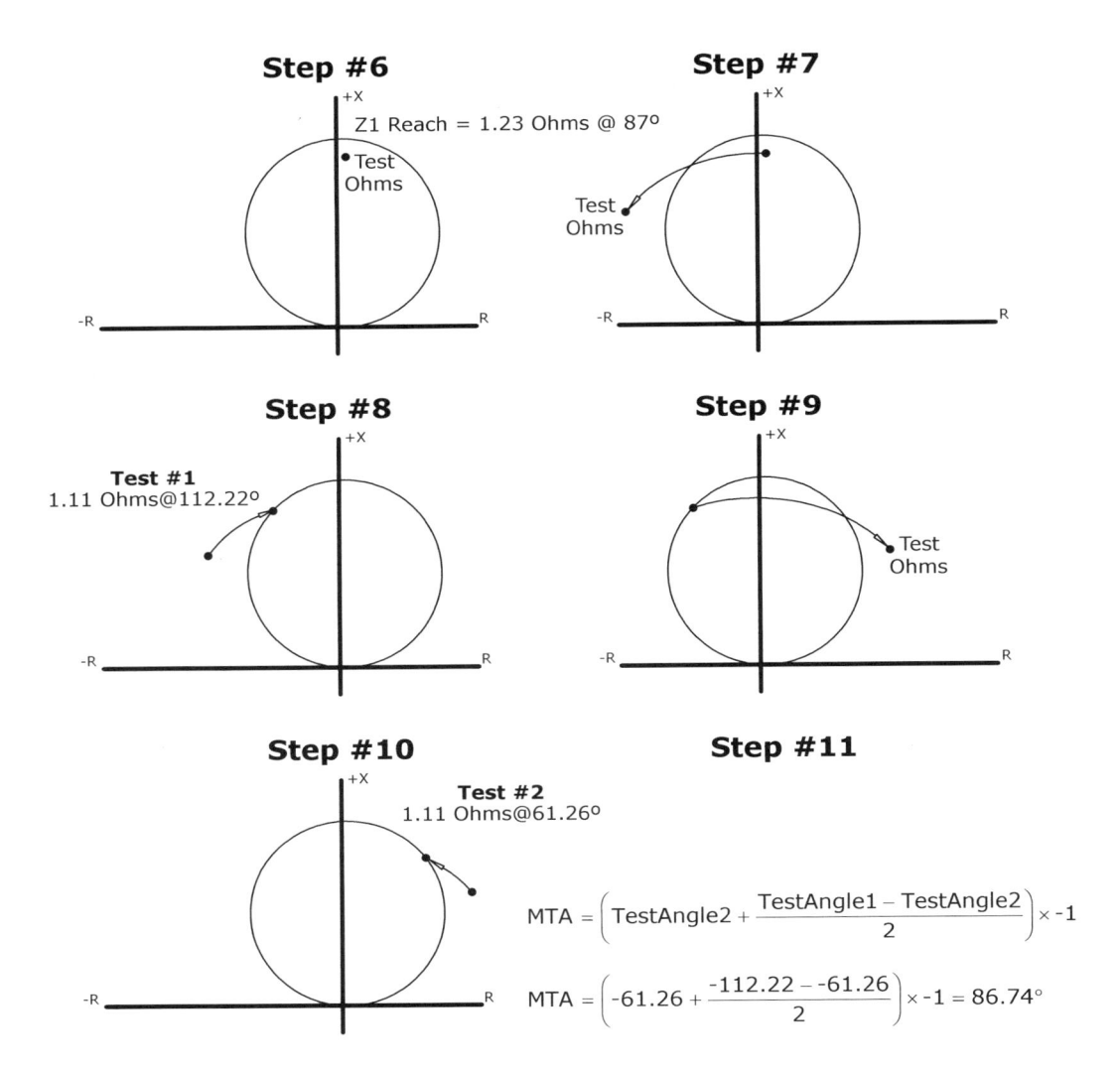

Figure 15-32: 3-Phase MTA Test Procedure on Impedance Diagram

12. Compare test results to manufacturer's specifications. There is typically no specific specification for the MTA test, but we can use the Phase Distance Reach (secondary Ω) specifications located in the GE D-60 manual as shown in the following figure which indicates a 5% tolerance.

PHASE DISTANCE

Characteristic:	mho (memory polarized or offset) or quad (memory polarized or non-directional), selectable individually per zone
Number of zones:	5
Directionality:	forward, reverse, or non-directional per zone
Reach (secondary Ω):	0.02 to 500.00 Ω in steps of 0.01
Reach accuracy:	±5% including the effect of CVT transients up to an SIR of 30

Figure 15-33: GE D-60 Phase Distance Specification

The test results measured an MTA of 87.225° and the expected result was 87°. Using the standard percent error calculation, we can determine the percent error and the test result passes. Record the percent error and pass/fail result on your test sheet.

$$\frac{\text{Actual Value - Expected Value}}{\text{Expected Value}} \times 100 = \text{Percent Error}$$

$$\frac{86.74° - 87.00°}{87.00°} \times 100 = \text{Percent Error}$$

-0.30% Error

13. Repeat the MTA test for all enabled 21-element zones.

3 PHASE DISTANCE TEST RESULTS (Ohms)											
	Z1 Function	**Enabled**	Z2 Function	**Enabled**	Z3 Function	**Enabled**	Z4 Function	**Disabled**			
	Z1 Reach	**1.23**	Z2 Reach	**1.85**	Z3 Reach	**3.1**	Z4 Reach				
	Z1 RCA	**87**	Z2 RCA	**87**	Z3 RCA	**87**	Z4 RCA				
	Z1 Delay	**0.00**	Z2 Delay	**0.30**	Z3 Delay	**1.00**	Z4 Delay				
	Z1 Direction	**F**	Z2 Direction	**F**	Z3 Direction	**F**	Z4 Direction				
MAXIMUM TORQUE ANGLE (MTA) TEST RESULTS (Ohms)											
	ZONE 1 TESTS			ZONE 2 TESTS			ZONE 3 TESTS			ZONE 4 TESTS		
	Ohms	PU1	PU2	Ohms	PU1	PU2	Ohms	PU1	PU2	Ohms	PU1	PU2
PU In deg	1.11	-112.2	-61.3	1.67	-112.8	-60.9	2.79	-112.8	-60.6			
	M1P	MFG	%ERR	M2P	MFG	%ERR	M3P	MFG	%ERR			
MTA	86.74	87.0	-0.30	86.84	87.0	-0.19	86.70	87.0	-0.34			

$$\text{MTA} = (\text{PU2} + ((\text{PU1-PU2})/2)) * (-1)$$

C) **Reach Test Procedure**

While the MTA test procedure can be considered optional, the reach test procedure is a mandatory test for the 21-element. The relay tester chooses one or more angles to perform the test (typically starting at the MTA angle) and then performs a standard pickup test. Some relay test procedures require multiple points around the MHO circle which can be considered unnecessary in a digital relay for several reasons.

- Multiple test points were required for electromechanical relays because it was possible that the relay was out of calibration and resistors and magnets could be adjusted to obtain a different result. No such adjustments exist on digital relays. Therefore, a digital relay 21-element test proves that the relay settings are applied correctly which can be achieved with one pickup test.
- Many digital relays change the MHO characteristic based on system conditions and the MHO circle may not be the static circle you are expecting. Multi-point tests on an MHO circle may produce results outside of the expected tolerance for a static circle but not for a dynamic circle.

Follow the steps below to perform a reach test.

1. Determine how you will monitor pickup and set the relay accordingly, if required. (Pickup indication by LED, output contact, front panel display, etc. See the *Relay Test Procedures* section starting on page 109 for details.)

2. Determine the Test Angle for the test. (MTA, MTA+n, MTA-n, 95° lag, etc.) and calculate the expected impedance at that angle. The following example will use the MTA angle. The expected result for any MHO circle with a 0.00Ω offset can be calculated using the formula:

$$\text{TestOhms} = \text{Z1Reach} \times \text{Cos}(\text{MTA}° - \text{Test}°)$$

Example calculations are shown below:

$\text{TestOhms} = \text{Z1Reach} \times \text{Cos}(\text{MTA}° - \text{Test}°)$	$\text{TestOhms} = \text{Z1Reach} \times \text{Cos}(\text{MTA}° - \text{Test}°)$
$\text{TestOhms} = 1.23 \times \text{Cos}(87° - 57°)$	$\text{TestOhms} = 1.23 \times \text{Cos}(87° - 117°)$
$\text{TestOhms} = 1.23 \times 0.866$	$\text{TestOhms} = 1.23 \times 0.866$
$\text{TestOhms} = 1.065Ω$	$\text{TestOhms} = 1.065Ω$

3. Calculate the test configuration using one of these test methods:

   a. If the Z1 Reach setting is greater than 0.66 Ohms, arbitrarily determine the maximum amount of current to be applied to the relay. A good rule of thumb for maximum continuous current for most relays is 10.0A. Calculate the 3-phase balanced voltage required to apply 95% of the Z1 Reach setting using the following formula based on Ohm's Law.

$$V = I \times R$$

$$FaultVolts = 0.95 \times FaultAmps \times Z1Reach$$

$$FaultVolts = 0.95 \times 9.0 \times 1.23$$

$$FaultVolts = 10.5165V$$

   b. If the Z1 Reach setting is less than 0.66Ω, use the traditional formula described in this step with a 5.0V voltage setting. If more than 10.0A is required, the test should be performed as quickly as possible to prevent equipment damage.

   Traditional test techniques reverse the formula in the previous step because electromechanical relays required a minimum voltage of approximately 20V to function reliably. Digital relays typically only require 3.0V to properly determine direction when a prefault voltage is applied. The traditional formula can also be used to calculate the test parameters in the first step, but you may need to make several calculations to keep the test current at reasonable levels.

$I = \dfrac{V}{R}$	$I = \dfrac{V}{R}$
$FaultAmps = \dfrac{FaultVolts}{0.95 \times Z1Reach}$	$FaultAmps = \dfrac{FaultVolts}{0.95 \times Z1Reach}$
$FaultAmps = \dfrac{20.0V}{0.95 \times 1.23}$	$FaultAmps = \dfrac{10.5165V}{0.95 \times 1.23}$
$FaultAmps = 17.11A$	$FaultAmps = 9.0A$

4. Configure a test with two states (prefault and fault) as shown in the following chart:

PREFAULT	FAULT
V1 = Nominal Volts @ 0°	V1 = FaultVolts @ 0°
V2 = Nominal Volts @ -120°	V2 = FaultVolts @ -120°
V3 = Nominal Volts @ 120°	V3 = FaultVolts @ 120°
I1 = 0.00A	I1 = FaultAmps @ V1° - Test°
I2 = 0.00A	I2 = FaultAmps @ V2° - Test°
I3 = 0.00A	I3 = FaultAmps @ V3° - Test°

Figure 15-34: 3-Phase Example MHO Distance MTA Test Configuration

The test configuration for the example settings is shown in the following chart:

PREFAULT	FAULT
V1 = 69.28V @ 0°	V1 = 10.52V @ 0°
V2 = 69.28V @ -120°	V2 = 10.52V @ -120°
V3 = 69.28V @ 120°	V3 = 10.52V @ 120°
I1 = 0.00A	I1 = 9.0A @ -87° (0° - 87°)
I2 = 0.00A	I2 = 9.0A @ 153°(-120° - 87°)
I3 = 0.00A	I3 = 9.0A @ 33°(120° - 87°)

5. Configure your test-set to change all three fault current magnitudes and/or voltage magnitudes simultaneously.

6. Apply the prefault condition to the relay and verify the metering is correct, if required.

7. Apply the fault condition and ensure the 21-element operates.

8. Increase all three voltage magnitudes OR decrease all three current magnitudes until the 21-element drops out.

9. Reverse the current or voltage changes until the 21-element picks up. Record the impedance at pickup. (1.22Ω in our example) $R = \frac{V}{I}$. Record the results on your test sheet.

Steps 7-9 are shown in the following figure:

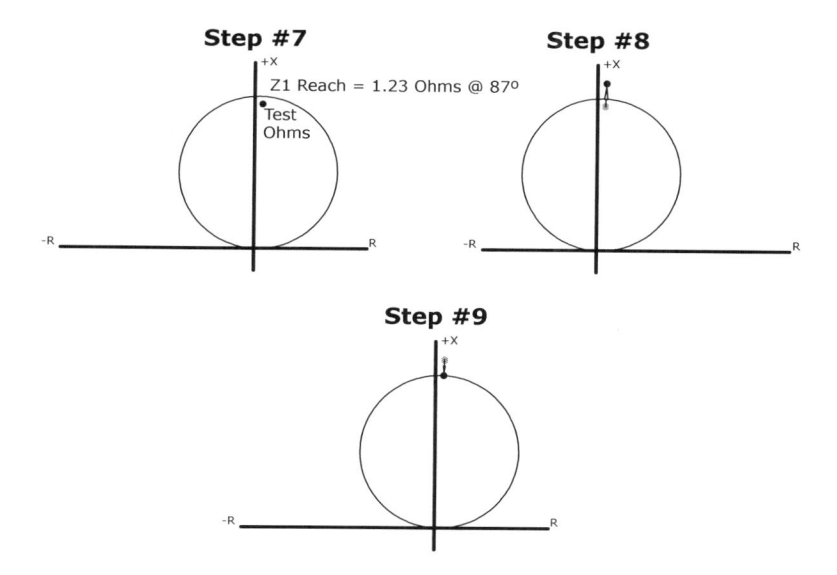

Figure 15-35: 3-Phase Example MHO Distance Reach Test Procedure

10. Compare test results to manufacturer's specifications. The "Phase Distance" "Reach (secondary Ω)" specifications located in the GE D-60 manual indicates a 5% tolerance.

PHASE DISTANCE

Characteristic:	mho (memory polarized or offset) or quad (memory polarized or non-directional), selectable individually per zone
Number of zones:	5
Directionality:	forward, reverse, or non-directional per zone
Reach (secondary Ω):	0.02 to 500.00 Ω in steps of 0.01
Reach accuracy:	±5% including the effect of CVT transients up to an SIR of 30

Figure 15-36: GE D-60 Phase Distance Specification

The test results measured a reach of 1.22Ω and the expected result was 1.23Ω. Using the standard percent error calculation, we can determine the percent error and the test result passes. Record the percent error or pass/fail result on your test sheet.

$$\frac{\text{Actual Value - Expected Value}}{\text{Expected Value}} \times 100 = \text{Percent Error}$$

$$\frac{1.22\Omega - 1.23\Omega}{1.23\Omega} \times 100 = \text{Percent Error}$$

-0.81% Error

11. Repeat Steps 2-10 for all desired test points.

12. Repeat Steps 2-11 for all enabled zones.

3 PHASE DISTANCE TEST RESULTS (Ohms)									
	Z1 Function	**Enabled**	Z2 Function	**Enabled**	Z3 Function	**Enabled**	Z4 Function	**Disabled**	
	Z1 Reach	**1.23**	Z2 Reach	**1.85**	Z3 Reach	**3.1**	Z4 Reach		
	Z1 RCA	**87**	Z2 RCA	**87**	Z3 RCA	**87**	Z4 RCA		
	Z1 Delay	**0.00**	Z2 Delay	**0.30**	Z3 Delay	**1.00**	Z4 Delay		
	Z1 Direction	**F**	Z2 Direction	**F**	Z3 Direction	**F**	Z4 Direction		

REACH TEST RESULTS (Ohms)												
	ZONE 1 TESTS			**ZONE 2 TESTS**			**ZONE 3 TESTS**			**ZONE 4 TESTS**		
TEST ANGLE	M1P	MFG	%ERR	M2P	MFG	%ERR	M3P	MFG	%ERR	M4P	MFG	%ERR
57	**1.10**	1.07	3.27	**1.68**	1.60	4.86	**2.78**	2.68	3.55			
87	**1.22**	1.23	-0.81	**1.86**	1.85	0.54	**3.12**	3.10	0.65			
117	**1.05**	1.07	-1.43	**1.59**	1.60	-0.76	**2.67**	2.68	-0.55			

ZONE 1 TESTS MFG =Z1Reach*COS(RADIANS(Z1RCA-TESTANGLE))

Zone 2 TESTS MFG =Z2Reach*COS(RADIANS(Z2RCA-TESTANGLE))

ZONE 3 TESTS MFG =Z3Reach*COS(RADIANS(Z3RCA-TESTANGLE))

NOTE: An MTA test can be incorporated into the reach test by performing two reach tests with the same angular difference from the MTA as shown in the test sheet above. If both test results have the same result, the MTA is correct.

D) Timing Test Procedure

The Timing test is relatively simple compared to the MTA and reach tests. Apply a prefault, then immediately transition to a fault at least 90% inside the circle, and record the time between fault and the relay operation.

Follow the steps below to perform a timing test.

1. Configure the test-set timer to start when the fault state is energized, and stop when the trip contact operates.

2. Determine the Test Angle for the test. (MTA, MTA+n, MTA-n, 95° lag, etc.) and calculate the expected impedance at that angle. The following example will use the MTA angle. The expected result for any MHO circle with a 0.00Ω offset can be calculated using the following formula:

$$TestOhms = Z1Reach \times Cos(MTA° - Test°)$$

3. Calculate the test configuration using one of these test methods:

    a. If the Z1 Reach setting is greater than 0.66 Ohms, arbitrarily determine the maximum amount of current to be applied to the relay. A good rule of thumb for maximum continuous current for most relays is 10.0A. Calculate the 3-phase balanced voltage required to apply 70% of the Z1 Reach setting using the following formula based on Ohm's Law.

$$V = I \times R$$

$$FaultVolts = 0.70 \times FaultAmps \times Z1Reach$$

$$FaultVolts = 0.70 \times 9.0 \times 1.23$$

$$FaultVolts = 7.749V$$

    b. If the Z1 Reach setting is less than 0.66Ω, use the traditional formula described in this step with a 5.0V voltage setting. If more than 10.0A is required, the test should be performed as quickly as possible to prevent equipment damage.

Traditional test techniques reverse the formula in the previous step because electromechanical relays required a minimum voltage of approximately 20V to function reliably. Digital relays typically only require 3.0V to properly determine direction when a prefault voltage is applied. The traditional formula can also be used to calculate the test parameters, but you may need to make several calculations to keep the test current at reasonable levels as shown in the following examples:

$$I = \frac{V}{R}$$

$$FaultAmps = \frac{FaultVolts}{0.70 \times Z1Reach}$$

$$FaultAmps = \frac{20.0V}{0.90 \times 1.23}$$

$$FaultAmps = 23.23A$$

$$I = \frac{V}{R}$$

$$FaultAmps = \frac{FaultVolts}{0.70 \times Z1Reach}$$

$$FaultAmps = \frac{7.749V}{0.70 \times 1.23}$$

$$FaultAmps = 9.0A$$

4. Configure a test with two states (prefault and fault) as shown in the following chart:

PREFAULT	FAULT
V1 = Nominal Volts @ 0°	V1 = FaultVolts @ 0°
V2 = Nominal Volts @ -120°	V2 = FaultVolts @ -120°
V3 = Nominal Volts @ 120°	V3 = FaultVolts @ 120°
I1 = 0.00A	I1 = FaultAmps @ V1° - Test°
I2 = 0.00A	I2 = FaultAmps @ V2° - Test°
I3 = 0.00A	I3 = FaultAmps @ V3° - Test°

Figure 15-37: 3-Phase Example MHO Distance MTA Test Configuration

The test configuration for the example settings are shown in the following chart:

PREFAULT	FAULT
V1 = 69.28V @ 0°	V1 = 7.749V @ 0°
V2 = 69.28V @ -120°	V2 = 7.749V @ -120°
V3 = 69.28V @ 120°	V3 = 7.749V @ 120°
I1 = 0.00A	I1 = 9.0A @ -87° (0° - 87°)
I2 = 0.00A	I2 = 9.0A @ 153°(-120° - 87°)
I3 = 0.00A	I3 = 9.0A @ 33°(120° - 87°)

5. Apply the prefault condition to the relay and verify the metering is correct, if required.

6. Apply the fault condition and ensure the 21-element operates within the specified time.

7. Compare test results to manufacturer's specifications. This step can be more complicated than normal as we have the normal tolerance with a "Timing accuracy: +/- 3% or 4 ms, whichever is greater" and the Phase Distance Operating Time Curves.

PHASE DISTANCE

Time delay:	0.000 to 65.535 s in steps of 0.001
Timing accuracy:	±3% or 4 ms, whichever is greater

Figure 15-38: GE D-60 Phase Distance Timing Specification

PHASE DISTANCE OPERATING TIME CURVES

The operating times are response times of a microprocessor part of the relay. See output contacts specifications for estimation of the total response time for a particular application. The operating times are average times including variables such as fault inception angle or type of a voltage source (magnetic VTs and CVTs).

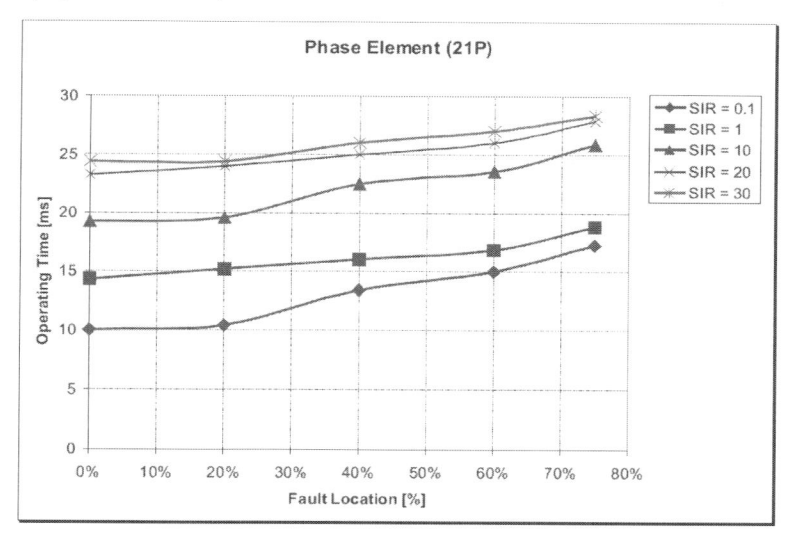

Figure 15-39: GE D-60 Phase Element (21P) Timing Specification

Using the worst case scenario of 27.5 ms (SIR = 0.1 at 70% fault location) +/- 3% or 4 ms, we can determine that the maximum expected time is:

$$27.5ms \times 3\% = 27.5ms \times 0.03 = 0.825ms$$

$$0.825ms < 4ms$$

$$ExpectedTime = 27.5ms + 4ms$$

$$ExpectedTime = 31.5ms$$

The measured time for our example was 31.0ms. The time is less than our expected time and is acceptable for service.

$$\frac{Actual\ Value - Expected\ Value}{Expected\ Value} \times 100 = Percent\ Error$$

$$\frac{31.0ms - 31.5ms}{31.5ms} \times 100 = Percent\ Error$$

-1.59% Error

8. Repeat Steps 2-7 for all enabled zones.

3 PHASE DISTANCE TEST RESULTS (Ohms)											
	Z1 Function	Enabled	Z2 Function	Enabled	Z3 Function	Enabled	Z4 Function	Disabled			
	Z1 Reach	1.23	Z2 Reach	1.85	Z3 Reach	3.1	Z4 Reach				
	Z1 RCA	87	Z2 RCA	87	Z3 RCA	87	Z4 RCA				
	Z1 Delay	0.00	Z2 Delay	0.30	Z3 Delay	1.00	Z4 Delay				
	Z1 Direction	F	Z2 Direction	F	Z3 Direction	F	Z4 Direction				
TIMING TESTS (in Seconds)											
MULT	M1PT	MFG	%ERR	M2PT	MFG	%ERR	M3P	MFG	%ERR	M4PT	MFG	%ERR
0.7 * PU	**0.031**	0.032	-1.59	**0.304**	0.30	1.33	**1.015**	1.0	1.50			
COMMENTS:											
RESULTS ACCEPTABLE:			✓ YES			☐ NO			☐ SEE NOTES		

5. Phase-to-Phase Line Distance Protection Testing

Phase-to-phase testing in the past was a simple procedure where a single-phase voltage and single-phase current was applied to the phases under test with the same test procedure as the 3-phase test. Modern relays are designed to prevent mis-operations caused by incorrect metering, such as a fuse-failure, and the test techniques of the past will not work without setting changes, if they will work at all, as relays become more sophisticated.

Modern test-sets allow you to simulate a more realistic fault so that you can test the phase-to-phase 21-element without changing settings. The first step in creating an acceptable fault for a relay is understanding what happens in a phase-to-phase fault. The following figure depicts a prefault and fault condition with a A-B fault. The following information describes the conditions required to create a realistic fault for modern relays in the absence of a system fault study.

Fault Voltages

- The C-phase voltage does not change because the load impedance to ground has not changed. (In a real fault, C-phase will change slightly compared to the A-phase and B-phase voltages, but the changes cannot be determined without a system model.)
- The A-B Phase voltage collapses because the fault resistance is significantly less than the load resistance. The collapsed voltage turns the equilateral triangle (all sides and angles are equal) in the prefault condition into an isosceles triangle (two angles and sides are equal) during the fault.
- The A-N voltage magnitude is equal to the B-N voltage magnitude because the line impedance to the fault should be nearly identical.
- The A-N and B-N magnitudes should be calculated to create maximum negative sequence voltage to provide a strong reference for the directional control inside the relay.
- The A-N and B-N magnitudes should be calculated to create minimum zero sequence voltage to prevent miscalculation inside the relay.
- The A-N and B-N angles should be equal if the N-C (C-N reversed) voltage is used as a reference to maintain the isosceles triangle.
- The faulted voltage angle should not change between prefault and fault to maintain a consistent reference.
  - A-B fault voltage @ 30°
  - B-C fault voltage @ 90°
  - C-A fault voltage @ 150°

Fault Currents

- The C-phase current does not change because the load impedance to ground has not changed. (In a real fault, C-phase will change slightly compared to the A-phase and B-phase voltages but the changes cannot be determined without a system model.)
- The fault current (FAULT AMPS) is the current flowing through the fault and is the vector sum of A PH AMPS and B PH AMPS. Notice that the A-phase and B-phase amps measured at the source terminals are 180° from each other. Therefore, the A-phase and B-phase currents are the same and equal to one-half of the fault current.
- The fault current angle is referenced to the fault voltage and not the origin. We always keep the A-B fault angle at 30° and the A-phase current will lag 30° by the fault angle. The B-phase current will be 180° from the A-phase current.

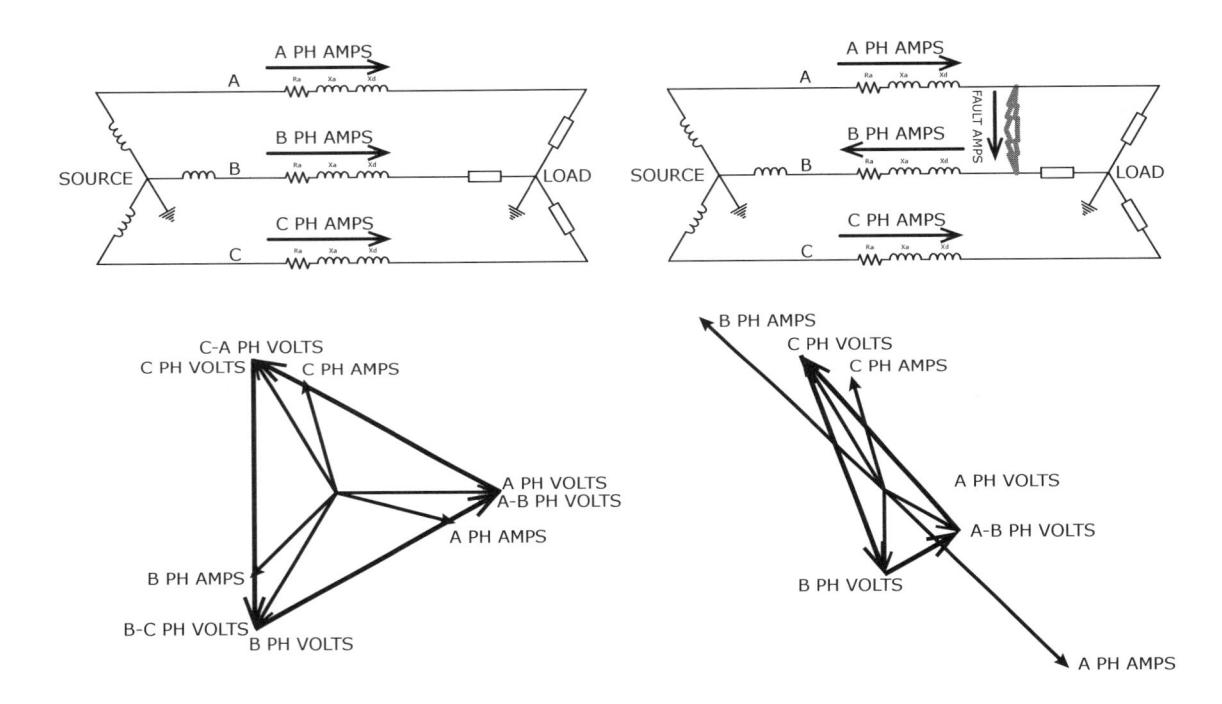

Figure 15-40: Phase-to-Phase Fault Characteristic

A) Phase-to-Phase Test Calculations

Some digital relays have additional complexity such as negative-sequence-impedance directional elements and source impedance measuring to prevent mis-operations. These additional elements to distance protection can make the phase-to-phase calculation quite involved, but the following procedure applied to a spreadsheet program should make the process relatively painless. We will use the same settings from the 3-phase test example described previously.

- Z1 Direction = F
- Z1 Shape = Mho
- Z1 Reach = 1.23
- Z1 RCA = 87 deg

1. Apply a balanced 3-phase nominal voltage within the prefault state. Ensure all currents are 0.00A in prefault.

CHANNEL	PREFAULT		FAULT	
	MAG	ANGLE	MAG	ANGLE
Vphase-to-phase	120V			
Ifault				
IPhase	0.000A			
V1	69.28V	0.00°		
V2	69.28V	-120.00°		
V3	69.28V	120.00°	69.28V	120.00°
I1	0.000A	0.00°	8.000A	
I2	0.000A	-120.00°	8.000A	
I3	0.000A	120.00°	0.000A	120.00°

2. Determine the maximum amount of current you wish to apply for the test. We'll use 8.0A for our example.

3. Calculate the fault voltage using the current from Step 2.

$$V_{phase-phase(Fault)} = 2 \times I_{Phase} \times TestOhms$$

$$V_{phase-phase(Fault)} = 2 \times 8A \times 1.23\Omega$$

$$V_{phase-phase(Fault)} = 19.68V$$

where

Vphase-to-phase(Fault) = fault voltage

IPhase = Applied current in each phase

TestOhms = Desired test impedance

4. Ensure that the fault voltage from Step 3 is greater than 5.0V and less than 90% of the prefault. If not, adjust the current accordingly or determine the test current with a preset voltage using the following formula:

$$I_{Phase} = \frac{V_{phase-phase(Fault)}}{2 \times TestOhms}$$

$$I_{Phase} = \frac{19.68V}{2 \times 1.23}$$

$$I_{Phase} = 8.0A$$

CHANNEL	PREFAULT		FAULT	
	MAG	ANGLE	MAG	ANGLE
Vphase-to-phase	120V		19.68	
Ifault			16.0A	87°
IPhase	0.000A		8.0A	
V1	69.28V	0.00°		
V2	69.28V	-120.00°		
V3	69.28V	120.00°	69.28V	120.00°
I1	0.000A	0.00°	8.000A	
I2	0.000A	-120.00°	8.000A	
I3	0.000A	120.00°	0.000A	120.00°

5. Apply A-N and B-N fault current magnitudes using I_{Phase} calculated in the previous steps. The A-phase angle equals the phase-to-phase Voltage angle (30°) minus the Test Angle. The B-phase angle equals the A-phase angle plus/minus 180°.

$I1 = I2 = I_{Phase}$ $I1 = I2 = 8.0A$	$I1Angle = V_{Phase-Phase(Fault)}Angle - TestAngle$ $I1Angle = 30° - 87°$ $I1Angle = -57°$	$I2Angle = I1Angle \pm 180°$ $I2Angle = -57° + 180$ $I2Angle = 123°$

	PREFAULT		FAULT	
CHANNEL	**MAG**	**ANGLE**	**MAG**	**ANGLE**
Vphase-to-phase	120V		19.68	
Ifault			16.0A	87°
IPhase	0.000A		8.0A	
V1	69.28V	0.00°		
V2	69.28V	-120.00°		
V3	69.28V	120.00°	69.28V	120.00°
I1	0.000A	0.00°	8.000A	-57.00°
I2	0.000A	-120.00°	8.000A	123.00°
I3	0.000A	120.00°	0.000A	120.00°

6. Calculate the A-N and B-N voltage magnitudes using the following formula:

$$V1_{Fault} = V2_{Fault} = \frac{\sqrt{V1_{Prefault}^2 + V_{Phase-Phase(Fault)}^2}}{2}$$

$$V1_{Fault} = V2_{Fault} = \frac{\sqrt{69.28V^2 + 19.68V^2}}{2}$$

$$V1_{Fault} = V2_{Fault} = 36.01V$$

	PREFAULT		FAULT	
CHANNEL	**MAG**	**ANGLE**	**MAG**	**ANGLE**
Vphase-to-phase	120V		19.68	
Ifault			16.0A	87°
IPhase	0.000A		8.0A	
V1	69.28V	0.00°	36.01V	
V2	69.28V	-120.00°	36.01V	
V3	69.28V	120.00°	69.28V	120.00°
I1	0.000A	0.00°	8.000A	-57.00°
I2	0.000A	-120.00°	8.000A	123.00°
I3	0.000A	120.00°	0.000A	120.00°

7. Calculate the A-N and B-N voltage angles using the following formulas:

$$V1_{FaultAngle} = V3_{FaultAngle} + 180 + \tan^{-1}\left(\frac{V_{Phase-Phase(Fault)}}{V3_{PrefaultMag}}\right)$$

$$V2_{FaultAngle} = V3_{FaultAngle} + 180 - \tan^{-1}\left(\frac{V_{Phase-Phase(Fault)}}{V3_{PrefaultMag}}\right)$$

$$V1_{FaultAngle} = 120 + 180 + \tan^{-1}\left(\frac{19.68}{69.28}\right)$$

$$V2_{FaultAngle} = 120 + 180 - \tan^{-1}\left(\frac{19.68}{69.28}\right)$$

$$V1_{FaultAngle} = 120 + 180 + 15.86$$

$$V2_{FaultAngle} = 120 + 180 - 15.86$$

$$V1_{FaultAngle} = 315.86° \text{ or } -44.14°$$

$$V2_{FaultAngle} = 284.14° \text{ or } -75.86°$$

CHANNEL	PREFAULT		FAULT	
	MAG	ANGLE	MAG	ANGLE
Vphase-to-phase	120V		19.68	
Ifault			16.0A	87°
IPhase	0.000A		8.0A	
V1	69.28V	0.00°	36.01V	-44.14°
V2	69.28V	-120.00°	36.01V	-75.86°
V3	69.28V	120.00°	69.28V	120.00°
I1	0.000A	0.00°	8.000A	-57.00°
I2	0.000A	-120.00°	8.000A	123.00°
I3	0.000A	120.00°	0.000A	120.00°

You can use the following template to create a spreadsheet to perform these calculations automatically:

A-B	PREFAULT (PRE)		FAULT (FLT)	
CHANNEL	MAG	ANGLE (ANG)	MAG	ANGLE(ANG)
Vphase-to-phase(Vpp)	120V		19.68	
Ifault (IFlt)			16.000A	87°
IPhase (IP)	0.000A		=IFltMag/2	
V1	Formula1	0.00°	Formula 2	Formula 3
V2	Formula1	=V1PreAng+240	Formula 2	Formula 4
V3	Formula1	=V1PreAng+120	=V3PreMag	=V3PreAng
I1	=IPPreMag	=V1PreAng-IPPreAng	=IPFltMag	Formula 5
I2	=IPPreMag	=I1PreAng+240	=IPFltMag	Formula 6
I3	=IPPreMag	=I1PreAng+120	=I3PreMag	=I3PreAng

Formula 1 =VppPreMag/sqrt(3)
Formula 2 =SQRT(V1PreMag^2+VppFltMag^2)/2
Formula 3 =MOD(V3FltAng+180+DEGREES(ATAN(VppFltMag/V1PreMag)),360)
Formula 4 =MOD(V3FltAng+180-DEGREES(ATAN(VppFltMag/V1PreMag)),360)
Formula 5 =MOD((30-IFltAng),360)
Formula 6 =MOD((I1FltAng+180),360)

Figure 15-41: A-B Fault Test-Set Configuration

8. Repeat Steps 5-7 for a B-C fault using the following template:

B-C	PREFAULT (PRE)		FAULT (FLT)	
CHANNEL	MAG	ANGLE (ANG)	MAG	ANGLE(ANG)
Vphase-to-phase(Vpp)	120V		19.68	
Ifault (IFlt)			16.000A	87°
IPhase (IP)	0.000A		=IFltMag/2	
V1	Formula1	0.00°	=V1PreMag	=V1PreAng
V2	Formula1	=V1PreAng+240	Formula 2	Formula 7
V3	Formula1	=V1PreAng+120	Formula 2	Formula 8
I1	=IPPreMag	=V1PreAng-IPPreAng	=I1PreMag	=I1PreAng
I2	=IPPreMag	=I1PreAng+240	=IPFltMag	Formula 9
I3	=IPPreMag	=I1PreAng+120	=IPFltMag	Formula 10

Formula 1 =VppPreMag/sqrt(3)
Formula 2 =SQRT(V1PreMag^2+VppFltMag^2)/2
Formula 7 =MOD(V1FltAng+180+DEGREES(ATAN(VppFltMag/V1PreMag)),360)
Formula 8 =MOD(V1FltAng+180-DEGREES(ATAN(VppFltMag/V1PreMag)),360)
Formula 9 =MOD((-90-IFltAng),360)
Formula 10 =MOD((I2FltAng+180),360)

Figure 15-42: B-C Fault Test-Set Configuration

9. Repeat Steps 5-7 for a C-A fault using the following template:

C-A	PREFAULT (PRE)		FAULT (FLT)	
CHANNEL	MAG	ANGLE (ANG)	MAG	ANGLE(ANG)
Vphase-to-phase(Vpp)	120V		19.68	
Ifault (IFlt)			16.000A	87°
IPhase (IP)	0.000A		=IFltMag/2	
V1	Formula1	0.00°	Formula 2	Formula 12
V2	Formula1	=V1PreAng+240	=V2PreMag	=V2PreAng
V3	Formula1	=V1PreAng+120	Formula 2	Formula 11
I1	=IPPreMag	=V1PreAng-IPPreAng	=IPFltMag	Formula 14
I2	=IPPreMag	=I1PreAng+240	=I2PreMag	=I2PreAng
I3	=IPPreMag	=I1PreAng+120	=IPFltMag	Formula 13

Formula 1 =VppPreMag/sqrt(3)

Formula 2 =SQRT(V1PreMag^2+VppFltMag^2)/2

Formula 11 =MOD(V2FltAng+180+DEGREES(ATAN(VppFltMag/V1PreMag)),360)

Formula 12 =MOD(V2FltAng+180-DEGREES(ATAN(VppFltMag/V1PreMag)),360)

Formula 13 =MOD((150-IFltAng),360)

Formula 14 =MOD((I3FltAng+180),360)

Figure 15-43: C-A Fault Test-Set Configuration

B) Test-Set Connections

The following test-set connections apply for each phase-to-phase fault.

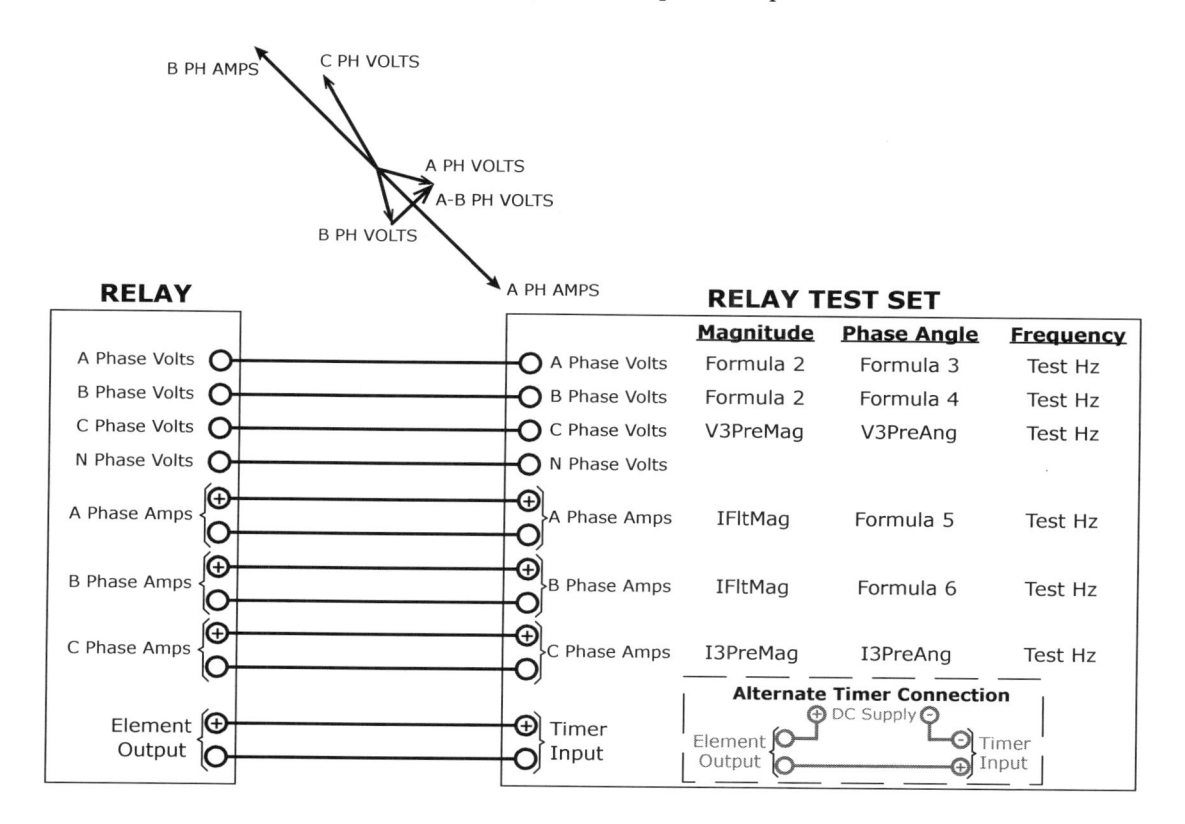

Figure 15-44: A-B Phase-to-Phase Test-Set Connections

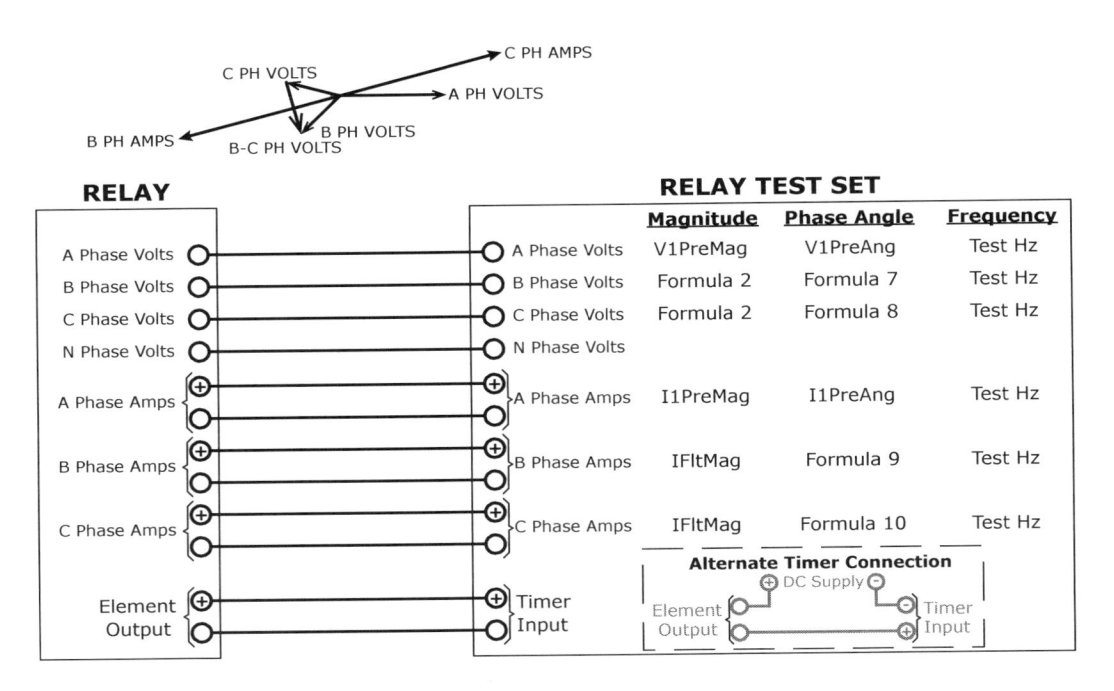

Figure 15-45: B-C Phase-to-Phase Test-Set Connections

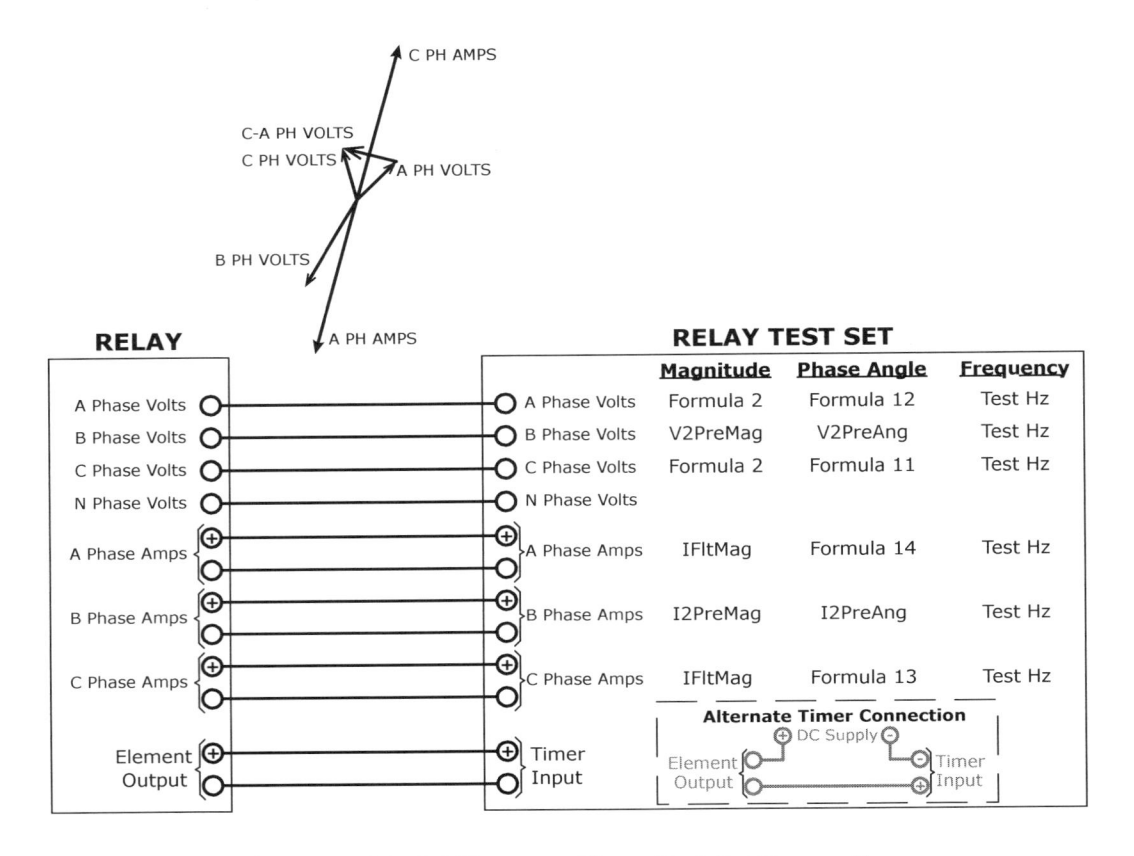

Figure 15-46: C-A Phase-to-Phase Test-Set Connections

C) MTA Test Procedure

The first test you can perform is the Maximum Torque Angle (MTA) test which verifies the RCA setting inside the GE relay. This test is nearly identical to the "MTA Test Procedure" description in the "3-Phase Line Distance Protection Testing" section earlier in this document which you can review for more details. We are going to change the phase angle references from $0\pm180°$ to $0\text{-}360°$ to make the calculations easier to understand.

Follow the steps below to perform a phase-to-phase MTA test.

1. Determine how you will monitor pickup and set the relay accordingly, if required. (Pickup indication by LED, output contact, front panel display, etc. See the *Relay Test Procedures* section starting on page 109 for details.)

2. Calculate the test configuration using the "Phase-to-Phase Test Calculations" description described earlier in this section. The test impedance should be 90% of the Z1 Reach Setting. $\text{ZTest} = 0.9 \times \text{Z1 Reach} = 0.9 \times 1.23 = 1.107\Omega$

$$V_{\text{phase-phase(Fault)}} = 2 \times I_{\text{Phase}} \times \text{TestOhms}$$

$$V_{\text{phase-phase(Fault)}} = 2 \times 8A \times 1.107\Omega$$

$$V_{\text{phase-phase(Fault)}} = 17.712V$$

A-B	PREFAULT (PRE)		FAULT (FLT)	
CHANNEL	MAG	ANGLE (ANG)	MAG	ANGLE(ANG)
Vphase-to-phase(Vpp)	120V		17.71V	
Ifault (IFlt)			16.000A	87°
IPhase (IP)	0.000A		=IFltMag/2	
V1	Formula1	0.00°	Formula 2	Formula 3
V2	Formula1	=V1PreAng+240	Formula 2	Formula 4
V3	Formula1	=V1PreAng+120	=V3PreMag	=V3PreAng
I1	=IPPreMag	=V1PreAng-IPPreAng	=IPFltMag	Formula 5
I2	=IPPreMag	=I1PreAng+240	=IPFltMag	Formula 6
I3	=IPPreMag	=I1PreAng+120	=I3PreMag	=I3PreAng
Formula 1 =VppPreMag/sqrt(3)				
Formula 2 =SQRT(V1PreMag^2+VppFltMag^2)/2				
Formula 3 =MOD(V3FltAng+180+DEGREES(ATAN(VppFltMag/V1PreMag)),360)				
Formula 4 =MOD(V3FltAng+180-DEGREES(ATAN(VppFltMag/V1PreMag)),360)				
Formula 5 =MOD((30-IFltAng),360)				
Formula 6 =MOD((I1FltAng+180),360)				

B-C	PREFAULT (PRE)		FAULT (FLT)	
CHANNEL	**MAG**	**ANGLE (ANG)**	**MAG**	**ANGLE(ANG)**
Vphase-to-phase(Vpp)	120V		17.71V	
Ifault (IFlt)			16.000A	87°
IPhase (IP)	0.000A		=IFltMag/2	
V1	Formula1	0.00°	=V1PreMag	=V1PreAng
V2	Formula1	=V1PreAng+240	Formula 2	Formula 7
V3	Formula1	=V1PreAng+120	Formula 2	Formula 8
I1	=IPPreMag	=V1PreAng-IPPreAng	=I1PreMag	=I1PreAng
I2	=IPPreMag	=I1PreAng+240	=IPFltMag	Formula 9
I3	=IPPreMag	=I1PreAng+120	=IPFltMag	Formula 10

Formula 1 =VppPreMag/sqrt(3)

Formula 2 =SQRT(V1PreMag^2+VppFltMag^2)/2

Formula 7 =MOD(V1FltAng+180+DEGREES(ATAN(VppFltMag/V1PreMag)),360)

Formula 8 =MOD(V1FltAng+180-DEGREES(ATAN(VppFltMag/V1PreMag)),360)

Formula 9 =MOD((-90-IFltAng),360)

Formula 10 =MOD((I2FltAng+180),360)

C-A	PREFAULT (PRE)		FAULT (FLT)	
CHANNEL	**MAG**	**ANGLE (ANG)**	**MAG**	**ANGLE(ANG)**
Vphase-to-phase(Vpp)	120V		17.71V	
Ifault (IFlt)			16.000A	87°
IPhase (IP)	0.000A		=IFltMag/2	
V1	Formula1	0.00°	Formula 2	Formula 12
V2	Formula1	=V1PreAng+240	=V2PreMag	=V2PreAng
V3	Formula1	=V1PreAng+120	Formula 2	Formula 11
I1	=IPPreMag	=V1PreAng-IPPreAng	=IPFltMag	Formula 14
I2	=IPPreMag	=I1PreAng+240	=I2PreMag	=I2PreAng
I3	=IPPreMag	=I1PreAng+120	=IPFltMag	Formula 13

Formula 1 =VppPreMag/sqrt(3)

Formula 2 =SQRT(V1PreMag^2+VppFltMag^2)/2

Formula 11 =MOD(V2FltAng+180+DEGREES(ATAN(VppFltMag/V1PreMag)),360)

Formula 12 =MOD(V2FltAng+180-DEGREES(ATAN(VppFltMag/V1PreMag)),360)

Formula 13 =MOD((150-IFltAng),360)

Formula 14 =MOD((I3FltAng+180),360)

3. Configure a test with two states (prefault and fault) as shown in the following chart and figure:

A-B	PREFAULT (PRE)		FAULT (FLT)	
CHANNEL	MAG	ANGLE (ANG)	MAG	ANGLE(ANG)
V1	69.28V	0.00°	35.75V	314.34°
V2	69.28V	240.00°	35.75V	285.66°
V3	69.28V	120.00°	69.28V	120.00°
I1	0.000A	0.00°	8.000A	303.00°
I2	0.000A	240.00°	8.000A	123.00°
I3	0.000A	120.00°	0.000A	120.00°

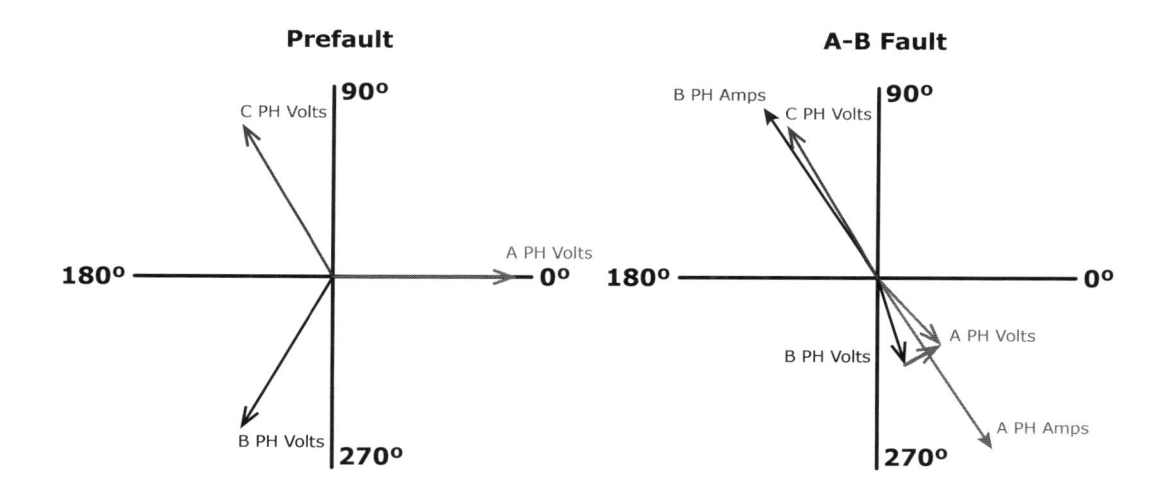

4. Configure your test-set to change all the faulted (I1 and I2 for an A-B fault) current angles equally and simultaneously.

5. Apply the prefault condition to the relay and verify the metering is correct, if required.

6. Apply the fault condition and ensure the 21-element operates.

7. Increase the faulted current phase angles equally until the 21-element drops out.

8. Decrease the faulted current phase angles until the 21-element picks up. Record the first faulted current phase angle (I1 for an A-B fault) at pickup (277.58° in our example).

9. Continue decreasing the faulted current phase angles until the 21-element drops out.

10. Increase the faulted current phase angles until the 21-element picks up. Record the first faulted current phase angle (I1 for an A-B fault) at pickup (328.65° in our example).

11. Calculate the average of the two angles and add the result to the smaller angle and the phase-to-phase reference (Remember that the A-B voltage is the reference for this test which is found at 30°). Multiply the result times -1 to convert the angle reference from a phasor reference to an impedance reference. Record the calculated value on your test sheet.

$$MTA = \left(\frac{TestAngle1 + TestAngle2}{2} - VoltageReference° - 360° \right) \times -1$$

$$MTA = \left(\frac{277.58 + 328.65}{2} - 30° - 360° \right) \times -1 = 86.885°$$

12. Compare test results to manufacturer's specifications. There is typically no direct specification for the MTA test, but we can use the Phase Distance Reach (secondary Ω) specifications located in the GE D-60 manual as shown in the following figure which indicates a 5% tolerance.

PHASE DISTANCE

Characteristic:	mho (memory polarized or offset) or quad (memory polarized or non-directional), selectable individually per zone
Number of zones:	5
Directionality:	forward, reverse, or non-directional per zone
Reach (secondary Ω):	0.02 to 500.00 Ω in steps of 0.01
Reach accuracy:	±5% including the effect of CVT transients up to an SIR of 30

Figure 15-47: GE D-60 Phase Distance Specification

The test results measured an MTA of 86.885° and the expected result was 87°. Using the standard percent error calculation, we can determine the percent error and the test result passes. Record the percent error and pass/fail result on your test sheet.

$$\frac{Actual\ Value - Expected\ Value}{Expected\ Value} \times 100 = Percent\ Error$$

$$\frac{86.885° - 87.00°}{87.00°} \times 100 = Percent\ Error$$

-0.13% Error

13. Repeat Steps 3-12 for a B-C fault. I2 and I3 are the faulted current phase angles and the I2 phase angle is the recorded angle. The B-C voltage reference is 270°.

B-C	PREFAULT (PRE)		FAULT (FLT)	
CHANNEL	MAG	ANGLE (ANG)	MAG	ANGLE(ANG)
V1	69.28V	0.00°	69.28V	0.00°
V2	69.28V	240.00°	35.75V	194.34°
V3	69.28V	120.00°	35.75V	165.66°
I1	0.000A	0.00°	0.000A	0.00°
I2	0.000A	240.00°	8.000A	183.00°
I3	0.000A	120.00°	8.000A	3.00°

$$MTA = \left(\frac{TestAngle1 + TestAngle2}{2} - VoltageReference° \right) \times -1$$

$$MTA = \left(\frac{157.33 + 208.99}{2} - 270° \right) \times -1 = 86.84°$$

14. Repeat Steps 3-12 for a C-A fault. I3 and I1 are the faulted current phase angles and the I3 phase angle is the recorded angle. The C-A voltage reference is 150°.

C-A	PREFAULT (PRE)		FAULT (FLT)	
CHANNEL	MAG	ANGLE (ANG)	MAG	ANGLE(ANG)
V1	69.28V	0.00°	35.75V	45.66°
V2	69.28V	240.00°	69.28V	240.00°
V3	69.28V	120.00°	35.75V	74.34°
I1	0.000A	0.00°	8.000A	243.00°
I2	0.000A	240.00°	0.000A	240.00°
I3	0.000A	120.00°	8.000A	63.00°

$$MTA = \left(\frac{TestAngle1 + TestAngle2}{2} - VoltageReference° \right) \times -1$$

$$MTA = \left(\frac{37.15 + 89.01}{2} - 150° \right) \times -1 = 86.92°$$

15. Repeat the MTA test for all enabled 21-element zones.

PHASE-PHASE DISTANCE TEST RESULTS (Ohms)								
	Z1 Function	Enabled	Z2 Function	Enabled	Z3 Function	Enabled	Z4 Function	Disabled
	Z1 Reach	1.23	Z2 Reach	1.85	Z3 Reach	3.1	Z4 Reach	
	Z1 RCA	87	Z2 RCA	87	Z3 RCA	87	Z4 RCA	
	Z1 Delay	0.00	Z2 Delay	0.30	Z3 Delay	1.00	Z4 Delay	
	Z1 Direction	F	Z2 Direction	F	Z3 Direction	F	Z4 Direction	

MAXIMUM TORQUE ANGLE (MTA) TEST RESULTS (Ohms)												
	ZONE 1 TESTS			ZONE 2 TESTS			ZONE 3 TESTS			ZONE 4 TESTS		
	Ohms	PU1°	PU2°	Ohms	PU1°	PU2°	Ohms	PU1°	PU2°	Ohms	PU1°	PU2°
A-B PU (°)	1.11	277.58	328.65	1.67	277.58	328.57	2.79	277.51	328.67			
B-C PU (°)	1.11	157.33	208.99	1.67	157.53	208.78	2.79	157.50	208.67			
C-A PU (°)	1.11	37.15	89.01	1.67	37.33	88.81	2.79	37.28	88.86			
	M1P	MFG	%ERR	M2P	MFG	%ERR	M3P	MFG	%ERR			
A-B MTA (°)	86.89	87.0	-0.13	86.93	87.0	-0.09	86.91	87.0	-0.10			
B-C MTA (°)	86.84	87.0	-0.18	86.85	87.0	-0.18	86.92	87.0	-0.10			
C-A MTA (°)	86.92	87.0	-0.09	86.93	87.0	-0.08	86.93	87.0	-0.08			

A-B ZONE 1 TESTS M1P =MOD((((ZONE1TESTSPU1°+ ZONE1TESTSPU2°)/2)-30)*(-1),360)

B-C ZONE 1 TESTS M2P =MOD((((ZONE1TESTSPU1°+ ZONE1TESTSPU2°)/2)-270)*(-1),360)

C-A ZONE 1 TESTS M3P =MOD((((ZONE1TESTSPU1°+ ZONE1TESTSPU2°)/2)-150)*(-1),360)

D) Reach Test Procedure

The reach test is similar to the 3-phase reach test with the phase-to-phase calculations described in the MTA test procedure. The relay tester chooses one or more angles to perform the test (typically starting at the MTA angle) and then performs a standard pickup test.

Follow the steps below to perform a reach test.

1. Determine how you will monitor pickup and set the relay accordingly, if required. (Pickup indication by LED, output contact, front panel display, etc. See the *Relay Test Procedures* section starting on page 109 for details.)

2. Determine the Test Angles for the test (MTA, MTA+n, MTA-n, 95° lag, etc.) The following example will use the MTA angle plus 30°. The expected result for any MHO circle with a 0.00Ω offset can be calculated using the formula:

$$TestOhms = Z1Reach \times Cos(MTA° - Test°)$$

Example calculations are shown below:

$$\text{TestOhms} = \text{Z1Reach} \times \text{Cos}(\text{MTA}° - \text{Test}°)$$

$$\text{TestOhms} = 1.23 \times \text{Cos}(87° - 117°)$$

$$\text{TestOhms} = 1.23 \times 0.866$$

$$\text{TestOhms} = 1.065 \Omega$$

3. Calculate the test configuration using the "Phase-to-Phase Test Calculations" description described earlier in this section. The test impedance should be 95% of the TestOhms calculation for the test angle. $\text{ZTest} = 0.95 \times \text{TestOhms} = 0.95 \times 1.065 = 1.012 \Omega$

$$V_{\text{phase-phase(Fault)}} = 2 \times I_{\text{Phase}} \times \text{TestOhms}$$

$$V_{\text{phase-phase(Fault)}} = 2 \times 8A \times 1.012 \Omega$$

$$V_{\text{phase-phase(Fault)}} = 16.192V$$

A-B	PREFAULT (PRE)		FAULT (FLT)	
CHANNEL	MAG	ANGLE (ANG)	MAG	ANGLE(ANG)
Vphase-to-phase(Vpp)	120V		16.192V	
Ifault (IFlt)			16.000A	117°
IPhase (IP)	0.000A		=IFltMag/2	
V1	Formula1	0.00°	Formula 2	Formula 3
V2	Formula1	=V1PreAng+240	Formula 2	Formula 4
V3	Formula1	=V1PreAng+120	=V3PreMag	=V3PreAng
I1	=IPPreMag	=V1PreAng-IPPreAng	=IPFltMag	Formula 5
I2	=IPPreMag	=I1PreAng+240	=IPFltMag	Formula 6
I3	=IPPreMag	=I1PreAng+120	=I3PreMag	=I3PreAng
Formula 1 =VppPreMag/sqrt(3)				
Formula 2 =SQRT(V1PreMag^2+VppFltMag^2)/2				
Formula 3 =MOD(V3FltAng+180+DEGREES(ATAN(VppFltMag/V1PreMag)),360)				
Formula 4 =MOD(V3FltAng+180-DEGREES(ATAN(VppFltMag/V1PreMag)),360)				
Formula 5 =MOD((30-IFltAng),360)				
Formula 6 =MOD((I1FltAng+180),360)				

B-C	PREFAULT (PRE)		FAULT (FLT)	
CHANNEL	MAG	ANGLE (ANG)	MAG	ANGLE(ANG)
Vphase-to-phase(Vpp)	120V		16.192V	
Ifault (IFlt)			16.000A	117°
IPhase (IP)	0.000A		=IFltMag/2	
V1	Formula1	0.00°	=V1PreMag	=V1PreAng
V2	Formula1	=V1PreAng+240	Formula 2	Formula 7
V3	Formula1	=V1PreAng+120	Formula 2	Formula 8
I1	=IPPreMag	=V1PreAng-IPPreAng	=I1PreMag	=I1PreAng
I2	=IPPreMag	=I1PreAng+240	=IPFltMag	Formula 9
I3	=IPPreMag	=I1PreAng+120	=IPFltMag	Formula 10

Formula 1 =VppPreMag/sqrt(3)

Formula 2 =SQRT(V1PreMag^2+VppFltMag^2)/2

Formula 7 =MOD(V1FltAng+180+DEGREES(ATAN(VppFltMag/V1PreMag)),360)

Formula 8 =MOD(V1FltAng+180-DEGREES(ATAN(VppFltMag/V1PreMag)),360)

Formula 9 =MOD((-90-IFltAng),360)

Formula 10 =MOD((I2FltAng+180),360)

C-A	PREFAULT (PRE)		FAULT (FLT)	
CHANNEL	MAG	ANGLE (ANG)	MAG	ANGLE(ANG)
Vphase-to-phase(Vpp)	120V		16.192V	
Ifault (IFlt)			16.000A	117°
IPhase (IP)	0.000A		=IFltMag/2	
V1	Formula1	0.00°	Formula 2	Formula 12
V2	Formula1	=V1PreAng+240	=V2PreMag	=V2PreAng
V3	Formula1	=V1PreAng+120	Formula 2	Formula 11
I1	=IPPreMag	=V1PreAng-IPPreAng	=IPFltMag	Formula 14
I2	=IPPreMag	=I1PreAng+240	=I2PreMag	=I2PreAng
I3	=IPPreMag	=I1PreAng+120	=IPFltMag	Formula 13

Formula 1 =VppPreMag/sqrt(3)

Formula 2 =SQRT(V1PreMag^2+VppFltMag^2)/2

Formula 11 =MOD(V2FltAng+180+DEGREES(ATAN(VppFltMag/V1PreMag)),360)

Formula 12 =MOD(V2FltAng+180-DEGREES(ATAN(VppFltMag/V1PreMag)),360)

Formula 13 =MOD((150-IFltAng),360)

Formula 14 =MOD((I3FltAng+180),360)

4. Configure a test with two states (prefault and fault) as shown in the following chart:

A-B	PREFAULT (PRE)		FAULT (FLT)	
CHANNEL	MAG	ANGLE (ANG)	MAG	ANGLE(ANG)
V1	69.28V	0.00°	35.57V	313.15°
V2	69.28V	240.00°	35.57V	286.85°
V3	69.28V	120.00°	69.28V	120.00°
I1	0.000A	0.00°	8.000A	273.00°
I2	0.000A	240.00°	8.000A	93.00°
I3	0.000A	120.00°	0.000A	120.00°

5. Configure your test-set to change the faulted (I1 and I2 for an A-B fault) current magnitudes equally and simultaneously.

6. Apply the prefault condition to the relay and verify the metering is correct, if required.

7. Apply the fault condition and ensure the 21-element operates.

8. Decrease both faulted current magnitudes simultaneously until the 21-element drops out. (You cannot change the voltage magnitudes for this test unless your test-set also calculates the appropriate angle changes instantaneously as well.)

9. Reverse the current changes until the 21-element picks up. Record the impedance at pickup. (1.061Ω in our example) $R = \dfrac{Vpp}{|I1|+|I2|} = \dfrac{16.192}{7.627+7.627} = 1.061\Omega$. Record the results on your test sheet.

10. Compare test results to manufacturer's specifications. The Phase Distance Reach (secondary Ω) specifications are located in the GE D-60 manual in the following figure which indicates a 5% tolerance.

PHASE DISTANCE

Characteristic:	mho (memory polarized or offset) or quad (memory polarized or non-directional), selectable individually per zone
Number of zones:	5
Directionality:	forward, reverse, or non-directional per zone
Reach (secondary Ω):	0.02 to 500.00 Ω in steps of 0.01
Reach accuracy:	±5% including the effect of CVT transients up to an SIR of 30

Figure 15-48: GE D-60 Phase Distance Specification

The test results measured 1.061Ω with an expected result of 1.065Ω. Using the standard percent error calculation, we can determine the percent error and the test result passes. Record the percent error and pass/fail result on your test sheet.

$$\frac{\text{Actual Value - Expected Value}}{\text{Expected Value}} \times 100 = \text{Percent Error}$$

$$\frac{1.061\Omega - 1.065\Omega}{1.065\Omega} \times 100 = \text{Percent Error}$$

-0.38% Error

11. Repeat Steps 2-10 for B-C Phase.

B-C	PREFAULT (PRE)		FAULT (FLT)	
CHANNEL	MAG	ANGLE (ANG)	MAG	ANGLE(ANG)
V1	69.28V	0.00°	69.28V	0.00°
V2	69.28V	240.00°	35.57V	193.15°
V3	69.28V	120.00°	35.57V	166.85°
I1	0.000A	0.00°	0.000A	0.00°
I2	0.000A	240.00°	8.000A	153.00°
I3	0.000A	120.00°	8.000A	333.00°

12. Repeat Steps 2-10 for C-A Phase.

C-A	PREFAULT (PRE)		FAULT (FLT)	
CHANNEL	MAG	ANGLE (ANG)	MAG	ANGLE(ANG)
V1	69.28V	0.00°	35.57V	46.85°
V2	69.28V	240.00°	69.28V	240.00°
V3	69.28V	120.00°	35.57V	73.15°
I1	0.000A	0.00°	8.000A	213.00°
I2	0.000A	240.00°	0.000A	240.00°
I3	0.000A	120.00°	8.000A	33.00°

13. Repeat Steps 2-12 for all enabled zones.

PHASE-PHASE DISTANCE TEST RESULTS (Ohms)											
Z1 Function	**Enabled**	Z2 Function	**Enabled**	Z3 Function	**Enabled**	Z4 Function	**Disabled**				
Z1 Reach	**1.23**	Z2 Reach	**1.85**	Z3 Reach	**3.1**	Z4 Reach					
Z1 RCA	**87**	Z2 RCA	**87**	Z3 RCA	**87**	Z4 RCA					
Z1 Delay	**0.00**	Z2 Delay	**0.30**	Z3 Delay	**1.00**	Z4 Delay					
Z1 Direction	**F**	Z2 Direction	**F**	Z3 Direction	**F**	Z4 Direction					

REACH TEST RESULTS (Ohms)											
	ZONE 1 TESTS			**ZONE 2 TESTS**			**ZONE 3 TESTS**			**ZONE 4 TESTS**		
TEST ANGLE	M1P	MFG	%ERR	M2P	MFG	%ERR	M3P	MFG	%ERR	M4P	MFG	%ERR
57	**1.06**	1.07	-0.49	**1.59**	1.60	-0.76	**2.67**	2.68	-0.55			
87	**1.22**	1.23	-0.81	**1.84**	1.85	-0.54	**3.09**	3.10	-0.32			
117	**1.06**	1.07	-0.49	**1.58**	1.60	-1.38	**2.68**	2.68	-0.17			

NOTE: An MTA test can be incorporated into the reach test by performing two reach tests with the same angular difference from the MTA as shown in the test sheet above. If both test results have the same result, the MTA is correct.

E) Timing Test Procedure

The Timing test is relatively simple compared to the MTA and reach tests above. Apply a prefault then transition to a fault at least 90% of the reach and record the time between fault and the relay operation.

Follow the steps below to perform a timing test.

1. Configure the test-set timer to start when the fault state is energized and stop when the trip contact operates.

2. Determine the Test Angles for the test. (MTA, MTA+n, MTA-n, 95° lag, etc.) The following example will use the MTA angle minus 30°. The expected result for any MHO circle with a 0.00Ω offset can be calculated using the formula:

$$TestOhms = Z1Reach \times Cos(MTA° - Test°)$$

Example calculations are shown below:

$$TestOhms = Z1Reach \times Cos(MTA° - Test°)$$

$$TestOhms = 1.23 \times Cos(87° - 57°)$$

$$TestOhms = 1.23 \times 0.866$$

$$TestOhms = 1.065Ω$$

Principles and Practice

3. Calculate the test configuration using one of these test methods:

a. If the Z1 Reach setting is greater than 0.66 Ohms, arbitrarily determine the maximum amount of current to be applied to the relay. A good rule of thumb for maximum continuous current for most relays is 10.0A. Calculate the 3-phase balanced voltage required to apply 70% of the Z1 Reach setting using the following formula based on Ohm's Law. $ZTest = 0.70 \times TestOhms = 0.70 \times 1.065 = 0.7455\Omega$

$$V_{phase-phase(Fault)} = 2 \times I_{Phase} \times TestOhms$$

$$V_{phase-phase(Fault)} = 2 \times 8A \times 0.7455\Omega$$

$$V_{phase-phase(Fault)} = 11.928V$$

b. If the Z1 Reach setting is less than 0.66Ω, use the traditional formula described in this step with a 5.0V voltage setting. If more than 10.0A is required, the test should be performed as quickly as possible to prevent equipment damage.

Traditional test techniques reverse the formula in the previous step because electromechanical relays required a minimum voltage of approximately 20V to function reliably. Digital relays typically only require 3.0V to properly determine direction when a prefault voltage is applied. The traditional formula can also be used to calculate the test parameters, but you may need to make several calculations to keep the test current at reasonable levels as shown in the following examples:

$I = \dfrac{V}{R}$	$I = \dfrac{V}{R}$
$FaultAmps = \dfrac{FaultVolts}{0.70 \times Z1Reach}$	$FaultAmps = \dfrac{FaultVolts}{0.70 \times Z1Reach}$
$FaultAmps = \dfrac{20.0V}{0.70 \times 1.065}$	$FaultAmps = \dfrac{11.928V}{0.70 \times 1.065}$
$FaultAmps(IFlt) = 26.83A$	$FaultAmps(IFlt) = 16.000A$
$PhaseAmpls(IP) = \dfrac{IFlt}{2} = \dfrac{26.83A}{2} = 13.42A$	$PhaseAmpls(IP) = \dfrac{IFlt}{2} = \dfrac{16.000A}{2} = 8.000A$

A-B	PREFAULT (PRE)		FAULT (FLT)	
CHANNEL	MAG	ANGLE (ANG)	MAG	ANGLE(ANG)
Vphase-to-phase(Vpp)	120V		11.93V	
Ifault (IFlt)			16.000A	57°
IPhase (IP)	0.000A		=IFltMag/2	
V1	Formula1	0.00°	Formula 2	Formula 3
V2	Formula1	=V1PreAng+240	Formula 2	Formula 4
V3	Formula1	=V1PreAng+120	=V3PreMag	=V3PreAng
I1	=IPPreMag	=V1PreAng-IPPreAng	=IPFltMag	Formula 5
I2	=IPPreMag	=I1PreAng+240	=IPFltMag	Formula 6
I3	=IPPreMag	=I1PreAng+120	=I3PreMag	=I3PreAng

Formula 1 =VppPreMag/sqrt(3)
Formula 2 =SQRT(V1PreMag^2+VppFltMag^2)/2
Formula 3 =MOD(V3FltAng+180+DEGREES(ATAN(VppFltMag/V1PreMag)),360)
Formula 4 =MOD(V3FltAng+180-DEGREES(ATAN(VppFltMag/V1PreMag)),360)
Formula 5 =MOD((30-IFltAng),360)
Formula 6 =MOD((I1FltAng+180),360)

B-C	PREFAULT (PRE)		FAULT (FLT)	
CHANNEL	MAG	ANGLE (ANG)	MAG	ANGLE(ANG)
Vphase-to-phase(Vpp)	120V		11.93V	
Ifault (IFlt)			16.000A	57°
IPhase (IP)	0.000A		=IFltMag/2	
V1	Formula1	0.00°	=V1PreMag	=V1PreAng
V2	Formula1	=V1PreAng+240	Formula 2	Formula 7
V3	Formula1	=V1PreAng+120	Formula 2	Formula 8
I1	=IPPreMag	=V1PreAng-IPPreAng	=I1PreMag	=I1PreAng
I2	=IPPreMag	=I1PreAng+240	=IPFltMag	Formula 9
I3	=IPPreMag	=I1PreAng+120	=IPFltMag	Formula 10

Formula 1 =VppPreMag/sqrt(3)
Formula 2 =SQRT(V1PreMag^2+VppFltMag^2)/2
Formula 7 =MOD(V1FltAng+180+DEGREES(ATAN(VppFltMag/V1PreMag)),360)
Formula 8 =MOD(V1FltAng+180-DEGREES(ATAN(VppFltMag/V1PreMag)),360)
Formula 9 =MOD((-90-IFltAng),360)
Formula 10 =MOD((I2FltAng+180),360)

C-A	PREFAULT (PRE)		FAULT (FLT)	
CHANNEL	MAG	ANGLE (ANG)	MAG	ANGLE(ANG)
Vphase-to-phase(Vpp)	120V		11.93V	
Ifault (IFlt)			16.000A	57°
IPhase (IP)	0.000A		=IFltMag/2	
V1	Formula1	0.00°	Formula 2	Formula 12
V2	Formula1	=V1PreAng+240	=V2PreMag	=V2PreAng
V3	Formula1	=V1PreAng+120	Formula 2	Formula 11
I1	=IPPreMag	=V1PreAng-IPPreAng	=IPFltMag	Formula 14
I2	=IPPreMag	=I1PreAng+240	=I2PreMag	=I2PreAng
I3	=IPPreMag	=I1PreAng+120	=IPFltMag	Formula 13

Formula 1 =VppPreMag/sqrt(3)
Formula 2 =SQRT(V1PreMag^2+VppFltMag^2)/2
Formula 11 =MOD(V2FltAng+180+DEGREES(ATAN(VppFltMag/V1PreMag)),360)
Formula 12 =MOD(V2FltAng+180-DEGREES(ATAN(VppFltMag/V1PreMag)),360)
Formula 13 =MOD((150-IFltAng),360)
Formula 14 =MOD((I3FltAng+180),360)

4. Configure a test with two states (prefault and fault) as shown in the following chart:

A-B	PREFAULT (PRE)		FAULT (FLT)	
CHANNEL	MAG	ANGLE (ANG)	MAG	ANGLE(ANG)
V1	69.28V	0.00°	35.15V	309.77°
V2	69.28V	240.00°	35.15V	290.23°
V3	69.28V	120.00°	69.28V	120.00°
I1	0.000A	0.00°	8.000A	333.00°
I2	0.000A	240.00°	8.000A	153.00°
I3	0.000A	120.00°	0.000A	120.00°

5. Apply the prefault condition to the relay and verify the metering is correct, if required.

6. Apply the fault condition and ensure the 21-element operates within the specified time.

7. Compare test results to manufacturer's specifications. This step can be more complicated than normal as we have the normal tolerance with a "Timing accuracy: +/- 3% or 4 ms, whichever is greater" and the Phase Distance Operating Time Curves.

PHASE DISTANCE

Time delay:	0.000 to 65.535 s in steps of 0.001
Timing accuracy:	±3% or 4 ms, whichever is greater

Figure 15-49: GE D-60 Phase Distance Timing Specification

PHASE DISTANCE OPERATING TIME CURVES

The operating times are response times of a microprocessor part of the relay. See output contacts specifications for estimation of the total response time for a particular application. The operating times are average times including variables such as fault inception angle or type of a voltage source (magnetic VTs and CVTs).

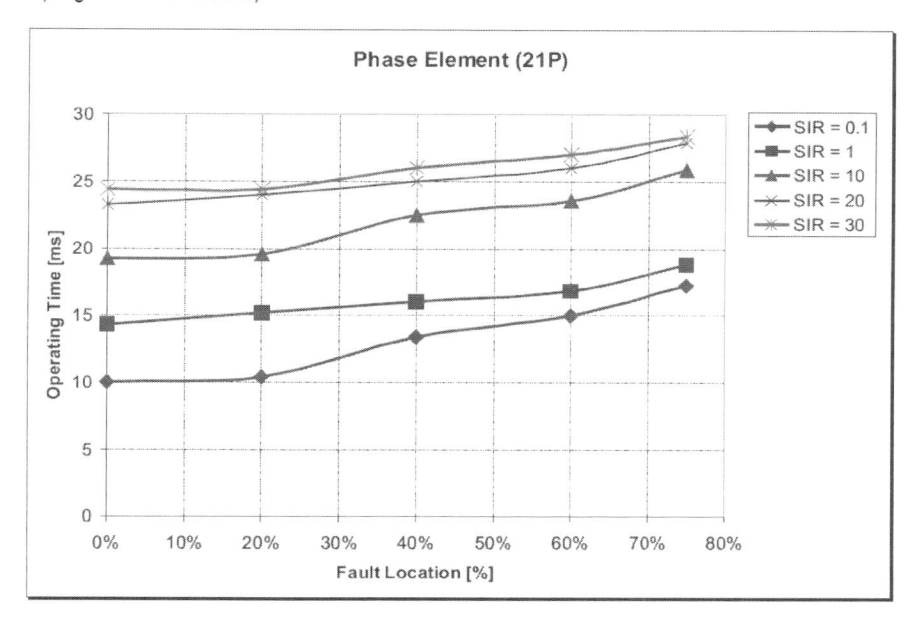

Figure 15-50: GE D-60 Phase Element (21P) Timing Specification

Using the worst case scenario of 27.5 ms (SIR = 0.1 at 70% fault location) +/- 3% or 4 ms, we can determine that the maximum expected time is:

$$27.5ms \times 3\% = 27.5ms \times 0.03 = 0.825ms$$

$$0.825ms < 4ms$$

$$ExpectedTime = 27.5ms + 4ms$$

$$ExpectedTime = 31.5ms$$

The measured time for our example was 31.0ms. The time is less than our expected time and is acceptable for service.

$$\frac{\text{Actual Value - Expected Value}}{\text{Expected Value}} \times 100 = \text{Percent Error}$$

$$\frac{30.5ms - 31.5ms}{31.5ms} \times 100 = \text{Percent Error}$$

-3.17% Error

8. Repeat Steps 2-7 for B-C Phase.

B-C	PREFAULT (PRE)		FAULT (FLT)	
CHANNEL	MAG	ANGLE (ANG)	MAG	ANGLE(ANG)
V1	69.28V	0.00°	69.28V	0.00°
V2	69.28V	240.00°	35.15V	189.77°
V3	69.28V	120.00°	35.15V	170.23°
I1	0.000A	0.00°	0.000A	0.00°
I2	0.000A	240.00°	8.000A	213.00°
I3	0.000A	120.00°	8.000A	33.00°

9. Repeat Steps 2-10 for C-A Phase.

C-A	PREFAULT (PRE)		FAULT (FLT)	
CHANNEL	MAG	ANGLE (ANG)	MAG	ANGLE(ANG)
V1	69.28V	0.00°	35.15V	50.23°
V2	69.28V	240.00°	69.28V	240.00°
V3	69.28V	120.00°	35.15V	69.77°
I1	0.000A	0.00°	8.000A	273.00°
I2	0.000A	240.00°	0.000A	240.00°
I3	0.000A	120.00°	8.000A	93.00°

10. Repeat Steps 2-9 for all enabled zones.

PHASE-PHASE DISTANCE TEST RESULTS

	Z1 Function	**Enabled**	Z2 Function	**Enabled**	Z3 Function	**Enabled**	Z4 Function	**Disabled**		
	Z1 Reach	**1.23**	Z2 Reach	**1.85**	Z3 Reach	**3.1**	Z4 Reach			
	Z1 RCA	**87**	Z2 RCA	**87**	Z3 RCA	**87**	Z4 RCA			
	Z1 Delay	**0.00**	Z2 Delay	**0.30**	Z3 Delay	**1.00**	Z4 Delay			
	Z1 Direction	**F**	Z2 Direction	**F**	Z3 Direction	**F**	Z4 Direction			

TIMING TESTS (in s)											
TEST	M1PT	MFG	%ERR	M2PT	MFG	%ERR	M3P	MFG	%ERR	M4PT	MFG	%ERR
A-B 0.7x	**0.03**	0.03	-1.61	**0.304**	0.30	1.33	**1.03**	1.00	3.00			
B-C 0.7x	**0.03**	0.03	-16.13	**0.303**	0.30	1.00	**1.028**	1.00	2.80			
C-A 0.7x	**0.03**	0.03	-9.68	**0.305**	0.30	1.67	**1.022**	1.00	2.20			

COMMENTS:

RESULTS ACCEPTABLE: ☑ YES   ☐ NO   ☑ SEE NOTES

6. Phase-to-Ground Line Distance Protection Testing

Phase-to-ground protection is often the easiest protective element to understand and test. This is not the case when performing phase-to-ground impedance tests. 3-phase and phase-to-phase faults involve transmission line or cable impedance with a fault impedance. The transmission line or cable can easily be modeled and a century of faults allows us to reasonably predict fault impedance characteristics. Phase-to-ground faults also include the ground impedance as the return path which must be added to our calculations. We like to pretend that all prefault conditions have 3-phase balanced conditions when writing textbooks, but the real world will never be balanced and there will always be some existing zero-sequence components in an electrical system. These two factors combine during phase-to-ground faults and can cause the measured impedance to be significantly different from the fault impedance. The difference between phase and neutral faults is commonly referred to as the K-factor, K-zero-factor, or zero-sequence factor and defined by the formula $\dfrac{\dfrac{Z0}{Z1}-1}{3}$.

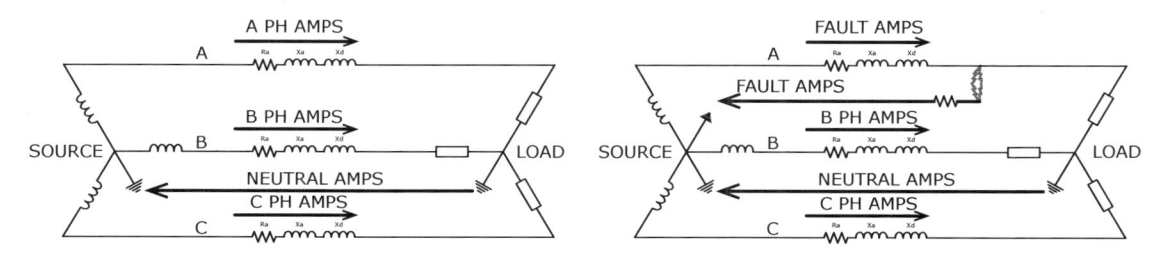

Figure 15-51: Phase-to-Ground Fault Characteristic

A) Phase-to-Ground Test Calculations

Most modern relays use the same pickup settings for phase and ground faults which doesn't appear to make sense knowing that the ground fault has additional impedances to account for. Ground distance protection often includes a correction factor which is applied to the measured impedance to determine the actual fault impedance. GE UR relays use the "Z0/Z1 Mag" and "Z0/Z1 Ang" settings to compensate for the ground return and source impedance as shown in the following information copied from the 3-phase test procedure.

RELAY SETTINGS	A-N TEST SETTINGS
Z1 Direction = F	V1 = 9.96V @ 0.00°
Z1 Shape = Mho	V2 = 69.28V @ -120°
Z1 Reach = 1.23	V3 = 69.28 @ 120°
Z1 RCA = 87 deg	I1 = 9.000A @ -87.00°
Z0/Z1 Mag = 3.93Ω	I2 = 0.000A @ -120°
Z0/Z1 Ang = -5°	I3 = 0.000A @ 120°

MEASURED IMPEDANCE	IMPEDANCE AFTER COMPENSATION
$$Z_{AN} = \frac{V1}{I1} =$$ $$Z_{AN} = \frac{9.96@0°}{9.000@-87°} = 1.11\Omega$$ $$Z_{AN} = 1.11\Omega @ 87°$$	$$ZC_{AN} = \frac{\left(\dfrac{V1}{I1}\right)}{\left(\vec{1} + \dfrac{(Z0/Z1\ Mag\ @\ Z0/Z1\ Ang) - \vec{1}}{3}\right)}$$ $$ZC_{AN} = \frac{\left(\dfrac{9.96V@0°}{9.000A@-87°}\right)}{\left(\vec{1} + \dfrac{(3.93\Omega@-5°) - \vec{1}}{3}\right)}$$ $$ZC_{AN} = \frac{1.11\Omega@87°}{\left(\vec{1} + \dfrac{2.94\Omega@-6.7°}{3}\right)}$$ $$ZC_{AN} = \frac{1.11\Omega@87°}{\left(\vec{1} + 0.98\Omega@-6.7°\right)}$$ $$ZC_{AN} = \frac{1.11\Omega@87°}{1.97\Omega@-3.31°}$$ $$ZC_{AN} = \left(\frac{1.11\Omega}{1.97\Omega}\right)@(87° - -3.31°)$$ $$ZC_{AN} = 0.563\Omega@90.31°$$

In the example above, an A-N fault has a measured impedance of $1.11\Omega@87°$ which would be a good starting point for a reach test. However, the relay would convert that 1.11Ω to $0.563\Omega@90.31°$. The converted number is significantly lower than the expected value, and the reach test results will be nowhere close to the expected results, if you do not apply K-factor. The following spreadsheet formulas will allow you to automatically apply the K-factor to your results.

ZAN Compensated Magnitude

=Ohms/(SQRT(((((Z0Z1Mag*COS(RADIANS(Z0Z1Ang)))-1)/3)+1)^2+((Z0Z1Mag*SIN(RADIANS(Z0Z1Ang)))/3)^2))

ZAN Compensated Angle

=Angle-DEGREES(TANH(((Z0Z1Mag*SIN(RADIANS(Z0Z1Ang)))/3)/((((Z0Z1Mag*COS(RADIANS(Z0Z1Ang)))-1)/3)+1)))

Where:
Ohms = Test Ohms
Angle = Test Angle

The following figure displays the difference between the Measured Impedance and the Compensated Impedance.

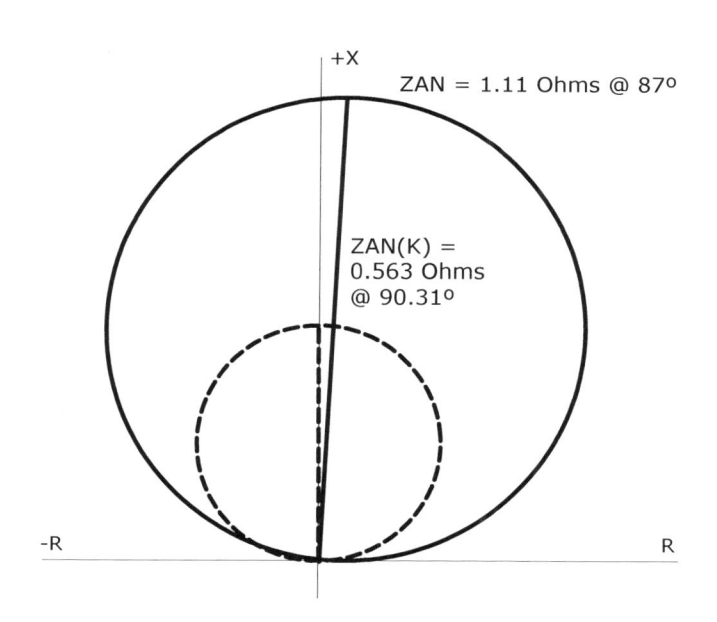

Different manufacturers calculate the ground correction factor in different ways. The following settings would apply to an SEL relay with identical line protection from the previous example.

RELAY SETTINGS	A-N TEST SETTINGS
Z1ANG = 87.00	V1 = 18.067V @ 0.00°
Z1MG = 1.23	V2 = 69.28V @ -120°
Z1D = 0.00	V3 = 69.28 @ 120°
k0M1 = 0.978	I1 = 9.000A @ -87.00°
k0A1 = -6.701	I2 = 0.000A @ -120°
	I3 = 0.000A @ 120°

MEASURED IMPEDANCE	IMPEDANCE AFTER COMPENSATION
$$Z_{AN} = \frac{V1}{I1} =$$ $$Z_{AN} = \frac{9.96@0°}{9.000@-87°} = 1.11\Omega$$ $$Z_{AN} = 1.11\Omega$$	$$ZC_{AN} = \frac{\left(\frac{V1}{I1}\right)}{(1 + k0M\angle k0A)}$$ $$ZC_{AN} = \frac{\left(\frac{9.96V@0°}{9.000A@-87°}\right)}{\left(1 + 0.978@-6.7°\right)}$$ $$ZC_{AN} = \frac{1.11\Omega@87°}{\left(1 + 0.978@-6.7°\right)}$$ $$ZC_{AN} = \frac{1.11\Omega@87°}{1.97\Omega@-3.31°}$$ $$ZC_{AN} = \left(\frac{1.11\Omega}{1.97\Omega}\right)@\left(87° - -3.31°\right)$$ $$ZC_{AN} = 0.563\Omega@90.31°$$

The compensation formula can be difficult to apply because it is a complex equation, but the following spreadsheet formula can help automate the calculation. Make sure you are applying the correct k0M(1) and k0A(1) settings that are appropriate for the zone under test.

ZAN Compensated Magnitude

=Ohms/SQRT((1+(k0M1*COS(RADIANS(k0A1))))^2+(k0M1*SIN(RADIANS(k0A1)))^2)

ZAN Compensated Angle

=Angle-(DEGREES(TANH((K0M1*SIN(RADIANS(K0A1)))/(1+(K0M1*COS(RADIANS(K0A1))))))))

Where:
Ohms = Test Ohms
Angle = Test Angle

B) Phase-to-Ground SEL Relay Settings

We will use an SEL-311C for the examples in this section. The following settings are the key settings for testing the Zone 2 21-element protection with any Schweitzer Engineering Laboratories relay from the 3xx or greater series.

Z1ANG = 78.60	This is the maximum torque angle of the distance protection. All tests should start at this angle.
Z2MG = 1.85	This is the phase-to-ground minimum pickup setting. This will not be the actual pickup setting because the K0M and K0A factors are applied to this setting in conjunction with Z1ANG determine the true phase-to-ground minimum pickup.
50L1 = 1.00	This is the minimum amount of phase current that must flow to enable the ground 21-element. This element can cause problems with very large impedance settings (>8Ω) and prevent the element from operating if the fault current falls below this setting.
50GZ1 = 0.50	This is the minimum amount of residual or unbalanced current that must flow to enable the ground 21-element. This element can cause problems with very large impedance settings (>8Ω) and prevent the element from operating if the fault current falls below this setting.
k0M = 0.978	This is the magnitude used in the zero compensation factors that must be applied during every phase-to-ground impedance test. Unfortunately, we could not find the correction factor formula in more recent publications. The following formula can be found in the Testing section of an SEL-321 Instruction Manual. $$Z_{AG} \cdot (1 + k0) = Z_{TEST} = Z2MG \cdot (1 + k0M \angle k0A)$$
k0A = -6.70	This is the angle used in the zero compensation factors that must be calculated during every phase-to-ground impedance test.
Z2GD = 18 cycles	This is the phase-to-ground time delay applied to Zone 2 ground protection.
Z2D = 0.00	Most zone timers have the same setting for phase-to-phase, 3-phase, and phase-to-ground faults. This setting applies to all Zone 2 faults to simplify the relay settings.

C) MTA Test Procedure

The first test you can perform is the Maximum Torque Angle (MTA) test which verifies the Z1ANG setting. This test is nearly identical to the "MTA Test Procedure" description in the "3-Phase Line Distance Protection Testing" section earlier in this document which you can review for more details. We are going to change the phase angle references from $0\pm180°$ to $0\text{-}360°$ to make the calculations easier to understand.

You should notice that SEL relays use a common setting (Z1ANG) for all impedance elements so it could be argued that only one test (3-P, P-P, or P-G) is required to verify the MTA. Don't be fooled by the "Z1" in the front of the setting. Z1 stands for positive sequence impedance and not Zone 1.

Follow the steps below to perform a phase-to-ground MTA test.

1. Determine how you will monitor pickup and set the relay accordingly, if required. (Pickup indication by LED, output contact, front panel display, etc. See the *Relay Test Procedures* section starting on page 109 for details.)

2. Calculate the measured impedance using the calculation described below. This calculation is the opposite of the calculations described earlier because, in this case, we are converting the relay-calculated impedance into a measured-impedance. The previous equations converted a measured-impedance into the relay-calculated impedance.

$$Z2MG = \frac{ZMeasured}{(1 + k0M\angle k0A)}$$

$$1.85\Omega @ 78.6° = \frac{ZMeasured}{\left(1 + 0.978 @ \text{-}6.7°\right)}$$

$$ZMeasured = 1.85\Omega @ 78.6° \times \left(1 + 0.978 @ \text{-}6.7°\right)$$

$$ZMeasured = 1.85\Omega @ 78.6° \times 1.975\Omega @ \text{-}3.31°$$

$$ZMeasured = \left(1.85\Omega \times 1.975\Omega\right) @ \left(78.6° + \text{-}3.31°\right)$$

$$ZMeasured = 3.653\Omega @ 75.29°$$

The Excel™ formula for this calculation is:

ZAN Compensated Magnitude

=Ohms*SQRT((1+(k0M1*COS(RADIANS(k0A1))))^2+(k0M1*SIN(RADIANS(k0A1)))^2)

ZAN Compensated Angle

=Angle+(DEGREES(TANH((K0M1*SIN(RADIANS(K0A1)))/(1+(K0M1*COS(RADI
ANS(K0A1))))))))

3. Calculate the test impedance using the calculation described below. The test impedance should not exceed 90% of the measured impedance.

$$ZTest = ZMeasured \times 90\%$$

$$ZTest = 3.653\Omega @ 75.29° \times 90\%$$

$$ZTest = 3.288\Omega @ 75.29°$$

The Excel™ formula for this calculation is:

ZAN Compensated Magnitude

=0.9*(Ohms*SQRT((1+(k0M1*COS(RADIANS(k0A1))))^2+(k0M1*SIN(RADIANS(
k0A1)))^2))

ZAN Compensated Angle

=Angle+(DEGREES(TANH((K0M1*SIN(RADIANS(K0A1)))/(1+(K0M1*COS(RADI
ANS(K0A1))))))))

4. Calculate the test configuration using one of these test methods:

   a. If the Z1 Reach setting is greater than 0.66 Ohms, arbitrarily determine the maximum amount of current to be applied to the relay. A good rule of thumb for maximum continuous current for most relays is 10.0A. Calculate the P-N voltage required to apply the test impedance using the following formula based on Ohm's Law.

$$V = I \times R$$

$$FaultVolts = FaultAmps \times ZTest$$

$$FaultVolts = 9.0A \times 3.288\Omega$$

$$FaultVolts = 29.59V$$

   b. If the Z1 Reach setting is less than 0.66Ω, use the traditional formula described in this step with a 5.0V voltage setting. If more than 10.0A is required, the test should be performed as quickly as possible to prevent equipment damage.

Traditional test techniques reverse the formula in the previous step because electromechanical relays required a minimum voltage of approximately 20V to function reliably. Digital relays typically only require 3.0V to properly determine direction when a prefault voltage is applied. The traditional formula can also be used to calculate the test parameters but you may need to make several calculations to keep the test current at reasonable levels.

$$I = \frac{V}{R}$$

$$FaultAmps = \frac{FaultVolts}{ZTest}$$

$$FaultAmps = \frac{20.0V}{3.288\Omega}$$

$$FaultAmps = 6.083A$$

$$I = \frac{V}{R}$$

$$FaultAmps = \frac{FaultVolts}{ZTest}$$

$$FaultAmps = \frac{29.59V}{3.288\Omega}$$

$$FaultAmps = 9.000A$$

5. Configure a test with two states (prefault and fault) as shown in the following chart:

PREFAULT	FAULT
V1 = Nominal Volts @ 0°	V1 = FaultVolts @ 0°
V2 = Nominal Volts @ -120°	V2 = Nominal Volts @ -120°
V3 = Nominal Volts @ +120°	V3 = Nominal Volts @ +120°
I1 = 0.00A	I1 = FaultAmps @ V1° - ZTestAngle
I2 = 0.00A	I2 = 0.00A
I3 = 0.00A	I3 = 0.00A

Figure 15-52: Phase-to-Neutral Example MHO Distance MTA Test Configuration

The test configuration for the example settings are shown in the following chart:

PREFAULT	FAULT
V1 = 69.28V @ 0°	V1 = 29.59V @ 0°
V2 = 69.28V @ -120°	V2 = 69.28V @ -120°
V3 = 69.28V @ 120°	V3 = 69.28V @ 120°
I1 = 0.00A	I1 = 9.000A @ 284.71° (360°-75.29°)
I2 = 0.00A	I2 = 0.00A
I3 = 0.00A	I3 = 0.00A

6. Configure your test-set to change the faulted current angle.

7. Apply the prefault condition to the relay and verify the metering is correct, if required.

8. Apply the fault condition and ensure the 21-element operates.

9. Increase the faulted phase angles equally until the 21-element drops out.

10. Decrease all three phase angles until the 21-element picks up. Record the angle at pickup. (259.50° in our example)

11. Continue decreasing all three phase angles until the 21-element drops out.

12. Increase all three phase angles until the 21-element picks up. Record the faulted angle at pickup. (311.62° in our example)

13. Calculate the average of the two angles and add the result to the smaller angle. Record the calculated value on your test sheet.

$$\text{MTA} = \left(\frac{\text{TestAngle1} + \text{TestAngle2}}{2} - \text{VoltageReference}° - 360° \right) \times \text{-1}$$

$$\text{MTA} = \left(\frac{259.5° + 311.62°}{2} - 0° - 360° \right) \times \text{-1}$$

$$\text{MTA} = \left(285.56° - 360° \right) \times \text{-1} = 74.44°$$

14. Compare test results to manufacturer's specifications. There is typically no direct specification for the MTA test but we can use the "Mho…Ground Distance Element" specifications located in the SEL-311C manual as shown in the following figure which indicates a 5% tolerance.

MHO AND QUADRILATERAL GROUND DISTANCE ELEMENT	
Accuracy:	±5% of Setting at line angle 30 ≤ SIR ≤ 60
	±5% of Setting at line angle SIR <30

Figure 15-53: SEL-311C Ground Distance Specification

The test results measured an MTA of 74.44° and the expected result was 75.29°. Using the standard percent error calculation, we can determine the percent error and the test result passes. Record the percent error and pass/fail result on your test sheet.

$$\frac{\text{Actual Value - Expected Value}}{\text{Expected Value}} \times 100 = \text{Percent Error}$$

$$\frac{74.44° - 75.29°}{75.29°} \times 100 = \text{Percent Error}$$

-1.13% Error

15. Repeat Steps 2-11 for a B-N fault.

PREFAULT	FAULT
V1 = Nominal Volts @ 0°	V1 = Nominal Volts @ 0°
V2 = Nominal Volts @ -120°	V2 = FaultVolts @ -120°
V3 = Nominal Volts @ +120°	V3 = Nominal Volts @ +120°
I1 = 0.00A	I1 = 0.00A
I2 = 0.00A	I2 = FaultAmps @ V2° - ZTestAngle I3 =
I3 = 0.00A	0.00A

$$\text{MTA} = \left(\frac{\text{TestAngle1} + \text{TestAngle2}}{2} - \text{VoltageReference}° - 360° \right) \times \text{-1}$$

$$\text{MTA} = \left(\frac{190.88° + 139.17°}{2} - \text{-120°} - 360° \right) \times \text{-1}$$

$$\text{MTA} = \left(285.025° - 360° \right) \times \text{-1} = 74.975°$$

16. Repeat Steps 2-11 for a C-N fault.

PREFAULT	FAULT
V1 = Nominal Volts @ 0°	V1 = Nominal Volts @ 0°
V2 = Nominal Volts @ -120°	V2 = Nominal Volts @ -120°
V3 = Nominal Volts @ +120°	V3 = FaultVolts @ +120°
I1 = 0.00A	I1 = 0.00A
I2 = 0.00A	I2 = 0.00A
I3 = 0.00A	I3 = FaultAmps @ V3° - ZTestAngle

$$\text{MTA} = \left(\frac{\text{TestAngle1} + \text{TestAngle2}}{2} - \text{VoltageReference}° \right) \times \text{-1}$$

$$\text{MTA} = \left(\frac{70.85° + 18.99°}{2} - 120° \right) \times \text{-1}$$

$$\text{MTA} = \left(-75.08° \right) \times \text{-1} = 74.975°$$

17. Repeat the MTA test for all enabled 21-element zones.

PHASE-NEUTRAL DISTANCE TEST RESULTS (Ohms)											
	Z1MG	1.23	Z2MG	1.85	Z3MG	3.1					
	Z1ANG	78.6									
	Z1GD	0.00	Z2GD	18.00	Z3GD	60.00					
	kOM1	0.978	kOM	0.978							
	kOA1	-6.70	kOA	-6.70							
	Z1 Direction	F	Z2 Direction	F	Z3 Direction	F					

MAXIMUM TORQUE ANGLE (MTA) TEST RESULTS (Ohms)											
	ZONE 1 TESTS			ZONE 2 TESTS			ZONE 3 TESTS			ZONE 4 TESTS		
	Ohms	PU1º	PU2º	Ohms	PU1º	PU2º	Ohms	PU1º	PU2º	Ohms	PU1º	PU2º
A-N PU (º)	2.19	259.29	310.21	3.28	259.50	311.62	5.51	258.98	310.67			
B-N PU (º)	2.19	190.30	139.59	3.28	190.88	139.17	5.51	190.71	139.29			
C-N PU (º)	2.19	70.24	19.48	3.28	70.85	18.99	5.51	70.67	18.99			
	Z1G	MFG	%ERR	Z2G	MFG	%ERR	Z3GD	MFG	%ERR			
A-N MTA (º)	75.25	75.3	-0.05	74.44	75.3	-1.13	75.18	75.3	-0.15			
B-N MTA (º)	75.06	75.3	-0.31	74.98	75.3	-0.42	75.00	75.3	-0.39			
C-N MTA (º)	75.14	75.3	-0.20	75.08	75.3	-0.28	75.17	75.3	-0.16			

A-N ZONE 1 TESTS Z1G =MOD((((ZONE1TESTSPU1º+ZONE1TESTSPU2º)/2)-0)*(-1),360)

B-N ZONE 1 TESTS Z2G =MOD((((ZONE1TESTSPU1º+ ZONE1TESTSPU2º)/2)+120)*(-1),360)

C-N ZONE 1 TESTS Z3G =MOD((((ZONE1TESTSPU1º+ ZONE1TESTSPU2º)/2)-120)*(-1),360)

D) Reach Test Procedure

The reach test is similar to the 3-phase reach test with the phase-to-neutral calculations described in the MTA test procedure of this section. The relay tester chooses one or more angles to perform the test (typically starting at the MTA angle) and then performs a standard pickup test.

Follow the steps below to perform a reach test.

1. Determine how you will monitor pickup and set the relay accordingly, if required. (Pickup indication by LED, output contact, front panel display, etc. See the *Relay Test Procedures* section starting on page 109 for details.)

2. Determine the Test Angles for the test. (MTA, MTA+n, MTA-n, 95° lag, etc.) The following example will use the MTA angle plus/minus 30°. The expected result for any MHO circle with a 0.00Ω offset can be calculated using the formula:

$$\text{TestOhms} = \text{Z1Reach} \times \text{Cos}(\text{MTA}° - \text{Test}°)$$

Example calculations are shown below:

$$ZReach = Z2MG \times (1 + k0M\angle k0A)$$

$$ZReach = 1.85\Omega@78.6° \times \overrightarrow{(1 + 0.978@-6.7°)}$$

$$ZReach = 1.85\Omega@78.6° \times 1.975\Omega@-3.31°$$

$$ZReach = (1.85\Omega \times 1.975\Omega)@(78.6° + -3.31°)$$

$$ZReach = 3.653\Omega@75.29°$$

$TestOhms = ZReach \times Cos(MTA° - Test°)$	$TestOhms = Z1Reach \times Cos(MTA° - Test°)$
$TestOhms = 3.653 \times Cos(75.29° - 105.29°)$	$TestOhms = 3.653 \times Cos(75.29° - 45.29°)$
$TestOhms = 3.653 \times 0.866$	$TestOhms = 3.653 \times 0.866$
$TestOhms = 3.164\Omega@105.29°$	$TestOhms = 3.164\Omega@45.29°$

3. Starting with a reach test at the MTA angle. Calculate the test impedance at 95% of the TestOhms calculation for the test angle.

$$ZTest = 0.95 \times TestOhms = 0.95 \times 3.653\Omega = 3.47\Omega@75.29°$$

ZAN Compensated Magnitude

=0.95*((Ohms*SQRT((1+(K0M1*COS(RADIANS(K0A1))))^2+(K0M1*SIN(RADIANS(K0A1)))^2))*COS(RADIANS(ZangleK-TESTANGLE)))

ZAN Compensated Angle (ZangleK)

=Angle+(DEGREES(TANH((K0M1*SIN(RADIANS(K0A1)))/(1+(K0M1*COS(RADIANS(K0A1)))))))

4. Calculate the test configuration using one of these test methods:

   a. If the Z1 Reach setting is greater than 0.66 Ohms, arbitrarily determine the maximum amount of current to be applied to the relay. A good rule of thumb for maximum continuous current for most relays is 10.0A. Calculate the 3-phase balanced voltage required to apply 90% of the Z1 Reach setting using the following formula based on Ohm's Law.

$$V = I \times R$$

$$FaultVolts = FaultAmps \times ZTest$$

$$FaultVolts = 9.0A \times 3.47\Omega$$

$$FaultVolts = 31.23V$$

b. If the Z1 Reach setting is less than 0.66Ω, use the traditional formula described in this step with a 5.0V voltage setting. If more than 10.0A is required, the test should be performed as quickly as possible to prevent equipment damage.

Traditional test techniques reverse the formula in the previous step because electromechanical relays required a minimum voltage of approximately 20V to function reliably. Digital relays typically only require 3.0V to properly determine direction when a prefault voltage is applied. The traditional formula can also be used to calculate the test parameters, but you may need to make several calculations to keep the test current at reasonable levels as shown in the following examples:

$$I = \frac{V}{R}$$ $$FaultAmps = \frac{FaultVolts}{ZTest}$$ $$FaultAmps = \frac{20.0V}{3.47\Omega}$$ $$FaultAmps = 5.764A$$	$$I = \frac{V}{R}$$ $$FaultAmps = \frac{FaultVolts}{ZTest}$$ $$FaultAmps = \frac{31.23V}{3.47\Omega}$$ $$FaultAmps = 9.000A$$

5. Configure a test with two states (prefault and fault) as shown in the following chart:

PREFAULT	FAULT
V1 = Nominal Volts @ 0°	V1 = FaultVolts @ 0°
V2 = Nominal Volts @ -120°	V2 = Nominal Volts @ -120°
V3 = Nominal Volts @ +120°	V3 = Nominal Volts @ +120°
I1 = 0.00A	I1 = FaultAmps @ V1° - ZTestAngle
I2 = 0.00A	I2 = 0.00A
I3 = 0.00A	I3 = 0.00A

Figure 15-54: Phase-to-Neutral Example MHO Distance MTA Test Configuration

The test configuration for the example settings are shown in the following chart:

PREFAULT	FAULT
V1 = 69.28V @ 0°	V1 = 31.23V @ 0°
V2 = 69.28V @ -120°	V2 = 69.28V @ -120°
V3 = 69.28V @ 120°	V3 = 69.28V @ 120°
I1 = 0.00A	I1 = 9.000A @ 284.71° (360°-75.29°)
I2 = 0.00A	I2 = 0.00A
I3 = 0.00A	I3 = 0.00A

6. Configure your test-set to change the faulted (I1 for an A-N fault) current magnitude.

7. Apply the prefault condition to the relay and verify the metering is correct, if required.

8. Apply the fault condition and ensure the 21-element operates.

9. Decrease the faulted current or increase the voltage magnitude until the 21-element drops out.

10. Reverse the magnitude changes until the 21-element picks up. Record the impedance at pickup. (3.65Ω in our example) $Z = \dfrac{FaultVolts}{FaultAmps} = \dfrac{31.23V}{8.557A} = 3.65\Omega$. Record the results on your test sheet.

11. Compare test results to manufacturer's specifications. The "Mho...Ground Distance Element" specifications are located in the SEL-311C manual and shown in the following figure which indicates a 5% tolerance.

MHO AND QUADRILATERAL GROUND DISTANCE ELEMENT	
Accuracy:	±5% of Setting at line angle 30 ≤ SIR ≤ 60
	±5% of Setting at line angle SIR <30

Figure 15-55: SEL-311C Ground Distance Specification

The test results measured 3.65Ω with an expected result of 3.653Ω. Using the standard percent error calculation, we can determine the percent error and the test result passes. Record the percent error and pass/fail result on your test sheet.

$$\frac{\text{Actual Value - Expected Value}}{\text{Expected Value}} \times 100 = \text{Percent Error}$$

$$\frac{3.65\Omega - 3.653\Omega}{3.653\Omega} \times 100 = \text{Percent Error}$$

-0.08% Error

12. Repeat Steps 2-12 for a B-N fault.

PREFAULT	FAULT
V1 = Nominal Volts @ 0°	V1 = Nominal Volts @ 0°
V2 = Nominal Volts @ -120°	V2 = FaultVolts @ -120°
V3 = Nominal Volts @ +120°	V3 = Nominal Volts @ +120°
I1 = 0.00A	I1 = 0.00A
I2 = 0.00A	I2 = FaultAmps @ V2° - ZTestAngle I3 =
I3 = 0.00A	0.00A

13. Repeat Steps 2-12 for a C-N fault.

PREFAULT	FAULT
V1 = Nominal Volts @ 0°	V1 = Nominal Volts @ 0°
V2 = Nominal Volts @ -120°	V2 = Nominal Volts @ -120°
V3 = Nominal Volts @ +120°	V3 = FaultVolts @ +120°
I1 = 0.00A	I1 = 0.00A
I2 = 0.00A	I2 = 0.00A
I3 = 0.00A	I3 = FaultAmps @ V3° - ZTestAngle

14. Repeat the reach test for all enabled 21-element zones.

PHASE-NEUTRAL DISTANCE TEST RESULTS (Ohms)									
	Z1MG	1.23	Z2MG	1.85	Z3MG	3.1			
	Z1ANG	78.6							
	Z1GD	0.00	Z2GD	18.00	Z3GD	60.00			
	k0M1	0.978	k0M	0.978	k0M	0.978			
	k0A1	-6.70	k0A	-6.70	k0A	-6.70			
	Z1 Direction	F	Z2 Direction	F	Z3 Direction	F			

A-N REACH TEST RESULTS (Ohms)												
	ZONE 1 TESTS			ZONE 2 TESTS			ZONE 3 TESTS			ZONE 4 TESTS		
TEST ANGLE	M1P	MFG	%ERR	M2P	MFG	%ERR	M3P	MFG	%ERR	M4P	MFG	%ERR
105.29	2.09	2.10	-0.63	3.12	3.16	-1.38	5.31	5.30	0.17			
75.29	2.42	2.43	-0.36	3.65	3.65	-0.08	6.11	6.12	-0.18			
45.29	2.11	2.10	0.31	3.19	3.16	0.83	5.32	5.30	0.35			

B-N REACH TEST RESULTS (Ohms)												
	ZONE 1 TESTS			ZONE 2 TESTS			ZONE 3 TESTS			ZONE 4 TESTS		
TEST ANGLE	M1P	MFG	%ERR	M2P	MFG	%ERR	M3P	MFG	%ERR	M4P	MFG	%ERR
105.29	2.08	2.10	-1.11	3.13	3.16	-1.06	5.31	5.30	0.17			
75.29	2.42	2.43	-0.36	3.65	3.65	-0.08	6.11	6.12	-0.18			
45.29	2.10	2.10	-0.16	3.18	3.16	0.51	5.30	5.30	-0.03			

C-N REACH TEST RESULTS (Ohms)												
	ZONE 1 TESTS			ZONE 2 TESTS			ZONE 3 TESTS			ZONE 4 TESTS		
TEST ANGLE	M1P	MFG	%ERR	M2P	MFG	%ERR	M3P	MFG	%ERR	M4P	MFG	%ERR
105.29	2.09	2.10	-0.63	3.15	3.16	-0.43	5.30	5.30	-0.02			
75.29	2.43	2.43	0.05	3.65	3.65	-0.08	6.11	6.12	-0.18			
45.29	2.10	2.10	-0.16	3.19	3.16	0.83	5.30	5.30	-0.03			

NOTE: An MTA test can be incorporated into the reach test by performing two reach tests with the same angular difference from the MTA as shown in the test sheet above. If both test results have the same result, the MTA is correct.

E) Timing Test Procedure

The Timing test is relatively simple compared to the MTA and reach tests described previously. Apply a prefault then transition to a fault at least 90% of the reach and record the time between fault and the relay operation.

Follow the steps below to perform a timing test.

1. Configure the test-set timer to start when the fault state is energized and stop when the trip contact operates.

2. Determine the Test Angles for the test. (MTA, MTA+n, MTA-n, 95° lag, etc.) Most timing tests are performed at the MTA angle. The expected result for any MHO circle with a 0.00Ω offset can be calculated using the formula:

$$\text{TestOhms} = \text{Z1Reach} \times \text{Cos}(\text{MTA}° - \text{Test}°)$$

Example calculations are shown below:

$$\text{ZReach} = \text{Z2MG} \times \overline{(1 + \text{k0M}\angle\text{k0A})}$$

$$\text{ZReach} = 1.85\Omega @ 78.6° \times \overline{\left(1 + 0.978 @ \text{-}6.7°\right)}$$

$$\text{ZReach} = 1.85\Omega @ 78.6° \times 1.975\Omega @ \text{-}3.31°$$

$$\text{ZReach} = \left(1.85\Omega \times 1.975\Omega\right) @ \left(78.6° + \text{-}3.31°\right)$$

$$\text{ZReach} = 3.653\Omega @ 75.29°$$

3. Starting with a reach test at the MTA angle. Calculate the test impedance at 80% of the TestOhms calculation for the test angle.

$$\text{ZTest} = 0.833 \times \text{TestOhms} = 0.8 \times 3.653\Omega = 2.922\Omega @ 75.29°$$

ZAN Compensated Magnitude

=0.80*((Ohms*SQRT((1+(K0M1*COS(RADIANS(K0A1))))^2+(K0M1*SIN(RADIANS(K0A1)))^2))*COS(RADIANS(ZangleK-TESTANGLE)))

ZAN Compensated Angle (ZangleK)

=Angle+(DEGREES(TANH((K0M1*SIN(RADIANS(K0A1)))/(1+(K0M1*COS(RADIANS(K0A1)))))))

4. Calculate the test configuration using one of these test methods:

   a. If the Z1 Reach setting is greater than 0.66 Ohms, arbitrarily determine the maximum amount of current to be applied to the relay. A good rule of thumb for maximum continuous current for most relays is 10.0A. Calculate the 3-phase balanced voltage required to apply 90% of the Z1 Reach setting using the following formula based on Ohm's Law.

$$V = I \times R$$

$$FaultVolts = FaultAmps \times ZTest$$

$$FaultVolts = 9.0A \times 2.922\Omega$$

$$FaultVolts = 26.30V$$

   b. If the Z1 Reach setting is less than 0.66Ω, use the traditional formula described in this step with a 5.0V voltage setting. If more than 10.0A is required, the test should be performed as quickly as possible to prevent equipment damage.

   Traditional test techniques reverse the formula in the previous step because electromechanical relays required a minimum voltage of approximately 20V to function reliably. Digital relays typically only require 3.0V to properly determine direction when a prefault voltage is applied. The traditional formula can also be used to calculate the test parameters, but you may need to make several calculations to keep the test current at reasonable levels as shown in the following examples:

$$I = \dfrac{V}{R}$$	$$I = \dfrac{V}{R}$$
$$FaultAmps = \dfrac{FaultVolts}{ZTest}$$	$$FaultAmps = \dfrac{FaultVolts}{ZTest}$$
$$FaultAmps = \dfrac{20.0V}{2.922\Omega}$$	$$FaultAmps = \dfrac{26.3V}{2.922\Omega}$$
$$FaultAmps = 6.845A$$	$$FaultAmps = 9.000A$$

5. Configure a test with two states (prefault and fault) as shown in the following chart:

PREFAULT	FAULT
V1 = Nominal Volts @ 0°	V1 = FaultVolts @ 0°
V2 = Nominal Volts @ -120°	V2 = Nominal Volts @ -120°
V3 = Nominal Volts @ +120°	V3 = Nominal Volts @ +120°
I1 = 0.00A	I1 = FaultAmps @ V1° - ZTestAngle
I2 = 0.00A	I2 = 0.00A
I3 = 0.00A	I3 = 0.00A

Figure 15-56: Phase-to-Neutral Example MHO Distance MTA Test Configuration

The test configuration for the example settings are shown in the following chart:

PREFAULT	FAULT
V1 = 69.28V @ 0°	V1 = 20.00V @ 0°
V2 = 69.28V @ -120°	V2 = 69.28V @ -120°
V3 = 69.28V @ 120°	V3 = 69.28V @ 120°
I1 = 0.00A	I1 = 6.845A @ 284.71° (360°-75.29°)
I2 = 0.00A	I2 = 0.00A
I3 = 0.00A	I3 = 0.00A

6. Apply the prefault condition to the relay and verify the metering is correct, if required.

7. Apply the fault condition and ensure the 21-element operates within the specified time.

8. Compare test results to manufacturer's specifications. The "Timer Specifications" and "Output Contact" specifications are located in the SEL-311C manual and shown in the following figure.

The test results measured 19.95 cycles with an expected result of 18 cycles. Using the standard percent error calculation, we can determine the percent error.

$$\frac{\text{Actual Value - Expected Value}}{\text{Expected Value}} \times 100 = \text{Percent Error}$$

$$\frac{19.95\text{cy - }18.00\text{cy}}{18.00\text{cy}} \times 100 = \text{Percent Error}$$

10.83% Error

10.83% error seems a little high for a digital relay. We should calculate the allowable percent error using the specification in Figure 15-57.

The SEL-311C specifications for Distance element timing are shown in the following figure:

MHO AND QUADRILATERAL GROUND DISTANCE ELEMENT	
Max. Operating Time:	See pickup and reset time curves in Figure 3.18 and Figure 3.19

TIMER SPECIFICATIONS	
Pickup / Dropout accuracy for all timers:	±0.25 cycle and ±0.1% of setting

OUTPUT CONTACTS	
Pickup time:	<5 ms

Figure 15-57: SEL-311C Timer and Output Specifications

If instantaneous overcurrent elements are made directional (with standard directional elements such as 32QF), the pickup time curve in SEL Figure 3.18 (Or our Figure 15-58 below) is adjusted as follows:

- multiples of pickup setting ≤4: add 0.25 cycle
- multiples of pickup setting >4: add 0.50 cycle

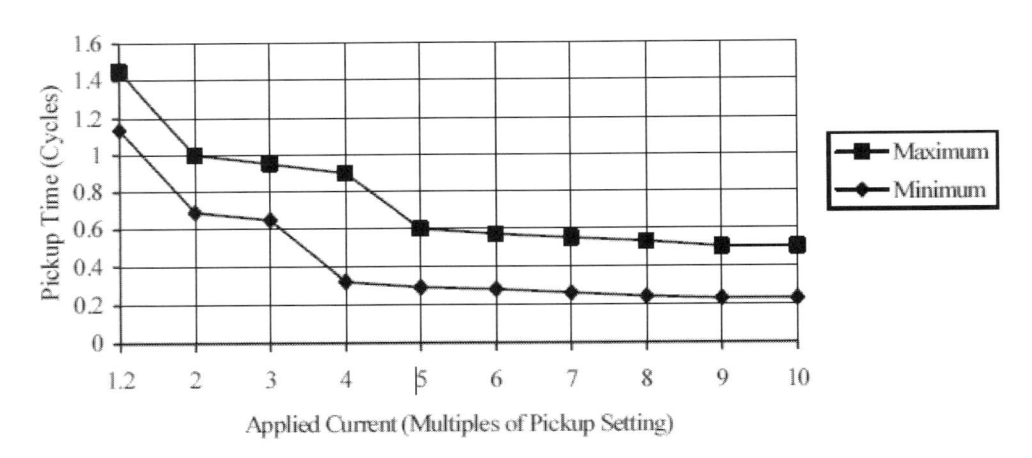

Figure 15-58: SEL-311C Timer and Output Graph (SELFigure 3.18)

We calculate the allowable percent error by adding all of the internal delays and then dividing by the expected value. The internal delays are:

- Max. operating time 1: The 21-element is directional. We add 0.25 cycles because our multiple of pickup is <4.
- Max. operating time 2: We are applying approximately 1.2 x the pickup setting and can add 1.4 cycles as per Figure 15-58 (SEL Figure 3-18).

- Timer specifications: add 0.25 cycles and 0.1% of the time delay setting.
- Output contacts. Add <5ms or 4.99ms.

$$AllowablePercentError = \frac{MaxOperatingTime+TimerSpecifications + OutputContacts}{ExpectedValue} \times 100$$

$$AllowablePercentError = \frac{0.25cycles+1.4cycles + 0.25\ cycles+(0.1\%\ of\ Setting) + (<5\ ms)}{18cycles} \times 100$$

$$AllowablePercentError = \frac{0.25cycles+1.4cycles+0.25cycles+(0.001 \times 18cycles)+(4.99ms \times 60)}{18cycles} \times 100$$

$$AllowablePercentError = \frac{0.25cycles+1.4cycles+0.25cycles+0.018cycles+0.2994cycles}{18cycles} \times 100$$

$$AllowablePercentError = \frac{2.2174cycles}{18cycles} \times 100$$

$$AllowablePercentError = 0.1231 \times 100$$

$$AllowablePercentError = 12.31\%$$

The test result percent error does fall within the maximum expected result. You can add the results to the test sheet. A 10+% error may create questions on a test sheet, and I often change the % Error field to a Pass/Fail or "OK".

9. Repeat Steps 5-8 for a B-N fault.

PREFAULT	FAULT
V1 = Nominal Volts @ 0°	V1 = Nominal Volts @ 0°
V2 = Nominal Volts @ -120°	V2 = FaultVolts @ -120°
V3 = Nominal Volts @ +120°	V3 = Nominal Volts @ +120°
I1 = 0.00A	I1 = 0.00A
I2 = 0.00A	I2 = FaultAmps @ V2° - ZTestAngle I3 =
I3 = 0.00A	0.00A

10. Repeat Steps 5-8 for a C-N fault.

PREFAULT	FAULT
V1 = Nominal Volts @ 0°	V1 = Nominal Volts @ 0°
V2 = Nominal Volts @ -120°	V2 = Nominal Volts @ -120°
V3 = Nominal Volts @ +120°	V3 = FaultVolts @ +120°
I1 = 0.00A	I1 = 0.00A
I2 = 0.00A	I2 = 0.00A
I3 = 0.00A	I3 = FaultAmps @ V3° - ZTestAngle

11. Repeat the reach test for all enabled 21-element zones.

PHASE-NEUTRAL DISTANCE TEST RESULTS

	Z1MG		**1.23**	Z2MG		**1.85**	Z3MG		**3.1**		
	Z1ANG		**78.6**								
	Z1GD		**0.00**	Z2GD		**18.00**	Z3GD		**60.00**		
	k0M1		**0.978**	k0M		**0.978**	k0M		**0.978**		
	k0A1		**-6.70**	k0A		**-6.70**	k0A		**-6.70**		
	Z1 Direction		**F**	Z2 Direction		**F**	Z3 Direction		**F**		

A-N TIMING TESTS (in cycles)

MULT	Z1G	MFG	Diff	Z2GT	MFG	%ERR	Z3GT	MFG	%ERR	Z4GT	MFG	%ERR
0.8 * PU	**2.00**	0.00	2.00	**19.95**	18.00	10.83	**61.86**	60.00	3.10			

B-N TIMING TESTS (in cycles)

MULT	Z1G	MFG	Diff	Z2GT	MFG	%ERR	Z3GT	MFG	%ERR	Z4GT	MFG	%ERR
0.8 * PU	**1.87**	0.00	1.87	**19.96**	18.00	10.89	**61.94**	60.00	3.23			

C-N TIMING TESTS (in cycles)

MULT	Z1G	MFG	Diff	Z2GT	MFG	%ERR	Z3GT	MFG	%ERR	Z4GT	MFG	%ERR
0.8 * PU	**1.86**	0.00	1.86	**19.83**	18.00	10.17	**61.78**	60.00	2.97			

COMMENTS: Z1 Maximum time delay = 2.2 cycles
Z2 Maximum Allowable Percent Error = 12.31%, Z3 Maximum Allowable Percent Error = 3.77%

RESULTS ACCEPTABLE: ☑ YES   ☐ NO   ☑ SEE NOTES

Chapter 16

Understanding Digital Logic

Every manufacturer uses a different method to display their logic. However, they all use the same basic logic elements. We will review the basic logic principles in this chapter then apply these principles to protective relays.

1. Understanding Logic

Basic logic can be separated into the following logic functions: OR, AND, NOT, NAND, NOR, XOR, and XNOR. A single logic function is called an "operator" or "gate." A logic gate has inputs and outputs that can be used in conjunction with other logic gates (logic functions) inside the relay. Logic is expressed in binary terms which means that an input/output can be in only one of two states, ON (1) (also called "Active" or "Enabled") or OFF (0). The number of inputs for any gate is limited by the manufacturer's interface, and all examples are shown with two inputs for simplicity's sake.

A) OR

The OR function is the most common gate used in relaying applications. An OR gate will be ON (1) if **ANY** of its inputs are in the ON (1) state. If all inputs are OFF (0), the output will be OFF (0). The electrical equivalent of an OR gate is normally-open contacts in parallel.

DESCRIPTOR	SYMBOL	MATRIX	ELECTRICAL EQUIVILANT
OR		1,1 OR—1; 1,0 OR—1; 0,0 OR—0	NORMALLY-OPEN CONTACTS IN PARALLEL

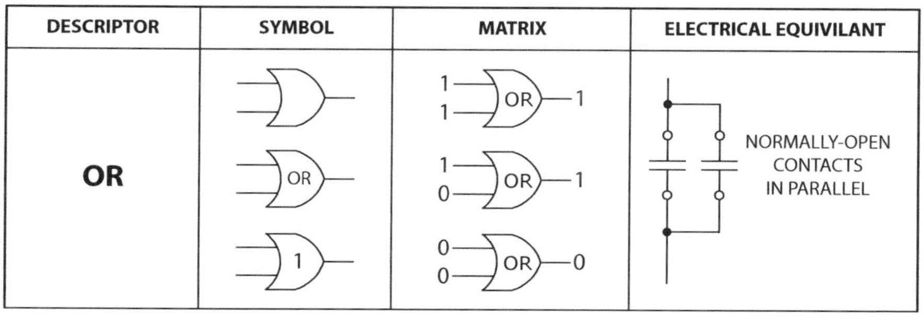

Figure 16-1: OR Gate Logic

B) AND

The AND function will be ON (1) if **ALL** its inputs are ON (1). If any input is OFF (0), the output will be OFF (0). The electrical equivalent of an AND gate is normally-open contacts in series.

DESCRIPTOR	SYMBOL	MATRIX	ELECTRICAL EQUIVILANT
AND			NORMALLY-OPEN CONTACTS IN SERIES

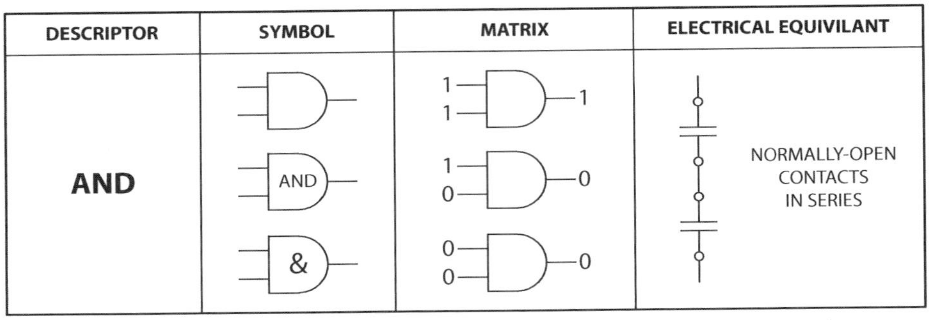

Figure 16-2: AND Gate Logic

C) NOT

The NOT gate reverses the state of any input/output and is usually represented as a circle connected to an input or output. If the input to a NOT is OFF (0), its output is ON (1) and visa-versa. The electrical equivalent of a NOT gate uses a relay with a normally-closed contact.

DESCRIPTOR	SYMBOL	MATRIX	ELECTRICAL EQUIVILANT
NOT		0 ——◯ 1 1 ——◯ 0	R1 RELAY WITH NORMALLY-CLOSED CONTACTS

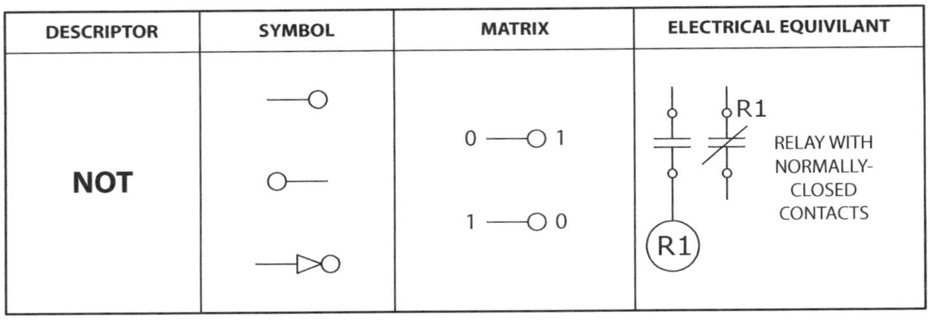

Figure 16-3: NOT Logic

474

D) NOR

NOR gates are not as common as OR, AND, or NOT gates and represent the opposite of an OR gate. Its symbol includes an OR gate with a NOT connected to the output. If any input is ON (1), the output is OFF (0). NOR gates only turn ON (1) when all inputs are OFF (0). The electrical equivalent of a NOR gate is normally-closed contacts in series.

DESCRIPTOR	SYMBOL	MATRIX	ELECTRICAL EQUIVILANT
NOR			NORMALLY-CLOSED CONTACTS IN SERIES

Figure 16-4: NOR Gate Logic

E) NAND

NAND gates are similar to NOR gates but they reverse the output of an AND gate instead of an OR gate. Therefore, the gate will be ON (1) if at least one input is OFF (0). The gate will be OFF (0) if all inputs are ON (1). The electrical equivalent of a NAND gate is normally-closed contacts in parallel.

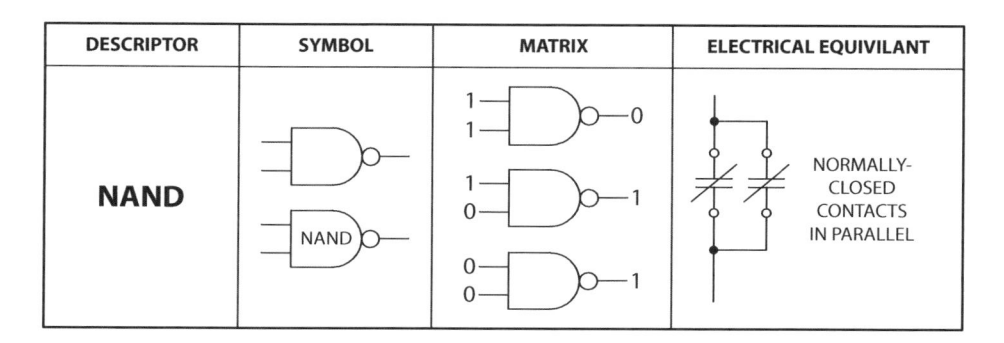

DESCRIPTOR	SYMBOL	MATRIX	ELECTRICAL EQUIVILANT
NAND			NORMALLY-CLOSED CONTACTS IN PARALLEL

Figure 16-5: NAND Gate Logic

F) XOR

The XOR gate is not available on most relay models and is infrequently used. The XOR gate is ON (1) if only one input is ON (1) and all other inputs are off. It is the electrical equivalent of an electrical-interlock like a forward/reverse motor control circuit.

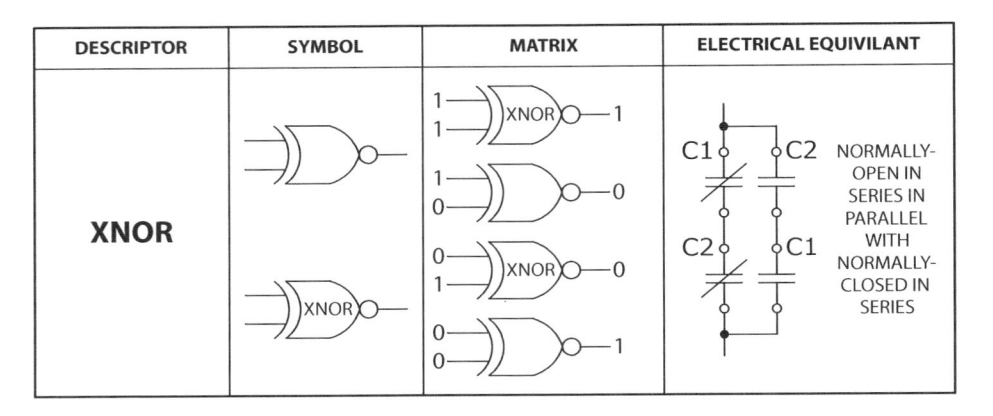

Figure 16-6: XOR Gate Logic

G) XNOR

The XNOR gate is not available on most relay models and is infrequently used. The XNOR gate is ON (1) if all inputs are in the same state. All inputs must be ON (1) or all inputs must be OFF (0) for the gate to be ON (1). Its electrical equivalent is normally-closed contacts in parallel with normally-open contacts.

Figure 16-7: XNOR Gate Logic

H) Comparator

The comparator is not a common logic gate but is included here because it is often found in a relay's internal logic schemes. A comparator compares an input to a setpoint or reference signal and turns ON (1) if the input is greater than the setpoint or reference. The input is typically an analog signal like a CT or PT and the comparator turns the analog signal into a digital signal to be used in the logic diagrams.

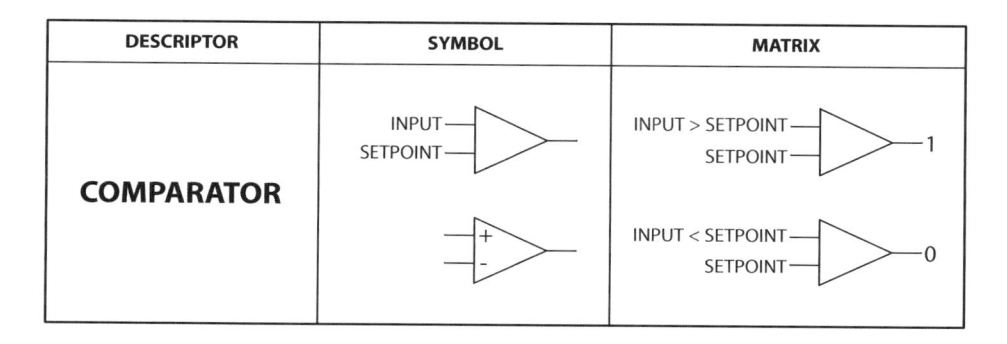

Figure 16-8: Comparator Logic

I) TIMER

Timers are not typically found in logic but are used extensively in relays. Timers can be on-delay and begin timing when its input is ON (1) or off-delay, and begin timing after the input has been removed. Time delays are defined by the user as a relay setpoint or can be pre-defined by the manufacturer. This logic device operates exactly like an equivalent electrical timer.

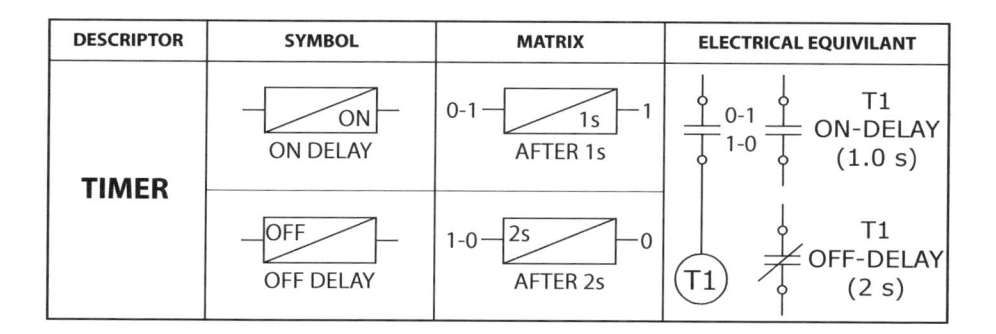

Figure 16-9: Timer Logic

J) Summary of Logic

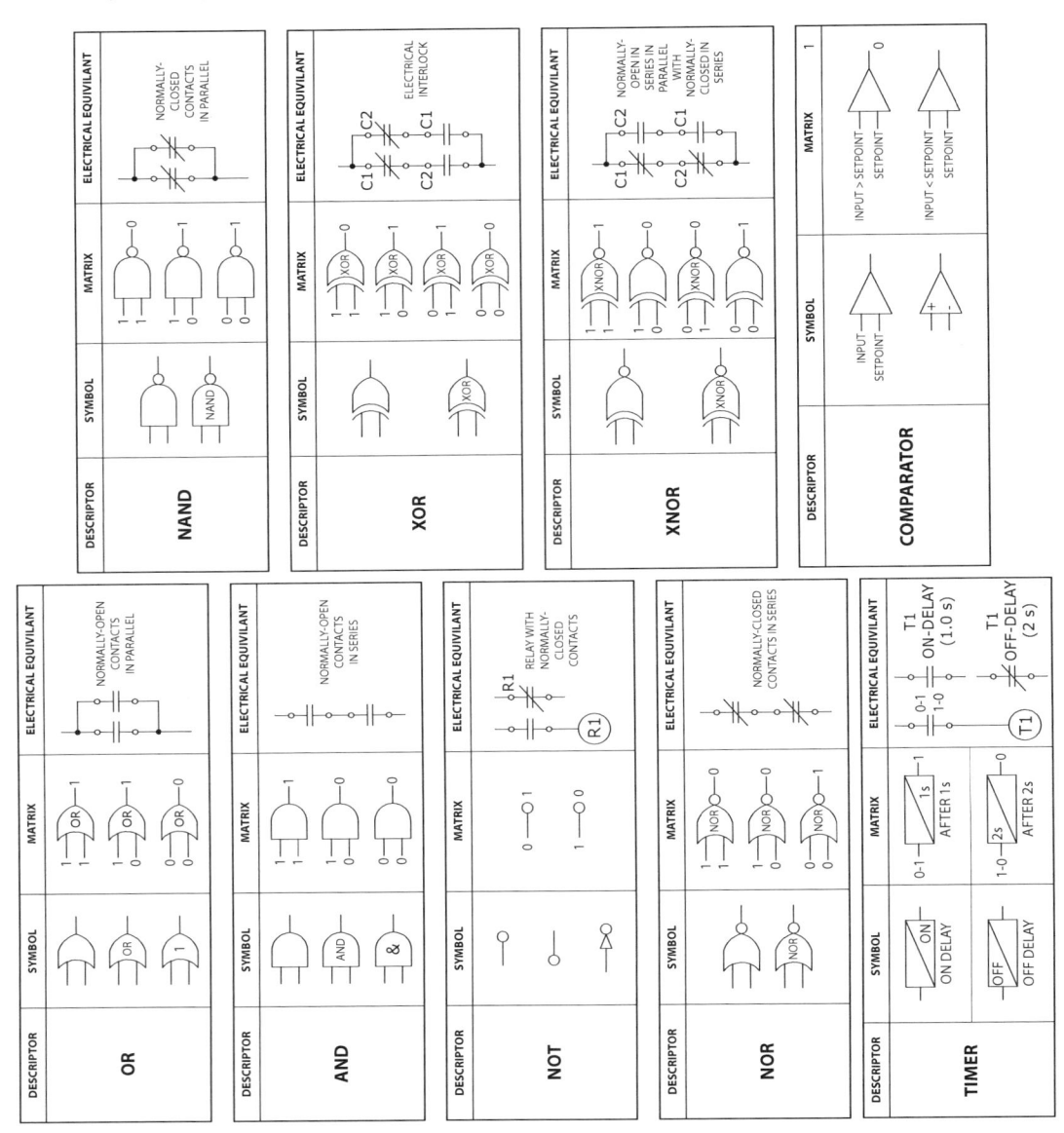

Figure 16-10: Summary of Logic Element Operation

2. Relay Labels

Before you can begin to understand any relay's logic scheme, you must understand what relay functions are available. The available relay functions will be called "labels" in this book for convenience because different manufacturers have different references for these elements. Every relay manufacturer has their own system for labels and there can be different systems between relay models from the same manufacturer. Simple, but catastrophic, errors can occur when you assume a label definition incorrectly. Here are just two examples that I have discovered in the past:

- A General Electric (GE) UR series relay was programmed with the label "87L DIFF OP A" instead of the correct label "87L DIFF OP." B-phase and C-phase differential trips would not operate the relay with this label selection.
- A Schweitzer Laboratories (SEL) relay had settings to operate the 51N elements and the trip logic included "51NT." These labels are used to provide overcurrent protection using the additional ground CT input to the relay that was not connected to anything. The application engineer assumed that 51N stood for residual neutral overcurrent instead of 51G which, in fact, represents residual neutral overcurrent. The relay would not have operated if a ground fault had occurred.

The first step in understanding the relay logic is to obtain a list of the available internal relay labels and their definitions. Review the following figures for examples of relay labels and their definitions from SEL and GE.

ROW	RELAY WORD BITS							
0	EN	TRP	TIME	COMM	SOTF	RCRS	RCLO	51
1	A	B	C	G	ZONE1	ZONE2	ZONE3	ZONE4
2	M1P	M1PT	Z1G	Z1GT	M2P	M2PT	Z2G	Z2GT
3	Z1T	Z1T	50P1	67P1	67P1T	50G1	67G1	67G1T
4	51G	51GT	51GR	LOP	ILOP	ZLOAD	ZLOUT	ZLIN
5	LB1	LB2	LB3	LB4	LB5	LB6	LB7	LB8

Figure 16-11: SEL-311C Relay Labels

ROW	BIT	DEFINITION	PRIMARY APPLICATION
3	Z1T	Zone 1 phase and/or ground distance, time delayed	
	Z2T	Zone 2 phase and/or ground distance, time delayed	
	50P1	Level 1 phase instantaneous overcurrent element (A, B, C) above pickup setting 50P1P	
	67P1	Level 1 torque controlled phase instantaneous overcurrent element (derived from 50P1)	
	67P1T	Level 1 phase definite-time overcurrent element 67P1T timed out (derived from 67P1)	
	50G1	Level 1 residual ground instantaneous overcurrent element (residual ground current above pickup setting 50G1P)	
	67G1	Level 1 torque controlled residual ground instantaneous overcurrent element (derived from 50G1)	
	67G1T	Level 1 residual ground definite-time overcurrent element 67P1T timed out (derived from 67P1)	

Figure 16-12: SEL-311C Relay Label Definitions

OPERAND TYPE	OPERAND SYNTAX	OPERAND DESCRIPTION
ELEMENT (Ground Distance)	GND DIST Z2 OP	Ground Distance Zone 2 is operated
ELEMENT (Phase IOC)	PHASE IOC1 PKP	At least one phase of IOC1 is picked up
ELEMENT (Phase Directional)	PH DIR1BLK	Phase Directional 1 Block
ELEMENT (Neutral IOC)	NTRL IOC1 PKP	Neutral IOC1 is picked up
ELEMENT (Neutral Directional)	NTRL DIR OC1 FWD NTRL DIR OC1 REV	Neutral Directional OC1 Forward is Operated Neutral Directional OC1 Reverse Operated

Figure 16-13: GE D-60 Relay Labels and Definitions

3. Internal Relay Control Schemes

All digital relays use a combination of predefined and user-defined logic schemes to operate that are included in the relay's algorithms (programming) or "firmware." The predefined logic schemes determine how the relay interprets the analog input signals (CTs and/or PTs) to operate the internal relay labels. The user-defined schemes (settings) determine what functions the relay will perform if any relay label becomes active. We will review the undervoltage (27) element for the SEL-311C and GE D-60 relays in this chapter to help understand relay logic functions.

A) SEL-311C Undervoltage Element

The following figures display the manufacturer's description of the undervoltage (27) element for the SEL-311C relay.

Figure 16-14 defines the undervoltage elements in a table from the manufacturer's literature. The labels 27A, 27B, 27C, and 3P27 are all controlled by the 27P setting and use the single-phase-to-neutral voltages V_A, V_B, and V_C. Therefore, the 27P settings define a single-phase-to-ground undervoltage (27) element that should not be confused with "27PP" settings which define a phase-to-phase undervoltage (27) element.

VOLTAGE ELEMENT (RELAY WORD BITS)	OPERATING VOLTAGE	PICKUP SETTING/RANGE
27A	V_A	27P
27B	V_B	OFF, 0.00—150.00 V Secondary
27C	V_C	
3P27	27A * 27B *27C	
27AB	V_{AB}	27PP
27BC	V_{BC}	OFF, 0.00—260.00 V secondary
27CA	V_{CA}	

Figure 16-14: SEL-311C Voltage Elements Settings and Setting Ranges

The next figure is another chart from the SEL manual which summarizes all of the available SEL logic labels that could be used in the relay settings. The labels in our example are located in TARGET ROW 42 and their real-time status could be viewed via the front-panel or software using the TAR 42 command in the software.

ROW	RELAY WORD BITS							
0	EN	TRP	TIME	COMM	SOTF	RCRS	RCLO	51
1	A	B	C	G	ZONE1	ZONE2	ZONE3	ZONE4
...	...	...	...	....	...	...	...	...
42	27A	27B	27C	59A	59B	59C	3P27	3P59

Figure 16-15: SEL-311C Relay Word Bits

SEL also provides another chart which defines each label as shown in the next figure.

ROW	BIT	DEFINITION	PRIMARY APPLICATION
42	27A	A-phase instantaneous undervoltage element (A-phase voltage below pickup setting 27P)	Control
	27B	B-phase instantaneous undervoltage element (B-phase voltage below pickup setting 27P)	Control
	27C	C-phase instantaneous undervoltage element (C-phase voltage below pickup setting 27P)	Control
	3P27	27A * 27B * 27C	Control

Figure 16-16: SEL-311C Relay Word Bit Definitions

It is usually easier to follow simple logic diagrams like the SEL version in the next figure by reviewing the diagram from the source(s) (left-hand side) to the outputs. (right-hand side) Follow the logic in Figure 16-17 from left to right to understand the operating characteristics of the undervoltage (27) element. There are three comparators used to determine which undervoltage element will operate. The first input to all three comparators is the "27P (Setting)" which is a setting entered by the user. The other input to each comparator is the analog input from the PTs for each phase. If the voltage input is less than the "27P" setting, the comparator will turn ON (1). For example if "27P" = 100V and the [VA] voltage fell to 95V so the "27A" relay label would be ON (1).

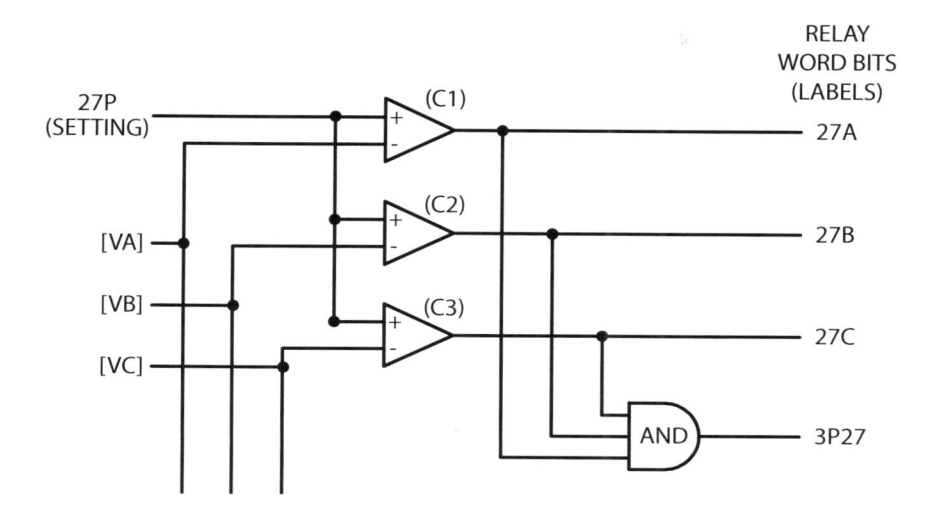

Figure 16-17: SEL-311C Single-Phase and Three-Phase Undervoltage (27) Logic

All three comparator outputs are also inputs into the AND gate. As discussed previously, an AND gate will turn ON only if **all** three inputs are ON. Therefore:

- [VA] must be lower the 27P setpoint; *and*
- [VB] must be lower the 27P setpoint; *and*
- [VC] must be lower the 27P setpoint

before the AND gate will turn ON (1). When the AND gate is ON, 3P27 turns ON (1). The electrical schematic of this logic is shown in the following figure where C1, C2, and C3 represent the three comparators:

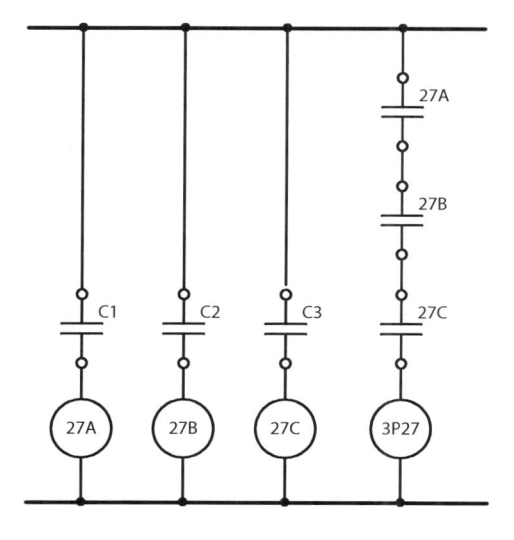

Figure 16-18: SEL-311C Undervoltage (27) Logic Electrical Schematic

B) SEL-311C Undervoltage (27) Logic Scenario #1

You can also add notations for a single scenario to help understand logic diagrams. Using the following conditions, see how adding notations can help you understand the SEL-311C undervoltage (27) logic diagram:

- 27P = 100V
- VA = 115V
- VB = 98V
- VC = 115V

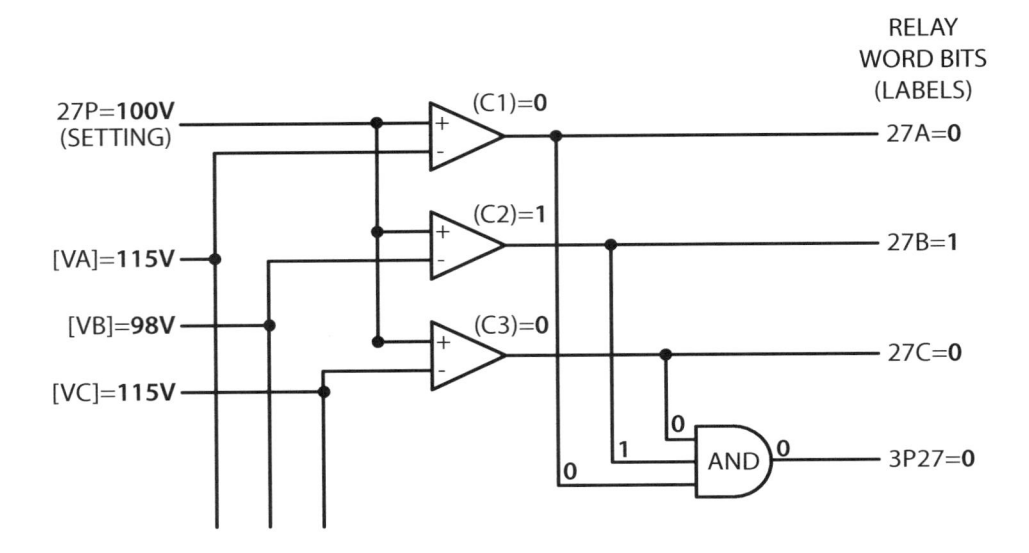

Figure 16-19: SEL-311C Undervoltage (27) Scenario #1 Logic

i) SEL-311C Undervoltage (27) Logic Scenario #2

Using the following conditions, see how adding notations can help understand the SEL-311C undervoltage (27) logic diagram with a different scenario:

- 27P = 100V
- VA = 80V
- VB = 80V
- VC = 80V

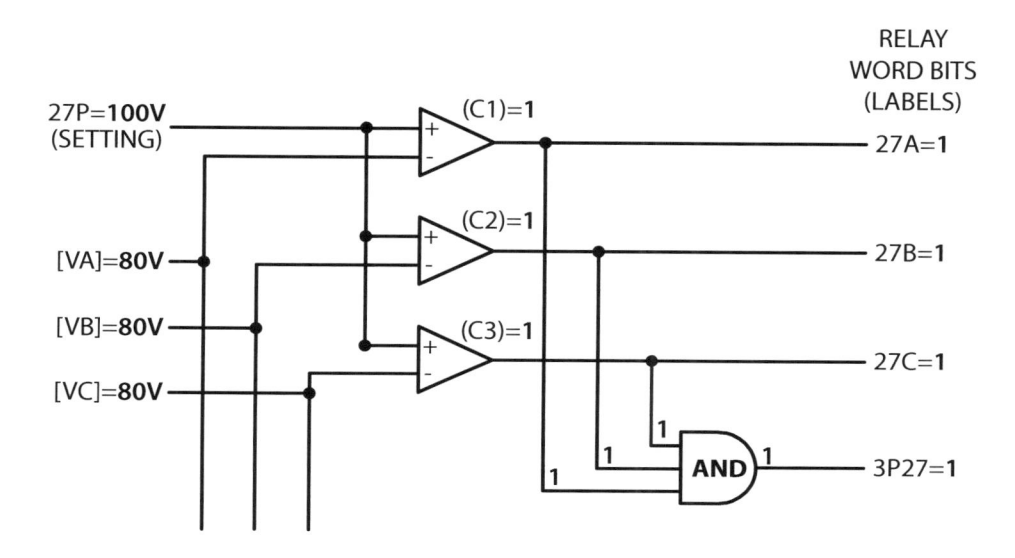

Figure 16-20: SEL-311C Undervoltage (27) Scenario #2 Logic

All of these undervoltage functions are performed automatically by the relay in its predefined logic controls. The user-defined logic determines what action the relay will perform if 27A, 27B, 27C, or 3P27 are ON (1) including fault record initiation or output contact operation. User-defined logic will be discussed in later sections of this chapter.

C) GE D-60 Undervoltage (27) Logic

The D-60 logic for the undervoltage (27) element is more complex. The following figures document information regarding the undervoltage (27) element from the manufacturer's instruction manual. They have been slightly modified to help understand the applied logic.

PHASE UNDERVOLTAGE	PHASE UV1 FUNCTION:	RANGE: DISABLED, ENABLED
	PHASE UV1 SIGNAL SOURCE: **SRC 1**	Range: SRC 1…, SRC 6
	PHASE UV1 MODE: Phase to Ground	Range: Phase to Ground, Phase to Phase
	PHASE UV1 PICKUP: **1.000 pu**	Range: 0.000 to 3.000 pu in steps of 0.001
	PHASE UV1 CURVE: **Definite Time**	Range: Definite Time, Inverse Time
	PHASE UV1 DELAY: **1.00s**	Range: 0.00 to 600.00s in steps of 0.01
	PHASE UV1 MINIMUM VOLTAGE: **0.100 pu**	Range: 0.000 to 3.000 pu in steps of 0.001
	PHASE UV1 BLOCK: **Off**	Range: FlexLogic™
	PHASE UV1 TARGET: **Self-Reset**	Range: Self-reset, Latched, Disabled
	PHASE UV1 EVENTS: **Disabled**	Range: Disabled, Enabled

Figure 16-21: GE D-60 Phase Undervoltage (27) Settings

OPERAND TYPE	OPERAND SYNTAX	OPERAND DESCRIPTION
ELEMENT (Phase UV)	PHASE UV1 PKP	At least one phase of UV1 is picked up
	PHASE UV1 OP	At least one phase of UV1 is operated
	PHASE UV1 DPO	At least one phase of UV1 is dropped out
	PHASE UV1 PKP A	Phase A of UV1 is picked up
	PHASE UV1 PKP B	Phase B of UV1 is picked up
	PHASE UV1 PKP C	Phase C of UV1 is picked up
	PHASE UV1 OP A	Phase A of UV1 is operated
	PHASE UV1 OP B	Phase B of UV1 is operated
	PHASE UV1 OP C	Phase C of UV1 is operated
	PHASE UV1 DPO A	Phase A of UV1 is dropped out
	PHASE UV1 DPO B	Phase B of UV1 is dropped out
	PHASE UV1 DPO C	Phase C of UV1 is dropped out

Figure 16-22: GE D-60 Undervoltage (27) Relay Labels and Definitions

Sometimes it is easier to follow more complex logic diagrams like the one in Figure 16-23 by choosing the output (right-hand side) you are trying to understand and working backwards to the inputs on the left-hand side of the drawing. The "Phase UV1 A PKP" output on the lower left hand side of the drawing is similar to the "27A" output in the previous SEL-311C example and we will try to see how that element operates. "Phase UV1 A PKP" is directly connected to the "S5 AP" output of the "Setting (S5)" module that is similar to the comparator module in the SEL-311C example. If the:

- "VAG" voltage signal from the PTs is lower than the "PHASE UV1 PICKUP" (Figure 16-21) setting, **and**
- the AND box (Figure 16-23) sends an ON (1) signal to the "RUN" inputs on the "Setting (S5)" module

the Phase UV1 A PKP label will operate and be ON (1).

The AND gate has three inputs and all three inputs must be ON (1) before the element will operate.

- The first input (S1) is the first setting of the element (Enabled / Disabled) and must be "ENABLED" or the input will be OFF (0) and the output "PHASE UV1 OP" will never operate.
- The second input (S2) is user-defined logic used to block the undervoltage from operating under certain conditions. For example, you could use the fuse failure (60) element to block the undervoltage (27) from operating until a new fuse could be installed. The input is connected to a NOT symbol which will reverse the state of the input so that an OFF (0) state will allow the element to operate. If you did not want to use this feature, the setting of "0" would be ON (1) to the AND gate and this would allow the element to operate.
- The third input is connected to the output of OR1 gate and any one of its inputs must be ON (1) to turn the AND gate input ON (1). The OR1 gates are connected to simple comparators similar to the ones shown in the SEL-311C example. If any of the voltages from the PTs in S3 is larger than the S4 settings for any phase, the related OR1 Gate input will be ON (1) and turn its output ON (1). Turning OR1 output ON (1) will allow the AND gate to operate if all its other inputs are also ON (1).

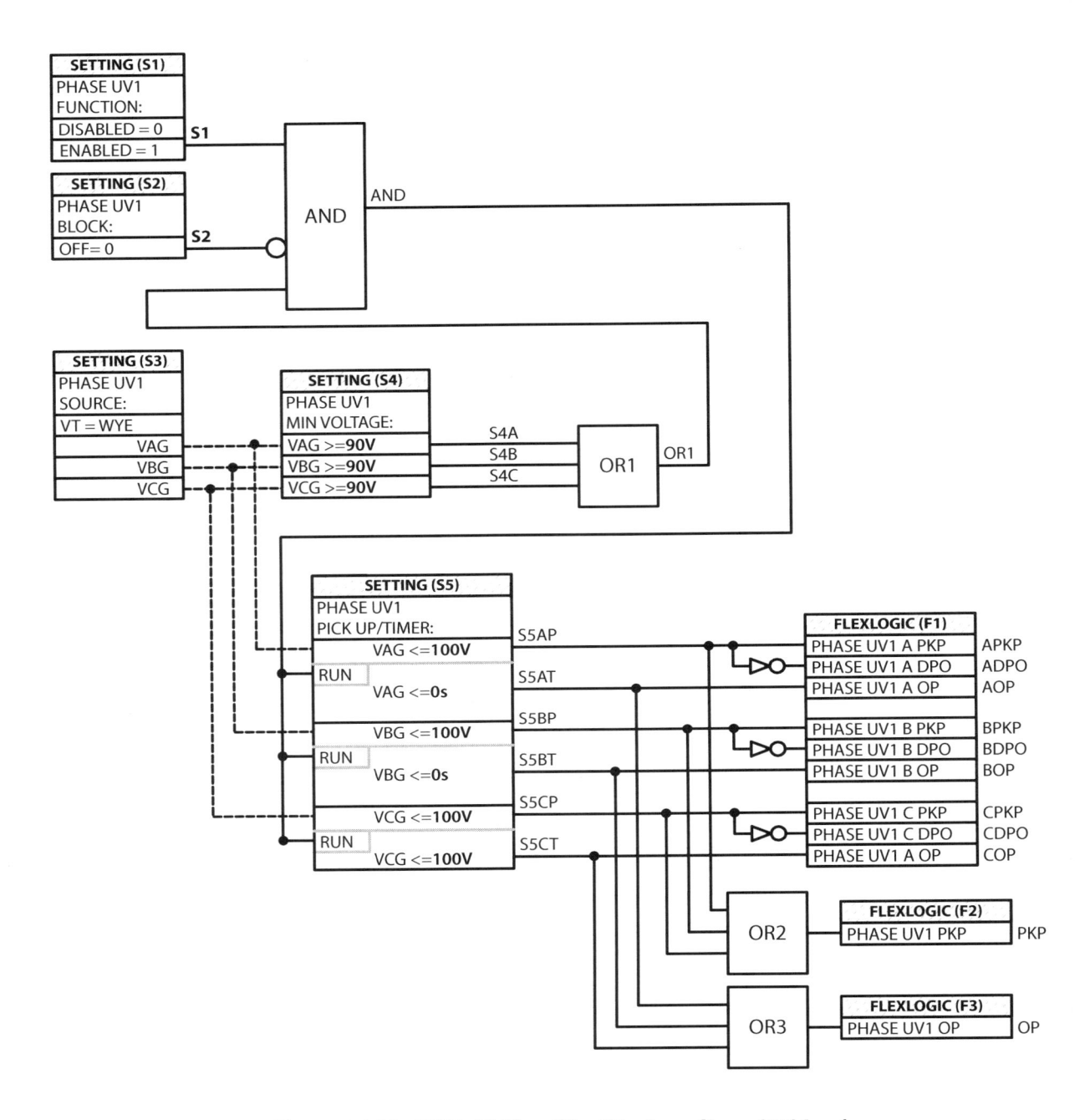

Figure 16-23: GE D-60 Simplified Undervoltage (27) Logic

Although both the SEL-311C and D-60 relays in the previous examples are designed to provide undervoltage protection, this is a great example of how two different manufacturers deal with the same protective element.

The following figure depicts the electrical equivalent of the GE D-60 logic diagram from Figure 16-23. See if you can follow along with the conversion from GE Logic diagrams to electrical Schematics.

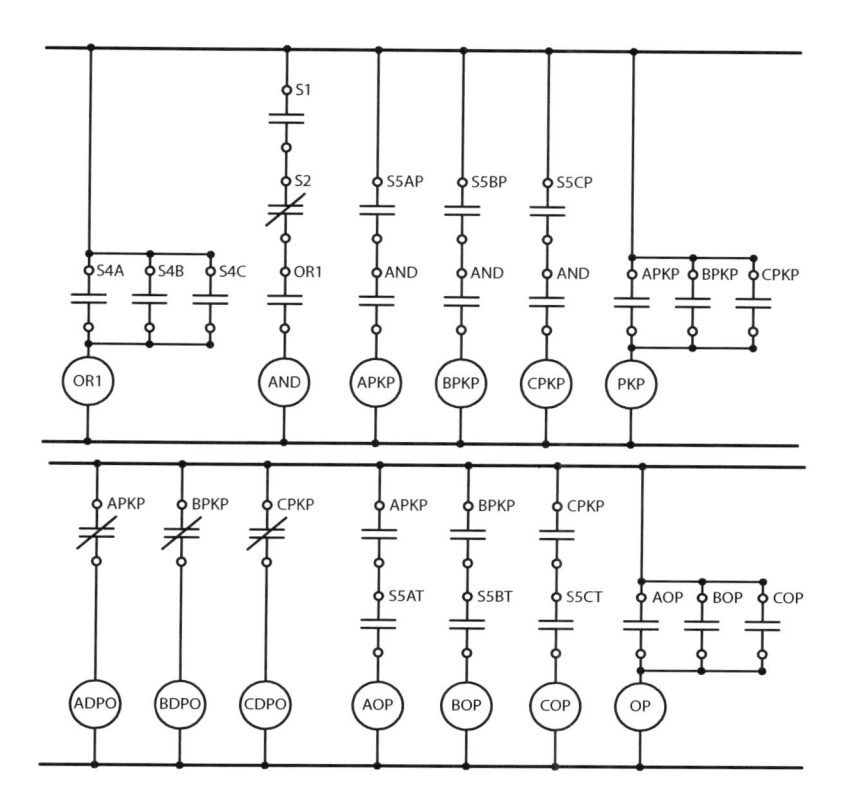

Figure 16-24: GE D-60 Electrical Schematic of Undervoltage (27) Logic

i)   GE-D60 Undervoltage (27) Logic Scenario Example #1

Using the following information, you can enter notations on the logic diagram to help understand what will happen under different conditions.

- VAG = **115V**
- VBG = **98V**
- VCG = **115V**
- S1 = Enabled
- S2 = Off
- S3 = Wye
- S4A, B, C = 90V
- S5PU = 100V
- S5TD = 0s

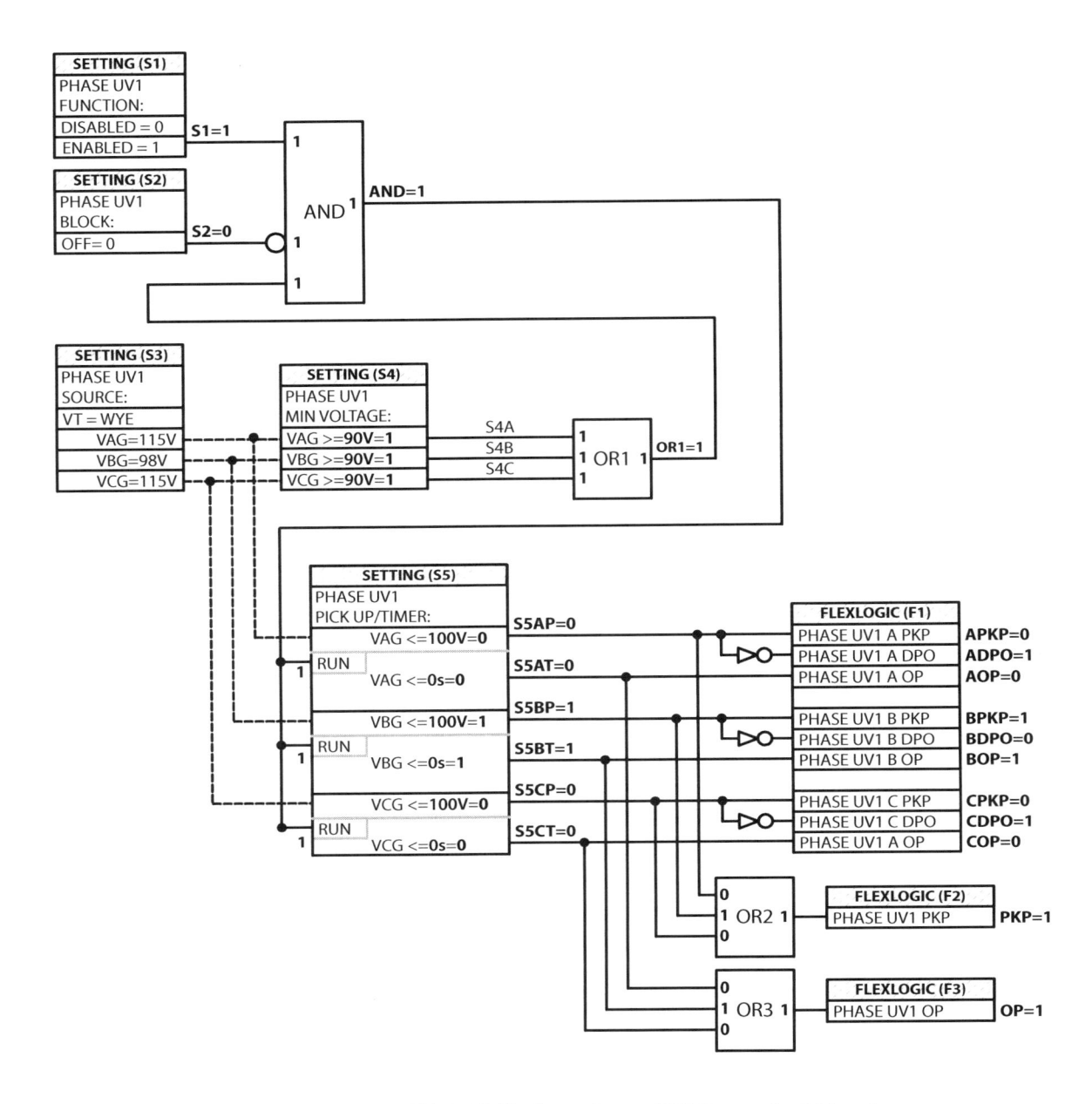

Figure 16-25: GE D-60 Undervoltage (27) Scenario #1 Logic

ii) GE-D60 Undervoltage (27) Logic Scenario Example #2

Using the following information, you can enter notations on the logic diagram to help understand what will happen under different conditions.

- VAG = **80V**
- VBG = **80V**
- VCG = **80V**

- S1 = Enabled
- S2 = Off
- S3 = Wye
- S4 = 90V
- S5PU = 100V
- S5TD = 0s

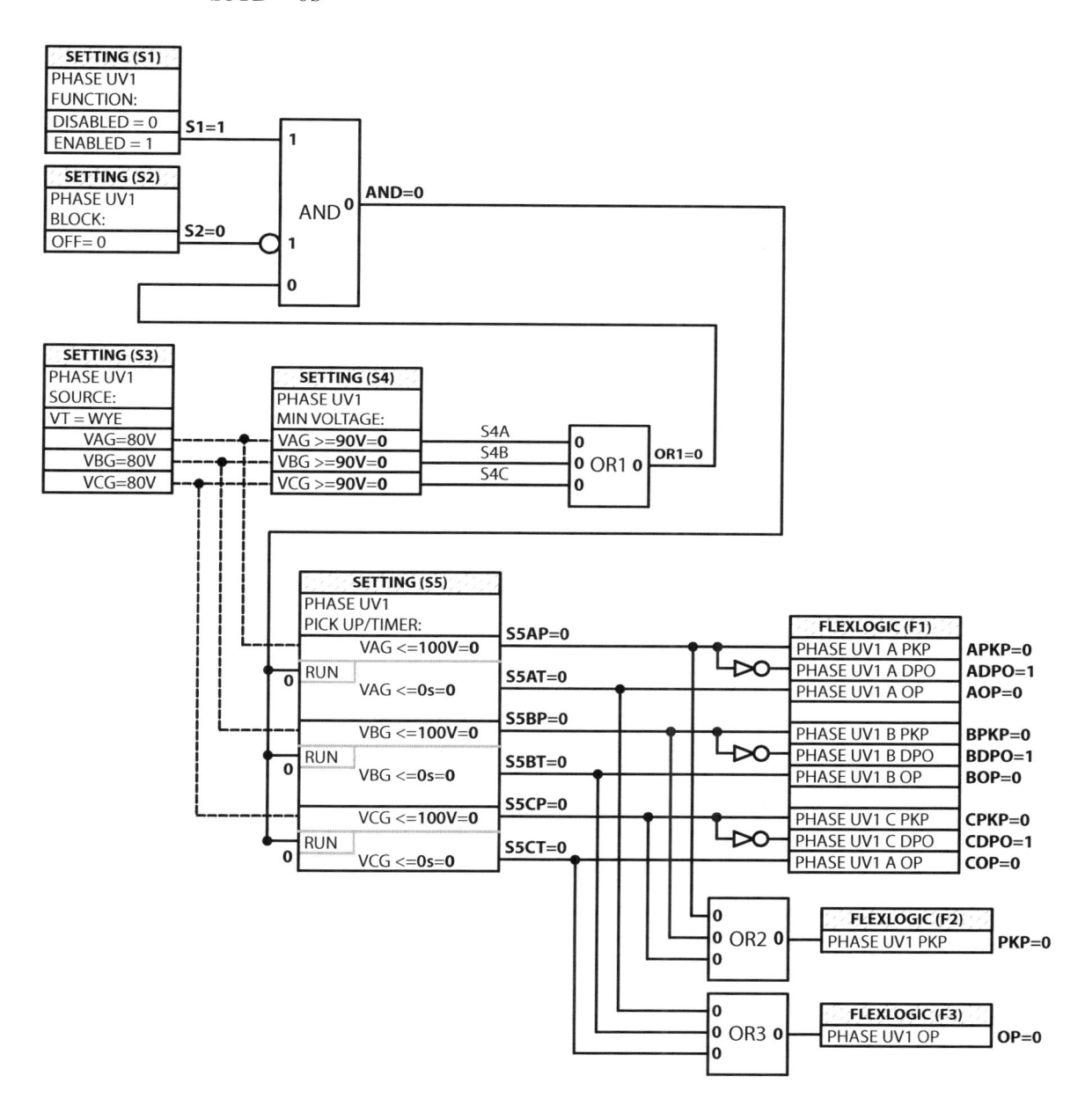

Figure 16-26: GE D-60 Undervoltage (27) Scenario #2 Logic

4. Individual Element Schemes

Individual element schemes are found in simpler, more straightforward digital relays where everything related to a protective element is located in one location or screen. Examples of these relays are the GE SR series and Beckwith Electric Relays. The element settings, output functions, and blocking elements are all set at once before moving on to the next element.

Some software programs will summarize the information for easy reference but most printouts can be more than ten pages long. It is a good idea to write down a summary of all element settings on one page for an easy reference to determine which elements interact with which inputs/outputs. The following figures are examples of individual element schemes:

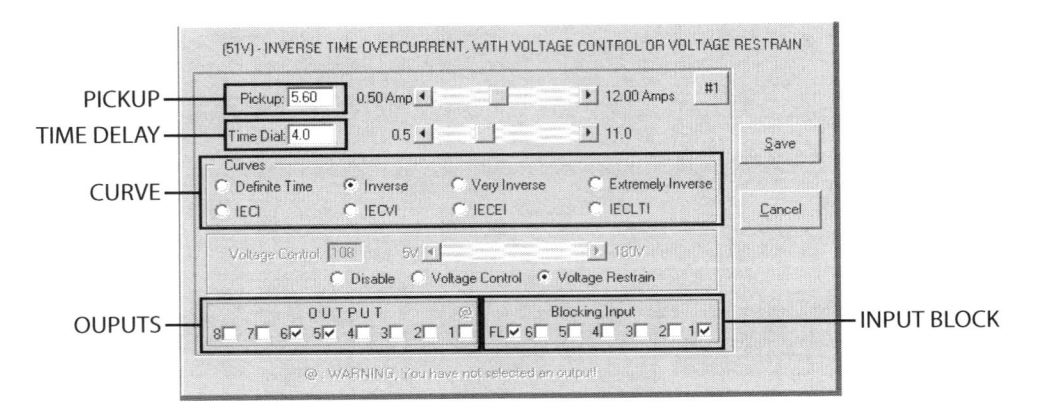

Figure 16-27: Example of Individual Element Scheme

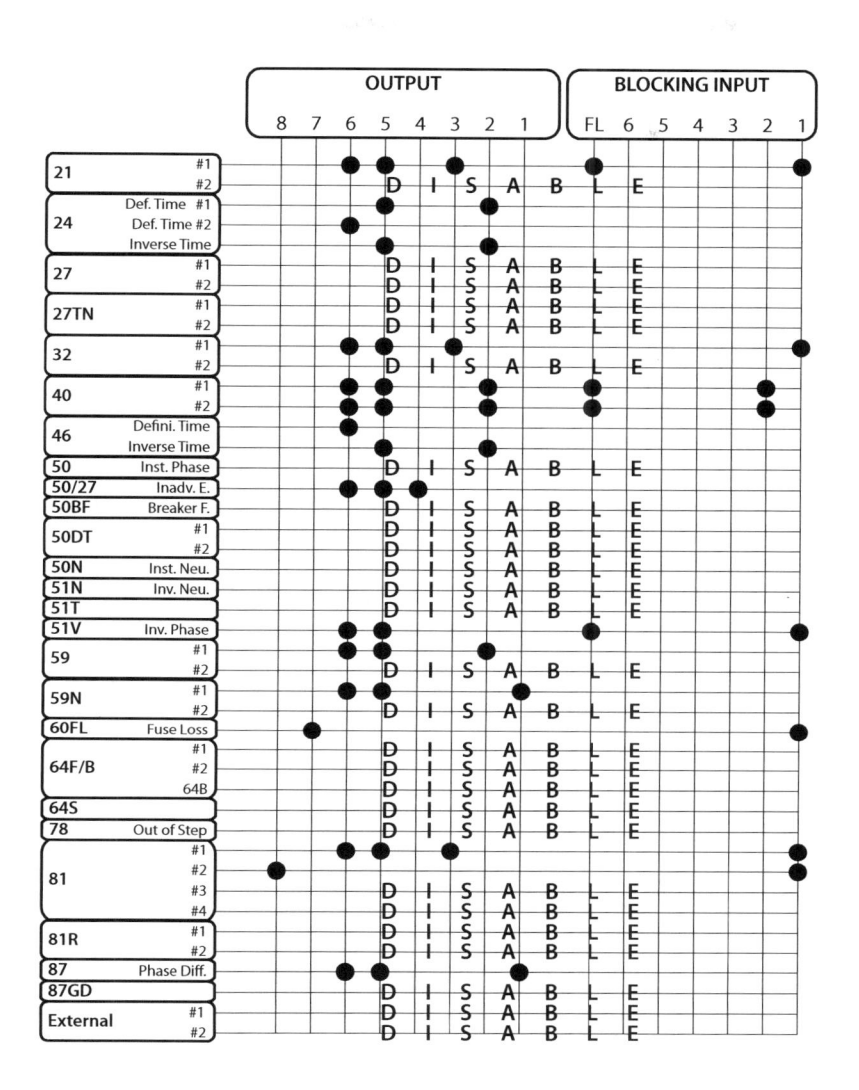

Figure 16-28: Summary Example for Individual Scheme

5. Binary Relays

Early digital relays, such as the Alstom K series, use a binary logic scheme to assign outputs and input functions. Actually, all digital relays use this logic scheme in some fashion but have an additional user interface to add labels and help the user. All entries in the binary scheme are ON (1) or OFF (0) and assigning an ON (1) in any cell turns that function on. The output relays or logic columns are typically arranged in multiples of eight and start with the number 0. Columns are arranged in order from right to left. An example of a binary scheme is shown in Figure 16-29. The functions to be assigned are usually arranged in columns with elements in rows. The element's function is assigned to the corresponding output in the "Relay Masks" row by entering ON (1) where the appropriate column and row intersect. In the following example,

- Output 0 will operate when elements to>>>, t>>>, or tV< operates
- Output 2 will operate when elements to>, to>>, tA>, tB>, tC>, or t>>operate.

RELAY MASKS	F	E	D	C	B	A	9	8	7	6	5	4	3	2	1	0
Io> Fwd	0	0	0	0	0	0	0	0	0	0	0	0	0	0	0	0
Io> Rev	0	0	0	0	0	0	0	0	0	0	0	0	0	0	0	0
to>	0	0	0	0	0	0	0	0	0	0	0	0	0	1	0	0
to>>	0	0	0	0	0	0	0	0	0	0	0	0	0	1	0	0
to>>>	0	0	0	0	0	0	0	0	0	0	0	0	0	0	0	1
I> Fwd	0	0	0	0	0	0	0	0	0	0	0	0	0	0	0	0
I> Rev	0	0	0	0	0	0	0	0	0	0	0	0	0	0	0	0
tA>	0	0	0	0	0	0	0	0	0	0	0	0	0	1	0	0
tB>	0	0	0	0	0	0	0	0	0	0	0	0	0	1	0	0
tC>	0	0	0	0	0	0	0	0	0	0	0	0	0	1	0	0
t>>	0	0	0	0	0	0	0	0	0	0	0	0	0	1	0	0
t>>>	0	0	0	0	0	0	0	0	0	0	0	0	0	0	0	1
CB Trip	0	0	0	0	0	0	0	0	0	0	0	0	0	0	0	0
CB Close	0	0	0	0	0	0	0	0	0	0	0	0	0	0	0	0
CB Fail	0	0	0	0	0	0	0	0	0	0	0	0	0	0	0	0
Aux 1	0	0	0	0	0	0	0	0	0	0	0	0	0	0	0	0
Aux 2	0	0	0	0	0	0	0	0	0	0	0	0	0	0	0	0
Aux 3	0	0	0	0	0	0	0	0	0	0	0	0	0	0	0	0
tV<	0	0	0	0	0	0	0	0	0	0	0	0	0	0	0	1
Level 1	0	0	0	0	0	0	0	0	0	0	0	0	0	0	0	0
Level 2	0	0	0	0	0	0	0	0	0	0	0	0	0	0	0	0
Level 3	0	0	0	0	0	0	0	0	0	0	0	0	0	0	0	0
thAlarm	0	0	0	0	0	0	0	0	0	0	0	0	0	0	0	0
thTrip	0	0	0	0	0	0	0	0	0	0	0	0	0	0	0	0
CB Alarm	0	0	0	0	0	0	0	0	0	0	0	0	0	0	0	0

Figure 16-29: Example of Simple Binary Logic Scheme

DISPLAY	STATUS	DESCRIPTION
to>	PWP	First stage time delayed earth fault output
to>>	PWP	Second stage time delayed output
to>>>	PWP	Third stage time delayed earth fault output
tA>	PWP	First stage time delayed overcurrent output for phase A
tB>	PWP	First stage time delayed overcurrent output for phase B
tC>	PWP	First stage time delayed overcurrent output for phase C
t>>	PWP	Second stage time delayed overcurrent output
t>>>	PWP	Third stage time delayed overcurrent output
tV<	PWP	Undervoltage time delayed output

Figure 16-30: Simple Binary Logic Relay Labels

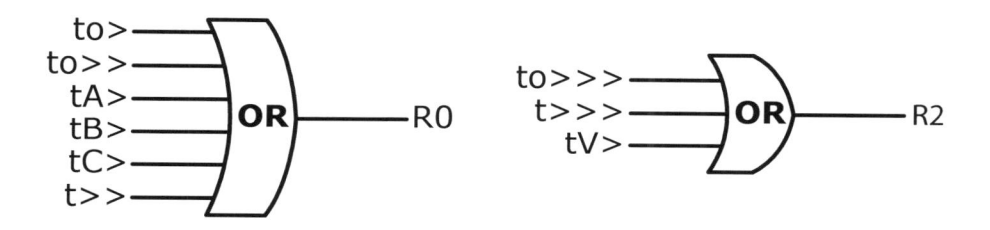

Figure 16-31: Simple Binary Logic Diagram

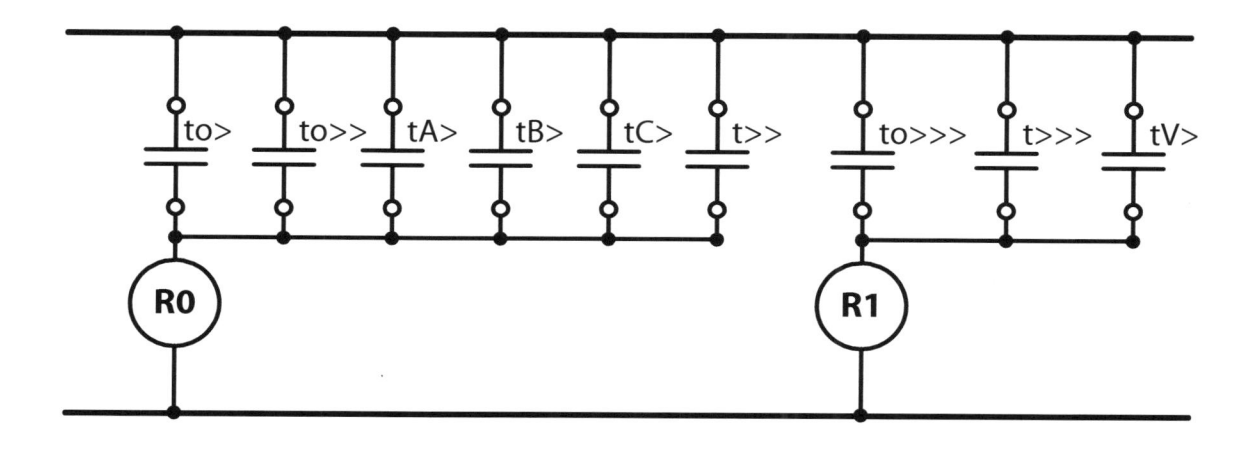

Figure 16-32: Simple Binary Logic Schematic

The following figures show how complex binary relays can perform full logic schemes. It is important to determine what logic function is performed in the rows and columns. In the following example for an Alstom LGPG Generator Protection Relay, each row in the "Logic Scheme" section represents one AND gate. Each output selected in the corresponding "Output Logic" row will be energized if the AND gate in the "Logic Scheme" section is ON (1).

Logic Scheme

	87G	51V	32R	32L	81O	81U-1	81U-2	27	59	46>	46>>	51N>	51N>>	59N-1	59N-2	67N	40	60	-60	60	I6	-I6	I7	-I7
L0	0	0	0	0	0	0	0	1	0	0	0	0	0	0	0	0	0	0	1	0	1	0	0	0
L1	0	0	1	0	0	0	0	0	0	0	0	0	0	0	0	0	0	0	1	0	1	0	0	0
L2	0	0	0	0	0	0	0	0	0	0	0	0	0	0	0	0	1	0	1	0	0	0	0	0
L3	0	0	0	0	0	0	0	0	0	0	0	1	0	0	0	0	0	0	0	0	0	0	0	0
L4	0	1	0	0	0	0	0	0	0	0	0	0	0	0	0	0	0	0	1	0	0	0	0	0
L5	0	0	0	0	0	0	0	1	0	0	0	0	0	0	0	0	0	0	0	0	0	0	0	0
L6	0	0	0	0	0	0	0	0	0	0	0	0	0	0	1	0	0	0	0	0	0	0	0	0
L7	0	0	0	0	0	0	0	0	0	0	0	0	0	0	0	0	1	0	0	0	0	0	0	0
L8	0	0	0	0	0	1	0	0	0	0	0	0	0	0	0	0	0	0	1	0	1	0	0	0
L9	0	0	0	0	0	0	1	0	0	0	0	0	0	0	0	0	0	0	1	0	1	0	0	0
L10	0	0	0	0	1	0	0	0	0	0	0	0	0	0	0	0	0	0	0	0	1	0	0	0
L11	1	0	0	0	0	0	0	0	0	0	0	0	0	0	0	0	0	0	0	0	0	0	0	0

AND ACROSS

Output Logic

	R15	R14	R13	R12	R11	R10	R9	R8	R7	R6	R5	R4	R3	R2	R1
L0	0	0	0	0	0	0	1	0	0	0	0	0	0	0	0
L1	0	0	0	0	0	0	0	1	0	0	0	0	0	0	1
L2	0	0	0	0	0	0	0	1	0	0	0	0	0	0	1
L3	0	0	0	0	0	0	0	1	0	0	0	0	0	0	1
L4	0	0	0	0	0	0	0	0	0	0	0	0	0	1	0
L5	0	0	0	0	0	0	0	0	0	0	0	0	0	1	0
L6	0	0	0	0	0	0	0	0	0	0	0	1	0	0	0
L7	0	0	0	0	0	0	0	1	0	0	0	0	0	0	0
L8	0	0	0	0	0	0	1	0	0	0	0	0	0	0	0
L9	0	0	0	0	0	0	1	0	0	0	0	0	0	0	0
L10	0	0	0	0	0	0	1	0	0	0	0	0	0	0	0
L11	0	0	0	0	0	0	0	0	0	0	0	0	0	1	0

OR DOWN

Figure 16-33: Example of Complex Binary Scheme

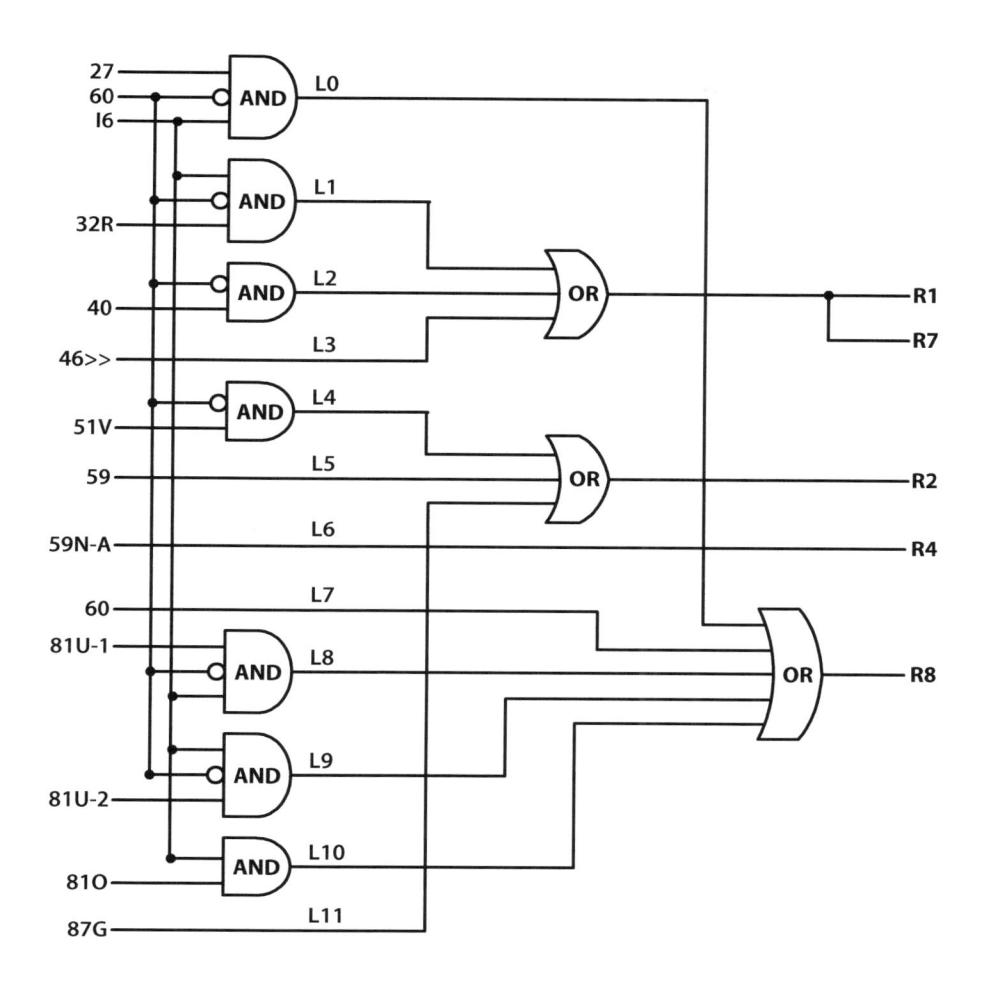

Figure 16-34: Complex Binary Example Logic Diagram

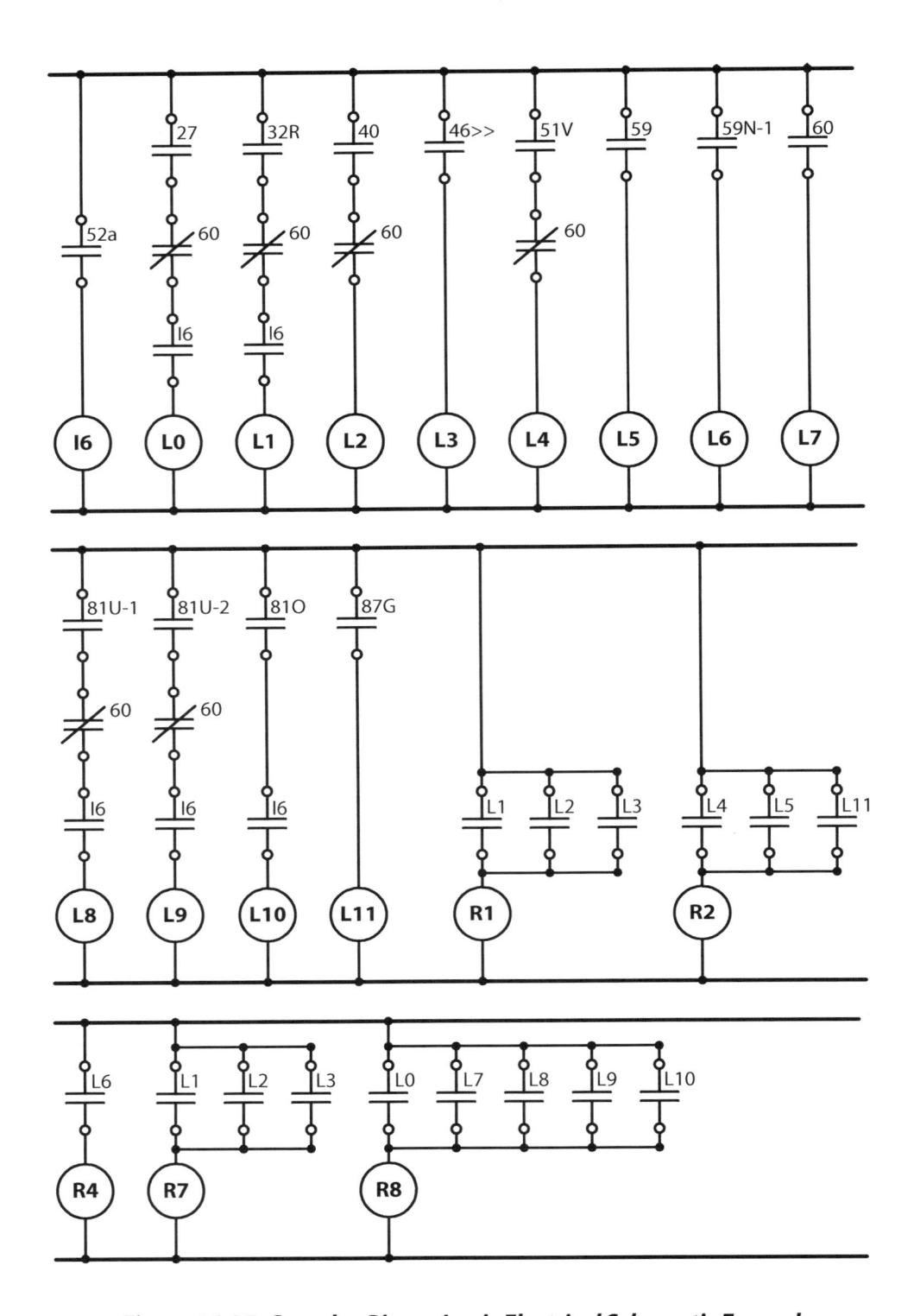

Figure 16-35: Complex Binary Logic Electrical Schematic Example

6. Arithmetic (Math) Scheme

Mathematical logic schemes can be the most flexible and most confusing methods for relay logic. Unfortunately, there is an almost inherent fear of algebra in modern society which hinders the use of this system. In math logic schemes, the relay logic is written as a mathematical equation using mathematical symbols and relay labels where each mathematical symbol represents a logic function. The chart in Figure 16-36 describes what logic function each symbol represents.

MATH SYMBOL	LOGIC FUNCTION	DESCRIPTION
/	Rising Edge Detect	Used as a pulse signal from an element. Only operates momentarily if the element is off and turns on. Typically used in event recorders.
\	Falling Edge Detect	Used as a pulse signal from an element. Only operates momentarily if the element is on and turns off. Typically used in event recorders.
!	NOT	Placing this symbol in front of an element reverses its state. !ON becomes OFF
( )	Parenthesis	Used to group parts of the equation to simplify data entry
*	AND	Both elements on either side of the * symbol must be enabled to cause the outputs to operate
+	OR	Operation of either element will cause the output to operate

Figure 16-36: Math Scheme Symbol Definitions in Order

The first step in understanding logic schemes is to understand that all equations must be analyzed in the correct order. Figure 16-36 displays all available logic functions in the order that they are processed. All the examples in this section will use the following labels:

ROW	RELAY WORD BITS							
0	EN	TRP	TIME	COMM	SOTF	RCRS	RCLO	51
1	A	B	C	G	ZONE1	ZONE2	ZONE3	ZONE4
2	M1P	M1PT	Z1G	Z1GT	M2P	M2PT	Z2G	Z2GT
3	Z1T	Z1T	50P1	67P1	67P1T	50G1	67G1	67G1T
4	51G	51GT	51GR	LOP	ILOP	ZLOAD	ZLOUT	ZLIN
...	...	...	...	...	...	...	...	...
23	*	*	IN106	IN105	IN104	IN103	IN102	IN101
24	ALARM	OUT107	OUT106	OUT105	OUT104	OUT103	OUT102	OUT101
...	...	...	...	...	...	...	...	...
42	27A	27B	27C	59A	59B	59C	3P27	3P59

ROW	BIT	DEFINITION	PRIMARY APPLICATION
2	M1P	Zone 1 phase distance, instantaneous	Tripping, Control
	M1PT	Zone 1 phase distance, time delayed	
	Z1G	Zone 1 mho and/or quad, distance, instantaneous	
	Z1GT	Zone 1 ground distance, time delayed	
	M2P	Zone 2 phase distance, instantaneous	
	M2PT	Zone 2 phase distance, time delayed	
	Z2G	Zone 2 mho and/or quad, distance, instantaneous	
	Z2GT	Zone 2 ground distance, time delayed	

ROW	BIT	DEFINITION	PRIMARY APPLICATION
3	Z1T	Zone 1 phase and/or ground distance, time delayed	
	Z2T	Zone 2 phase and/or ground distance, time delayed	
	50P1	Level 1 phase instantaneous overcurrent element (A, B, C) above pickup setting 50P1P	
	67P1	Level 1 torque controlled phase instantaneous overcurrent element (derived from 50P1)	
	67P1T	Level 1 phase definite-time overcurrent element 67P1T timed out (derived from 67P1)	
	50G1	Level 1 residual ground instantaneous overcurrent element (residual ground current above pickup setting 50G1P)	
	67G1	Level 1 torque controlled residual ground instantaneous overcurrent element (derived from 50G1)	
	67G1T	Level 1 residual ground definite-time overcurrent element 67P1T timed out (derived from 67P1)	

ROW	BIT	DEFINITION	PRIMARY APPLICATION
4	Z1T	Zone 1 phase and/or ground distance, time delayed	
	Z2T	Zone 2 phase and/or ground distance, time delayed	
	50P1	Level 1 phase instantaneous overcurrent element (A, B, C) above pickup setting 50P1P	
	67P1	Level 1 torque controlled phase instantaneous overcurrent element (derived from 50P1)	
	67P1T	Level 1 phase definite-time overcurrent element 67P1T timed out (derived from 67P1)	
	50G1	Level 1 residual ground instantaneous overcurrent element (residual ground current above pickup setting 50G1P)	
	67G1	Level 1 torque controlled residual ground instantaneous overcurrent element (derived from 50G1)	
	67G1T	Level 1 residual ground definite-time overcurrent element 67P1T timed out (derived from 67P1)	

ROW	BIT	DEFINITION	PRIMARY APPLICATION
23	*		
	*		
	IN106	Optoisolated input IN106 asserted	Relay input
	IN105	Optoisolated input IN105 asserted	status,
	IN104	Optoisolated input IN104 asserted	Control via
	IN103	Optoisolated input IN103 asserted	Optoisolated
	IN102	Optoisolated input IN102 asserted	inputs
	IN101	Optoisolated input IN101 asserted	

ROW	BIT	DEFINITION	PRIMARY APPLICATION
24	*		
	OUT107	Output contact OUT107 asserted	Relay output
	OUT106	Output contact OUT106 asserted	status,
	OUT105	Output contact OUT105 asserted	Control
	OUT104	Output contact OUT104 asserted	
	OUT103	Output contact OUT103 asserted	
	OUT102	Output contact OUT102 asserted	
	OUT101	Output contact OUT101 asserted	

ROW	BIT	DEFINITION	PRIMARY APPLICATION
42	27A	A-phase instantaneous undervoltage element (A-phase voltage below pickup setting 27P)	Control
	27B	B-phase instantaneous undervoltage element (B-phase voltage below pickup setting 27P)	
	27C	C-phase instantaneous undervoltage element (C-phase voltage below pickup setting 27P)	
	59A	A-phase instantaneous overvoltage element (A-phase voltage above pickup setting 59P)	
	59B	B-phase instantaneous overvoltage element (B-phase voltage above pickup setting 59P)	
	59C	C-phase instantaneous overvoltage element (C-phase voltage above pickup setting 59P)	
	3P27	27A * 27B * 27C	
	3P59	59A * 59B * 59C	

Figure 16-37: SEL Relay Label Definitions

A) Example #1 Parameters

In this example, Input 101 (IN101) is connected to a 52b circuit breaker contact. Output 101 (OUT101) is connected to the undervoltage alarm input. Output 101 (OUT101) should close when the circuit breaker is closed and an undervoltage (27) condition exits.

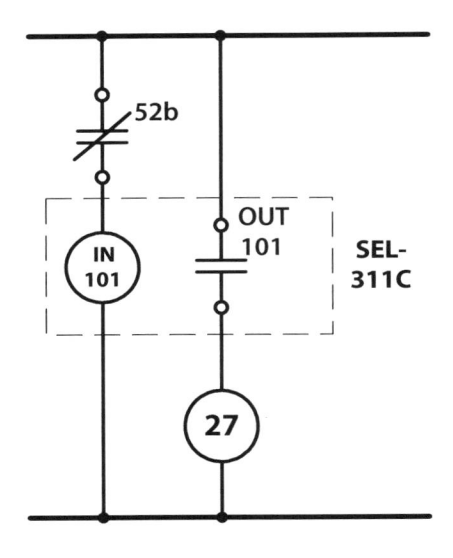

Figure 16-38: Relay Input/Output Schematic

The first step in designing the logic equation is to create a logic diagram to make sure we understand the application. Figure 16-39 starts with an electric schematic and logic diagram of the output requirements.

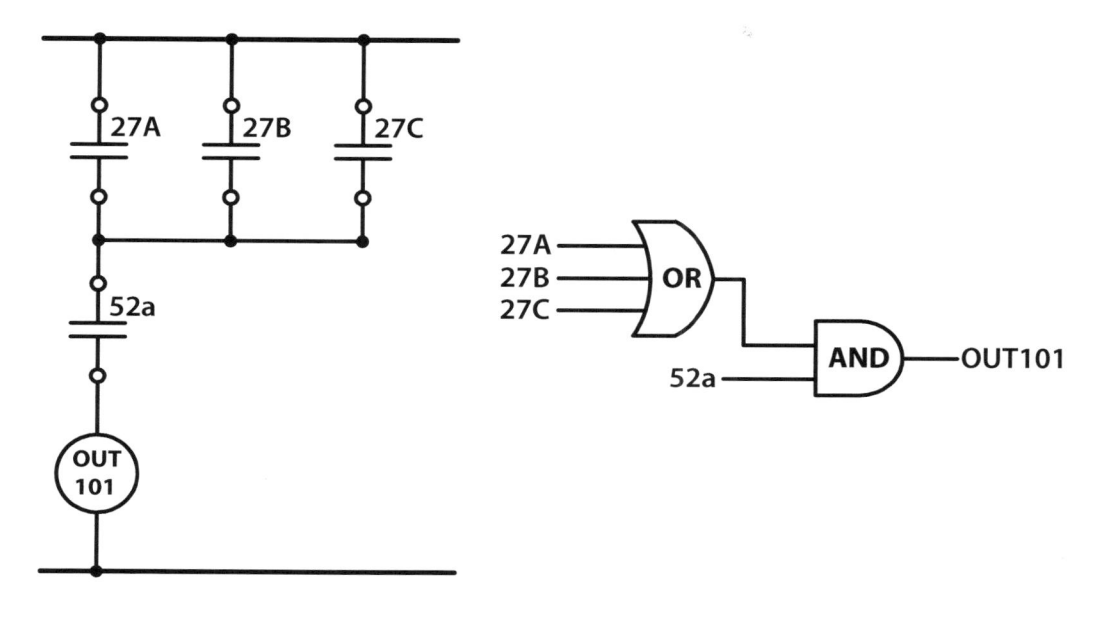

Figure 16-39: Example #1 Initial Logic

By focusing on the simpler input to the AND gate (52a), a problem immediately presents itself because there is no 52a contact connected to IN101 as shown in Figure 16-38. The 52a and 52b contacts are always opposite states, so we can still use the 52b contact by reversing its state with a NOT (!) function.

Therefore, the first part of our equation is "!IN101," and our logic changes as per the next section.

B) NOT (!)

The NOT (1) function changes any value to its opposite and is the first processed function in SEL relays. An ON (1) becomes OFF (0) and visa versa. The symbol is applied by placing an exclamation point (!) in front of the element you wish to NOT.

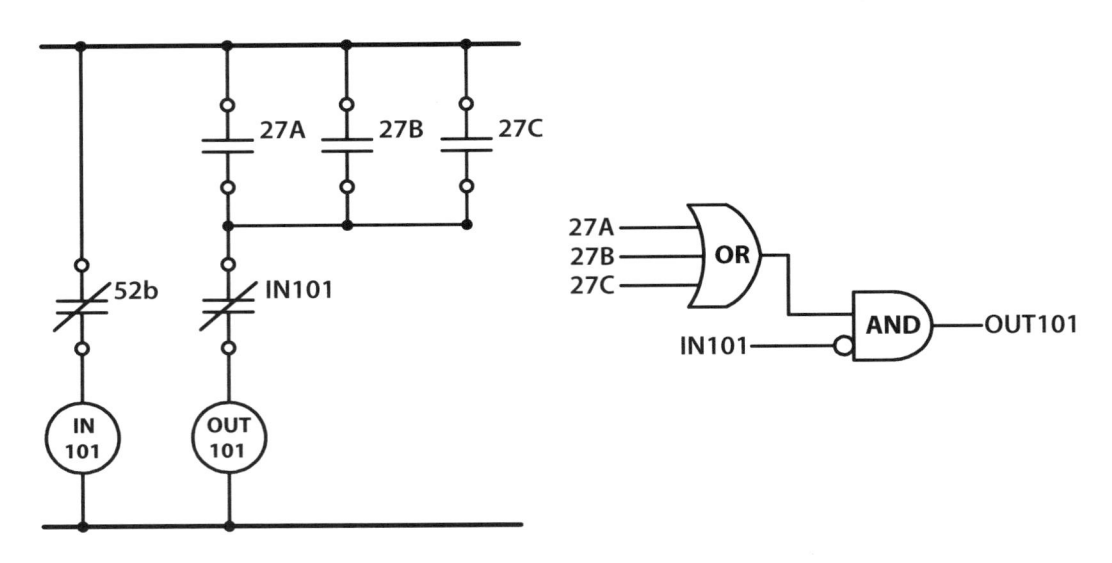

Figure 16-40: Arithmetic NOT Example Final Logic

C) OR (+)

The OR (+) arithmetic function is the most common function used and causes a lot of grief when trying to understand logic equations. Although your brain may automatically think "and" when it sees the "+" symbol, get in the habit of saying "OR" out loud when reading logic equations. For example, if you read a normal algebraic equation "x + y + z" it might sound like "x and y and z." When using math logic, the equation should sound like "x **or** y **or** z." Reading logic equations out loud may make you feel silly, but it is very helpful when working through equations. When developing equations with OR (+) gates, count the number of inputs and subtract one to determine the number of OR (+) symbols required. For example an OR (+) gate with two inputs would look like "x + y."

We need an OR (+) gate with three inputs to continue with our example. So our initial equation will be "x + y + z" to represent an OR (+) gate with three inputs. Substituting 27A, 27B, and 27C for x, y, and z gives us the equation "27A + 27B + 27C" which would be read aloud "twenty-seven A **or** twenty-seven B **or** twenty-seven C."

Add notations above the labels to represent a moment in time when evaluating math logic equations. Using our example, let us imagine the following conditions exist to evaluate our logic equation "27A + 27B + 27C":

- 27P = 100V
- 27A = VA < 27P = 120V < 100V = **0**
- 27B = VA < 27P = 90V < 100V = **1**
- 27C = VA < 27P = 120V < 100V = **0**

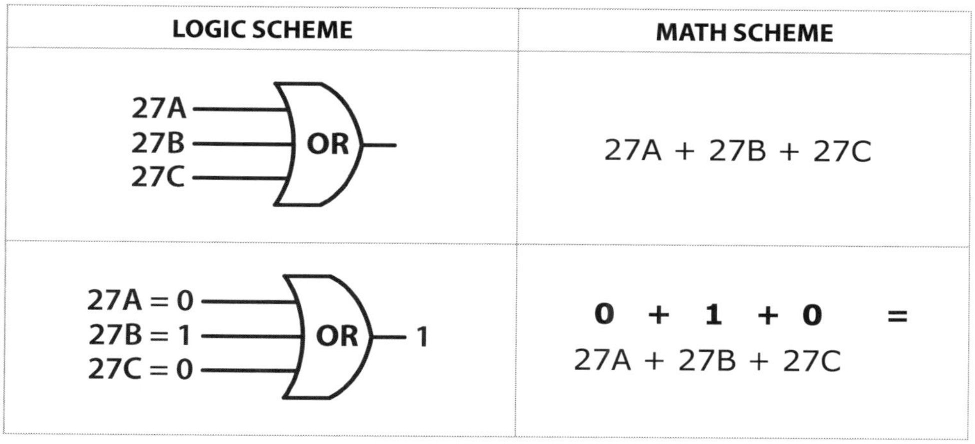

LOGIC SCHEME	MATH SCHEME
27A — 27B — OR — 27C	27A + 27B + 27C
27A = 0 — 27B = 1 — OR — 1 27C = 0	**0 + 1 + 0 =** 27A + 27B + 27C

Figure 16-41: Evaluating Math Logic Example

D) Parentheses ()

Parentheses are used to organize equations and make sure the SEL relay evaluates the equation in the order determined by the application. After the SEL relay reverses all NOT (!) gates, it looks to solve any equation in parentheses. Parentheses can also help understand complex equations by separating them into smaller groups to make them easier to understand. They can also group OR (+) functions together when used in conjunction with AND (*) functions to make sure equations are evaluated in the correct order. The relay always evaluates equations in the same order [NOT (!), Parentheses (), AND (*), and OR (+)] and sometimes equations will not work correctly unless parentheses are added. We have a combination of OR (+) and AND (*) gates in our example so we should add parentheses around our OR (+) logic equation. Therefore, "27A + 27B + 27C" becomes "(27A + 27B + 27C)."

E) AND (*)

AND (*) gates are ON (1) when both labels on each side of the AND (*) symbol are ON (1). When developing equations with AND (*) gates, count the number of inputs and subtract one to determine how many AND (*) symbols your need. Start building your equation by adding parentheses to all AND (*) gates to help keep track of inputs. For example "(x) * (y) * (z)"

In our example, we use an AND (*) gate with two inputs so our equation will use one AND (*) gate symbol. Our equation starts as "(x) * (y)." First we add the OR (+) gate equation to our example for "(27A +27B + 27C) * (y)." Now we add the breaker status to finish our equation "(27A +27B + 27C) * (!IN101)." The "!IN101" part of the equation is pretty simple, so let's remove the parentheses to save a little bit of typing and simplify the equation to "(27A +27B + 27C) * !IN101."

Now let's make sure our equation works by creating the following situation, and see what happens if the parentheses were missed in the equation

F) Example #1, Scenario #1

- Circuit Breaker is open (52b is ON (1)) therefore IN101 = 1
- 27P = 100V
- All three voltages are Zero, therefore 27A = 1, 27B = 1, 27C = 1

CORRECT EQUATION				INCORRECT EQUATION			
(27A +	27B +	27C) *	!IN101	27A +	27B +	27C *	!IN101
(1 +	1 +	1) *	!1	1 +	1 +	1 *	!1
NOT (!) gates are evaluated first				NOT (!) gates are evaluated first			
(1 +	1 +	1) *	0	1 +	1 +	1 *	0
Elements in parentheses next							
		1 *	0				
AND (*) Gates are evaluated next				AND (*) Gates are evaluated next			
			0	1 +	1 +	0	
				OR (+) Gates are evaluated Last			
				1			
Output 101 will not operate				**Output 101 will operate incorrectly**			

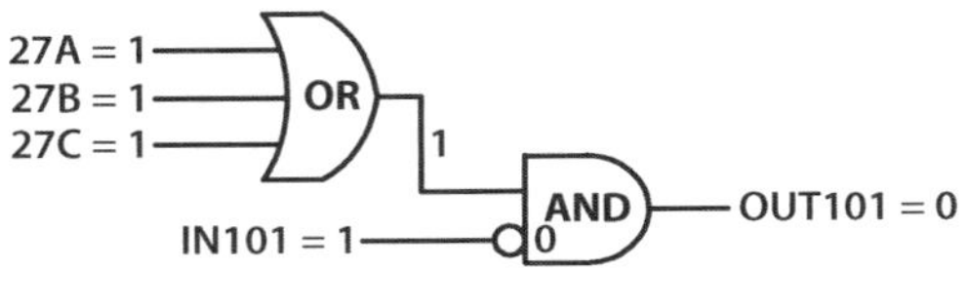

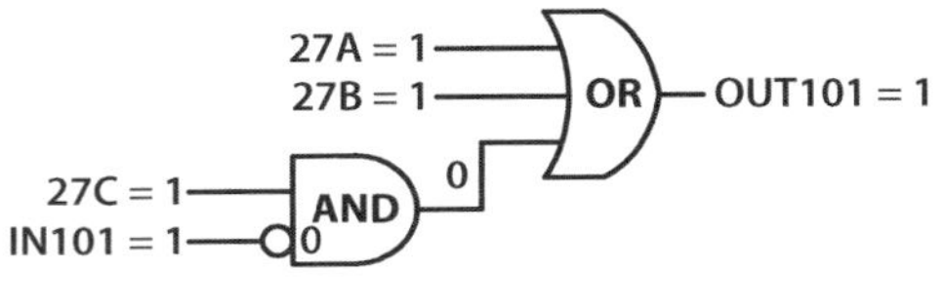

Figure 16-42: Evaluation of Example #1, Scenario #1 Logic

Let's try another scenario:

G) Example #1, Scenario #2

- Circuit Breaker is Closed (52b is OFF (0)) therefore IN101 = 0
- 27P = 100V
- All three voltages are 120V therefore 27A = 0, 27B = 0, 27C = 0

CORRECT EQUATION				INCORRECT EQUATION			
(27A +	27B +	27C) *	!IN101	27A +	27B +	27C *	!IN101
(0 +	0 +	0) *	!0	0 +	0 +	0 *	!0
NOT (!) gates are evaluated first				NOT (!) gates are evaluated first			
(0 +	0 +	0) *	1	0 +	0 +	0 *	1
Elements in parentheses next							
		0 *	1				
AND (*) Gates are evaluated next				AND (*) Gates are evaluated next			
			0	0 +	0 +	0	
				OR (+) Gates are evaluated Last			
				0			
Output 101 will not operate				**Output 101 will not operate**			

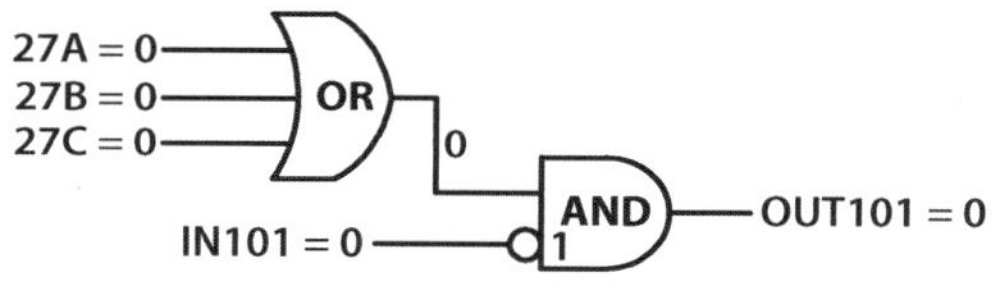

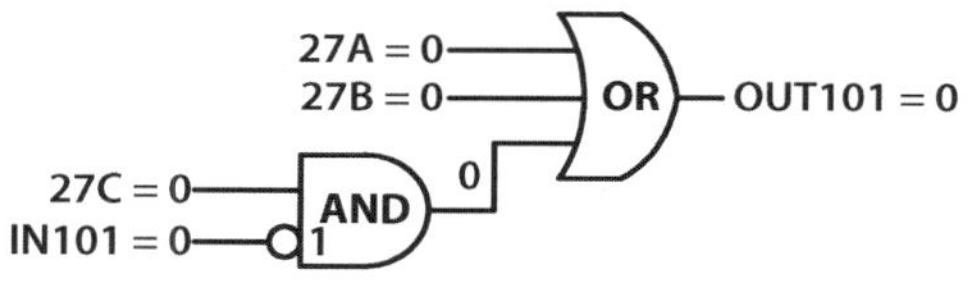

Figure 16-43: Evaluation of Example #1, Scenario #2 Logic

Here is another scenario:

H) Example #1, Scenario #3

- Circuit Breaker is Closed (52b is OFF (0)) therefore IN101 = 0
- 27P = 100V
- All three voltages are 90V therefore 27A = 1, 27B = 1, 27C = 1

CORRECT EQUATION				INCORRECT EQUATION			
(27A +	27B +	27C) *	!IN101	27A +	27B +	27C *	!IN101
(1 +	1 +	1) *	!0	1 +	1 +	1 *	!0
NOT (!) gates are evaluated first				NOT (!) gates are evaluated first			
(1 +	1 +	1) *	1	1 +	1 +	1 *	1
Elements in parentheses next							
		1 *	1				
AND (*) Gates are evaluated next				AND (*) Gates are evaluated next			
			1	1 +	1 +	1	
				OR (+) Gates are evaluated Last			
				1			
Output 101 will operate				**Output 101 will operate**			

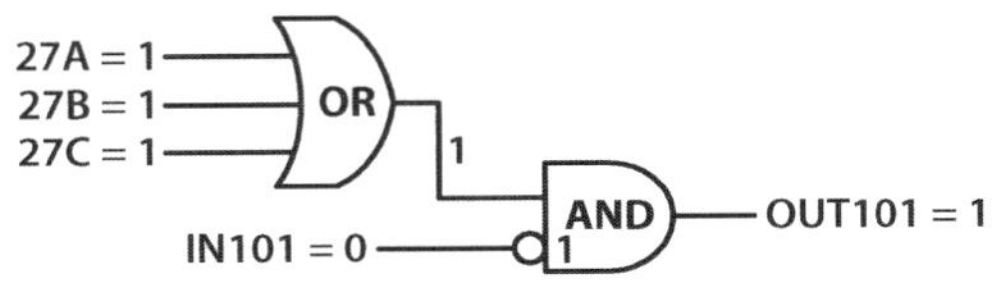

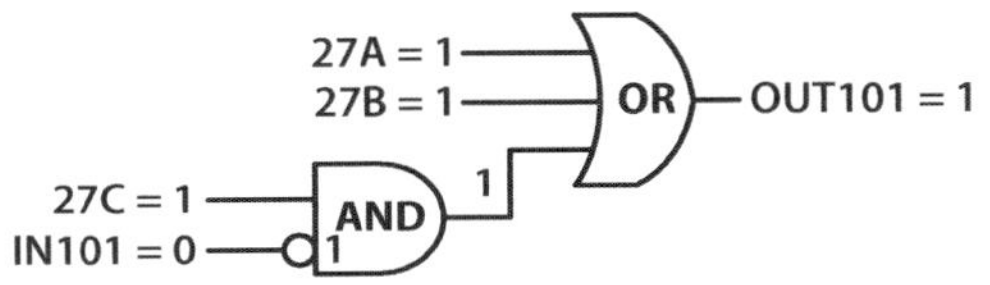

Figure 16-44: Evaluation of Example #1, Scenario #3 Logic

508

I) Example #2

Earlier in this chapter, we reviewed the internal logic for the undervoltage (27) element in SEL-311C and GE D-60 relays. The logic schemes were significantly different and we will create additional SEL logic and elements in this example to create an undervoltage scheme as close as possible to the GE D-60 scheme. First, a review of the GE D-60 Scheme:

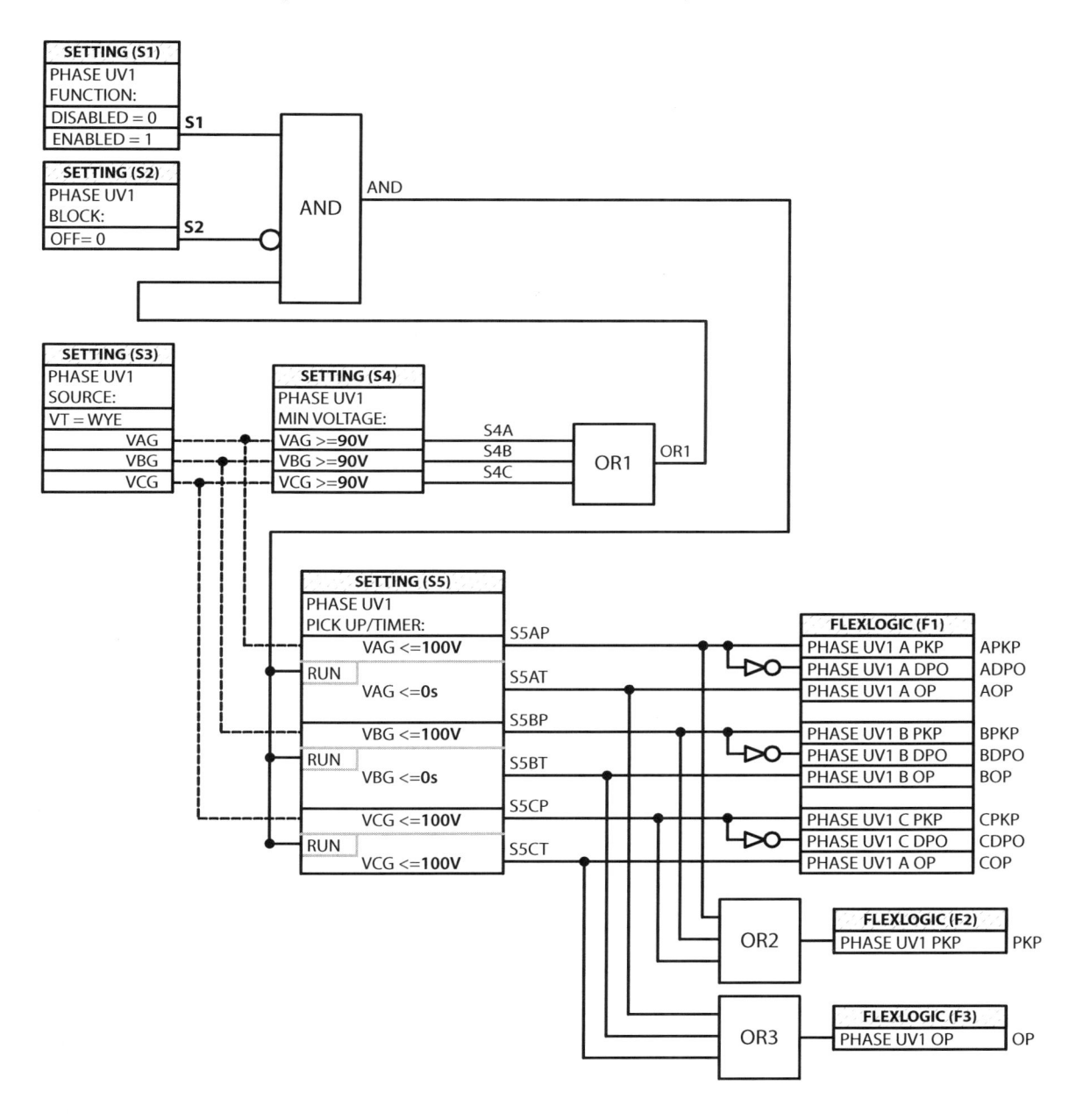

Figure 16-45: GE D-60 Simplified Undervoltage (27) Logic

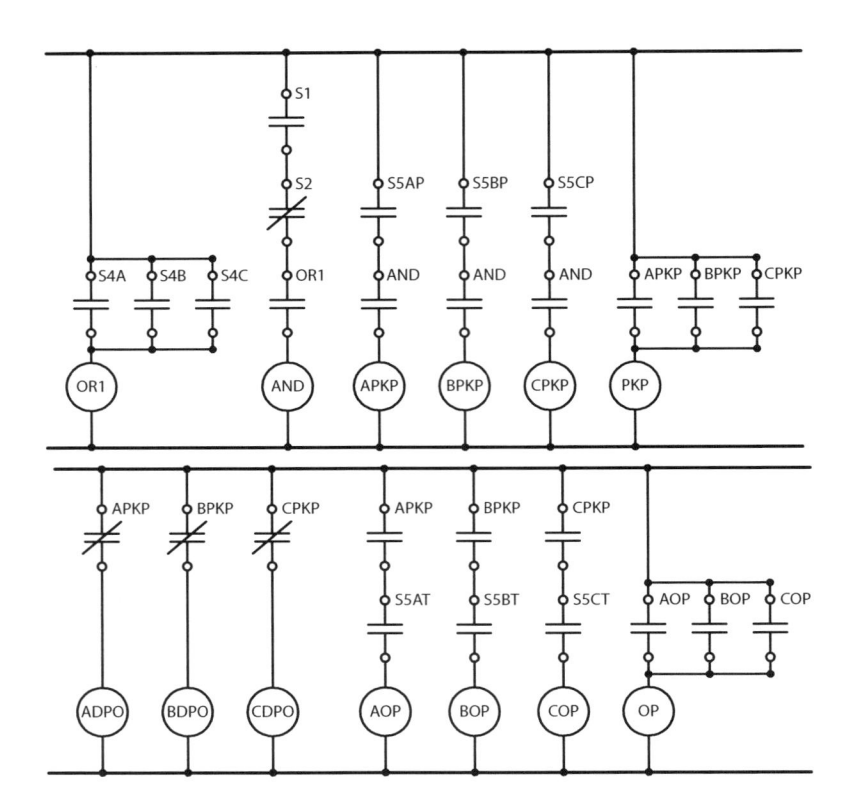

Figure 16-46: GE D-60 Undervoltage (27) Logic Electrical Schematic

Now a review of the SEL-311C undervoltage (27) logic:

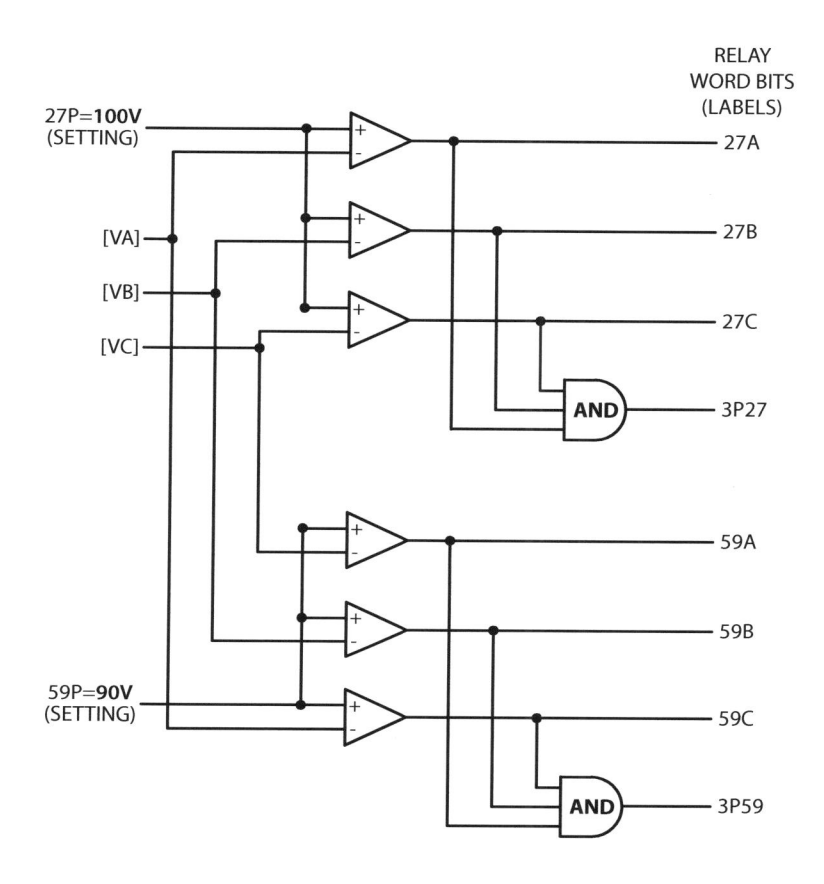

Figure 16-47: SEL-311C Undervoltage (27) and Overvoltage (59) Logic

VOLTAGE ELEMENT (RELAY WORD BITS)	OPERATING VOLTAGE	PICKUP SETTING/RANGE
27A	VA	27P
27B	VB	OFF, 0.00—150.00 V
27C	VC	Secondary
3P27	27A * 27B *27C	
27AB	VAB	27PP
27BC	VBC	OFF, 0.00—260.00 V
27CA	VCA	secondary
59A	VA	59P
59B	VB	OFF, 0.00—150.00 V
59C	VC	Secondary
3P59	59A * 59B *59C	

Figure 16-48: SEL-311C Voltage Elements Settings and Setting Ranges

ROW	RELAY WORD BITS							
0	EN	TRP	TIME	COMM	SOTF	RCRS	RCLO	51
1	A	B	C	G	ZONE1	ZONE2	ZONE3	ZONE4
…	…	…	…	….	…	…	…	…
11	SV1	SV2	SV3	SV4	SV1T	SV2T	SV3T	SV4T
…	…	…	…	….	…	…	…	…
42	27A	27B	27C	59A	59B	59C	3P27	3P59

Figure 16-49: SEL-311C Relay Word Bits

ROW	BIT	DEFINITION	PRIMARY APPLICATION
11	SV1	SELogic control equation variable timer input SV1 asserted	Control
	SV2	SELogic control equation variable timer input SV2 asserted	Control
	SV3	SELogic control equation variable timer input SV3 asserted	Control
	SV4	SELogic control equation variable timer input SV4 asserted	Control
	SV1T	SELogic control equation variable timer output SV1T asserted	Control
	SV2T	SELogic control equation variable timer output SV2T asserted	Control
	SV3T	SELogic control equation variable timer output SV3T asserted	Control
	SV4T	SELogic control equation variable timer output SV4T asserted	Control

ROW	BIT	DEFINITION	PRIMARY APPLICATION
42	27A	A-phase instantaneous undervoltage element (A-phase voltage below pickup setting 27P)	Control
	27B	B-phase instantaneous undervoltage element (B-phase voltage below pickup setting 27P)	Control
	27C	C-phase instantaneous undervoltage element (C-phase voltage below pickup setting 27P)	Control
	3P27	27A * 27B * 27C	Control
	59A	A-phase instantaneous overvoltage element (A-phase voltage above pickup setting 59P)	Control
	59B	B-phase instantaneous undervoltage element (B-phase voltage above pickup setting 59P)	Control
	59C	C-phase instantaneous undervoltage element (C-phase voltage above pickup setting 59P)	Control
	3P59	59A * 59B * 59C	Control

Figure 16-50: SEL-311C Relay Word Bit Definitions

The first step in developing this logic scheme is to create a complete logic diagram of the problem. We use the information above and steps below to create the logic diagram in Figure 16-51.

The SEL enabled/disabled function is built into its pickup settings by entering "OFF" to disable the setting and any valid pickup to enable the setting. Setting the 27P pickup to 100V is equivalent to an enabled input and it is not required in our logic.

- 27P = 100V

The SEL does not have a built-in undervoltage (27) blocking input like "S2" in the GE D-60. The SEL does have customizable virtual timers called "SELogic control equations variables/timers" that we can use in place of the setting. A virtual timer can be set to operate using any combination of available relay elements and can be a virtual relay by setting the time delays to zero. We will use timer "SV1" as the blocking function for our logic with the following settings. SV1 will always be "0" unless we define it.

- SV1 = 0
- SV1PU = 0s
- SV1DO = 0s

The SEL does not have a built-in undervoltage (27) minimum voltage setting like the GE D-60 but it does have an overvoltage function that we can use in its place with the following setting

- 59P = 90V

The SEL does not have built-in undervoltage (27) timers but does have up to 16 virtual timers that we can use. We will designate the following timers as undervoltage (27) timers

- SV2 = Phase A
- SV2PU = 10s
- SV2DO = 0s
- SV3 = Phase B
- SV3PU = 10s
- SV3DO = 0s
- SV4 = Phase C
- SV4PU = 10s
- SV4DO = 0s

Using these substitutions, we can start to apply the logic diagram in Figure 16-51.

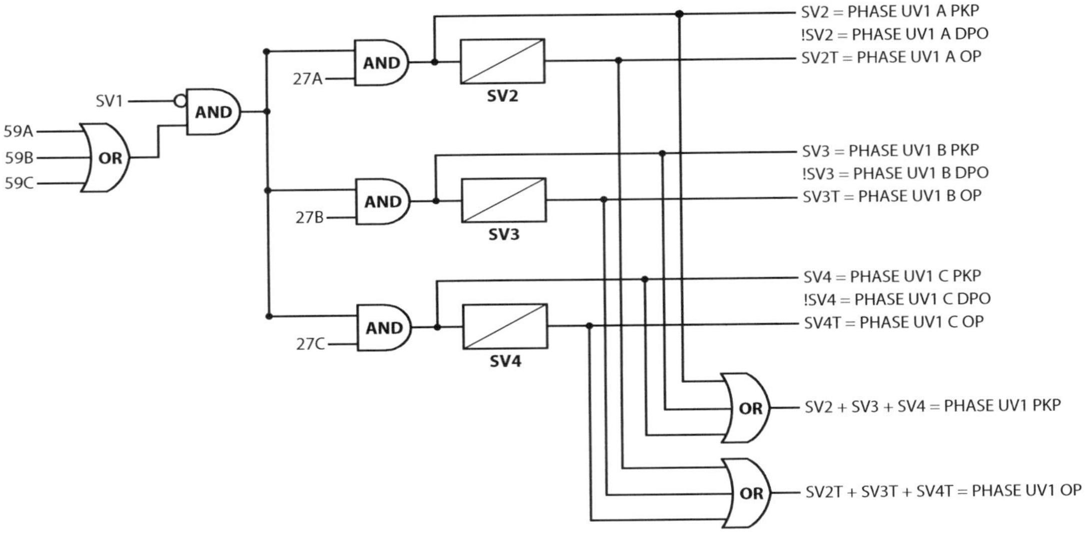

Figure 16-51: Example #2 Logic Diagram

The first logic element for our equations is the OR gate connected to the 59 elements. This is a 3-input OR gate so we will need 2 OR (+) symbols to create:

59A + 59B + 59C

The next element is the first AND gate, which has two inputs, so our equation starts with "(x) * (y)." Substituting the OR gate for x and the NOT SV1 for y and we have the following equation to represent the AND gate.

(59A + 59B + 59C) * !SV1

The input to timer SV2 is an AND gate with two inputs so we start with the equation "(x) * (y)."

SV2 = ((59A + 59B + 59C) * !SV1) * 27A

Our first problem has occurred. SEL relays will not allow you to enter parentheses inside another set of parentheses. This is called nesting and we must remove the inner parentheses without causing another problem as we saw in the previous example when the parentheses was intentionally dropped in an AND equation.

We have at least two possible solutions available to us. We can assign the first AND gate equation to another timer and edit our equations as shown below:

SV5 = (59A + 59B + 59C) * !SV1
SV2 = SV5 * 27A

Or, by taking advantage of the flexibility of this kind of logic scheme, we can re-draw our logic diagram so we don't need the additional relay.

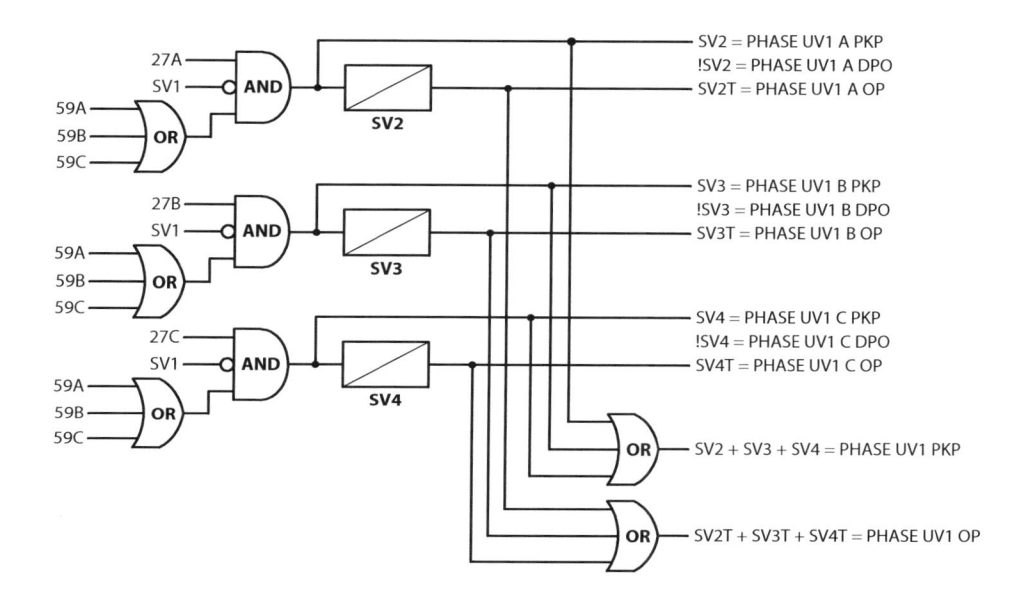

Figure 16-52: Example #2 Revised Logic Diagram

Using the revised logic drawing, we observe that all the OR gates are identical and have not changed so we can use our previous equation for all the input OR gates

59A + 59B + 59C

The first AND gate has three inputs so we need two AND (*) symbols like "(x) * (x) * (z)."

The first input is "27A," the second input is NOT SV1 (!SV1), and the third input is the OR gate we have previously defined as "59A + 59B + 59C" to give us the following formula:

SV2 = 27A * !SV1 * (59A + 59B + 59C)

The next AND gates are identical to the first AND gates except for the 27-element so the following formulas apply.

SV3 = 27B * !SV1 * (59A + 59B + 59C)
SV4 = 27C * !SV1 * (59A + 59B + 59C)

We now are ready to apply the following settings and check our logic using a couple of different scenarios.

• 27P	= 100V	• 59P = 90V
• SV1 = 0		• SV2 = 27A * !SV1 * (59A + 59B + 59C)
• SV1PU = 0s		• SV2PU = 10s
• SV1DO = 0s		• SV2DO = 0s
• SV3 = 27B * !SV1 * (59A + 59B + 59C)		• SV4 = 27C * !SV1 * (59A + 59B + 59C)
• SV3PU = 10s		• SV4PU = 10s
• SV3DO = 0s		• SV4DO = 0s

J) Example #2, Scenario #1

SEL MODIFIED UNDERVOLTAGE (27) LOGIC	GE UNDERVOLTAGE (27) ORIGINAL LOGIC
Settings:	Settings:
SV1 = 0	S1 = Enabled
	S2 = Off
59P = 90V	S3 = Wye
27P = 100V	S4 = 90V
SV2PU = 0s	S5PU = 100V
SV3PU = 0s	S5TD = 0s
SV4PU = 0s	
VAG = 115V	VAG = 115V
VBG = 98V	VBG = 98V
VCG = 115V	VCG = 115V

SOLVE FOR 27A						
27A =	VA <		27P			
27A =	115 <		100			
27A =	**0**					

SOLVE FOR PHASE UV1 MINIMUM VOLTAGE A			PHASE UV1 MINIMUM
59A =	VA >	59P	VOLTAGE A = 1
59A =	115 >	90	(From previous example)
59A =	**1**		

SOLVE FOR PHASE UV1 A PKP						PHASE UV1 A PKP = 0
SV2=	27A*	!SV1*	(59A+	59B+	59C)	(From previous example)
SV2=	0 *	!0 *	(1 +	1 +	1)	

SV2=	0 *	1 *	(1 +	1 +	1)	
SV2=	0 *	1 *	1			
SV2=	0 *	1				
SV2=	**0**					

SOLVE FOR PHASE UV1 A DPO				**PHASE UV1 A DPO = 1**	
!SV2 =	!SV2			(From previous example)	
!SV2 =	!0				
!SV2 =	**1**				

SOLVE FOR PHASE UV1 A OP					**PHASE UV1 A OP = 0**	
SV2T=	27A*	!SV1*	(59A+	59B+	59C)	(From previous example)
SV2T=	0 *	!0 *	(1 +	1 +	1)	
SV2T=	0 *	1 *	(1 +	1 +	1)	
SV2T=	0 *	1 *	1			
SV2T=	0 *	1				
SV2T=	**0**					

SOLVE FOR 27B			
27B =	VB <	27P	
27B =	98 <	100	
27B =	1		

SOLVE FOR PHASE UV1 MINIMUM VOLTAGE B			**PHASE UV1 MINIMUM**
59B =	VB >	59P	VOLTAGE B = 1
59B =	98 >	90	(From previous example)
59B =	1		

SOLVE FOR PHASE UV1 B PKP					**PHASE UV1 B PKP = 1**	
SV3=	27B*	!SV1*	(59A+	59B+	59C)	(From previous example)
SV3=	1 *	!0 *	(1 +	1 +	1)	
SV3=	1 *	1 *	(1 +	1 +	1)	
SV3=	1 *	1 *	1			
SV3=	**1**					

SOLVE FOR PHASE UV1 B DPO				**PHASE UV1 B DPO = 0**	
!SV3 =	!SV3			(From previous example)	
!SV3 =	!1				
!SV3 =	**0**				

SOLVE FOR PHASE UV1 B OP					**PHASE UV1 B OP = 1**	
SV3T=	27B*	!SV1*	(59A+	59B+	59C)	(From previous example)
SV3T=	1 *	!0 *	(1 +	1 +	1)	

SV3T=	1 *	1 *	(1 +	1 +	1)	
SV3T=	1 *	1 *	1			
SV3T=	**1**					

SOLVE FOR 27C

27C =	VC <	27P				
27C =	115 <	100				
27C =	**0**					

SOLVE FOR PHASE UV1 MINIMUM VOLTAGE C						PHASE UV1 MINIMUM
59C =	VC >	59P				VOLTAGE C = 1
59C =	115 >	90				(From previous example)
59C =	**1**					

SOLVE FOR PHASE UV1 C PKP						PHASE UV1 C PKP = 0
SV4=	27C*	!SV1*	(59A+	59B+	59C)	(From previous example)
SV4=	0 *	!0 *	(1 +	1 +	1)	
SV4=	0 *	1 *	(1 +	1 +	1)	
SV4=	0 *	1 *	1			
SV4=	**0**					

SOLVE FOR PHASE UV1 C DPO						PHASE UV1 C DPO = 1
!SV4 =	!SV4					(From previous example)
!SV4 =	!0					
!SV4 =	**1**					

SOLVE FOR PHASE UV1 C OP						PHASE UV1 C OP = 0
SV4T=	27C*	!SV1*	(59A+	59B+	59C)	(From previous example)
SV4T=	0 *	!0 *	(1 +	1 +	1)	
SV4T=	0 *	1 *	(1 +	1 +	1)	
SV4T=	0 *	1 *	1			
SV4T=	**0**					

SOLVE FOR PHASE UV1 PKP				PHASE UV1 PKP = 1
=	SV2 +	SV3 +	SV4	(From previous example)
=	0 +	1 +	0	
=	**1**			

SOLVE FOR PHASE UV1 OP				PHASE UV1 OP = 1
=	SV2T +	SV3T +	SV4T	(From previous example)
=	0 +	1 +	0	
=	**1**			

Figure 16-53: Evaluation of Example #2, Scenario #1 Logic

Based on this scenario, our logic appears to operate exactly like the GE D-60 logic. Success! Now let's try another scenario:

K) Example #2, Scenario #2

27P	= 100V		59P	= 90V
SV1 = 0			SV2	= 27A * !SV1 * (59A + 59B + 59C)
SV1PU = 0s			SV2PU	= 10s
SV1DO = 0s			SV2DO	= 0s
SV3 = 27B * !SV1 * (59A + 59B + 59C)			SV4	= 27C * !SV1 * (59A + 59B + 59C)
SV3PU = 10s			SV4PU	= 10s
SV3DO = 0s			SV4DO	= 0s

SEL MODIFIED UNDERVOLTAGE (27) LOGIC	GE UNDERVOLTAGE (27) ORIGINAL LOGIC
Settings:	Settings:
SV1 = 0	S1 = Enabled
	S2 = Off
59P = 90V	S3 = Wye
27P = 100V	S4 = 90V
SV2PU = 0s	S5PU = 100V
SV3PU = 0s	S5TD = 0s
SV4PU = 0s	
VAG = 80V	VAG = 80V
VBG = 80V	VBG = 80V
VCG = 80V	VCG = 80V

SOLVE FOR 27A							
27A =	VA <		27P				
27A =	80 <		100				
27A =	**1**						
SOLVE FOR PHASE UV1 MINIMUM VOLTAGE A				**PHASE UV1 MINIMUM**			
59A =	VA >		59P		VOLTAGE A = 0		
59A =	80 >		90		(From previous example)		
59A =	**0**						
SOLVE FOR PHASE UV1 A PKP					**PHASE UV1 A PKP = 0**		
SV2=	27A*	!SV1*	(59A+	59B+	59C)	(From previous example)	
SV2=	1 *	!0 *	(0 +	0 +	0)		

SV2=	1 *	1 *	(0 +	0 +	0)	
SV2=	1 *	1 *	0			
SV2=	1 *	0				
SV2=	**0**					

SOLVE FOR PHASE UV1 A DPO					PHASE UV1 A DPO = 1	
!SV2 =	!SV2				(From previous example)	
!SV2 =	!0					
!SV2 =	**1**					

SOLVE FOR PHASE UV1 A OP					PHASE UV1 A OP = 0	
SV2T=	27A*	!SV1*	(59A+	59B+	59C)	(From previous example)
SV2T=	1 *	!0 *	(0 +	0 +	0)	
SV2T=	1 *	1 *	(0 +	0 +	0)	
SV2T=	1 *	1 *	0			
SV2T=	1 *	0				
SV2T=	**0**					

SOLVE FOR 27B						
27B =	VB <		27P			
27B =	80 <		100			
27B =	**1**					

SOLVE FOR PHASE UV1 MINIMUM VOLTAGE B					PHASE UV1 MINIMUM	
59B =	VB >		59P		VOLTAGE B = 0	
59B =	80 >		90		(From previous example)	
59B =	**0**					

SOLVE FOR PHASE UV1 B PKP					PHASE UV1 B PKP = 0	
SV3=	27B*	!SV1*	(59A+	59B+	59C)	(From previous example)
SV3=	1 *	!0 *	(0 +	0 +	0)	
SV3=	1 *	1 *	(0 +	0 +	0)	
SV3=	1 *	1 *	0			
SV3=	**0**					

SOLVE FOR PHASE UV1 B DPO					PHASE UV1 B DPO = 1	
!SV3 =	!SV3				(From previous example)	
!SV3 =	!0					
!SV3 =	**1**					

SOLVE FOR PHASE UV1 B OP					PHASE UV1 B OP = 0	
SV3T=	27B*	!SV1*	(59A+	59B+	59C)	(From previous example)
SV3T=	1 *	!0 *	(0 +	0 +	0)	

SV3T=	1 *	1 *	(0 +	0 +	0)	
SV3T=	1 *	1 *	0			
SV3T=	**0**					

SOLVE FOR 27C			
27C =	VC <	27P	
27C =	80 <	100	
27C =	**1**		

SOLVE FOR PHASE UV1 MINIMUM VOLTAGE C			**PHASE UV1 MINIMUM**
59C =	VC >	59P	VOLTAGE C = 0
59C =	80 >	90	(From previous example)
59C =	**0**		

SOLVE FOR PHASE UV1 C PKP					**PHASE UV1 C PKP = 0**
SV4=	27C*	!SV1*	(59A+	59B+	59C)	(From previous example)
SV4=	1 *	!0 *	(0 +	0 +	0)
SV4=	1 *	1 *	(0 +	0 +	0)
SV4=	1 *	1 *	0		
SV4=	**0**				

SOLVE FOR PHASE UV1 C DPO		**PHASE UV1 C DPO = 1**
!SV4 =	!SV4	(From previous example)
!SV4 =	!0	
!SV4 =	**1**	

SOLVE FOR PHASE UV1 C OP					**PHASE UV1 C OP = 0**
SV4T=	27C*	!SV1*	(59A+	59B+	59C)	(From previous example)
SV4T=	1 *	!0 *	(0 +	0 +	0)
SV4T=	1 *	1 *	(0 +	0 +	0)
SV4T=	1 *	1 *	0		
SV4T=	**0**				

SOLVE FOR PHASE UV1 PKP				**PHASE UV1 PKP = 1**
=	SV2 +	SV3 +	SV4	(From previous example)
=	0 +	0 +	0	
=	**0**			

SOLVE FOR PHASE UV1 OP				**PHASE UV1 OP = 1**
=	SV2T +	SV3T +	SV4T	(From previous example)
=	0 +	0 +	0	
=	**0**			

Figure 16-54: Evaluation of Example #2, Scenario #1 Logic

Another success! We have successfully simulated the GE D-60 undervoltage (27) logic.

L) Latch Control Switches

SELogic also has pre-built logic to use in place of traditional latching relays. Like the traditional latching relay, the Latch Control switch has a set and reset function, but it has unlimited output contacts that can be used in any SELogic equation. The following figures provide additional information regarding SELogic Latch Control Switches:

ROW	RELAY WORD BITS							
0	EN	TRP	TIME	COMM	SOTF	RCRS	RCLO	51
1	A	B	C	G	ZONE1	ZONE2	ZONE3	ZONE4
…	…	…	…	…	…	…	…	…
9	LT1	LT2	LT3	LT4	LT5	LT6	LT7	LT8

ROW	BIT	DEFINITION	PRIMARY APPLICATION
9	LT1	Latch Bit 1 asserted	Latched Control -
	LT2	Latch Bit 2 asserted	Replacing Traditional
	LT3	Latch Bit 3 asserted	Latching Relays
	LT4	Latch Bit 4 asserted	
	LT5	Latch Bit 5 asserted	
	LT6	Latch Bit 6 asserted	
	LT7	Latch Bit 7 asserted	
	LT8	Latch Bit 8 asserted	

Figure 16-55: SEL-311C Relay Word Bits

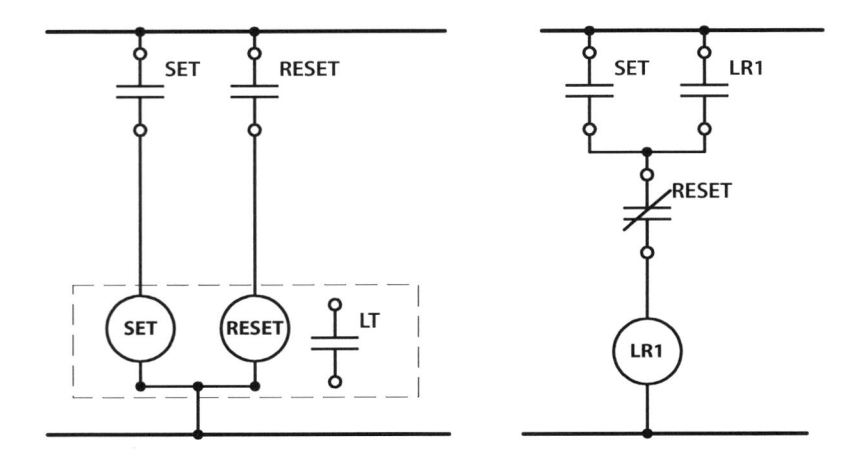

Figure 16-56: Electrical Schematic of Latching Relay

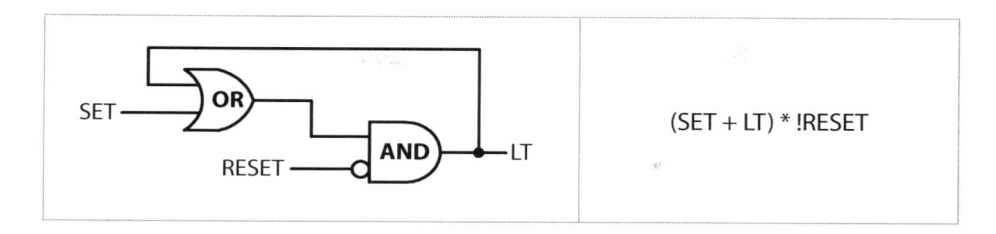

Figure 16-57: Logic Schemes for Latching Relays

M) Rising Edge (/) and Falling Edge (\)

Rising edge and falling edge logic detectors are primarily used to trigger fault history records, events, or oscillography records. The rising-edge detector turns ON (1) for one processing cycle when its input turns ON (1) and will not turn ON (1) again until its input has made the OFF (0)/ON (1) transition again as the Figure 16-58 displays. The space between dashed lines equals one processing cycle. If we did not use this kind of logic detector, one fault could cause hundreds of events and make event recording meaningless due to the sheer volume of data.

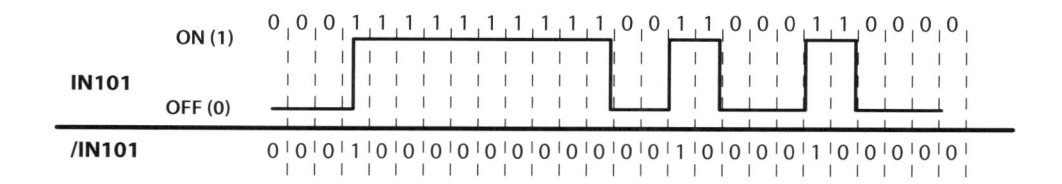

Figure 16-58: Illustration of Rising Edge Operation

Falling-edge detectors work exactly opposite of rising-edge detectors and turns ON (1) for one processing cycle when its input has turned OFF (0) and it will not turn ON (1) again until another ON (1)/OFF (0) transition as Figure 16-59 displays. The space between dashed lines equals one processing cycle.

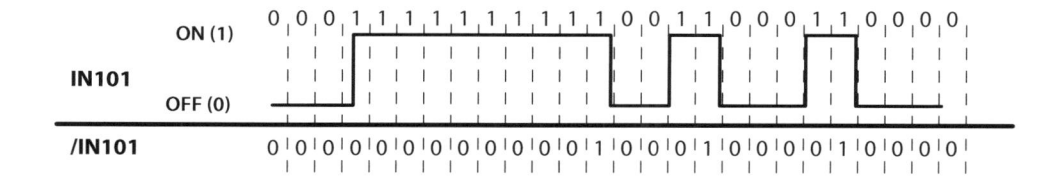

Figure 16-59: Illustration of Falling Edge Operation

7. General Electric FlexLogic™

FlexLogic™ incorporates logic gates and symbols as described in the understanding logic part of this chapter. The primary difference between manufacturers is how the logic is entered into the relay and how it is displayed. FlexLogic™ entry can be daunting but it is simple once you understand the rules. We will be dealing with FlexLogic™ data entry into the GE UR style relays here, but the other GE FlexLogic™ data entry styles are similar.

Unlike other relays, you are unable to assign multiple relay functions directly to an output in the GE UR series relays. All logic schemes must be assigned to virtual outputs, which are then used to operate other functions like output contacts. Virtual outputs are like the SELogic control equations/timers discussed previously.

FlexLogic™ is entered via the "FlexLogic Equation Editor" screen and each element is entered in rows and selected via pull down menus. When applying logic gates, all the logic inputs are entered in the rows directly above the gate entry. You must specify the number of previous rows that will be applied as inputs in the Syntax column. All logic function outputs must be assigned to perform some function like a virtual output or an input to another function or the relay will not accept the scheme. The first column narrows the selections in the final column by selecting the type of element. Possible choices are:

TYPE	DESCRIPTION
End of list	This is a very important setting and is used to indicate that the logic scheme is complete. If this selection is misplaced, the entire logic scheme may not operate.
NOT	This is a standard NOT gate. Selecting this type will reverse the state of the previous row's entry. All NOT gates must have three rows: 1. The value or function to be reversed 2. The NOT gate 3. What the NOT output is connected to
XOR	This is the standard XOR logic gate described earlier in the chapter. This gate is rarely used and uses the previous two rows as inputs to the XOR gate. All XOR gates must have four rows: 1. Input #1 2. Input #2 3. The XOR gate 4. What the XOR output is connected to
Latch	The Latch function represents a latching relay with set and reset inputs defined by the preceding two rows. All Latch functions must have four entry rows: 1. The SET input 2. The RESET input 3. The Latch gate 4. What the XOR output is connected to

TYPE	DESCRIPTION
OR	This is a standard OR gate with up to 16 inputs. All OR gates must have (number of inputs plus two) entry rows: 1. The OR inputs (One per row) 2. The OR selection with the correct number of inputs selected in the Syntax column 3. What the OR gate output is connected to
AND	This is a standard AND gate with up to 16 inputs. All AND gates must have (number of inputs plus two) entry rows: 1. The AND inputs (One per row) 2. The AND selection with the correct number of inputs selected in the Syntax column 3. What the AND gate output is connected to
NOR	This is a standard NOR gate with up to 16 inputs. All NOR gates must have (number of inputs plus two) entry rows: 1. The NOR inputs 2. The NOR selection with the correct number of inputs selected in the Syntax column 3. What the NOR gate output is connected to
NAND	This is a standard NAND gate with up to 16 inputs. All NAND gates must have (number of inputs plus two) entry rows: 1. The NAND inputs 2. The NAND selection with the correct number of inputs selected in the Syntax column 3. What the NAND gate output is connected to
TIMER	This timer function can be used as an output row for logic functions. There are 32 available timers controlled in the "FlexLogic Timers" entry screen. Timers must have two rows. What will turn the timer ON or OFF (logic gate or other function) The Timer with the specific timer selected in the Syntax column.
Write Virtual Output (Assign)	This function uses the preceding row state and assigns it as a virtual output that can be used elsewhere. A virtual output is a control relay inside the software that can be ON (1) or OFF (0) as controlled by its previous row, such as a logic gate or other function. Write Virtual Output (Assign) functions require two inputs: 1. What will turn the virtual output ON or OFF (logic gate or other function) 2. The Write Virtual Output function with the designated output selected in the Syntax column.

TYPE	DESCRIPTION
Positive One Shot	The Positive One Shot is similar to the Rising Edge Detect symbol described earlier in the SELogic section of this chapter. If the input to this function changes from OFF (0) to ON (1), the Positive One Shot output will be ON (1) for one processing cycle and will remain OFF (0) until the Input makes the OFF (0)/ON (1) transition again. Positive One Shot functions need 3 rows to operate: 1. What will initiate the Positive One Shot (logic gate or other function) 2. The Positive One Shot function. 3. What the Positive One Shot output is connected to
Negative One Shot	The Negative One Shot is similar to the Falling Edge Detect symbol described earlier in the SELogic section of this chapter. If the input to this function from ON (1) to OFF (0), the Negative One Shot output will be ON (1) for one processing cycle and will remain OFF (0) until the Input changes makes the ON (1)/OFF (0) transition again. Negative One Shot functions need 3 rows to operate: 1. What will initiate the Negative One Shot (logic gate or other function) 2. The Negative One Shot function. 3. What the Negative One Shot output is connected to
Dual One Shot	The Dual One Shot is like an OR gate between a Rising Edge Detect and Falling Edge Detect symbols described earlier in the SELogic section of this chapter. If the input to this function changes from OFF (0) to ON (1) OR ON (1) to OFF(1), the Dual One Shot output will be ON (1) for one processing cycle and will remain OFF (0) until the Input changes makes another OFF (0)/ON (1) OR ON(1)/OFF (0) transition. Dual One Shot functions need 3 rows to operate: 1. What will initiate the Dual One Shot (logic gate or other function) 2. The Dual One Shot function. 3. What the Dual One Shot output is connected to
OFF	A permanent OFF (0) state used in inputs to logic functions. Using OFF and ON functions is a good way to make allowances for future elements in logic so that new elements replace the ON (1) or OFF functions instead of re-entering all logic below the addition.
ON	A permanent ON (0) state used in inputs to logic functions. Using OFF and ON functions is a good way to make allowances for future elements in logic so that new elements replace the ON (1) or OFF functions instead of re-entering all logic below the addition.
Contact Inputs On	This function will return an ON (1) state if the physical digital input selected in the Syntax column is ON (1).
Contact Inputs Off	The function returns an ON (1) state if the physical input selected in the Syntax column is OFF (1). This function is used instead of "NOT Contacts input On."
Read Virtual Inputs	Virtual inputs are software commands issued via communications or keypads that are rarely used. This function monitors the Virtual Input status designated in the Syntax column and will be ON (1) if the virtual input is ON (1).

TYPE	DESCRIPTION
Contact Output Voltage On	Some contacts use electronic components to detect if there is a voltage across the open contact and can be used as a trip coil monitor. If the voltage across the contact is higher than a threshold voltage, this function will be ON (1). This style of output contacts must not be used as inputs to other electronic devices like SCADA systems as they may cause incorrect states through the internal diodes.
Contact Output Voltage Off	This status represents "NOT Contact Output Voltage ON."
Contact Output Current ON	Like Contact Output Voltage ON, this status input monitors the current flowing through the output contact to monitor the status of the output circuit. If the Current flowing through the contact is higher than the contact threshold current, the status will be ON (1).
Contact Output Current OFF	This status represents NOT Contact Output Current ON.
Remote Inputs On	Many GE UR series relays can communicate with each other via communication ports to share information and work as a system instead of a lonely island in the protection stream. If such a communication network is in place, this status input monitors the contact inputs of another relay and will turn ON (1) if the remote relay's input is ON (1). The syntax column selects which input on which relay is monitored.
Remote Inputs Off	This status represents NOT Remote Inputs On and the syntax column selects which remote relay's input will be monitored.
Remote Devices On	When a communication network is set up between GE UR series relays and information is shared between the relays, it is often imperative to know whether the remote relay is still communicating with the local relay. This status will be ON (1) if the remote relay selected in the syntax column is still communicating with the local relay.
Remote Devices Off	This logical input represents NOT Remote Devices On and the remote relay is selected in the Syntax column.
Protective Element	This logical input can be any of the relay labels selected in the syntax column. It is important to select the correct element from the text and understand its function.

Figure 16-60: Description of FlexLogic Functions

As you can see, the GE UR FlexLogic™ has an incredible amount of flexibility and overwhelming number of selections. While 90 +% of the available functions may never be used, it can be very cumbersome and confusing when entering FlexLogic™ equations. The "View" button at the top of the columns is a useful function that should be used often to prove equations and look for errors. Clicking this button automatically checks the logic for errors and creates a logic diagram for you to view.

Here are some real-world logic scenarios to demonstrate how FlexLogic™ is entered.

A) FlexLogic™ Example—The Situation

Our client has a D-60 relay with the following connections:

- Output Relay H1 connected to the trip coil of CB999
- Output Relay H2 connected to initiate a circuit breaker reclose
- Output Relay H3 connected to block a circuit breaker reclose
- Output Relay H4 connected to a SCADA alarm point
- Input H5 is connected to an external Transfer Trip relay

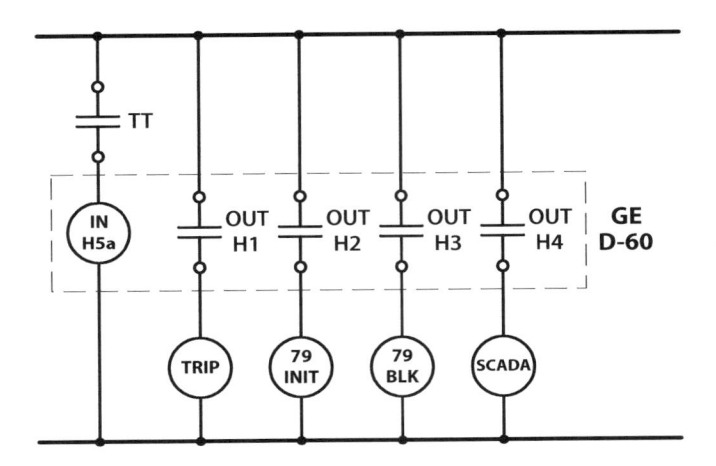

Figure 16-61: FlexLogic™ Example Relay Schematic

The design engineer has determined that the following functions should trip the CB999:

- Zone 1 phase distance trip
- Zone 2 phase distance time trip
- Zone 3 phase distance time trip IF the load encroachment element is OFF (0)
- Zone 1 ground distance trip
- Zone 2 ground distance time trip
- Zone 3 ground distance time trip IF the load encroachment element is OFF (0)
- Line Pickup time trip
- External Transfer Trip

The following functions should initiate a circuit breaker reclose:

- Zone 1 phase distance trip
- Zone 2 phase distance time trip
- Zone 1 ground distance trip
- Zone 2 ground distance time trip

The following functions should block the circuit breaker reclose:

- Zone 3 phase distance time trip IF the load encroachment element is OFF (0)
- Zone 3 ground distance time trip IF the load encroachment element is OFF (0)
- Line Pickup time trip

The SCADA is connected to a normally-closed alarm signal that will alarm if:

- any of the trip elements operate, or
- a Fuse failure is detected, or
- the relay powers down

This utility requires that all signals to external relays must seal in for 80ms to ensure that the relay will operate.

The following figures include the relevant manufacturer's information:

OPERAND TYPE	STATE	EXAMPLE FORMAT	CHARACTERISTICS [INPUT IS "1" (=ON) IF...]
Contact Input	On	Cont Ip On	Voltage is presently applied to input (external contact closed).
	Off	Cont Ip Off	Voltage is presently not applied to the input (external contact open).
Element (Analog)	Pickup	PHASE TOC1 PKP	The tested parameter is presently above the pickup setting of an element which responds to rising values or below the pickup setting of an element that responds to falling values.
	Dropout	PHASE TOC1 DPO	The operand is the logical inverse of the above PKP operand.
	Operate	PHASE TOC1 OP	The tested parameter has been above/below the pickup setting of the element for the programmed delay time, or has been at logic 1 and is now at logic 0 but the reset timer has not finished timing.
Element Digital)	Pickup	Dig Element 1 PKP	The input operand is at logic 1.
	Dropout	Dig Element 1 DPO	This operand is the logical inverse of the above pickup operand.
	Operate	Dig Element 1 OP	The input operand has been at Logic 1 for the programmed pickup delay time, or has been at logic 1 for this period and is now at logic 0 but the reset timer has not finished timing.

Figure 16-62: GE - UR FlexLogic™ Operand Types

OPERAND TYPE	OPERAND SYNTAX	OPERAND DESCRIPTION
ELEMENT Ground Distance	GND DIST Z1 PKP	Ground Distance Zone 1 has picked up
	GND DIST Z1 OP	Ground Distance Zone 1 has operated
	GND DIST Z1 OP A	Ground Distance Zone 1 Phase A has operated
	GND DIST Z1 OP B	Ground Distance Zone 1 Phase B has operated
	GND DIST Z1 OP C	Ground Distance Zone 1 Phase C has operated
	GND DIST Z1 PKP A	Ground Distance Zone 1 Phase A has picked up
	GND DIST Z1 PKP B	Ground Distance Zone 1 Phase B has picked up
	GND DIST Z1 PKP C	Ground Distance Zone 1 Phase C has picked up
	GND DIST Z1 SUPN IN	Ground Distance Zone 1 neutral is supervising
	GND DIST Z1 DPO A	Ground Distance Zone 1 Phase A has dropped out
	GND DIST Z1 DPO B	Ground Distance Zone 1 Phase B has dropped out
	GND DIST Z1 DPO C	Ground Distance Zone 1 Phase C has dropped out
	GND DIST Z2 DIR SUPN	Ground Distance Zone 2 directional is supervising
	GND DIST Z2 to Z4	Same set of operands as shown for GND DIST Z1
ELEMENT Line Pickup	LINE PICKUP OP	Line Pickup has operated
	LINE PICKUP PKP	Line Pickup has picked up
	LINE PICKUP DPO	Line Pickup has dropped out
	LINE PICKUP I<A	Line Pickup detected Phase A current below 5% of nom
	LINE PICKUP I<B	Line Pickup detected Phase B current below 5% of nom
	LINE PICKUP I<C	Line Pickup detected Phase C current below 5% of nom
	LINE PICKUP UV PKP	Line Pickup Undervoltage has picked up
	LINE PICKUP LEO PKP	Line Pickup Line End Open has picked up
	LINE PICKUP RCL TRIP	Line Pickup operated from overreaching Zone 2 when reclosing the line
ELEMENT Load Encroachment	LOAD ENCHR PKP	Load Encroachment has picked up
	LOAD ENCHR OP	Load Encroachment has operated
	LOAD ENCHR DPO	Load Encroachment has dropped out
ELEMENT Phase Distance	PH DIST Z1 PKP	Phase Distance Zone 1 has picked up
	PH DIST Z1 OP	Phase Distance Zone 1 has operated
	PH DIST Z1 OP AB	Phase Distance Zone 1 Phase AB has operated
	PH DIST Z1 OP BC	Phase Distance Zone 1 Phase BC has operated
	PH DIST Z1 OP CA	Phase Distance Zone 1 Phase CA has operated
	PH DIST Z1 PKP AB	Phase Distance Zone 1 Phase AB has picked up
	PH DIST Z1 PKP BC	Phase Distance Zone 1 Phase BC has picked up
	PH DIST Z1 PKP CS	Phase Distance Zone 1 Phase CA has picked up
	PH DIST Z1 SUPN IAB	Phase Distance Zone 1 AB IOC is supervising
	PH DIST Z1 SUPN IBC	Phase Distance Zone 1 BC IOC is supervising
	PH DIST Z1 SUPN ICA	Phase Distance Zone 1 CA IOC is supervising
	PH DIST Z1 DPO AB	Phase Distance Zone 1 Phase AB has dropped out
	PH DIST Z1 DPO BC	Phase Distance Zone 1 Phase BC has dropped out
	PH DIST Z1 DPO CA	Phase Distance Zone 1 Phase CA has dropped out

OPERAND TYPE	OPERAND SYNTAX	OPERAND DESCRIPTION
	PH DIST Z2 to Z4	Same set of operands as shown for PH DIST Z1
ELEMENT VTFF	SRC1 VT FF OP	Source 1 Fuse Failure detector has operated
	SRC1 VT FF DPO	Source 1 Fuse Failure detector has dropped out
	SRC1 VT FF VOL LOSS	Source 1 has lost voltage signals (V2 above 25% or V1 below 70% of nominal)
FIXED OPERANDS	Off	Logic = 0. Does nothing and may be used as a delimiter in an equation list, used as 'disable by other features. Logic = 1. Can be used as a test setting.
	ON	
INPUTS / OUTPUTS Contact Inputs	Cont Ip 1 On	Logic = 1
	Cont Ip 1 Off	Logic = 0

Figure 16-63: GE D-60 FlexLogic™ Operands

B) FlexLogic™ Example—CB999 Trip Logic

The first step in developing the scheme logic is to diagram the design engineer's requirements to help us visualize what the logic scheme will look like. We'll start with the requirements for tripping CB999.

By reviewing the drawing, we see that H1 is connected to the trip coil. Therefore, H1 will use the following labels from the previous figures in the instruction manual:

- Zone 1 phase distance trip (PH DIST Z1 PKP) Note: "PKP" is selected because no time delay is specified
- Zone 2 phase distance time trip (PH DIST Z2 OP)
- Zone 3 phase distance time trip IF the load encroachment element is OFF (0) ("PH DIST Z3" *AND* NOT "LOAD ENCHR OP")
- Zone 1 ground distance time trip (GND DIST Z1 PKP) Note: "PKP" is selected because no time delay is specified
- Zone 2 ground distance time trip (GND DIST Z2 OP)
- Zone 3 ground distance time trip IF the load encroachment element is OFF (0) ("GND DIST Z3 OP" *AND* NOT "LOAD ENCHR OP")
- Line Pickup time trip (LINE PICKUP OP)
- External Transfer Trip (Cont Ip H5a On)
- 80ms drop out timer (Timer 1)

An electrical schematic of this logic would look like:

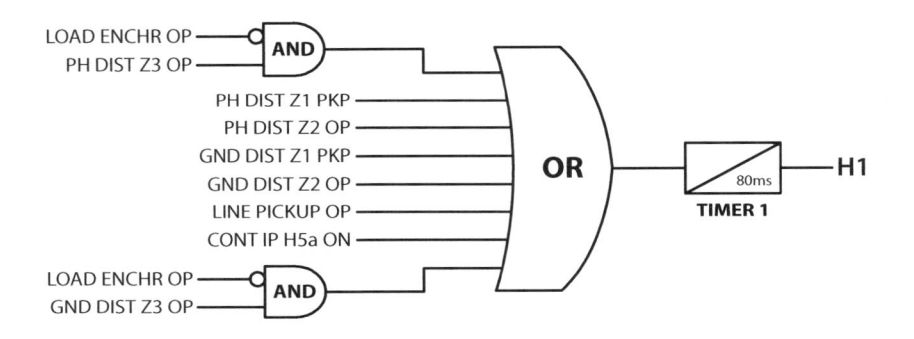

Figure 16-64: GE D-60 Logic Example Schematic Output H1

The logic diagram would look like:

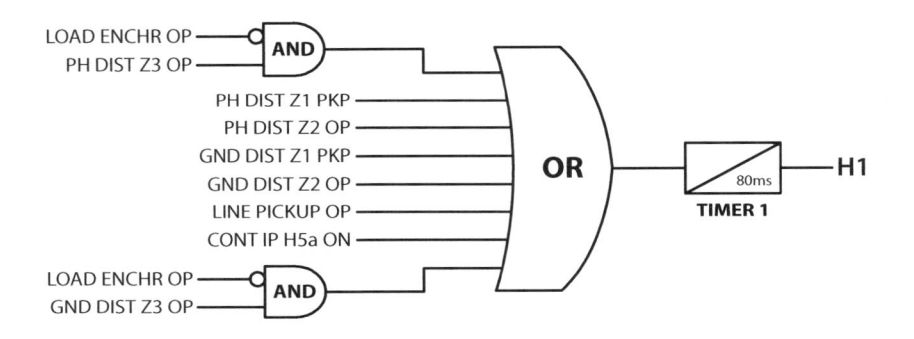

Figure 16-65: GE D-60 Logic Example Output H1

It's time to start entering logic into the D-60 relay. Start by opening the FlexLogic™ - Equation Editor Window via the software. The logic is entered by selecting a function type in the "Type" column and selecting the correct element in the "Syntax" column. You will be amazed by the number of available choices that are sorted in alphabetical order. You can skip ahead to the element you want by typing the element's first character in the pull-down menus.

Enter the elements with no associated logic first (one element per row) as shown in Figure 16-66 FLEXLOGIC ENTRY 1 to FLEXLOGIC ENTRY 6. The more complicated logic is entered next by selecting:

- PH DIST Z3 OP
- LOAD ENCHR OP
- NOT to reverse LOAD ENCHR OP in the row above
- AND with "2 Input" to select PH DIST Z3 and NOT LOAD ENCHR OP as its inputs

The GND DIST Z3 OP logic scheme is entered next by selecting almost identical information as follows:

- GND DIST Z3 OP
- LOAD ENCHR OP
- NOT to reverse LOAD ENCHR OP in the row above
- AND with "2 Input" to select GND DIST Z3 and NOT LOAD ENCHR OP as its inputs

Select OR with "8 input" to OR all the elements above.

Although this logic is designed for output H1, we cannot directly assign it to Output H1 and must create a virtual output to store the state of this logic. We will assign virtual output #1 (V01) by selecting "Write Virtual Output (Assign)" and "Virt Op 1 (VO1)."

Timer 1 is assigned after we assign Virt Op 1 gate to include the 80ms Drop Out delay necessary to seal in the output relay for 80ms. First, we must add "Read Virt Op On" Virt Op 1 (VO1) to provide an input for the timer. The Timer must be assigned an output, so another virtual output is assigned, VO2. The reason for this additional output will become apparent when we move on to the SCADA output.

If the logic was finished here, it would be vital that the next line be "End of List" to tell the relay that no more logic will be entered. Any other entry after "End of List" would be invalid and prevent the logic from operating.

The logic for tripping CB999 looks like Figure 16-66 after it has been entered.

FLEXLOGIC ENTRY	TYPE	SYNTAX
VIEW GRAPHIC	VIEW	VIEW
FLEXLOGIC ENTRY 1	Protection Element	PH DIST Z1 PKP
FLEXLOGIC ENTRY 2	Protection Element	PH DIST Z2 OP
FLEXLOGIC ENTRY 3	Protection Element	GND DIST Z1 PKP
FLEXLOGIC ENTRY 4	Protection Element	GND DIST Z2 OP
FLEXLOGIC ENTRY 5	Protection Element	LINE PICKUP OP
FLEXLOGIC ENTRY 6	Contact Inputs On	Cont Ip 1 ON (H5a)
FLEXLOGIC ENTRY 7	Protection Element	PH DIST Z3 OP
FLEXLOGIC ENTRY 8	Protection Element	LOAD ENCHR OP
FLEXLOGIC ENTRY 9	NOT	1 Input
FLEXLOGIC ENTRY 10	AND	2 Input
FLEXLOGIC ENTRY 11	Protection Element	GND DIST Z3 OP
FLEXLOGIC ENTRY 12	Protection Element	LOAD ENCHR OP
FLEXLOGIC ENTRY 13	NOT	1 Input
FLEXLOGIC ENTRY 14	AND	2 Input
FLEXLOGIC ENTRY 15	OR	8 Input
FLEXLOGIC ENTRY 16	Write Virtual Output (Assign)	= Virt Op 1 (VO1)
FLEXLOGIC ENTRY 17	Read Virtual Output On	Virt Op 1 On (VO1)
FLEXLOGIC ENTRY 18	Timer	Timer 1
FLEXLOGIC ENTRY 19	Write Virtual Output (Assign)	= Virt Op 2 (VO2)

Figure 16-66: GE D-60 H1 Output Logic Entry Example

This data entry screen can make it hard to understand the logic. A graphical representation of the logic is visible by selecting either of the two "VIEW" buttons at the top of the screen to check the entered logic against our logic drawings. This feature will also verify that our logic entries are compatible with the software and a warning message will appear if there is a problem.

If "VIEW" was clicked now, Timer 1 would not display a time delay. The time delay for Timer 1 is set by opening the "FlexLogic—FlexLogic Timers" screen and selecting "milliseconds" in the "Type" entry box and "80" in the drop out time entry box. Any changes made are not permanent until you tell the software to save them. Click the small lightning bolt icon or select "Action" and then "Save Settings" from the menu bar to save the timer settings.

Now you can return to the FlexLogic Equation Editor Screen and click "View."

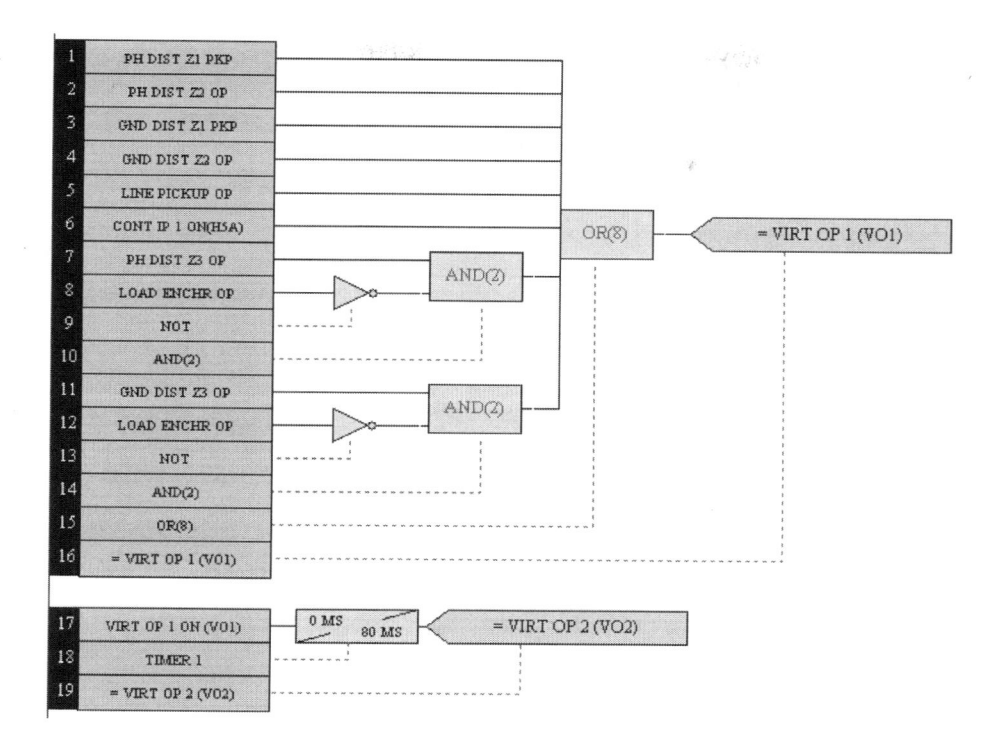

Figure 16-67: GE D-60 H1 Output Logic Display Example

Now we're ready to move on to the reclose initiate output. It's a good idea to save the changes in the FlexLogic Equation Editor at this point in case you make an error later.

C) FlexLogic™ Example—Reclose Initiate

Let's start with the reclose initiate requirements.

By reviewing the drawing, we see that H2 is connected to the reclose initiate coil. Therefore, H2 will use the following labels from the previous figures in the instruction manual:

- Zone 1 phase distance trip (PH DIST Z1 PKP) Note: "PKP" is selected because no time delay is specified
- Zone 2 phase distance time trip (PH DIST Z2 OP)
- Zone 1 ground distance time trip (GND DIST Z1 PKP) Note: "PKP" is selected because no time delay is specified
- Zone 2 ground distance time trip (GND DIST Z2 OP)

An electrical schematic of this logic would look like:

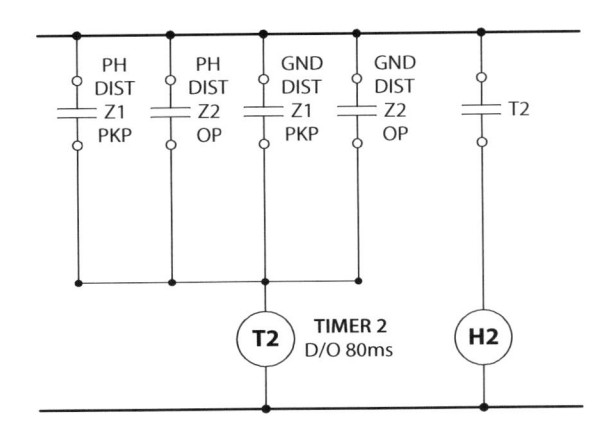

Figure 16-68: GE D-60 H2 Output Logic Example Schematic

The logic diagram would look like:

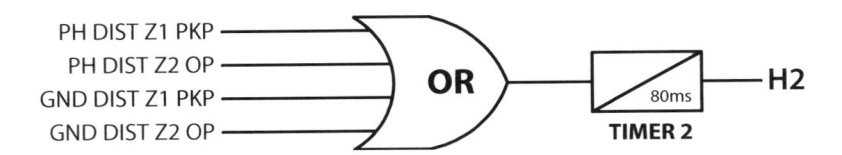

Figure 16-69: GE D-60 H2 Output Logic Example

The logic to initiate a reclose looks like Figure 16-70 after it has been entered.

FLEXLOGIC ENTRY	TYPE	SYNTAX
VIEW GRAPHIC	VIEW	VIEW
FLEXLOGIC ENTRY 20	Protection Element	PH DIST Z1 PKP
FLEXLOGIC ENTRY 21	Protection Element	PH DIST Z2 OP
FLEXLOGIC ENTRY 22	Protection Element	GND DIST Z1 PKP
FLEXLOGIC ENTRY 23	Protection Element	GND DIST Z2 OP
FLEXLOGIC ENTRY 24	OR	4 Input
FLEXLOGIC ENTRY 25	Timer	Timer 1
FLEXLOGIC ENTRY 26	Write Virtual Output (Assign)	= Virt Op 3 (VO3)

Figure 16-70: GE D-60 H2 Output Logic Entry Example

Don't forget to change the timer settings for Timer 2.

Now you can return to the FlexLogic Equation Editor Screen and click "View"

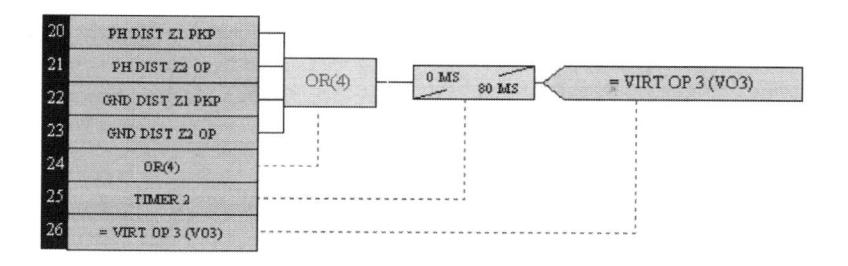

Figure 16-71: GE D-60 H2 Output Logic Display Example

It's a good idea to save the changes in the FlexLogic Equation Editor at this point in case you make an error later.

D) FlexLogic™ Example—Reclose Block

Let's start with the requirements for the reclose initiate requirements.

A review of the drawing shows that H3 is connected to the reclose block coil. Therefore, H3 will use the following labels from the previous figures in the instruction manual:

- Zone 3 phase distance time trip IF the load encroachment element is OFF (0) ("PH DIST Z3" AND NOT "LOAD ENCHR OP")
- Zone 3 ground distance time trip IF the load encroachment element is OFF (0) ("GND DIST Z3 OP" AND NOT "LOAD ENCHR OP")
- Line Pickup time trip (LINE PICKUP OP)

Let's simplify the logic a little bit to save us a little bit of typing. A simplified electrical schematic of this logic would look like:

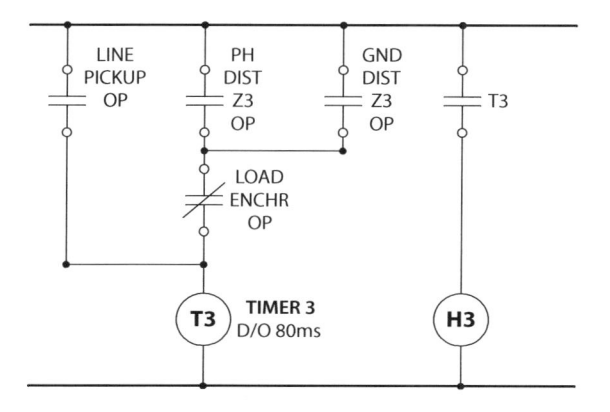

Figure 16-72: GE D-60 H3 Output Logic Example Schematic

The logic diagram would look like:

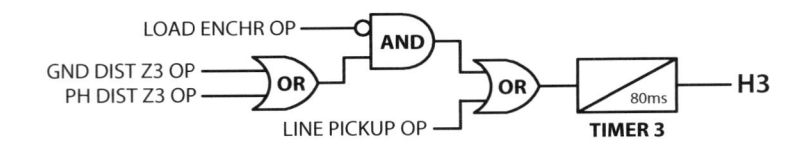

Figure 16-73: GE D-60 H3 Output Logic Example

The logic to block any reclose attempts looks like Figure 16-74 after it has been entered.

FLEXLOGIC ENTRY	TYPE	SYNTAX
VIEW GRAPHIC	VIEW	VIEW
FLEXLOGIC ENTRY 27	Protection Element	GND DIST Z3 OP
FLEXLOGIC ENTRY 28	Protection Element	PH DIST Z3 OP
FLEXLOGIC ENTRY 29	OR	2 Input
FLEXLOGIC ENTRY 30	Protection Element	LOAD ENCHR OP
FLEXLOGIC ENTRY 31	NOT	1 Input
FLEXLOGIC ENTRY 32	AND	2 Input
FLEXLOGIC ENTRY 33	Protection Element	LINE PICKUP OP
FLEXLOGIC ENTRY 34	OR	2 Input
FLEXLOGIC ENTRY 35	Timer	Timer 3
FLEXLOGIC ENTRY 36	Write Virtual Output (Assign)	= Virt Op 4 (VO4)

Figure 16-74: GE D-60 H3 Output Logic Entry Example

Don't forget to change the timer settings for Timer 3.

Now you can return to the FlexLogic Equation Editor Screen and click "View"

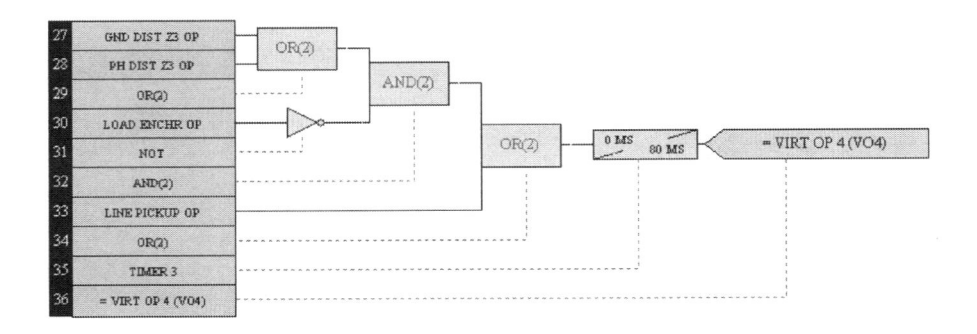

Figure 16-75: GE D-60 H3 Output Logic Display Example

It's a good idea to save the changes in the FlexLogic Equation Editor at this point in case you make an error later.

E) FlexLogic™ Example—SCADA Logic

The first step in developing the scheme logic is to diagram the design engineer's requirements to help visualize what the logic scheme will look like. We'll start with the requirements for SCADA logic.

A review of Figure 16-61 indicates that H4 is connected to SCADA. SCADA uses the following labels from the instruction manual as described in previous figures:

- Zone 1 phase distance trip (PH DIST Z1 PKP) Note: "PKP" is selected because no time delay is specified
- Zone 2 phase distance time trip (PH DIST Z2 OP)
- Zone 3 phase distance time trip IF the load encroachment element is OFF (0) ("PH DIST Z3" AND NOT "LOAD ENCHR OP")
- Zone 1 ground distance time trip (GND DIST Z1 PKP) Note: "PKP" is selected because no time delay is specified
- Zone 2 ground distance time trip (GND DIST Z2 OP)
- Zone 3 ground distance time trip IF the load encroachment element is OFF (0) ("GND DIST Z3 OP" AND NOT "LOAD ENCHR OP")
- Line Pickup time trip (LINE PICKUP OP)
- External Transfer Trip (Cont Ip H5a On)
- Fuse Failure (SRC1 FUSE FAIL OP)

An electrical schematic of this logic would look like:

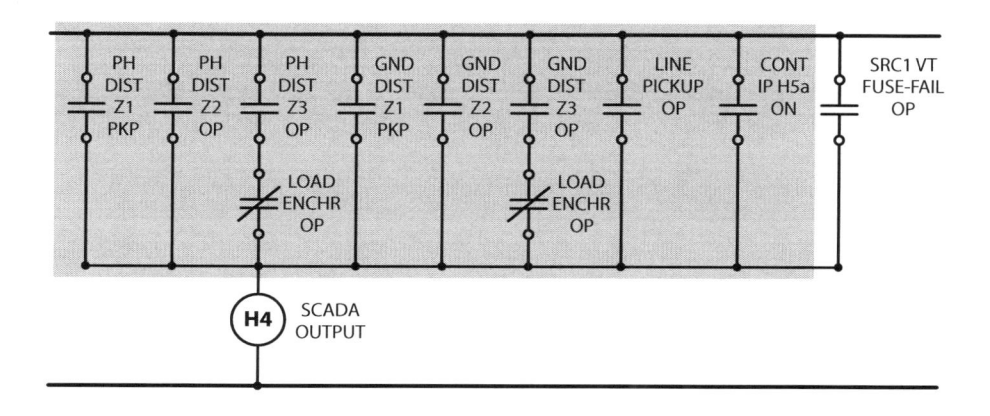

Figure 16-76: GE D-60 H4 Output Logic Example Schematic

If you look closely at the boxed section of the schematic diagram, it should look familiar. We have already entered all this logic and assigned it to Virtual Output VO1. Instead of entering all those rows over again, VO1 can be assigned as one input into the OR gate as shown in Figure 16-77.

The logic diagram would look like:

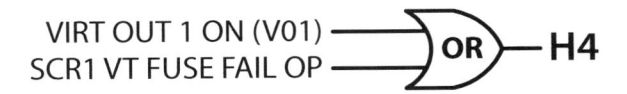

Figure 16-77: GE D-60 H4 Output Logic Example

The SCADA output logic looks like Figure 16-78 after it has been entered.

FLEXLOGIC ENTRY	TYPE	SYNTAX
VIEW GRAPHIC	VIEW	VIEW
FLEXLOGIC ENTRY 37	Read Virtual Output On	Virt Op 1 On (VO1)
FLEXLOGIC ENTRY 38	Protection Element	SRC1 VT FUSE FAIL OP
FLEXLOGIC ENTRY 39	OR	2 Input
FLEXLOGIC ENTRY 40	Write Virtual Output(Assign)	= Virt Op 4 (VO4)

Figure 16-78: GE D-60 H4 Output Logic Entry Example

Now you can return to the FlexLogic Equation Editor Screen and click "View"

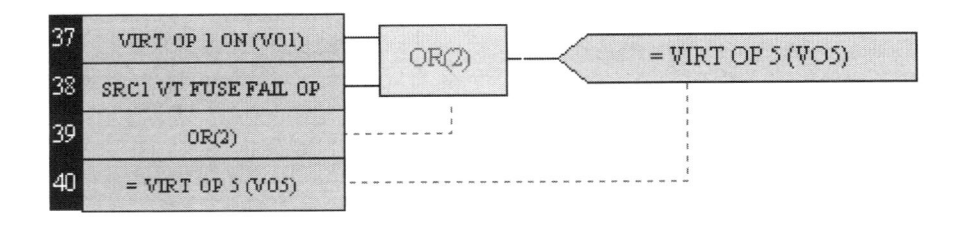

Figure 16-79: GE D-60 H4 Output Logic Display Example

We're not finished yet!

F) FlexLogic™ Example—Naming Logic

It can be difficult to keep track of all the virtual inputs and outputs and there is a way to re-label this information to help understand what is going on without referring to your notes. Open the Virtual Output window in Inputs/Outputs and assign names to your virtual outputs like Figure 16-80. The Events row tells the relay to save an event every time the output changes state. Don't forget to save your data.

PARAMETER	VIRT OP 1	VIRT OP 2	VIRT OP 3	VIRT OP 4	VIRT OP 5
ID	TRIP INIT	TRIP CB999	79 INIT	79 BLOCK	SCADA
Events	Disabled	Disabled	Disabled	Disabled	Disabled

Figure 16-80: GE D-60 Virtual Output Labels

G) FlexLogic™ Example—Assigning Output Functions

The relay will not operate after all the FlexLogic™ has been entered and saved until the outputs have been assigned to operate. Output functions are assigned in the Inputs/ Outputs then Contact Output window. You can add a custom label to the output like the virtual outputs in the ID row and assign its function by selecting an element from the pull-down menus in the operate rows. Selecting "ON" in the Seal-In row will cause the output to latch when operated until a reset command is received by the relay. Enabling events will tell the relay to record an event every time the output turns ON (1) or OFF (0).

PARAMETER	H1	H2	H3	H4
ID	TRIP CB999	79 INIT	79 BLOCK	SCADA
Operate	Trip CB999 On(V02)	79 INIT On(V03)	79 BLOCK On(V04)	SCADA On(V05)
Seal-In	OFF	OFF	OFF	OFF
Events	Enabled	Enabled	Enabled	Enabled

Figure 16-81: GE D-60 Contact Output Assignments

Congratulations! If you understood what happened in this chapter, you should be ready for almost any relay logic you're likely to find in the real world.

Chapter 17

Review the Application

1. Introduction

Test plans determine what will be tested and in what order. This is where we use all of the knowledge discussed in this book and apply it to test relay applications. Test plans should include all of the steps discussed in the next few chapters and can be performed manually or automatically. A summary of these steps are as follows:

1) Review the Relay Application

Collect the single-line drawings, three-line drawings, DC schematics, coordination study, and other relay related drawings to compare to the relay manufacturer's manual and customer-supplied relay settings. Pay particular attention to the following items:

- CT polarity and location
- CT and PT input connections
- CT and PT ratios
- CT and PT test points
- Output connections, descriptions and logic assignments
- Input connections, descriptions, and logic assignments
- Power supply voltage
- Settings match the coordination study and make sense

2) Establish Communications with the Relay and Apply Settings

After the application has been verified and all discrepancies have been resolved, you're ready to start applying settings. Power up the relay, connect your communication cable, and apply the relay settings. Sometimes this can be the hardest step when relay testing.

3) Connect Your Relay Test-Set

Connect your test-set to the relay to simulate the actual CTs and PTs. This can often require some creative thinking depending on the application and the make/model of your test-set. Connect the relay outputs/inputs that will be tested to the test-set.

4) Creating the Test Plan

Make a list of elements that affect all output contact, LED, or communication output operation. Review the list and test to ensure the elements are operating correctly and will operate the correct outputs. Try to test without changing relay settings. If you must change settings, minimize the relay setting changes and make sure that any disabled elements are tested after they are enabled. (Do not test an element, disable it to test another element, and then enable it again without verifying it still operates correctly.)

5) Perform Relay Self-Check and Verify Inputs/Outputs

Modern relays perform continuous self-checks to help detect in-service failures. Run a self-check to see if any problems are detected. Most relay failures occur with the analog/digital inputs that cannot be tested with the self-check feature. Apply currents and voltages to the appropriate inputs and verify that the relay metering is correct. Pulse or operate output contacts and verify operation with an ohmmeter or your test-set's contact-status detector. Apply the correct input voltage to each input and make sure the relay status point changes state.

6) Test Individual Elements for Pickup and Timing

Test the pickup and timing for each enabled element and compare to manufacturer's tolerances.

7) Final Output Tests

If the pickup and timing functions for elements are tested individually, verify all relay output logic functions to ensure there are no errors. This includes each output contact, LED, and communication output. Apply input currents/voltages/frequencies necessary to enable an element and make sure the correct output signals are created. You should be able to simulate each element in a relay whose settings have been correctly designed and applied. If you can't make every element operate during this step, there are redundant settings that should be questioned.

This step can be combined with Step #6 to minimize setting changes.

8) Final Report

Document all test results and save a record of all as-left settings. Document all changes, special procedures, and/or questions on the first page of the test sheet for easy review. The settings should be in a format that does not require any special software so that anyone who needs the information can access it.

2. Collecting Information

Collect the single-line drawing, three-line drawings, schematic drawings, relay settings, and coordination study with a pad of paper to write down key items.

A) Single-Line Drawings

Figure 17-1 depicts an example single line drawing. You can follow the lines from any device to find its source or what devices are connected to any source. As drawings become more complex, they can be split into different drawings because of their size or crossing lines that can make it nearly impossible to read. In these cases, symbols are often added on each side of a split line. The hexagon connected to CT's 4-5-6 indicate the CTs are connected to RLY-5 on Drawing #2 and Hexagon B indicates the second CT input will come from CTs 24-26 on Drawing #2. You can draw a mental line between Hexagons C on RLY-1 and VT2 as shown by the light dashed line.

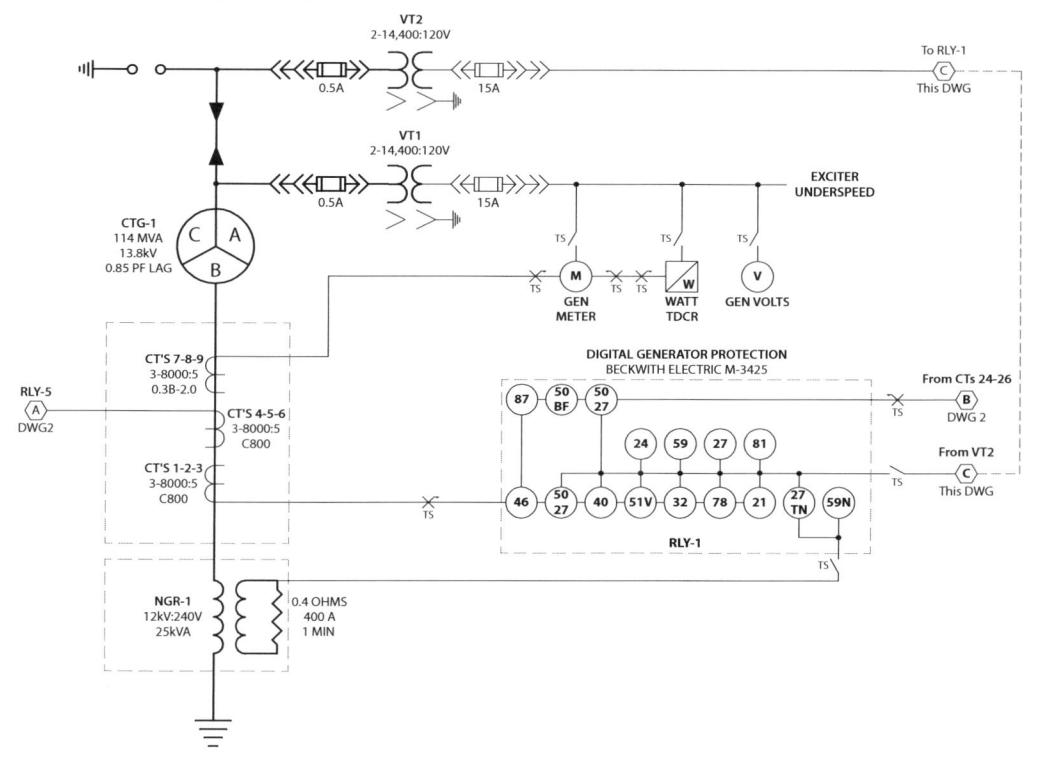

Figure 17-1: Complex Single-Line Drawings

Single-line drawings can become quite complex and it's important to follow the lines correctly. Use the following figures and instructions to learn what information a single line drawing can provide.

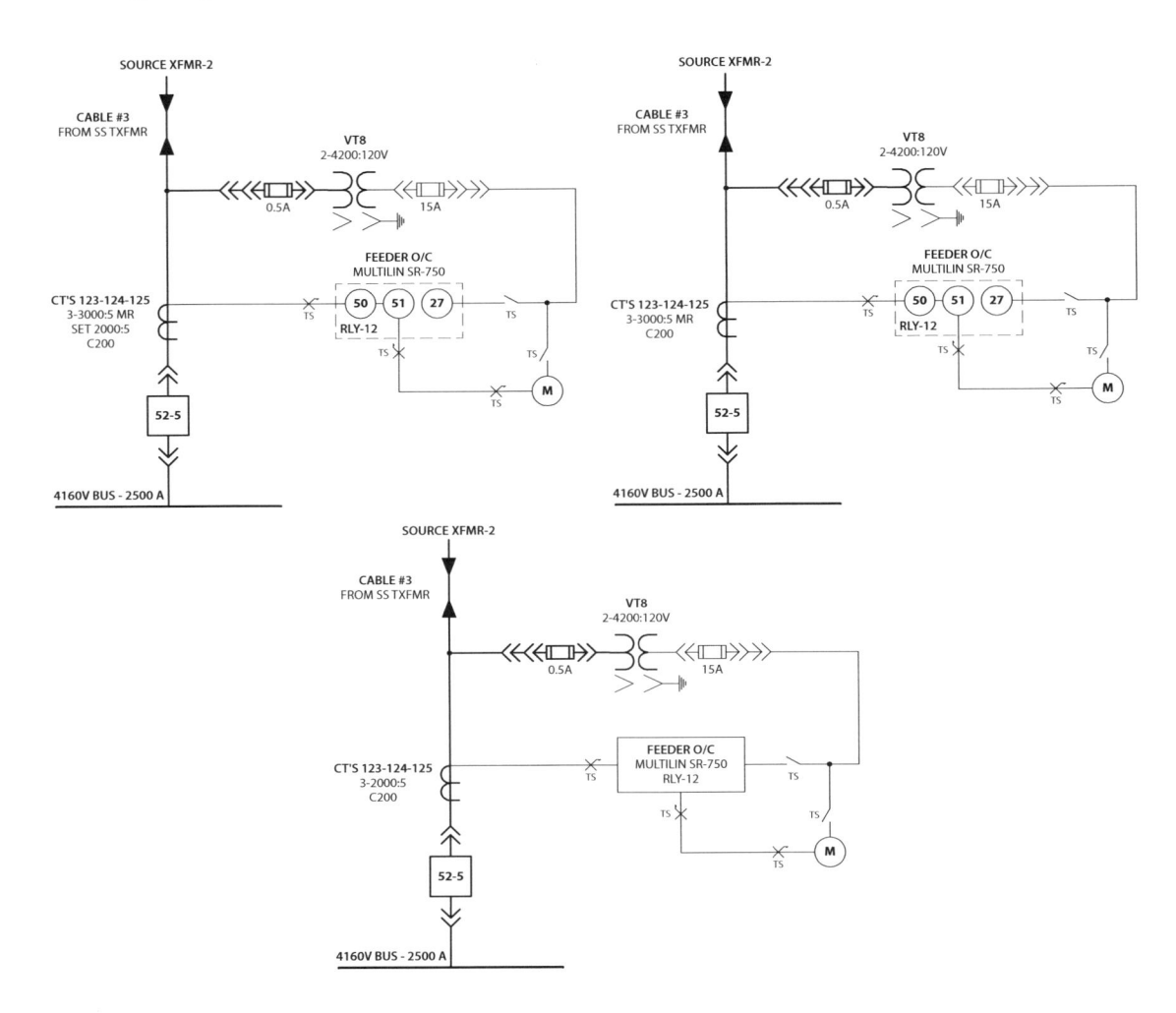

Figure 17-2: Simple Single-Line Drawings

Use Figure 17-2 to learn the following information and note it down for our test plan:

1. The relay should be a Multilin SR-750 (Make sure you have the relay manufacturer's bulletin) and its designation is RLY-12. All relays should have a unique designation as there could be many SR-750 relays at this project. This is commonly missed at sites and on drawings. You can make one up or use the designation from the coordination study that will likely have a unique identifier.

2. The CT is a multi-ratio CT and should be connected with a 2000:5 ratio. (Note: While I believe that the single-line should have the CT ratio listed, many companies only list the maximum ratio as shown on the top-right drawing. Find out what the standard is and make sure it's correctly shown.)

3. The CT polarity is connected to the cable and current flows from the cable to the bus.

4. The PT ratio is 4200:120V and is connected open-delta.

5. Test switches are installed for testing. (Good news for relay testers.)

6. 50, 51, and 27 elements should be enabled in the relay. (Some companies only show relays as a blank box or circle without any element information like the bottom drawing.)

7. There doesn't appear to be any ground or neutral protection.

Single-line drawings can also provide information regarding the logic settings and output contacts. Any information should be checked against the DC control schematics and the relay settings. Some examples of DC logic on single-line drawings are shown in Figure 17-3. Various styles of drawings give different amounts of information as you can see from our various examples.

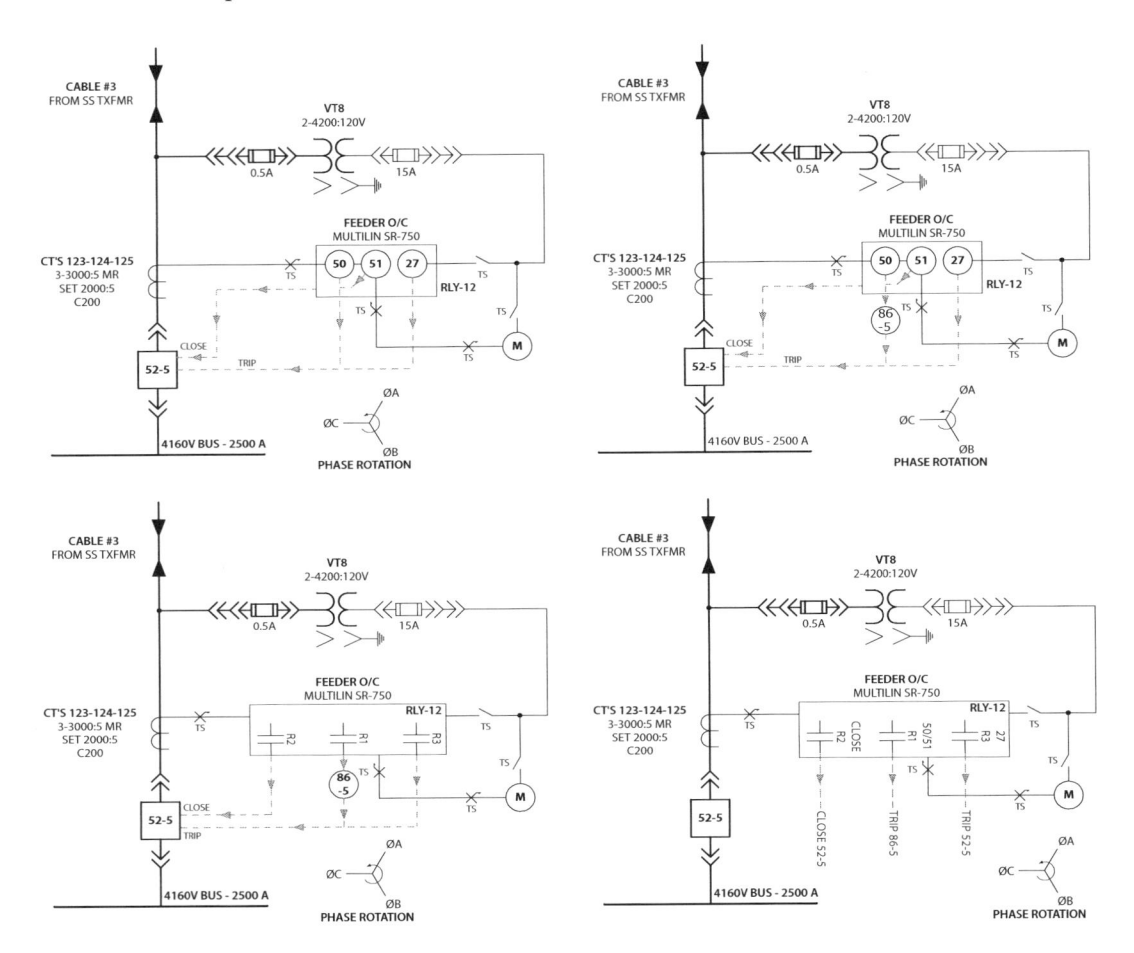

Figure 17-3: Single-Line Drawings with DC Logic and Settings

Use the following checklist to record information from the single-line drawing:

	SINGLE-LINE	THREE-LINE	DC SCHEMATIC	SETTINGS
Relay Designation	RLY-12			
Phase Rotation	A-B-C			
CT1 Location	Cable-Breaker			
CT1 Ratio	2000:5			
CT1 Polarity	Cable			
PT1 Location	Incoming			
PT1 Primary Volts	4200			
PT1 Secondary Volts	120			
PT1 Ratio	35:1			
PT1 Nominal Pri Volts	4160 V			
PT1 Nominal Sec Volts	118.86 V			
PT1 Connection	Open Delta			
Out1 Initiate	R1 = 50 + 51			
Out1 Connected	Trip 86-5			
Out2 Initiate	R2 = Close			
Out2 Connected	52-5 Close			
Out3 Initiate	R3 = 27			
Out3 Connected	Trip 52-5			

Figure 17-4: Example Relay Information Checklist #1

B) Three-Line Drawings

Three-line drawings are the most important drawings for the A/C connections to the relay. These drawings should show all of the connections throughout the A/C system from the CT secondary terminals, which determine the CT ratio and polarity to the relay terminals to ensure correct polarity. The following is a three-line drawing for our example:

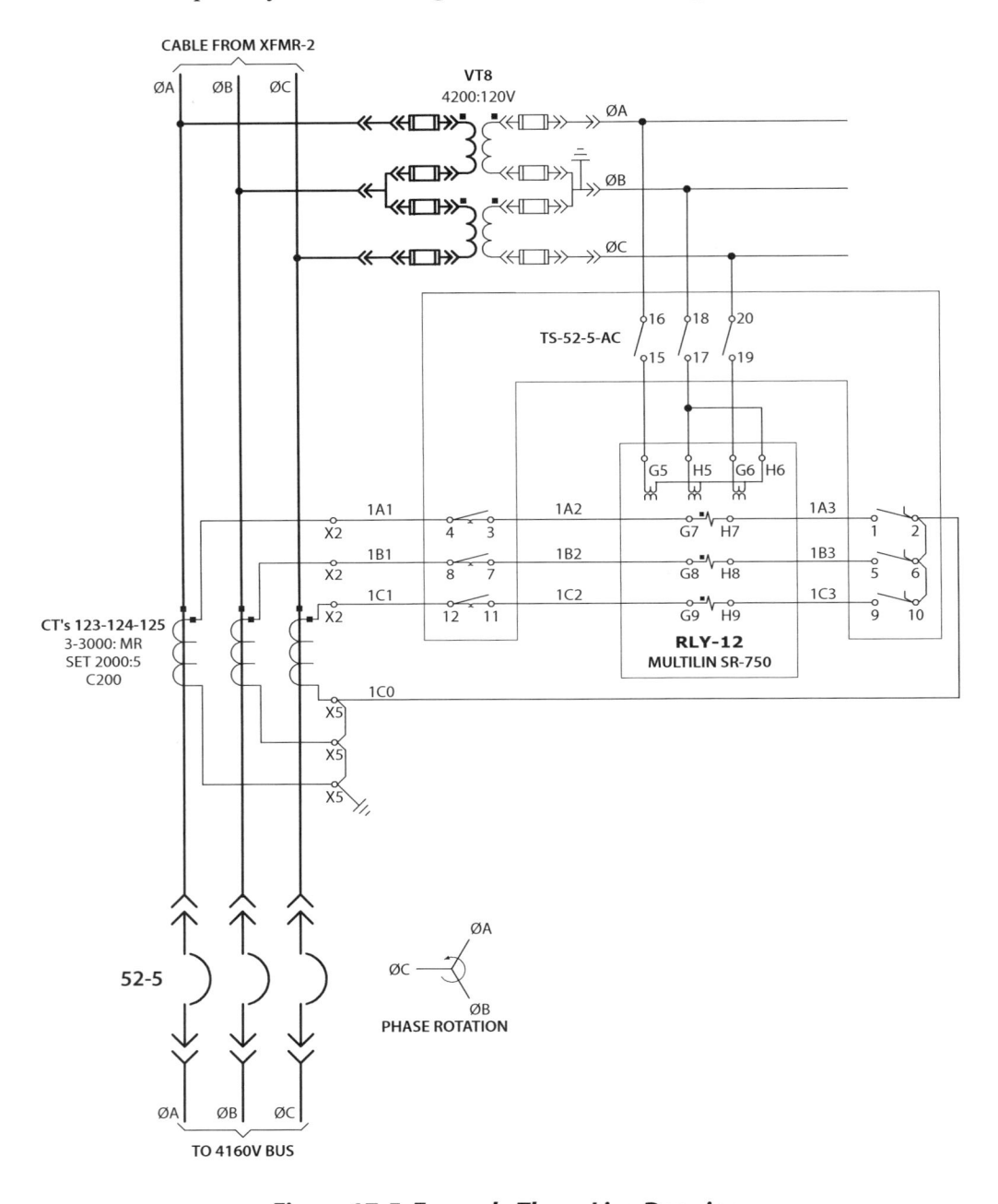

Figure 17-5: Example Three-Line Drawing

Carefully inspect the three-line drawing and continue filling out the checklist. Compare any differences between the three-line and single-line drawings. If there is a difference, find out which drawing is correct before testing the relay.

	SINGLE-LINE	THREE-LINE	DC SCHEMATIC	SETTINGS
Relay Designation	RLY-12	**RLY-12**		
Phase Rotation	A-B-C	**A-B-C**		
CT1 Location	Cable-Breaker	**Cable-Breaker**		
CT1 Ratio	2000:5	**2000:5**		
CT1 Polarity	Cable	**Cable**		
PT1 Location	Incoming	**Incoming**		
PT1 Primary Volts	4200	**4200**		
PT1 Secondary Volts	120	**120**		
PT1 Ratio	35:1	**35:1**		
PT1 Nominal Pri Volts	4160 V	**4160 V**		
PT1 Nominal Sec Volts	118.86 V (4160 / 35)	**118.86 V (4160 / 35)**		
PT1 Connection	Open Delta	**Open Delta**		
Out1 Initiate	R1 = 50 + 51	**NA**		
Out1 Connected	Trip 86-5	**NA**		
Out2 Initiate	R2 = Close	**NA**		
Out2 Connected	52-5 Close	**NA**		
Out3 Initiate	R3 = 27	**NA**		
Out3 Connected	Trip 52-5	**NA**		

Figure 17-6: Example Relay Information Checklist #2

CT connections can be confusing for some people, but the connections can be correctly interpreted using this method. Power flows from the transformer to the bus in this application so we can trace the flow of power from the A-phase transformer cable to the A-phase CT. Current flows into the CT polarity mark and out the CT secondary polarity mark (H1 & X1). Follow the line from the polarity mark to the test switch and record the number of the CT test switch connected to the relay. The test-switch terminal connected to the relay must always be an odd number or the shorting feature will not work correctly. (More on test switches later in the *Test Switches* section of this chapter) Keep following the line to the relay, and record the first relay terminal number. Keep following the terminal through the relay and record the outgoing relay terminal. Keep following the line to the test switch and record the test switch terminal number. Keep following the line and make sure it eventually makes its way back to the CT. Repeat this step for the other two phases and ground, if applicable.

The three-line drawing adds additional information required for relay testing and we can add another connection checklist to record this information. Using the example three-line drawing, follow the A-phase primary bus through the polarity marks of the PT, write down the test switch number, and note whether the odd or even number is connected to the relay. Continue following the line to the relay terminal and record this information. Repeat for the other two phases and ground, if installed.

Repeat the previous PT and CT steps on the manufacturer's drawings and record on the Connection Checklist.

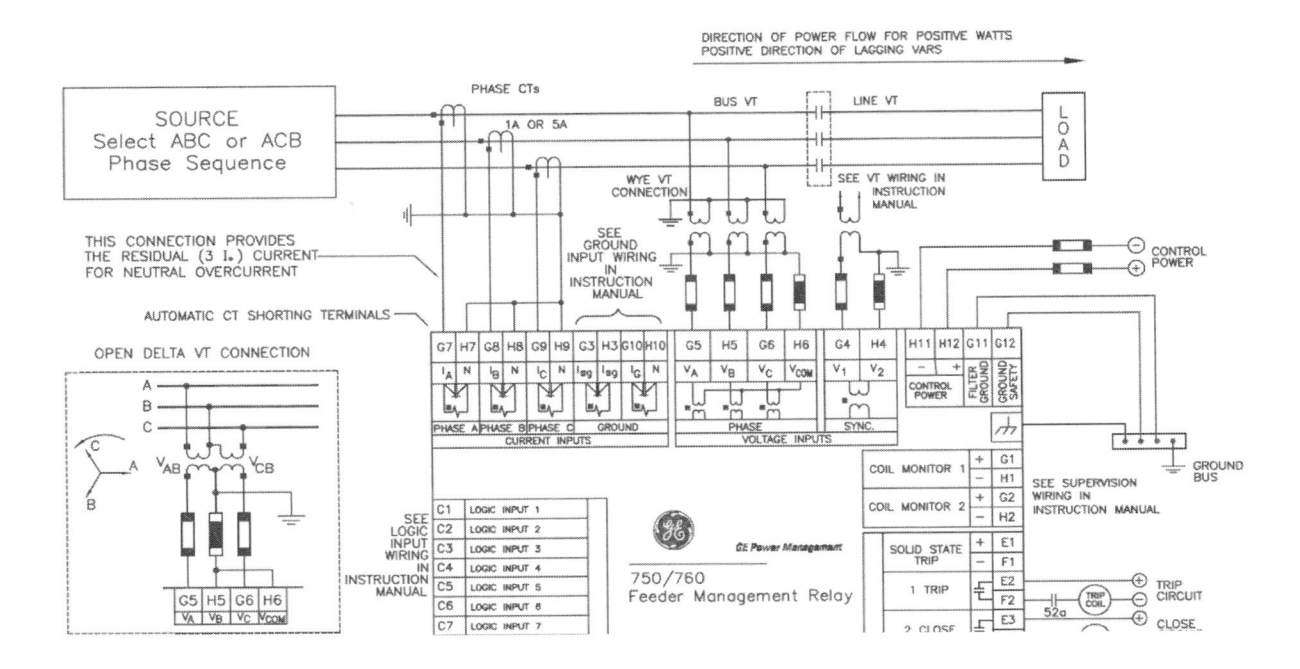

Figure 17-7: GE/Multilin Manufacturer's Typical Wiring Diagram

	SINGLE-LINE	THREE-LINE	RELAY SPECS
Relay Designation	RLY-12	RLY-12	**RLY-12**
Relay Model	SR-750	SR-750	**SR-750**
CT1 Location	Cable-Breaker	Source-Breaker	**Source-Load**
CT1 Ratio	2000:5	2000:5	**N/A**
CT1 Polarity	Cable	Cable (Source)	**Source**
CT1 A-Phase Test Switch	NA	TS-52-5-AC (3-1)	**N/A**
CT1 A-Phase Relay Connection	NA	RLY-12 (G7-H7)	**G7-H7**
CT1 B-Phase Test Switch	NA	TS-52-5-AC (7-5)	**N/A**
CT1 B-Phase Relay Connection	NA	RLY-12 (G8-H8)	**G8-H8**
CT1 C-Phase Test Switch	NA	TS-52-5-AC (11-9)	**N/A**
CT1 C-Phase Relay Connection	NA	RLY-12 (G9-H9)	**G9-H9**
CT1 G Phase Test Switch	NA	N/A	**N/A**
CT1 G Phase Relay Connection	NA	NA	**G10-H10**
PT1 Location	Incoming	Incoming	**Incoming**
PT1 Connection	Open Delta	Open delta	**Open Delta**
VT1 A-phase Test Switch	NA	TS-52-5-AC (15)	**N/A**
VT1 A-phase Relay Connection	NA	RLY-12 (G5)	**G5**
VT1 B-Phase Test Switch	NA	TS-52-5-AC (17)	**N/A**
VT1 B-Phase Relay Connection	NA	RLY-12 (H5 & H6)	**H5 & H6**
VT1 C-Phase Test Switch	NA	TS-52-5-AC (19)	**N/A**
VT1 C-Phase Relay Connection	NA	RLY-12 (G6)	**G6**
VT1 N Phase Test Switch	NA	N/A	**N/A**
VT1 N Phase Relay Connection	NA	N/A	**N/A**
VTS Phase Test Switch	NA	N/A	**N/A**
VTS Phase Relay Connection	NA	NA	**G4**
VTS N Test Switch	NA	N/A	**N/A**
VTS N Relay Connection	NA	NA	**G4**

Figure 17-8: Example Relay Connection Checklist

So far everything is working fine. Notice that the manufacturer included a separate window for open delta connections and we used this connection instead of the wye connection. The most common error for PT connections involves the jumper between V_B and V_N. It should be installed on nearly every relay when open-delta connected PTs are used.

Let's look at some examples where CT connections do not look like the drawings but will still operate correctly. Follow the power flow into and out of the CT polarity marks into the relay and compare to the checklist to see how the following connections are also correct.

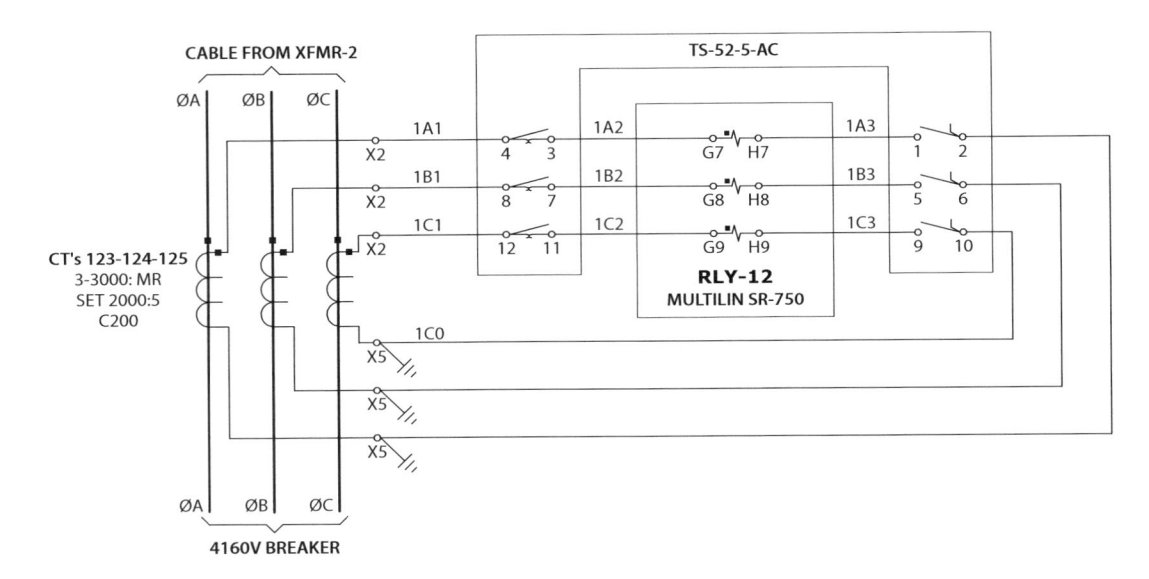

Figure 17-9: Alternate CT Connection #1

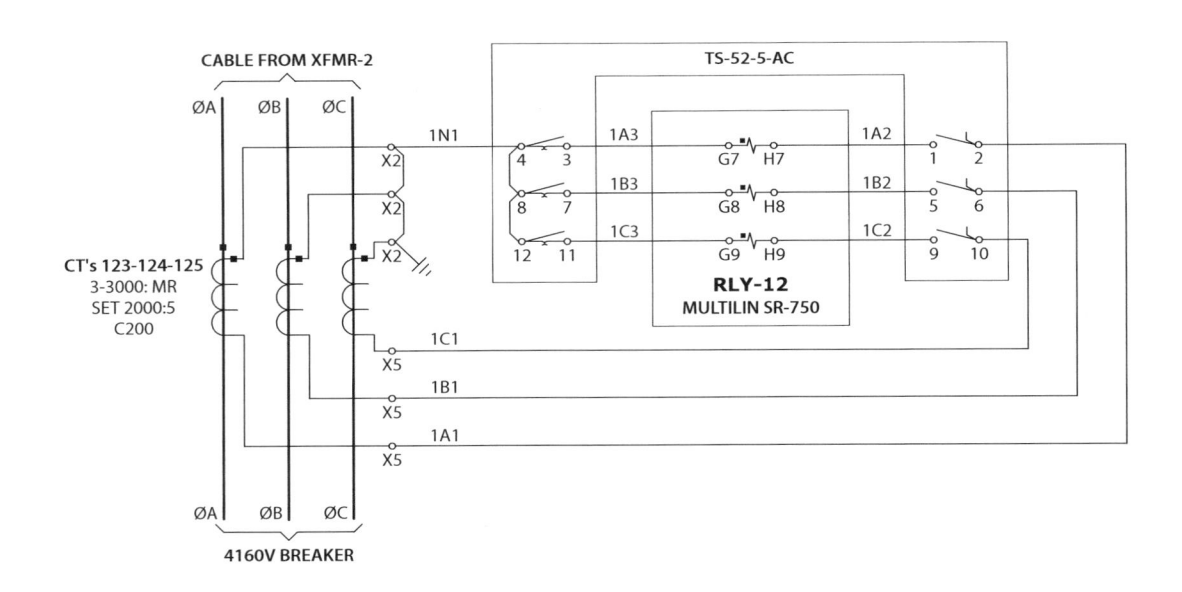

Figure 17-10: Alternate CT Connection #2

The following drawing illustrates why polarity mark location is less important that actual CT position and connections. Power flows into the non-polarity mark, so you must start tracing the flow of power in the secondary circuit from the non-polarity of the CT secondary.

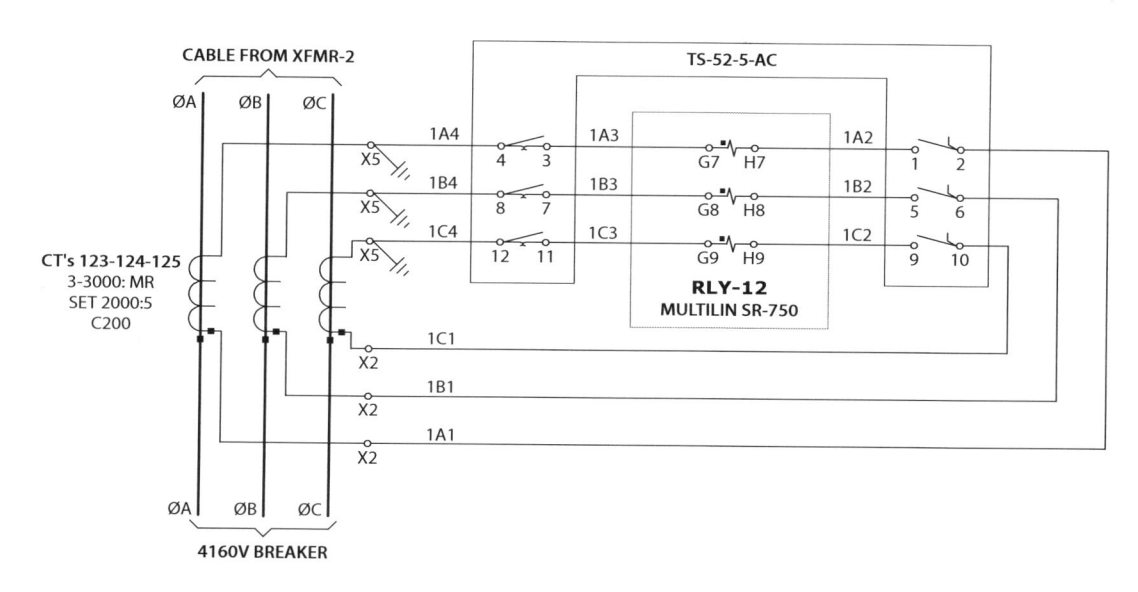

Figure 17-11: Alternate CT Connection #3

It's unlikely you'll ever see this CT configuration in the real world, but it demonstrates why polarity marks are important. Follow the flow of power into the CT polarity and out the secondary polarity mark.

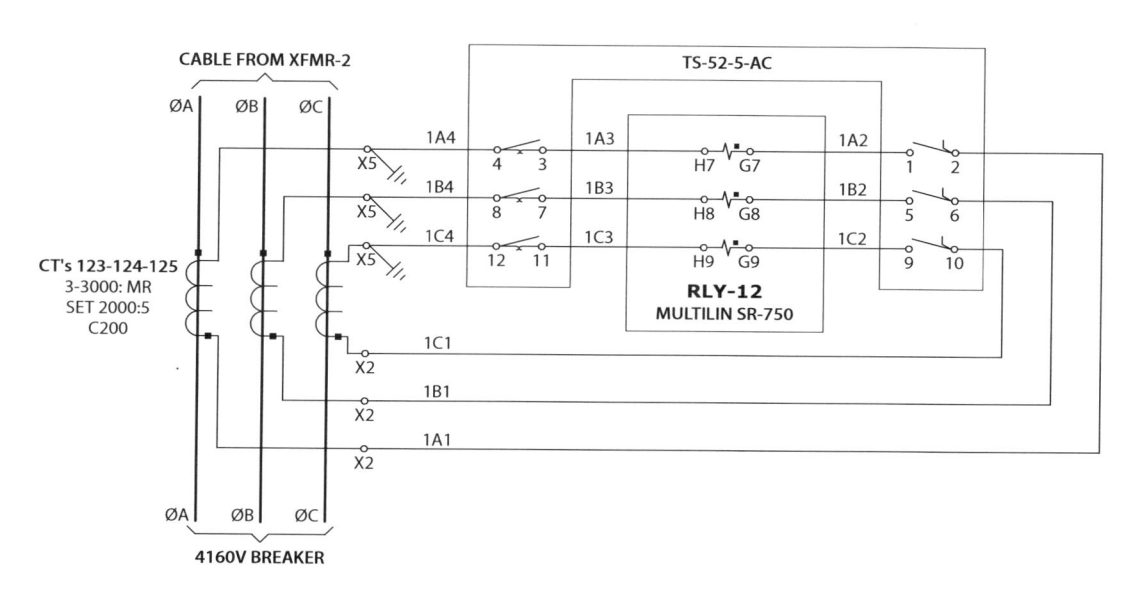

Figure 17-12: Alternate CT Connection #4

Some companies use A-C-B rotation and A-C-B connections with an A-C-B configured relay. The following connection is used in these cases. Pay careful attention to the system and relay's rotation when reviewing three-line drawings.

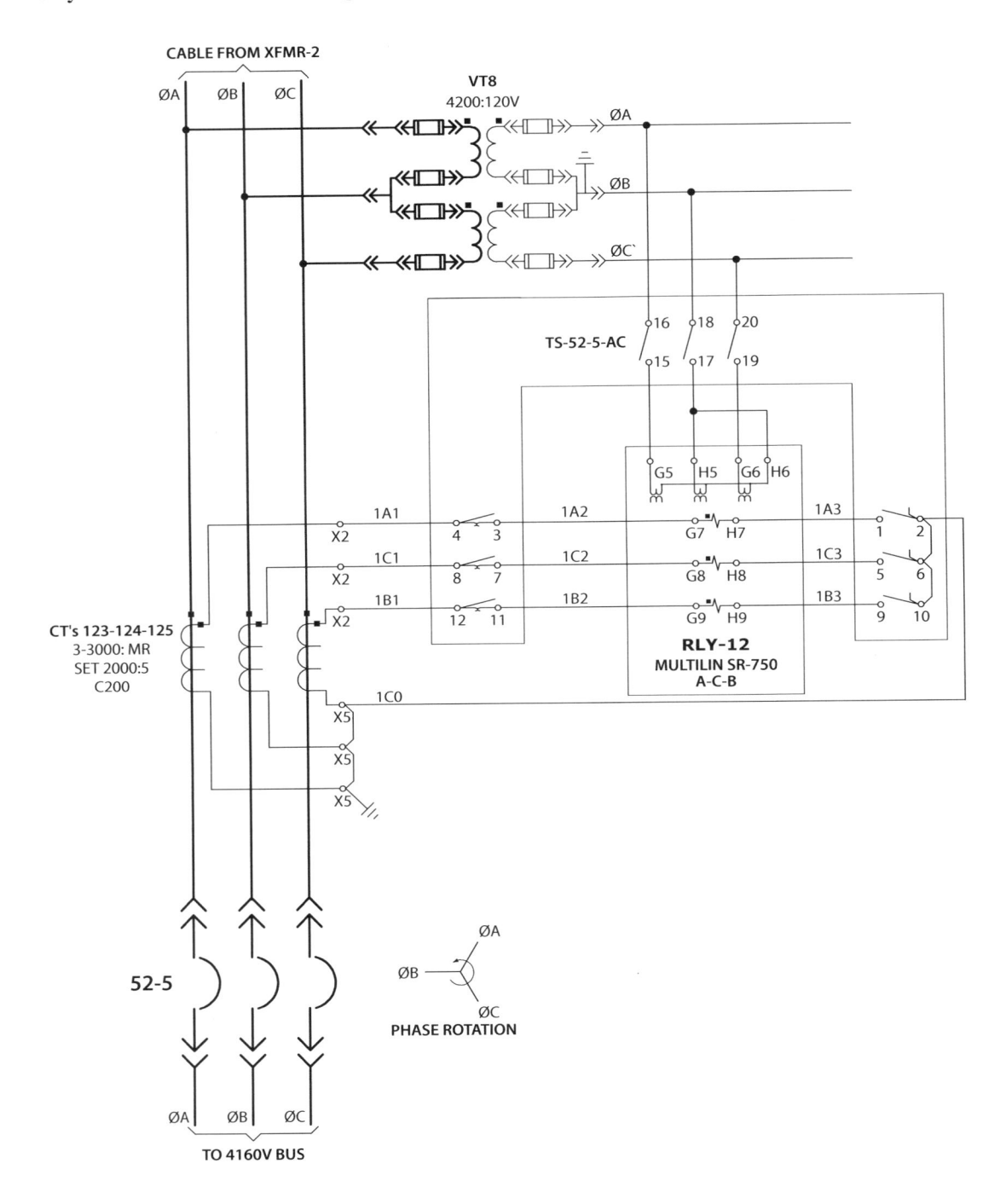

Figure 17-13: Using A-C-B Rotation with an A-C-B Relay

The following three-line drawing demonstrates how you can use an A-B-C configured relay in an A-C-B configured system. Both CT and PT connections must be swapped in the same manner otherwise the relay will not operate on any phase-related elements and may not correctly measure Watts, VARs, Power Factor, etc.

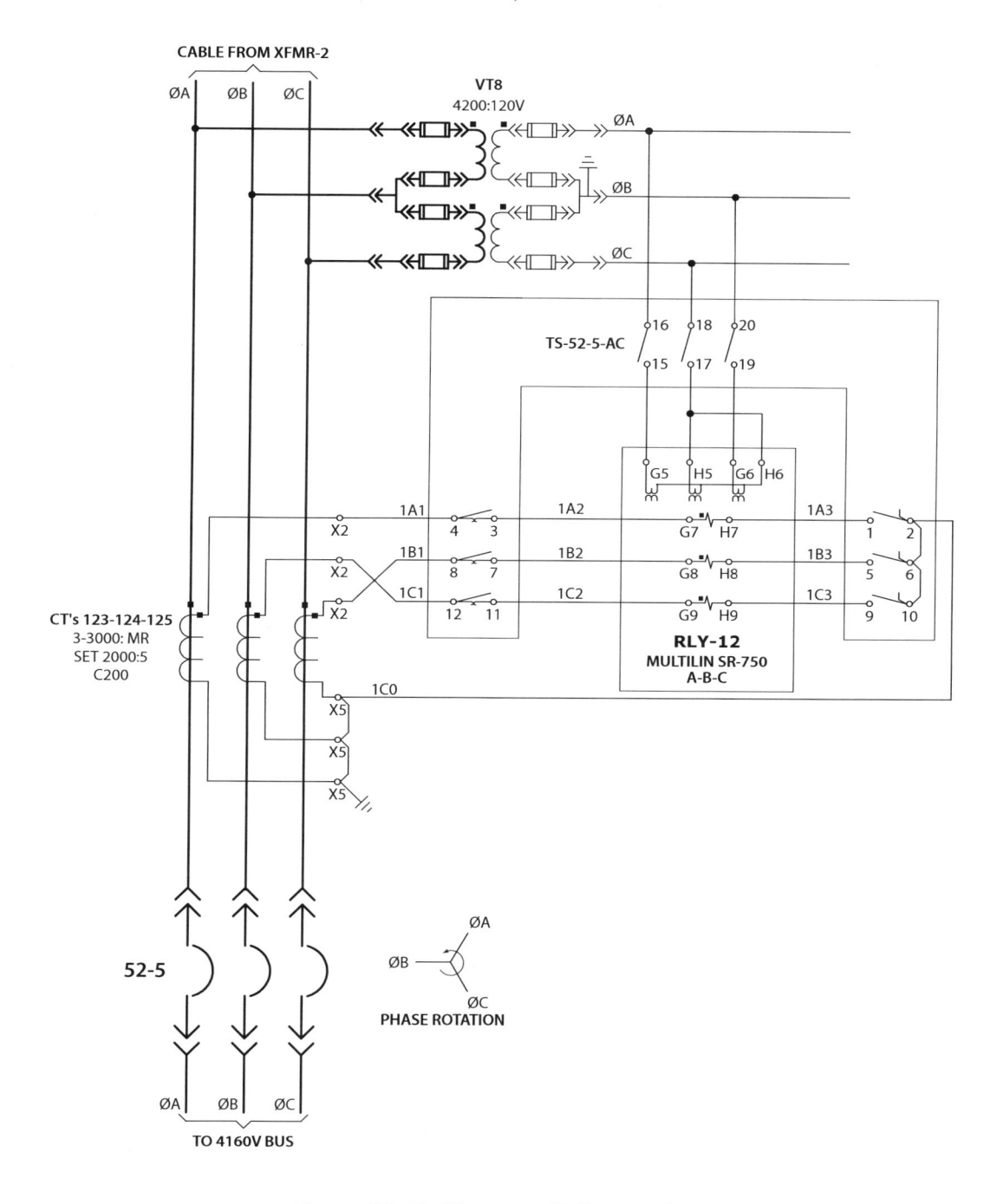

Figure 17-14: Alternate CT Connection

C) Test Switches

Test switches are installed between devices to isolate equipment without removing wiring. You will find test switches in nearly every utility environment, but they create an additional expense and are not as common in industrial environments. The most common test switch used in North America is based on the old Westinghouse "FT" style test switch. Each test block has ten (or more) switches and each switch can be individually configured to be a simple knife-blade style disconnect switch or; two, side-by-side, switches can be configured for CT connections. When looking from the front as shown in Figure 17-15, the terminal numbering starts from the top-left (1), the next number is bottom left (2), moving to the top of the next switch to the right (3). Odd numbers are always on top and even numbers are on the bottom. The switches are also labeled "A" to "J" from left to right. Test switch handles can be nearly every color and some companies color-code the switches for CTs, PTs, and DC connections. If color coding is applied, make sure it's consistent throughout the site.

Simple knife-blade style switches can be used in voltage circuits to isolate PTs; or in DC circuits to isolate trip outputs or relay inputs. Every company has a different standard when connecting test switches, and it is important not to make assumptions based on what you are used to. Some companies always connect the source of AC circuits (CTs and PTs) on the bottom of the test switch and the relays are always connected to the top. Other companies connect voltage sources on the topside of the switch to prevent the exposed blades from being energized if something metallic is dropped across open blades or to prevent unintentional shocks. It is important that the design is consistent throughout the site. If some test switches are supplied from the bottom and others are supplied from the top, make a note on your test sheet and bring it to your client's attention to prevent problems in the future. Figure 17-15 displays a FT test block configured with all voltage switches. All of the switches are closed except switch "C" (terminals 5-6) which is open. Obviously there is no connection between Terminals 5 and 6 when the switch is open

Figure 17-15: Voltage FT Test Switches

CT switches are a little more complicated and are always installed in pairs when used to short CT circuits. The right-side switch is installed with clips that will allow normal current flow when opened. A probe can be inserted between the upper and lower clips to directly measure the current in the circuit. An isolation device is inserted between the upper and lower clips when testing to prevent back feeds into the CTs and remove any parallel paths. You can use specially designed test plugs, test probes with open-circuits, wide tie-wraps, or (my personal favorite) store membership cards to isolate the CT circuits. Anything that will separate the upper and lower plug without permanently bending the clips will work in a pinch. The pair's left-hand switch is a shorting switch. When the switch is opened, the bottom of the switch makes contact with a metal bar that is installed between the two switches. When the left-hand switch is fully open, the bottom two (or even numbered) contacts are shorted together and current should flow into the bottom of one switch and out the bottom of the other. This is why it is very important that CTs are always connected to the bottom of the test switch. Incorrectly installed CT secondary connections can cause open-circuits that can create dangerously high CT secondary voltages. Current can flow into either switch and it's important to note which switch is which for test-set connections.

Test switches can be installed in nearly every configuration and incorrectly configured switches is a very common problem at new installations. Switch configurations can be quickly checked by looking for the clips and shorting bars and comparing them against the drawing terminal numbers. Figure 17-16 is a close up picture of a shorting style test block. If you look for the clips and shorting bars, you'll see that there could be three sets of CTs connected to this switch using switches:

- A-Phase = A(1-2) and B(3-4)
- B-Phase = C(5-6) and D(7-8)
- C-Phase = E(9-10) and F(11-12)

Figure 17-16: Current FT Test Switches

The previous test switch example has a very unique configuration because switch "G" has CT test plug clips but no corresponding shorting switch. This test switch is incorrectly configured unless the CT return path flows through this switch as shown in Figure 17-17 to allow for easy neutral current measurements.

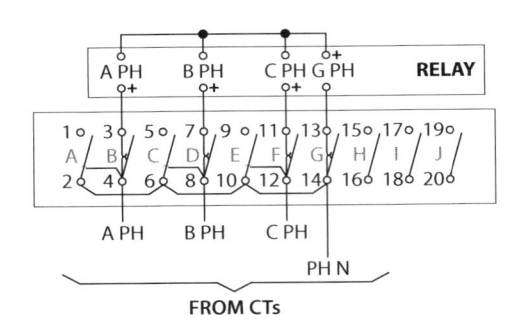

Figure 17-17: Current FT Test Switch Connection Drawing

The following drawings are examples of the most common test switch configurations.

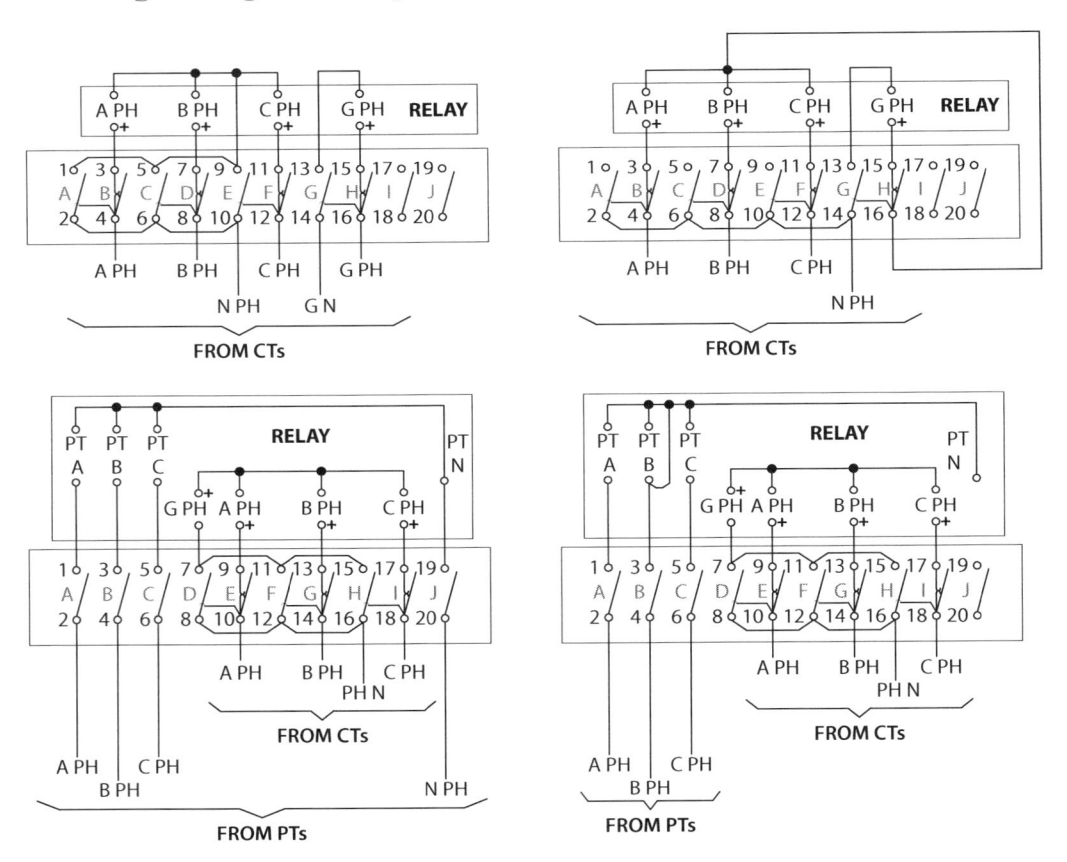

Figure 17-18: Common FT Test Switch Configurations

General Electric also produces a "PK" test block that has a very different principle of operation and is not common in newer installations. The test block uses a housing mounted in the panel and a keyed plug is inserted into the housing, connecting the top and bottom terminals during normal operation. The test block can be used for voltage and current applications depending on the metal jumpers installed in the housing as shown in Figure 17-19.

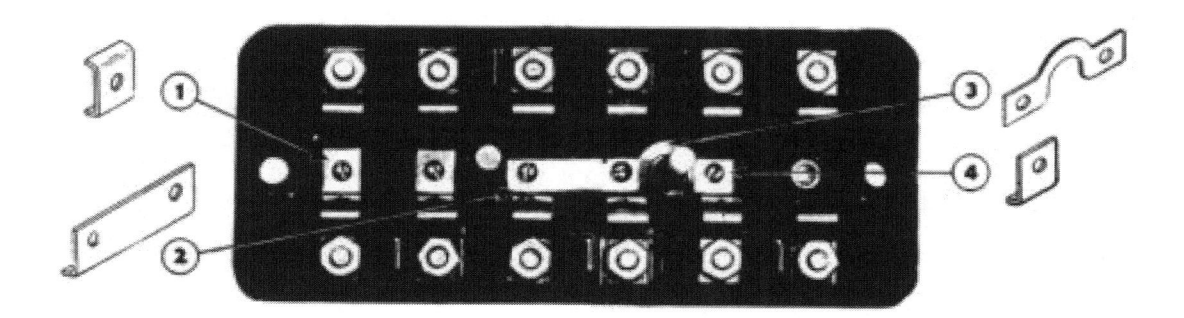

Figure 17-19: General Electric "PK" Style Test Block

The two sets of terminals on the left would be used for voltage circuits. When the plug is removed, the shorting link beside "1" connects the top and bottom terminals through spring clips installed on each terminal. When the normal plug is inserted, the spring clips separate from the link and the two terminals are connected through the plug. Special test plugs can be configured to separate the top from the bottom, or any insulating material can be inserted between the spring clip and shorting link for testing. The shorting clip is not necessary for correct operation of the test block and may not be desirable. The voltage source can be connected to the top or bottom depending on the site requirements.

Current test terminals use the horizontal shorting link to short the phase and return terminals connected to the CTs. One horizontal link can be used for each CT, or all of the horizontal links can be connected together with the circular shorting links designed to wrap around the alignment bars. The CT can be connected to the top or bottom of the test block but the shorting bars must be connected on the same side.

These test blocks can be dangerous because you cannot see the shorting link configuration until after the plug has been removed. If the configuration is incorrect, dangerously high open-circuit CT secondary voltages can be created. Always wear the appropriate personal protective equipment when opening an unproven PK style test block, and be prepared to immediately re-insert the plug if the configuration is incorrect.

D) Schematic Drawings

Schematic drawings show how the different control and protective devices interact to operate a device; or signal to display status. When relay testing, our primary focus will be the trip and close schematics for isolating devices (circuit breakers, contactors, etc.), SCADA or remote monitoring system points, and relay layout drawings if they exist. Collect all of these drawings to record information to continue our ongoing checklist.

The first drawing to look for is the relay layout drawing, if it exists. This is an extremely useful drawing for relay testing but can often be left out of a project as another way to cut costs. The drawing is usually associated with the relay's power supply if it has its own circuit and can look like the following figure. Check the terminals against the manufacturer's drawings and pay particular attention to input connections. The relay can easily be damaged by incorrect input connections. Use the example drawing to fill in the connection checklist we started earlier. Our example drawing has the ideal amount of information, but many real world drawings will be incomplete. Fill in as much information as possible on the checklist from this drawing.

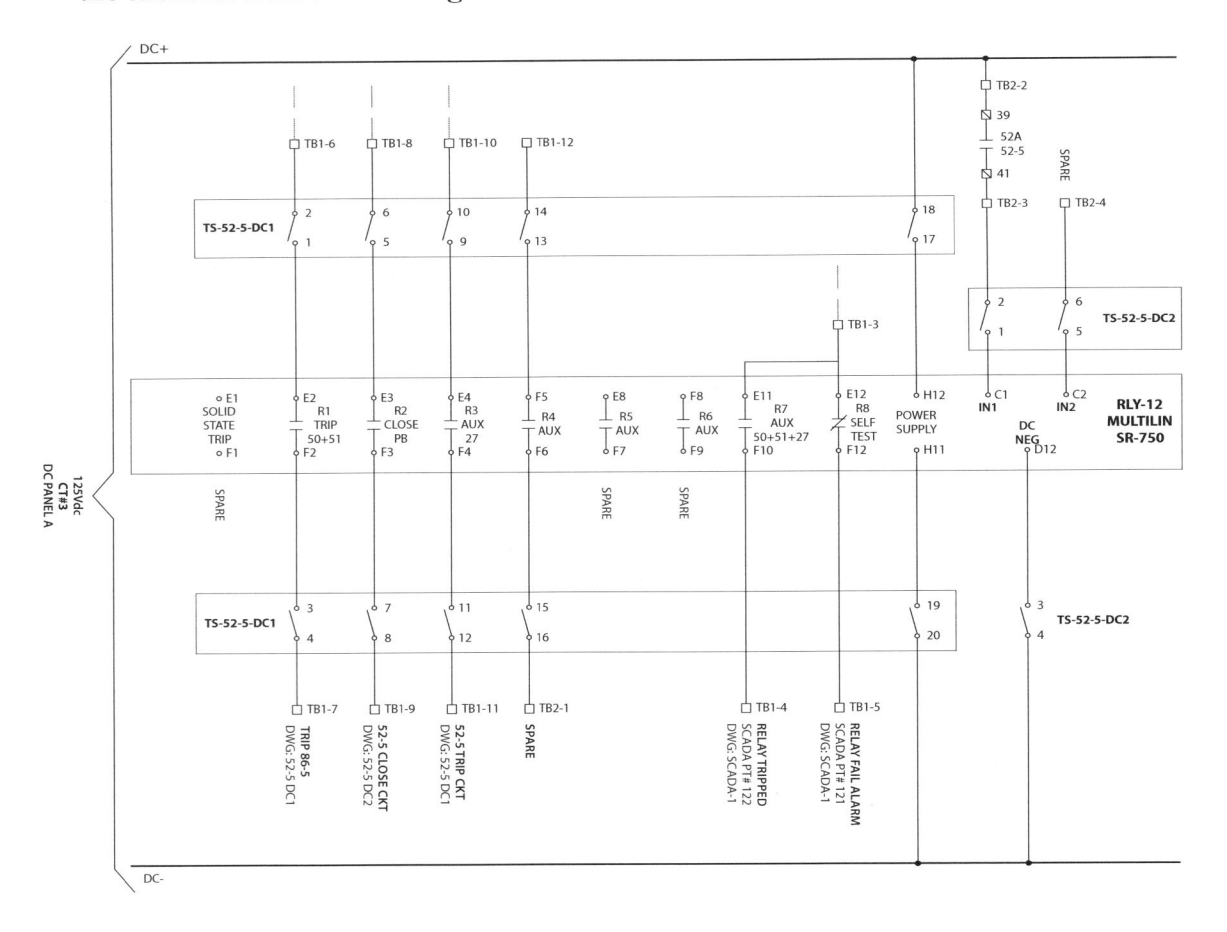

Figure 17-20: Example Relay Schematic Drawing

The following figures depict some examples of relay input connections from manufacturer's instruction manuals.

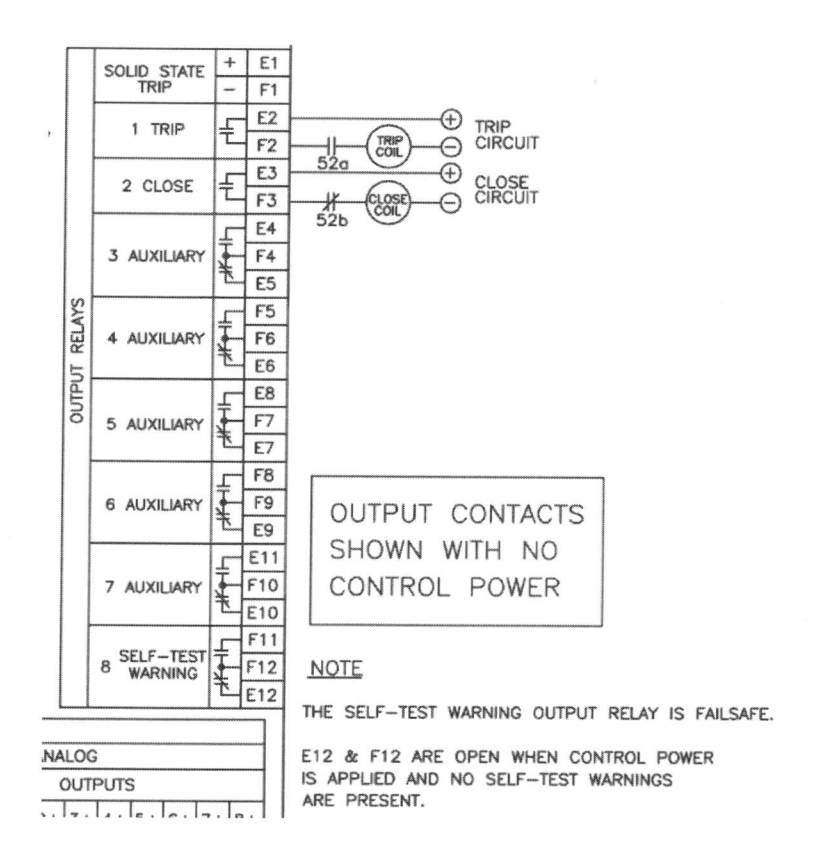

Figure 17-21: Example DC Output Connections

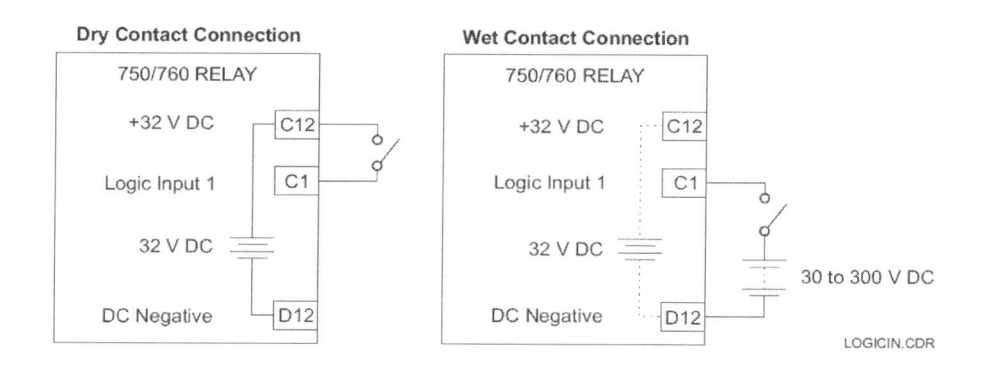

Figure 17-22: Example DC Input Connections

	SINGLE-LINE	3 LINE	DC SCHEMATIC	SETTINGS / NAMEPLATE
Relay Designation	RLY-12	RLY-12	**RLY-12**	
CT1 Location	Cable-Breaker	Cable-Breaker	**N/A**	
CT1 Ratio	2000:5	2000:5	**N/A**	
CT1 Polarity	Cable	Cable	**N/A**	
PT1 Location	Incoming	Incoming	**N/A**	
PT1 Primary Volts	4200	4200	**N/A**	
PT1 Secondary Volts	120	120	**N/A**	
PT1 Ratio	35:1	35:1	**N/A**	
PT1 Nominal Pri Volts	4160V	4160V	**N/A**	
PT1 Nominal Sec Volts	118.86V (4160 / 35)	118.86V (4160 / 35)	**N/A**	
PT1 Connection	Open Delta	Open Delta	**N/A**	
Relay Power	N/A	N/A	**125VDC**	
Out1 Initiate	R1 = 50 + 51	N/A	**50+51**	
Out1 Connected	TRIP 86-5	N/A	**TRIP 86-5**	
Out2 Initiate	R2 = CLOSE	N/A	**PB**	
Out2 Connected	52-5 CLOSE	N/A	**52-5 CLOSE**	
Out3 Initiate	R3 = 27	N/A	**27**	
Out3 Connected	Trip 52-5	N/A	**TRIP 52-5**	
Out4 Initiate	N/A	N/A	**SPARE**	
Out4 Connected	N/A	N/A	**SPARE**	
Out5 Initiate	N/A	N/A	**SPARE**	
Out5 Connected	N/A	N/A	**SPARE**	
Out6 Initiate	N/A	N/A	**SPARE**	
Out6 Connected	N/A	N/A	**SPARE**	
Out7 Initiate	N/A	N/A	**50+51+27**	
Out7 Connected	N/A	N/A	**SCADA**	
Out8 Initiate	N/A	N/A	**RELAY FAIL**	
Out8 Connected	N/A	N/A	**SCADA**	
IN1 Initiate	N/A	N/A	**52-A / 52-5**	
IN1 Operate	N/A	N/A	**N/A**	
IN2 Initiate	N/A	N/A	**SPARE**	
IN2 Operate	N/A	N/A	**N/A**	

Figure 17-23: Example Relay Information Checklist #3

Sometimes the relay layout drawing is a copy of the manufacturer's drawing with site specific information added that makes it easier to find the information you need. An example of this kind of drawing follows:

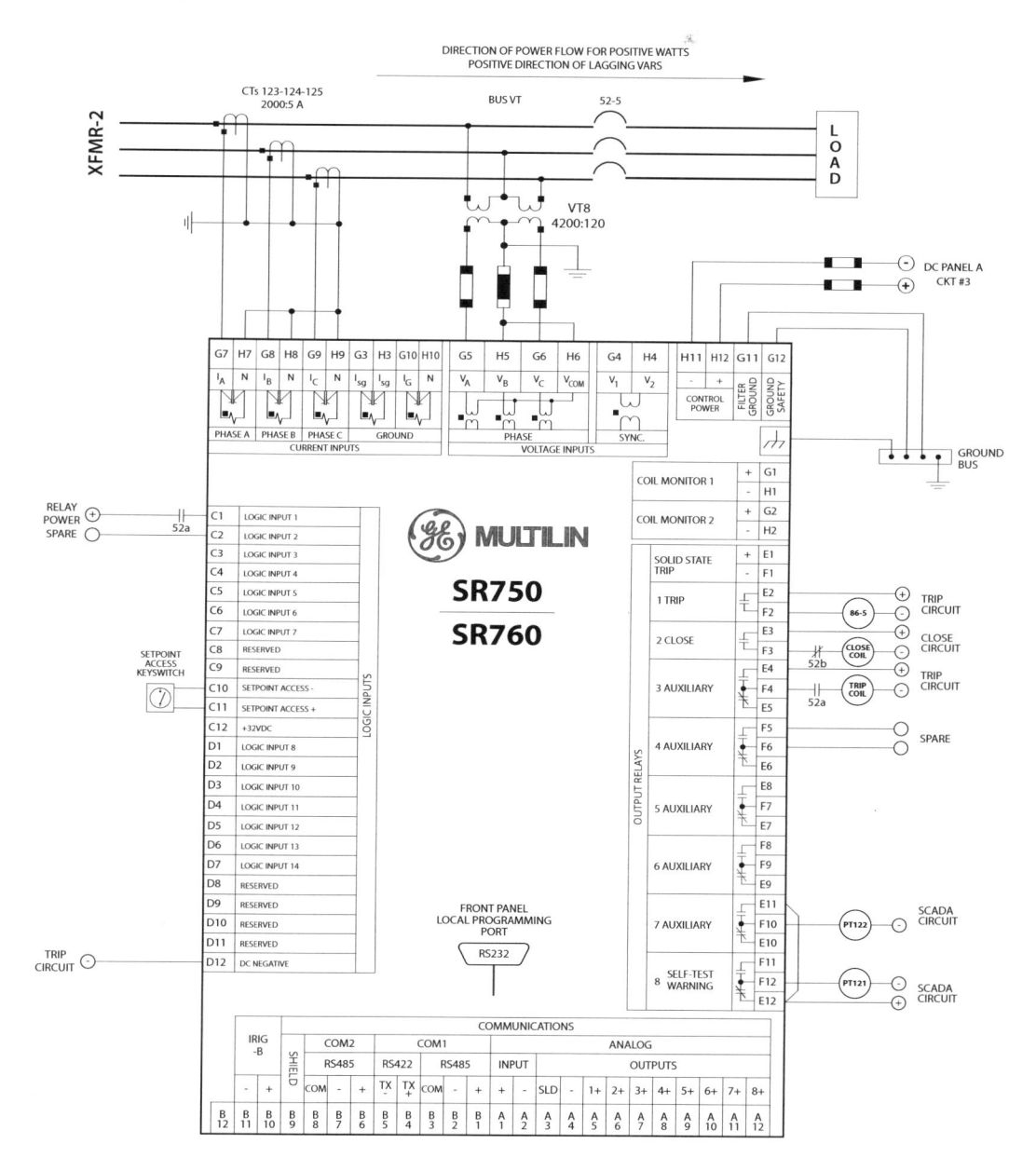

Figure 17-24: Example Manufacturer's Connection Diagram

Use the information from the relay layout drawing to quickly locate the relay inputs and outputs on the related schematic drawings. Compare each schematic to the checklist and ensure that everything matches.

If the relay layout schematic does not exist, look through each schematic until you locate all of the inputs and outputs related to the application. This can be tricky or difficult as the drawings become more complex. Sometimes it's necessary to look at the back of the relay to determine which outputs and inputs are connected to external wiring to find out how many are used. Our example application for RLY-12 uses five outputs as shown in the relay layout diagram on Figure 17-20. Three of these outputs can be found on drawing 52-5 DC1 as shown by the shaded areas in Figure 17-25. Without the relay layout drawings, you would have to look at all of the DC schematics related to the SCADA system and 52-5 to find all five outputs. The relay layout drawing helps us by specifying drawings 52-5 DC1, 52-5 DC2, and SCADA-1.

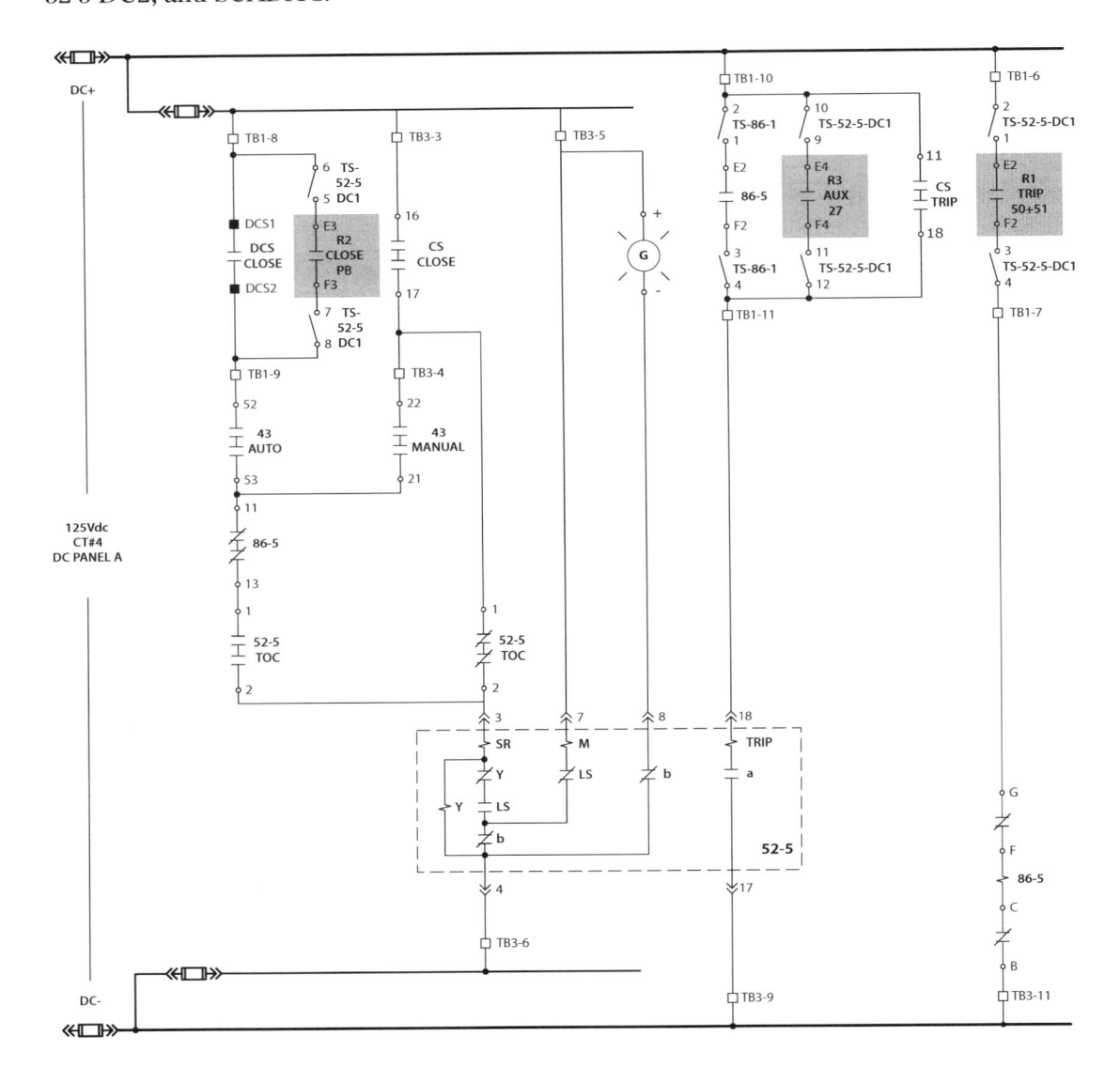

Figure 17-25: Example Trip / Close Schematic

E) Settings

Most of the problems found during relay testing occur when comparing the settings to all of the information we've gathered so far. The most common problems are listed below and should be carefully reviewed. Sometimes it seems that the drawings are not even reviewed when settings are issued!

- CT / PT ratios and connections don't match the drawings
- Outputs are assigned in the relay but not connected to anything
- Outputs are connected to circuits and not assigned inside the relay

Settings can be provided in their native software, as Microsoft Word™, Adobe Acrobat™ (pdf) documents, or in any combination. Print the settings (if necessary), collect the coordination study (if provided), and complete the checklist. I prefer to wait until I'm sitting in front of the relay and am entering the settings into the relay for two reasons:

First, relays are constantly evolving and manufacturers keep adding new features, correcting mistakes, and creating new revisions of the relay's operating software. The provided settings could have been created using an older version or incorrect relay model and some settings might be missing. Sometimes the provided relay settings are outside acceptable ranges and this problem won't be noticed until the settings are applied to the relay. By waiting until the settings are applied, I can hopefully find all of the problems at once.

The second reason is more a personal preference. I find it difficult to catch mistakes when they are simply printed on a page. Especially, when relays have dozens of setting pages for simple functions and 90% of the settings do not apply. Some relays will automatically block unused settings from view when an element is disabled but will print every setting making it difficult to separate the wheat from the chaff. Most people pay more attention when doing instead of reviewing, and this helps me find mistakes.

On the other hand, waiting till the last minute can delay testing as you try to resolve the discrepancies.

You are inevitably going to discover discrepancies between all of the information that you've collected and already know what needs to be done to correct it. It's important that you realize that you are not the engineer responsible for this project and should not unilaterally make changes based on the information you have. It's common to be provided drawings that are out of date that will make your changes obsolete. There may also be bigger-picture reasons for settings that you are unaware of. Summarize all of the problems, suggest corrections, and submit them to the engineer in charge. Document all changes and submit the final settings to the appropriate parties.

The following figure contains the pertinent information for the Multilin SR-750 relay in our example.

S2 System Setup

Current Sensing

Phase CT Primary	2000 A
Ground CT Primary	50 A
Sensitive Gnd CT Primary	1000 A

Power System

Nominal Frequency	60Hz
Phase Sequence	ABC
Cost of Energy	5.0 ¢/kWh

Bus VT Sensing

Type	Delta
Nominal VT Sec. Voltage	118.9 V
Ratio	35.0:1

Line VT Sensing

Type	Vab
Nominal VT Sec. Voltage	118.9 V
Ratio	35.0:1

S3 Logic Inputs

Logic Inputs Setup

Input1 Asserted Logic	Contact Close
Input2 Asserted Logic	Contact Close

Logic Inputs

Input1 Name	52-5 52a
Input2 Name	Logic Input 2

S4 Output Relays

Relay 1 Trip

Seal In Time	0.04 s

Relay 2 Close

Seal In Time	0.04 s

Relay 3 Auxiliary

Name	27 Trip
Non-operated State	De-energized
Output Type	Self-resetting

Relay 4 Auxiliary

Name	AUXILIARY
Non-operated State	De-energized
Output Type	Self-resetting

Relay 5 Auxiliary

Name	AUXILIARY
Non-operated State	De-energized
Output Type	Self-resetting

Relay 6 Auxiliary

Name	AUXILIARY
Non-operated State	De-energized
Output Type	Self-resetting

Relay 7 Auxiliary

Name	SCADA
Non-operated State	De-energized
Output Type	Self-resetting

S5 Protection-Phase Current

Phase Time Overcurrent 1

Function	Trip
Relays	----7
Pickup	1.00 X CT
Curve	Normally Inverse
Multiplier	3.00
Reset	Instantaneous
Direction	Disabled
Voltage Restraint	Disabled

Phase Time Overcurrent 2

Function	Disabled

Phase Instantaneous Overcurrent 1

Function	Trip
Relays	----7
Pickup	9.00 X CT
Delay	0.10 s
Phases Required for Operation	Any One
Direction	Disabled

Phase Instantaneous Overcurrent 2

Function	Disabled

Phase Directional

Function	Disabled

S5 Protection-Voltage

Bus Undervoltage 1

Function	Control
Relays	3---7
Pickup	0.90 X VT
Curve	Definite Time
Delay	60.0 s
Phases Required for Operation	Any One
Minimum Operating Voltage	0.30 X VT

Bus Undervoltage 2

Function	Control
Relays	3---7
Pickup	0.85 X VT
Curve	Definite Time
Delay	20.0 s
Phases Required for Operation	Any One
Minimum Operating Voltage	0.30 X VT

Figure 17-26: Example Setting Printout

	SINGLE-LINE	3 LINE	DC SCHEMATIC	SETTINGS / NAMEPLATE
Relay Designation	RLY-12	RLY-12	RLY-12	**RLY-12**
CT1 Location	Cable-Breaker	Cable-Breaker	N/A	**N/A**
CT1 Ratio	2000:5	2000:5	N/A	**2000:5**
CT1 Polarity	Cable	Cable	N/A	**N/A**
PT1 Location	Incoming	Incoming	N/A	**N/A**
PT1 Primary Volts	4200	4200	N/A	**N/A**
PT1 Secondary Volts	120	120	N/A	**N/A**
PT1 Ratio	35:1	35:1	N/A	**35:1**
PT1 Nominal Pri Volts	4160V	4160V	N/A	**N/A**
PT1 Nominal Sec Volts	118.86V (4160 / 35)	118.86V (4160 / 35)	N/A	**118.9 V**
PT1 Connection	Open Delta	Open Delta	N/A	**Delta**
Relay Power	N/A	N/A	125VDC	**90-300VDC From Nameplate**
Out1 Initiate	R1 = 50 + 51	N/A	50+51	**Trip = 50+51**
Out1 Connected	TRIP 86-5	N/A	TRIP 86-5	**N/A**
Out2 Initiate	R2 = CLOSE	N/A	PB	**Close**
Out2 Connected	52-5 CLOSE	N/A	52-5 CLOSE	**N/A**
Out3 Initiate	R3 = 27	N/A	27	**Bus27-1 + Bus27-2**
Out3 Connected	Trip 52-5	N/A	TRIP 52-5	**N/A**
Out4 Initiate	N/A	N/A	SPARE	**N/A**
Out4 Connected	N/A	N/A	SPARE	**N/A**
Out5 Initiate	N/A	N/A	SPARE	**N/A**
Out5 Connected	N/A	N/A	SPARE	**N/A**
Out6 Initiate	N/A	N/A	SPARE	**N/A**
Out6 Connected	N/A	N/A	SPARE	**N/A**
Out7 Initiate	N/A	N/A	50+51+27	**50 + 51 + Bus27-1 + Bus27-2**
Out7 Connected	N/A	N/A	SCADA	**N/A**
Out8 Initiate	N/A	N/A	RELAY FAIL	**N/A**
Out8 Connected	N/A	N/A	SCADA	**N/A**
IN1 Initiate	N/A	N/A	52-A / 52-5	**52a**
IN1 Operate	N/A	N/A	N/A	**N/A**
IN2 Initiate	N/A	N/A	SPARE	**N/A**
IN2 Operate	N/A	N/A	N/A	**N/A**

Figure 17-27: Example Relay Information Checklist #4

There were 25 pages of information in the original printout of settings from the relay that was narrowed to the information presented previously. I prefer connecting to the relay and opening every option screen in the software to find out what settings are enabled. The following screen captures show the same settings through the relay communication software.

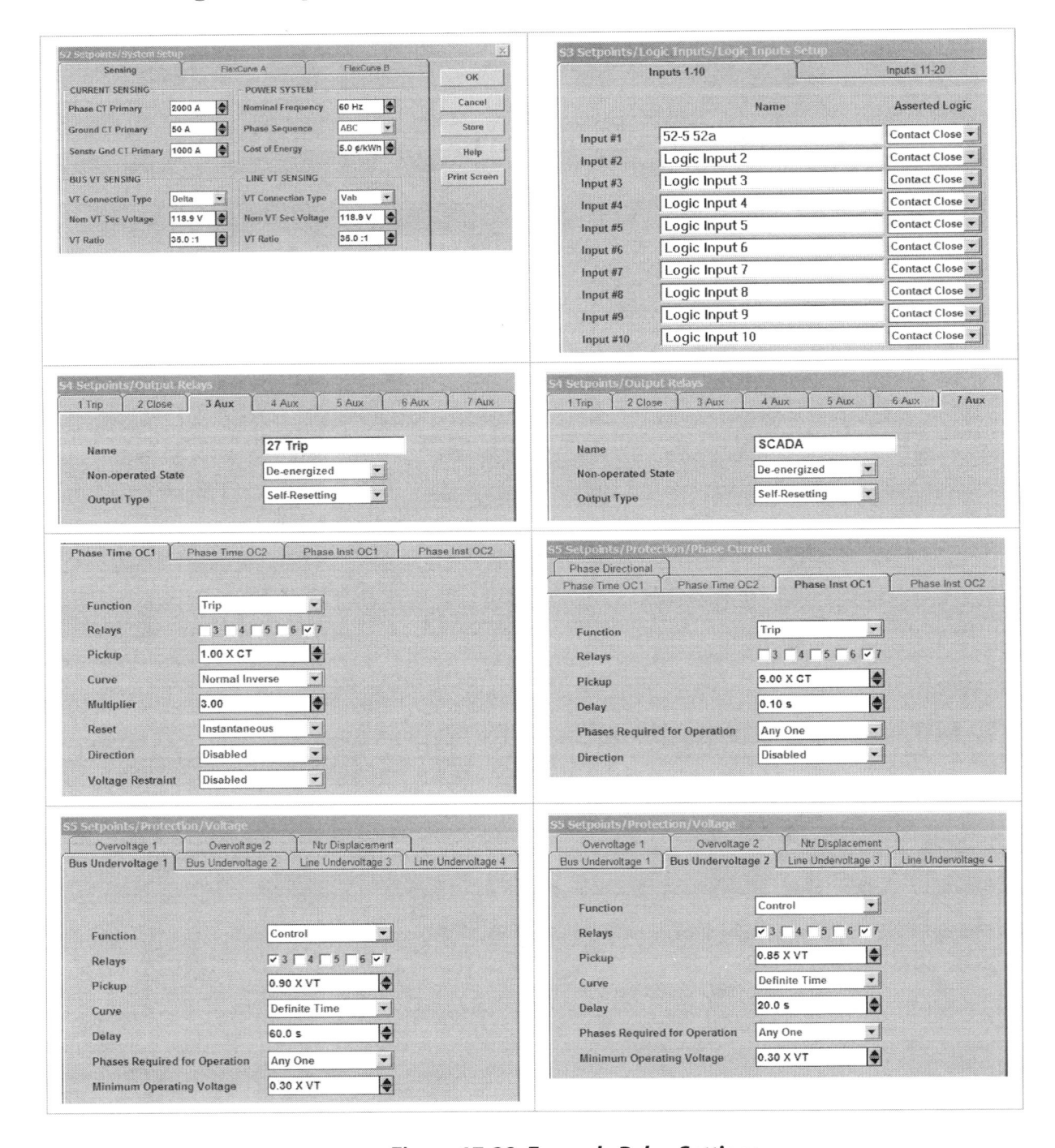

Figure 17-28: Example Relay Settings

Common sense should always prevail when reviewing settings. If a setting just doesn't make sense in the application, make a note of it and include it in your comments. Many relay errors are never detected until it's too late because relays are only designed to operate (hopefully) under rare circumstances.

F) Coordination Studies

Coordination studies (See *Time Coordination Curves (TCC) and Coordination* starting on page 73) display the overcurrent protection settings for all of the relays in TCC curves to ensure that relays will coordinate with each other. You should have this study when relay testing, but that doesn't happen very often. If you are provided with a copy, find the relay under test on a TCC curve and make sure that the settings in the description box match the relay settings. Figure 17-29 is an example of a simple TCC drawing.

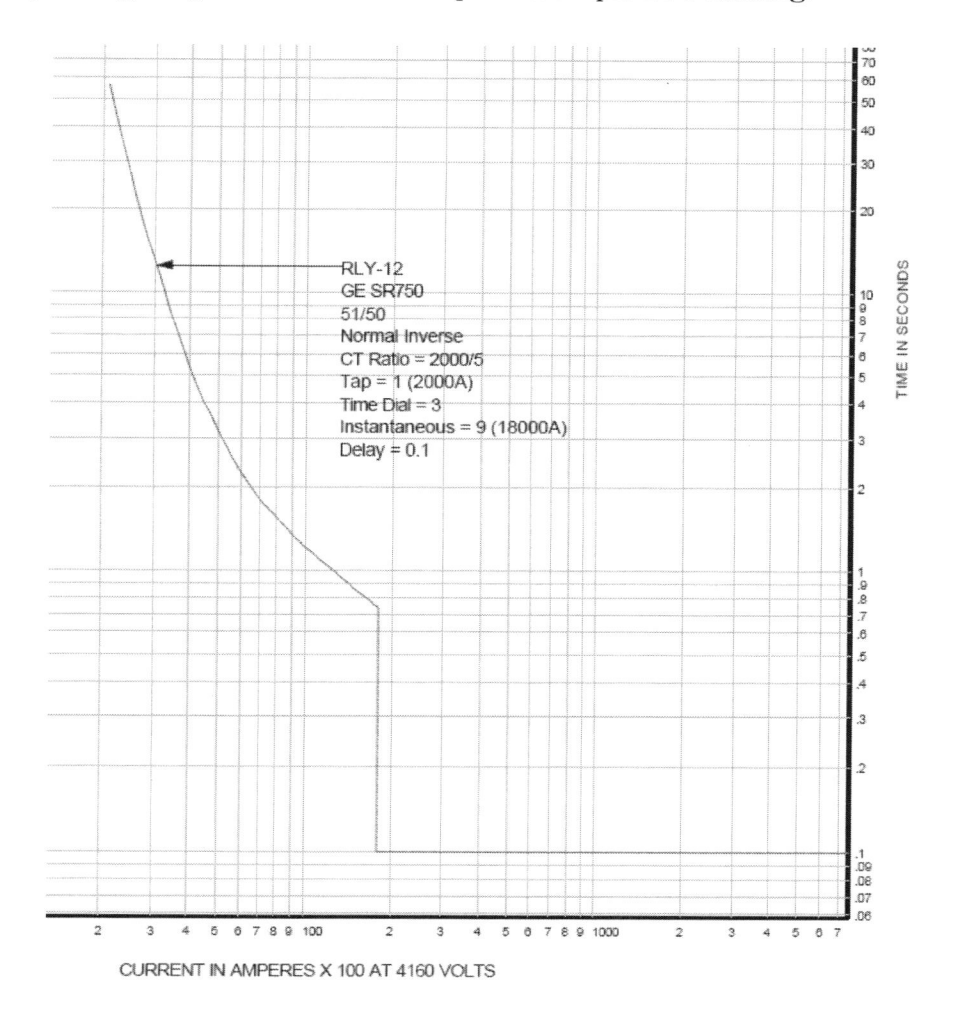

Figure 17-29: Example TCC Drawing

In the absence of a coordination study, check the pickup values, time dials, and CT ratios of the relays upstream and downstream of the relay under test. Calculate the primary pickup values using the relay pickup and CT ratio and make sure the upstream relay is greater than the relay under test pickup and the downstream relay pickup is less than the relay under test. If the pickup settings are close, make sure the time dial is lower than the upstream relay and higher than the downstream relay. This is particularly important in instantaneous and ground fault relay applications to prevent upstream relays from operating before downstream relays during a fault. the *Time Coordination Curves (TCC) and Coordination* section starting on page 73 details the calculations for different voltage levels. Do not compare ground fault settings at different voltage levels because the transformer should provide isolation between the two voltage levels.

Chapter 18

Preparing to Test

1. Prepare to Test

Always ensure that the relay is completely isolated from the rest of the electrical system before changing settings or connecting your test-set unless the relay cannot adversely affect the system. The trip circuit should be disabled first before isolating the AC circuits. Reconnect the AC circuits first and ensure that no trip signals are present before returning the trip and close circuits to service.

2. Establish Communication

While most relays will allow you to enter information via the front panel, this is a tedious and often unreliable process that should be avoided at all costs. It is usually easier to enter the settings using the relay manufacturer's software. Another benefit to remote communication is the ability to save the as-left settings. This information is vital to protect yourself if someone else changes the settings after you, and the relay does not operate correctly. While relay communication usually makes our life easier when it is working correctly, the most painful part of relay testing is often establishing communications with an unfamiliar relay.

When two computers communicate (and today's relays are just computers with analog to digital conversions, digital inputs, digital outputs, and programming), information is sent in packets called bytes. As each packet is received, additional information is sent to provide error checking and signal the start or stop of each byte. If the error correction bit matches the information sent, the computer accepts the information and sends its own data or continues to receive new data.

Here are some definitions and tips to help you along:

A) Port

Some relays have several communication ports to allow multiple devices to communicate simultaneously. It is important to make sure the settings of the port you will be using are the same as the final settings because when you send settings to the relay all at once, and the port settings are different, any changes after the communication setpoints are saved will not be reliably applied to the relay.

Your laptop computer probably doesn't have a RS-232 computer port, and you may need to use a USB/Serial port converter. Sometimes it is difficult to determine what computer port you're using. You can find and change the computer port number by opening the control panel in Microsoft Window XP™ and:

- Double-click the "system" folder/icon.
- Select the "Hardware" tab and click the "Device Manager" button.
- Find and expand the "Ports (COM & LPT)" directory and right click the description of your USB/Serial port adapter.
- Select "Properties" and look through the tabs. The port designation setting should be obvious at this point.

B) Port Configuration

Most communication ports are pre-configured from the factory. However, some relay port configurations can be changed via a jumper inside the relay. Some of the common configurations include:

1. **RS-232**—This is the most common configuration and is used to connect any two devices via a communication cable or modem. The communication port can be a 9 or 25 pin, but Pins 2 and 3 are normally used to transfer data. A straight through cable is wired so that Pin 2 and Pin 3 are the same on both ends. A "null modem" cable or adapter crosses Pins 2 and 3, so Pin 2 on one end is connected to Pin 3 on the other end. The null modem connection simulates a modem and is necessary to communicate to some relays. The following manufacturers usually use straight through cables for communication with most models:

   - Newer generation General Electric (D-60, L90, etc.)
   - GE /Multilin
   - Alstom
   - Basler Electric

These manufacturers usually require null modem connections for communication:

- Schweitzer Laboratories
- Beckwith Electric
- Older generation General Electric (DGP, LPS, etc.)

2. **RS-485**—The RS-485 configuration allows for multiple devices to be connected to a network that communicates over two or four wires. The network is only limited by the length of wire and the maximum number of device address numbers. With this network, different devices from different manufacturers can be connected together and communicate with a master device that collects all of the information, such as a SCADA system. However, all of the devices must all have the same communication settings with different addresses and speak the same language. The two most common languages are MODBUS and DNP. It is possible to connect a computer to a RS-485 port using an external RS-485 to RS-232 converter. Be wary when using self-powered converters as some models do not transfer all of the required data, and they may not work.

3. **Network**—Some newer relays will have a network port like the network port on your computer. While there are several different protocols possible for communication, the most common is TCP/IP. TCP/IP uses four number addresses separated by decimals such as "192.168.000.001." The most common problem when trying to communicate with relays that use the TCP/IP address system is that the computer IP address is incompatible with the relay address. The first 3 groups of IP address numbers should be identical between the computer and device when troubleshooting communication problems. Also, make sure each device on the connected network has a unique address. In Microsoft Windows XP™, change the network address by:

   - Selecting "Network Connections" from the control panel, right clicking on the port used (usually "Local Area Connection") and select "Properties."
   - Scroll down and select "Internet Protocol (TCP/IP)" then push the "Properties" button.
   - Select "Use the following IP address" button and enter the first three numbers from the relay with a fourth number that is unique to the network. The subnet mask must also be identical.
   - Click "OK" to save the settings.
   - You may also need to re-boot your computer to enable the new settings depending on your computer. Don't be surprised if you can't connect to your other networks or the internet until the changes have been reversed.

C) Protocol

Some relays allow you to change the protocol or "language" that is expected from the port. This setting allows different devices to retrieve information from the relay. The most common protocols are:

1. **MODBUS**—This language is used by SCADA systems to poll information from different devices via a RS-485 or TCP/IP network connection. Each relay will have a unique address and a map of available information that can be retrieved using the HEX numbering system. The remote device communicates to the relay using the MODBUS language, and it retrieves information that is stored in the selected memory locations for information such as real time metering, settings, last trip data, etc.

2. **DNP**—This is another language with the same application as MODBUS described above. Any information to be shared is set up with a unique address and scaling factors are applied to provide uniform information to the data collector.

D) Baud Rate

This setting determines the rate of data transfer. Make sure that your computer program communication settings match the relay's settings. Baud rates are typically 1200, 2400, 4800, 9600, 19200, 38400, etc. A higher baud rate will transfer data quickly, but higher baud rates can create errors due to interference. If you are having problems communicating and suspect interference from surrounding equipment, try a 2400 baud rate. The most common problem with communications is incorrect baud rates. Be wary when buying your next laptop because newer models do not have a serial port and you must purchase external USB or PC Card serial ports which may not work with DOS based programs. If the program and relay baud rates are the same and the relay still will not communicate, check the hardware port setting for the operating system you are using. These settings can be found by selecting "System" from the control panel, selecting the "Hardware" tab, clicking on the "Device Manager" button, opening the "Ports & LP1" directory, and double clicking the appropriate port. The following setting can also be included with the baud rate settings:

1. **Bit**—This setting determines the number of data bits that are sent and expected in each packet of information.

2. **Parity**—This setting determines whether or not an additional bit is added to the end of the data packet for error checking. If "odd" is selected, the parity bit will be a 0 or 1 to make the number of 1's in the byte an odd number. Conversely, an "even" setting will cause the parity bit to change to make the number of 1s in the byte even. If the two devices have different parity settings, all data transfers could be rejected as errors.

3. **Stop**—When a byte is transferred between devices there is a start bit, the data bits, and stop bits. This setting defines how many stop bits will be used.

4. **Time Out**—This setting limits the amount of time the port remains open after it has been idle. This is to prevent unauthorized personnel from communicating with the relay without the required passwords after you have disconnected from the relay without logging out.

3. Apply Settings

Use the manufacturer's software to install the supplied settings. Hopefully, the engineer has supplied a digital copy of the settings but you may have to enter all the settings in by hand.

The supplied settings are often out of date because the setting engineer used an older version of the relay or similar model to create the settings. These slight differences can cause the relay to reject settings or round up or down to the next acceptable value. Carefully document each discrepancy and communicate any required changes to the design engineer. After all of the settings have been accepted by the relay, save a copy of the file and print the settings. Compare every setting to the supplied setting to ensure they are identical. If changes are made, save and print the file and review it again. Repeat this process until the supplied and applied settings are identical.

If the settings require significant changes, STOP and do not proceed until the new setting changes have been approved by the powers that be.

4. Connect Your Relay Test-Set

A) Relay Output Connections

It is important to disable the connected device (circuit breaker, lockout relay, etc.) before beginning any relay testing. Carefully review the drawing to make sure all output contacts are accounted for and open any test switches, panel circuit breakers, fuses, etc. necessary to prevent unintended equipment operation. The circuit breaker position, relay operation, or metering values applied during testing could have unforeseen consequences in an external plant-wide logic controller that could cause embarrassing and expensive outages if appropriate measures are not taken.

There are several ways to connect relay test output contacts depending on your test-set and the field connections. The simplest connection applies your test-set input contacts directly across the relay's output contacts. This is a simple task with test switches as shown in Figure 18-1. TS-52-5-DC1 switches A and B are opened and the test-set input is connected at the test switches or on the relay itself. Test switches are nice but not always available, and you can probably connect your test-set input across the contact without test switches as shown on the right side of Figure 18-1. Check with your test-set manufacturer before attempting this connection, however. Some relay manufacturer inputs are polarity sensitive and you may need to reverse the polarity if the test-set senses the contact is closed when it is actually open.

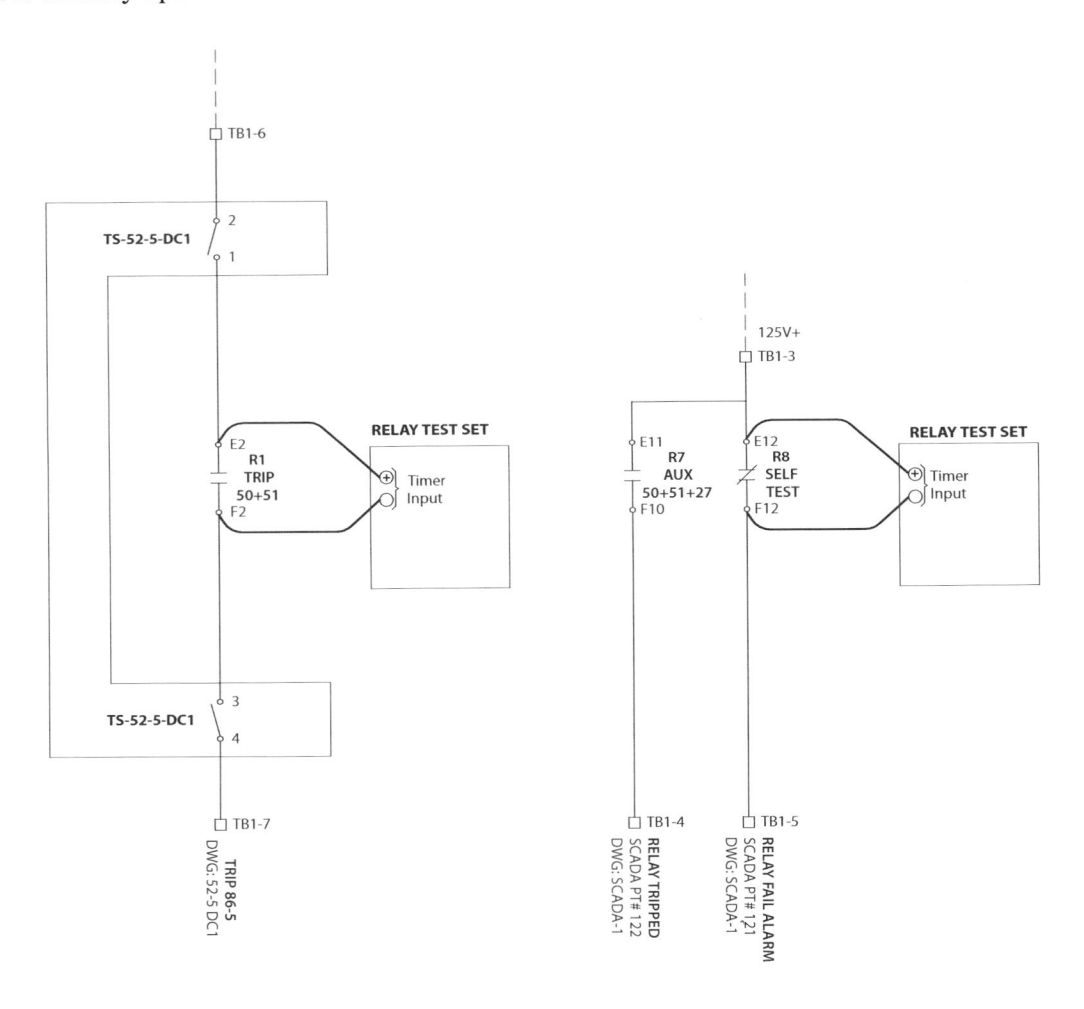

Figure 18-1: Simple Test-Set Input Connections

If test switches are not provided or are closed, another part of the circuit could be shorted in parallel with the output contact under test and cause a false operation. The relay "R2 Close" contact in Figure 18-2 is connected in parallel with the "DCS close" contact. If the DCS contact is closed when our test switches are closed, the relay input will sense that the contact is closed. We can solve this easily in our example by opening either of the test switches, but we must remove one wire when no test switches are provided. Figure 18-2 displays the different options when contacts are connected in parallel.

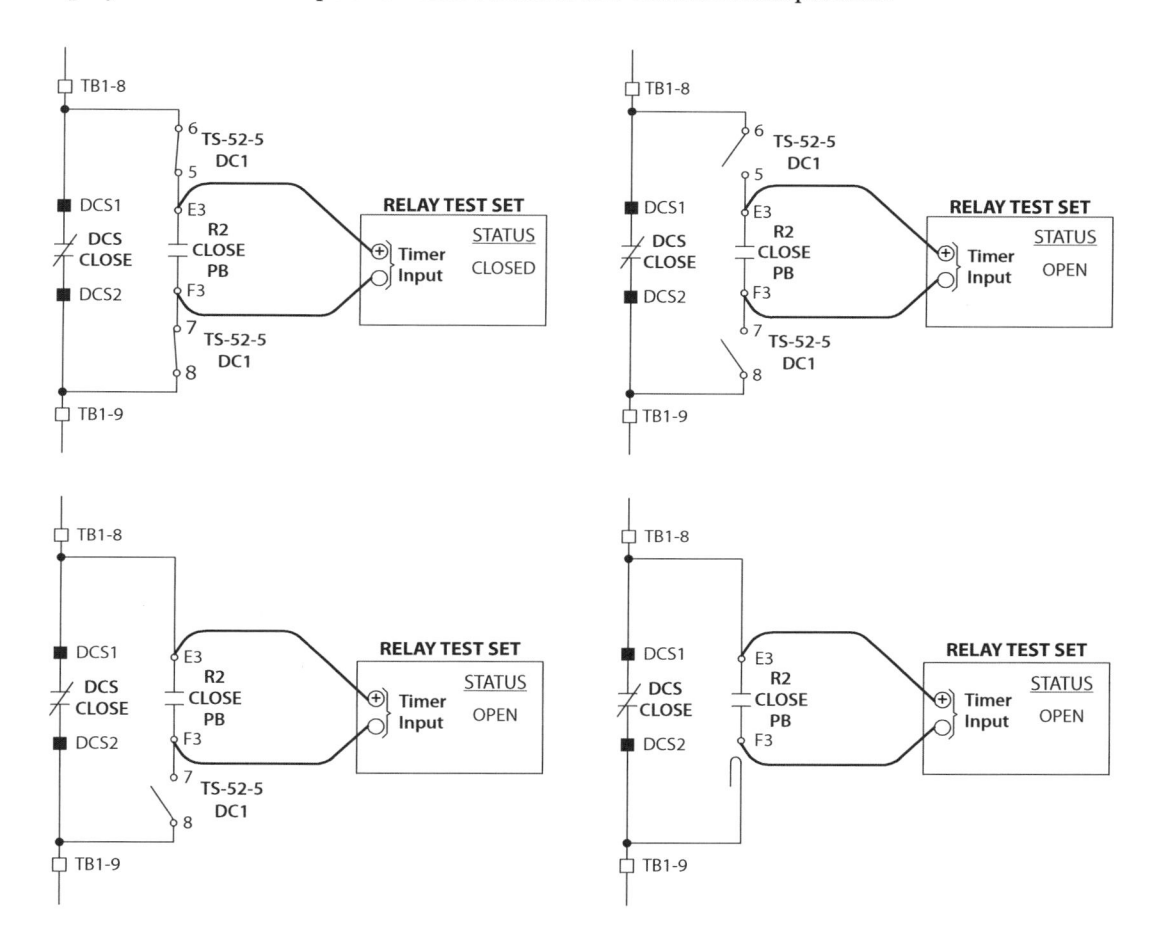

Figure 18-2: Test-Set Input Connections with Contact in Parallel

Most test-set manufacturers also allow voltage-monitoring inputs to reduce wiring changes when testing. Instead of monitoring whether a contact is closed or open, the voltage-monitoring option determines that the contact is closed when the measured voltage is above the test-set defined setpoint. The test-set assumes the contact is open if the measured voltage between the terminals is below the setpoint.

When the test switches on either side of the output contact are open as shown in Figure 18-1, there cannot be any voltage on the output. The test-set will determine that the output is open no matter what state the contact is in. The second scenario in our first example has a voltage applied and the test-set will detect output operation but in the opposite configuration. When the contact is closed, zero volts will be measured across the contact, and then the test-set will incorrectly determine that the output is open. When the output is open, the relay will detect $125V_{DC}$ across the contact and determine that the output is closed.

Another connection is required when using voltage-detecting test-set inputs when the correct contact state is necessary for testing. Any of the test-set connections can be used when voltage is applied to the circuit in Figure 18-3.

Starting from the left, The R2 timer is connected between TB1-9 and TB3-6 (circuit negative) with closed output contact test switches. When R2 and "DCS close" are open, the voltage between the two terminals should be negligible and the relay will detect an open contact. When R2 or "DCS close" closes, the relay will detect $125V_{DC}$ across the contacts and the test-set will detect a closed contact. Be wary of this connection because the circuit breaker will close if the circuit is complete!

The R3 timer is connected between Terminal 11 of the open test switch and circuit negative. This is a safe connection as the test-set will detect the correct contact position and the circuit breaker will not trip when the contact is operated.

The simplest connection uses the R1 timer with the R1 timer connected between Terminal 3 of the test switch and ground. (I often use the test switch cover screw as the ground that works in most applications) Obviously this connection will only work when the DC system is grounded at the midpoint, as most DC systems are. This connection is also safe as the connected 86-5 lockout will not operate when the contact is operated.

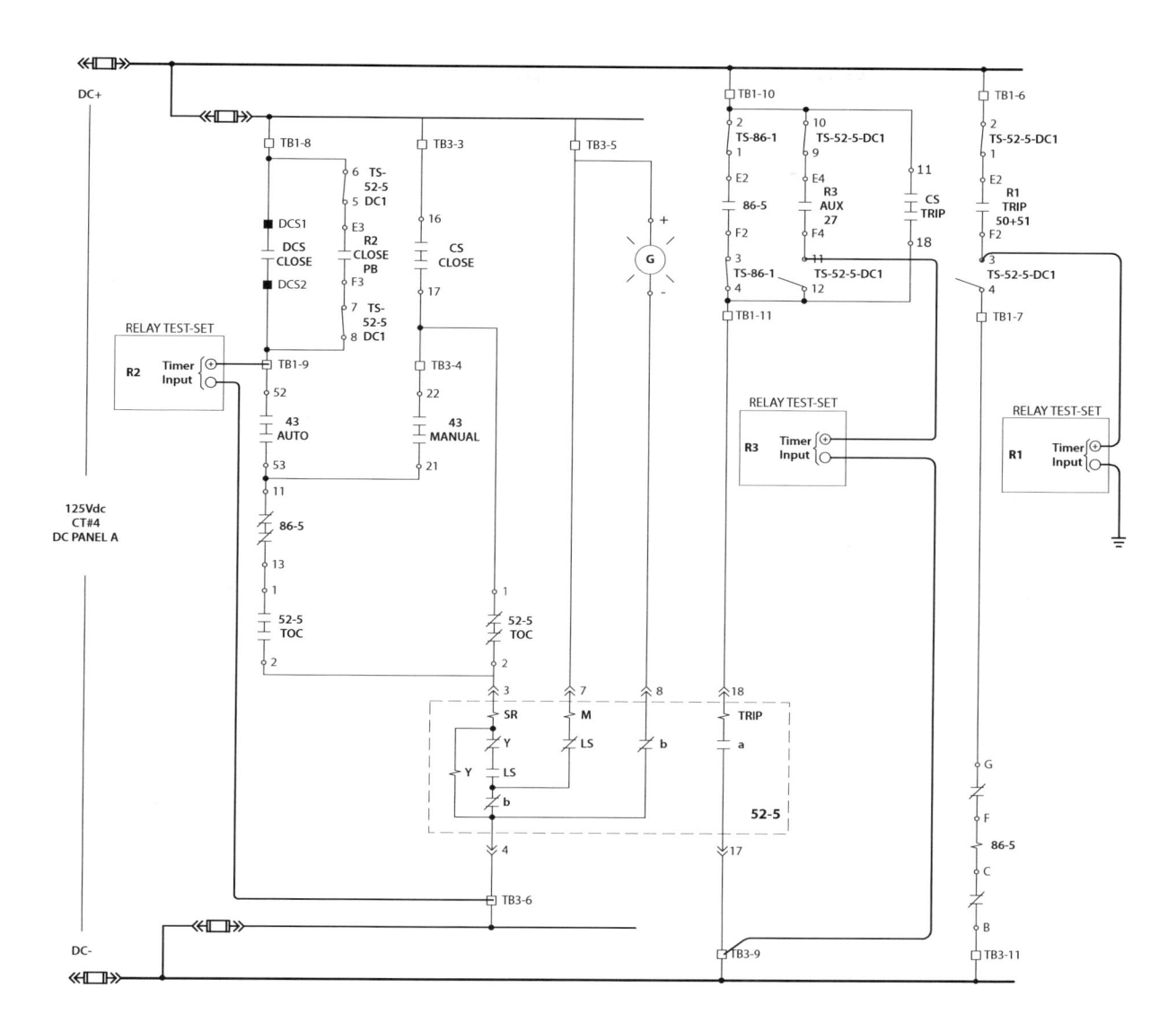

Figure 18-3: Test-Set Input Connections in DC Circuit

NEVER apply the following connection in a trip circuit unless you are absolutely sure that there will be no negative results if the circuit is completed and operates. Some test-set sensing contacts have a very low impedance and complete the circuit.

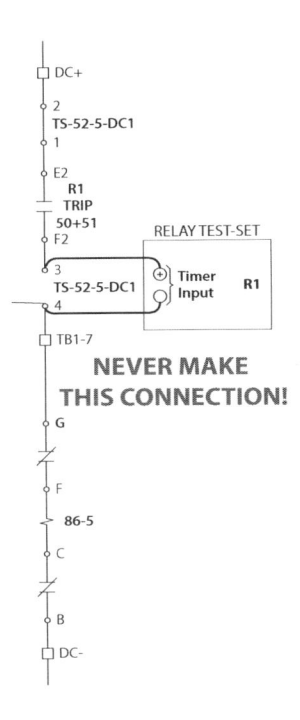

Figure 18-4: Dangerous Test-Set Input Connection in Trip Circuit

B) Relay Input Connections

Relay input connections should be tested as a part of the acceptance testing process and sometimes are integral to correct relay operation. Always review the manufacturer's literature when performing digital input acceptance testing because relays can be unforgiving when not correctly connected and cause some embarrassing and expensive smoke to be released. These connections should also be carefully compared to the application to ensure they are connected properly before applying power to the relay circuit. Figures 2-5, 2-6, and 2-9 show some typical examples of input connections from different relay manufacturers.

Figure 18-5 from the Beckwith Electric M-3310 manufacturer's bulletin displays the connections for the M-3310 relay input connections. The field contact is "dry" and the sensing voltage is supplied by the relay itself. Any external voltage connected in this circuit could damage the relay. The test-set dry output contact or jumper would be connected between Terminals 10 and 11 to simulate an IN 1 input.

Figure 18-6 from the SEL-311C manufacturer's bulletin shows that this relay requires "wet" inputs. An external voltage must be connected before the relay will detect input operation. Always check the input voltage to make sure it matches the source voltage.

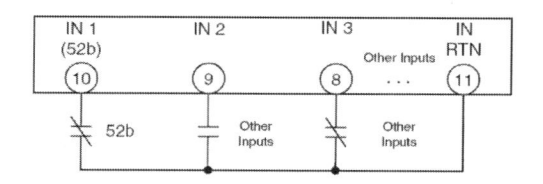

Figure 18-5: M-3310 Relay Input Connections

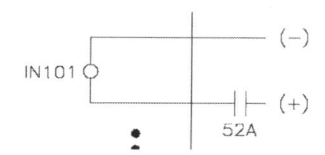

Figure 18-6: SEL-311C Input Connections

The GE Multilin SR-750 relay from our example can have "wet" or "dry" contacts connected as shown in Figure 18-7 from the manufacturer's bulletin. Relays that can accept both styles of input contacts are more prone to damaging connection errors and the site and manufacturer's drawings should be compared to ensure no errors have been made.

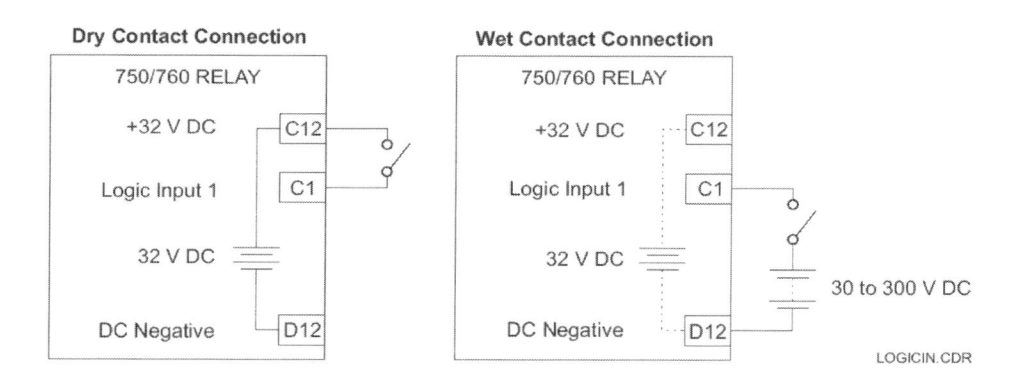

Figure 18-7: GE/Multilin SR-750 Input Connections

A fused jumper or closed contact should be connected between each IN terminal and IN RTN (11) on the M-3310 relay while monitoring input status to ensure all inputs operate correctly. The SEL-311C relay inputs are tested by monitoring input status and applying a test voltage, or source voltage with jumpers at the nameplate rated voltage across each input.

When testing, the test-set output is connected across the actual contact used in the circuit to prevent unintentional damage by applying incorrect test voltages. If the in-service contact is closed, you will have to remove wiring or open test switches to isolate the contact and allow you to test both input states.

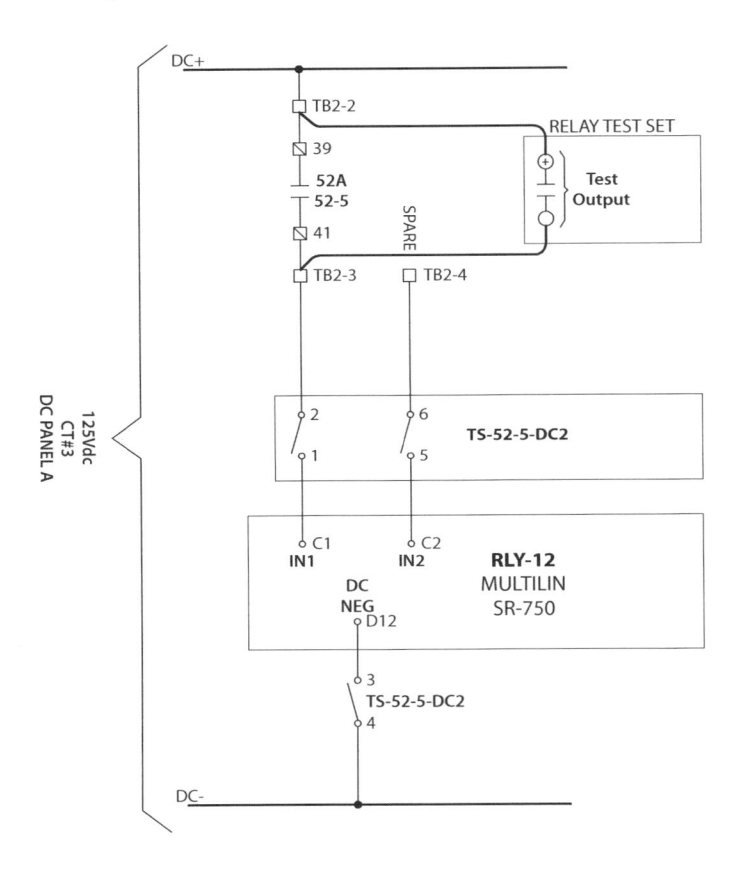

Figure 18-8: Test-Set Output Connections in DC Circuit

C) AC Connections

For the purposes of this publication, it is assumed that you are using a 3-phase test-set. Protective relays today use positive, negative, and zero-sequence components in their pick up calculations and these components can only be accurately simulated using 3-phase currents and voltages. While it is possible to test most elements using single-phase test procedures, most of your testing time is often spent trying to fool the relay into thinking that 3-phase inputs exist.

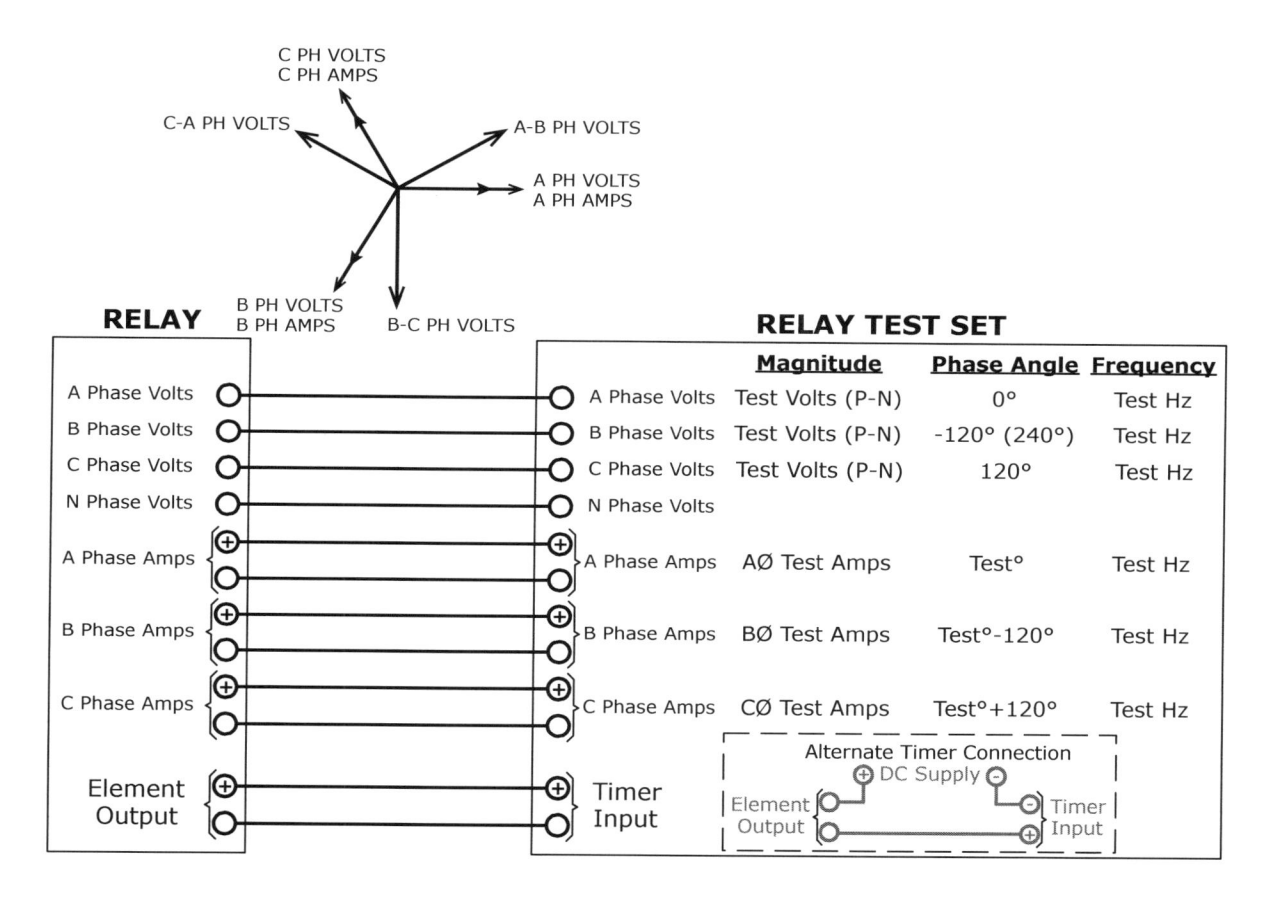

Figure 18-9: Relay Test-Set Connections

It is possible to simulate 3-phase testing with two current and two voltage channels as shown in Figure 18-10.

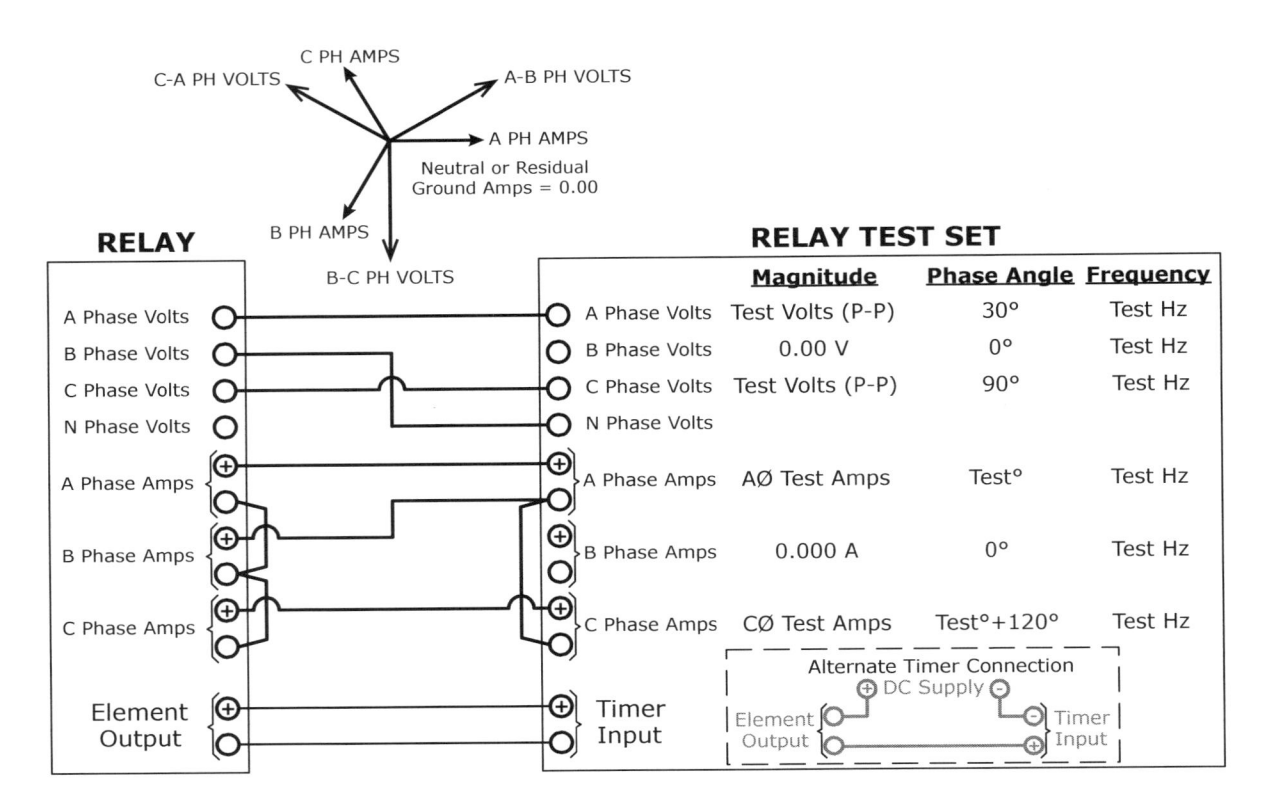

Figure 18-10: 3-Phase Test-Set Connections Using Two Phases

Connect your test-set to simulate the CT and PT inputs. Our goal is to simulate real life conditions as best as possible. Figure 18-11 shows the AC connections from our example. All CT test switches have been opened to short the CT inputs, and an isolating device has been inserted between the CT clips to isolate the top from the bottom.

Always pay attention to the PT connections and triple check to make sure that your test-set is connected to the relay side of the test switch. If you are connected to the wrong side, you could back-feed the connected PTs and apply a dangerous voltage to the high-side of the PTs.

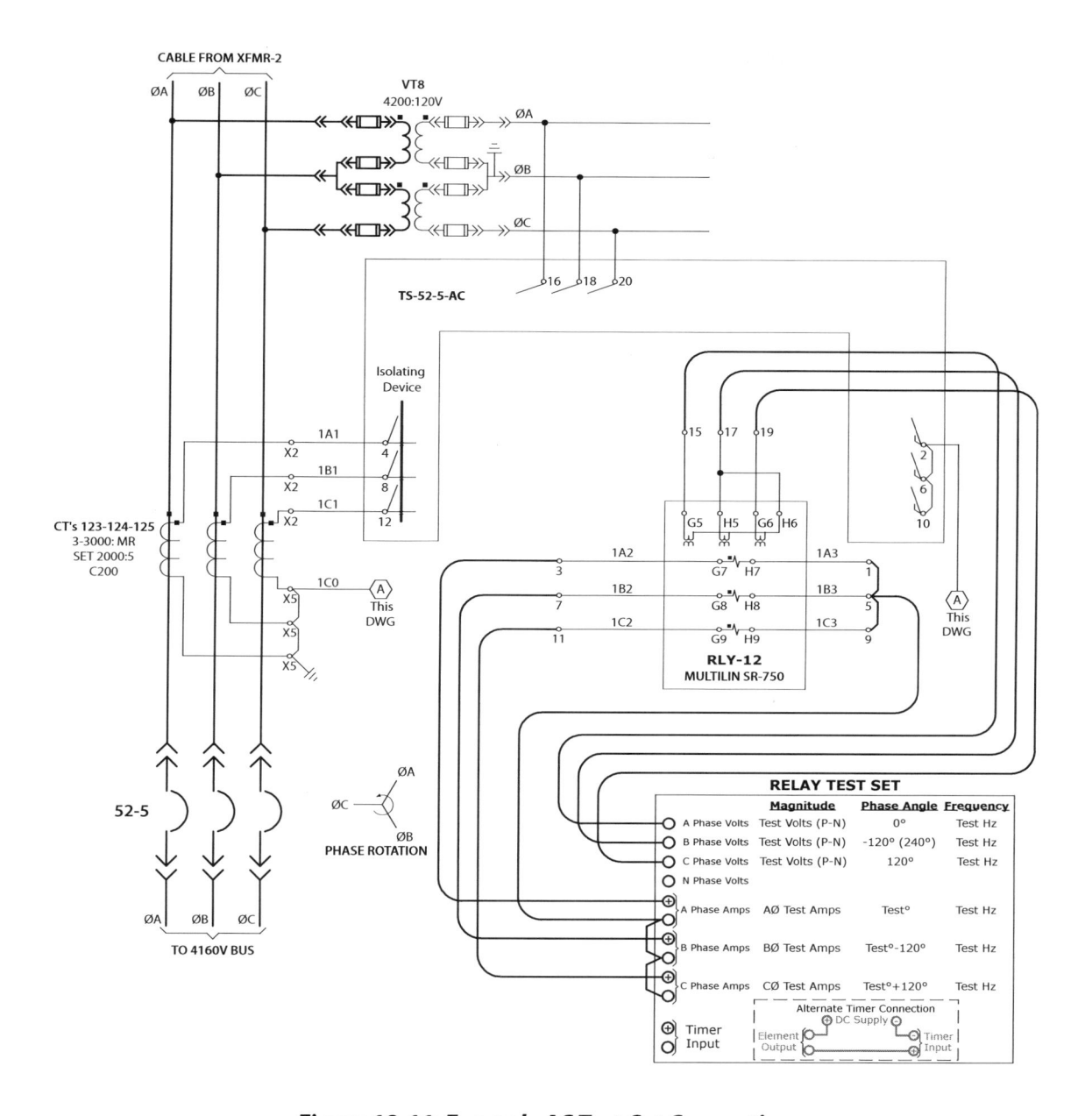

Figure 18-11: Example AC Test-Set Connections

If test switches are not available, the wiring will have to be removed to test the relay. Label each wire and write down all connections before removing any wiring. Replace the wiring after testing, and check the terminations against your notes. It is always preferable to have CT/PT loop tests completed after relay testing to ensure the connections are correct. Always check the online metering after energization to ensure the wiring has been replaced correctly. It is a good idea to carry a checklist of each wire removed to ensure every wire is returned to service.

5. Creating the Test Plan

The test plan determines what elements will be tested and the order of testing. Ideally, the testing should be completed without changing any of the final, as-left settings. Your goal is to move from element to element with the least amount of connection and setting changes.

Before you start creating your test plan, it is a good idea to step back and ask yourself "Why am I testing this relay?" If your answer is to put numbers onto a test sheet, than the traditional test techniques and even some of the advanced test techniques can help you get those numbers and you can skip to the next header to learn those techniques. However, if your answer is "To make sure that the relay operates correctly when it should", the traditional test techniques described in this book probably aren't giving you the answer you are looking for and even the advanced procedures in this book only increase the odds of a properly configured relay.

As we've mentioned previously in this book, the relay tester's instinct of looking to the relay settings for answers can actually be counter-productive. The relay will perform exactly as it is programmed and when we look to the relay settings to create our test plans, we are really testing the relay manufacturer's algorithms which is closer to performing type-tests instead of commissioning. Almost all problems with modern protective relaying are actually simple mistakes such caused by human error when entering the relay settings, cut-and-paste errors when using templates to create the relay settings, or an incorrect understanding of the relays settings. If the design engineer creates flawed settings, asking the relay to tell you what it is supposed to do will only tell you what it was programmed to do and not what is *intended* to do.

Sometimes you can use common sense and experience to spot a setting mistake; such as the 51G element enabled in the element settings but 51N is assigned in the output logic. Sometimes applying a logic test or a combination test will also find a mistake; such as an engineer assigning "XFMR PCNT DIFF OP A" to an output which will only trip during an AØ fault instead of the "XFMR PCNT DIFF OP" label that will trip on all phases. Sometimes performing tests using in-service status inputs instead of using a simulation will find that the 52a contact that the design intended was really a 52b contact which caused the differential protection to be disabled whenever the equipment was in service. These are just a few examples of real-life problems that were discovered by moving beyond the traditional test techniques into more advanced testing but they are really accidental discoveries. We really need to ask the design engineers what they intended when they created the settings.

This is a relatively simple task when reviewing overcurrent protection because most installations have a coordination study performed and the relay tester can review that coordination study and pick out the appropriate pickup settings and time delays to use in their test plan. This is a much more effective test as it proves that the relay is set correctly and operating correctly. Asking the relay for its settings and testing those settings only proves that the relay is functioning correctly. Unfortunately, engineering studies beyond overcurrent protection are few and far between so we should ask design engineers for descriptions of operation *when they create the settings*. Asking for that description months or years later will probably not help as the relay will be long forgotten and the design engineer will likely just look at the settings to create the description the same way a relay tester does now. But if you get that description of operation before the settings are even created, this can be a valuable tool that allows the relay tester to test that the design engineer provided the correct settings and the relay operated. Here is an example description of operation for the example relay in the next section:

1. If the bus voltage drops below 102V secondary, Outputs R3 and R7 will operate in 20s, and the relay should indicate a Bus Undervoltage trip.

2. If the bus voltage drops below 108V secondary but is greater than 102V secondary, Outputs R3 and R7 will operate in 60s, and the relay should indicate a Bus Undervoltage trip.

3. The relay overcurrent protection will not operate below 5.0A secondary and will trip the TRIP and R7 outputs at the following test points:

    a. 10A = 5.296s
    b. 20A = 1.538s
    c. 30A = 1.033s
    d. >45A = 0.1s

You can test any relay from any manufacturer with this information to ensure that it is working correctly and will operate the correct outputs at the correct time.

I realize that getting description of operations from overworked design engineers is an ideal situation that is unlikely to be realized in most cases, but it usually doesn't hurt to ask so that you can make your relay testing easier, more effective, and more efficient.

The following practical tips will help you create a test plan for any situation regardless of the test technique you wish to employ.

A) Traditional Pickup and Timing Tests

Collect all of the charts created in previous steps and determine what elements or inputs are enabled and assigned to an output. Outputs can be LEDs on the front of the relay, status points transferred between relays via communication, and the standard relay output contacts. It is sometimes easier to reverse engineer this step by starting with the output settings. The SR-750 relay uses the individual element approach to relay settings and there are no independent output settings so we must look at every element setting in the example. A summary of our example settings can be found in the following chart, but I strongly recommend using values from the coordination study whenever possible to ensure that the relay will trip when the coordination study requires it to and not when the relay is programmed to operate. It is very possible that the two times do not match due to a setting error.

ELEMENT	PICKUP	CURVE	TIME DIAL	RELAYS
51 (TOC)	1xCT (5.0A)	Normally Inverse (NI)	3.00	Trip & R7
50 (IOC) #1	9xCT (45A)	Definite Time (DT)	0.10 s	Trip & R7
27 (BUV) #1	0.9xVT (108V)	DT	60.0s	R3 & R7
27 (BUV) #2	0.85xVT (102V)	DT	20.0s	R3 & R7

Figure 18-12: Example Relay Settings

Review the settings and look for overlapping protection that will interfere with testing. The overcurrent settings are straightforward. This relay has one pickup LED for all elements that can be used for 51-element pickup testing without interference from other elements. Timing tests can be performed up to 9x pickup without interference from the 50-element. The bus voltage during both tests must be greater than 108V to ensure the voltage protection does not interfere with the overcurrent tests.

The 50-element tests require a little more finesse. The pickup LED can only be used for pickup testing if the 51-element is disabled. If the 51-element is disabled for pickup testing, it must be tested after the 50-element testing is complete. You could also test 50-element pickup using the relay trip (R2 or R7) contacts without any setting changes. Using the manufacturer's formula, the 51-element time delay for a Normally Inverse curve at the 50-element pickup level is 0.73s. This is far greater than the 50-elements 0.1s time delay. The 50-element pickup is tested by performing timing tests and increasing the current in 0.2A increments starting from 5% less than the 50-element pickup. The test current that causes the time delay to drop to approximately 0.1s is the pickup current. This method is preferred when such large currents are involved to prevent damage to the relay. The timing test is performed at 1.1x the pickup value to ensure correct operation.

The 27-1-element pickup testing is also straightforward using the front LED to test for pickup as long as the test current is less than 5.0A. I always apply a current (typically 1.0 A) below any overcurrent pickup value when testing 27-elements because some relays will block 27-element operation when no current is flowing to prevent nuisance trips. Timing tests are usually performed at 10% below the pickup (97.2V) but this value would cause the 27-2-element to operate first. The timing test should be performed at a level significantly higher than the 27-2-element and would be performed at 103V. A second test could be justified to prove a definite time element and both tests should be performed between the two pickup voltages. A prefault voltage above 108V is required for 27-element testing.

The 27-2 pickup testing cannot be tested with the pickup LED when the 27-1-element is operational. The 27-1-element could be disabled before testing the 27-2-element pickup, but the 27-1 pickup tests must be performed afterwards to ensure it is left enabled. 27-2-element pickup tests could also use the method described above for the 50-element testing. This method could take some time while waiting more than 20s between tests and isn't practical unless time is not an issue.

The third method that can be used to test the pickup of the 27-2-element uses the communication software between the computer and relay. This method is not as accurate as those described previously for this relay brand but can be used. Use the software to open the status screen and change between a prefault nominal voltage and a pickup voltage. Count out the time delay between the voltage change and the computer target display. Start the test voltage 5% higher than the pickup and lower the voltage in small voltage increments. Wait for the previously counted time delay to elapse between increments and stop when the 27-2 pickup target is displayed. Timing tests for this element are very straightforward and are tested at 10% below the pickup (91.8V)

Calculate the expected tolerances as discussed in the *Pickup Test Procedure* section starting on page 110 to determine a pass or fail test result.

B) Combining Pickup, Timing, and Logic Tests

A microprocessor relay element does not fall out of calibration…it either works correctly or it doesn't. Using this principal, the pickup and timing test can be combined into one test. Logic testing can also be added to the test by connecting all of the outputs to the test-set inputs and applying multiple timers. We can apply this test type to each element using the following steps:

1. Configure your test-set to stop if any assigned output for the element under test operates and ignore any output not assigned to the relay element under test.

2. Apply a normal, prefault voltage to the relay for a few seconds to simulate an in-service normal condition.

3. Apply a fault condition that is just outside the relay's pickup setting and manufacturer's accuracy specifications.

4. Apply a fault condition that is just inside the relay's pickup setting and manufacturer's accuracy specifications. Start a timer for each relay-output assigned to the element. Stop the timers when the appropriate output operates.

The following specifications are from the GE/Multilin SR-750 relay instruction bulletin:

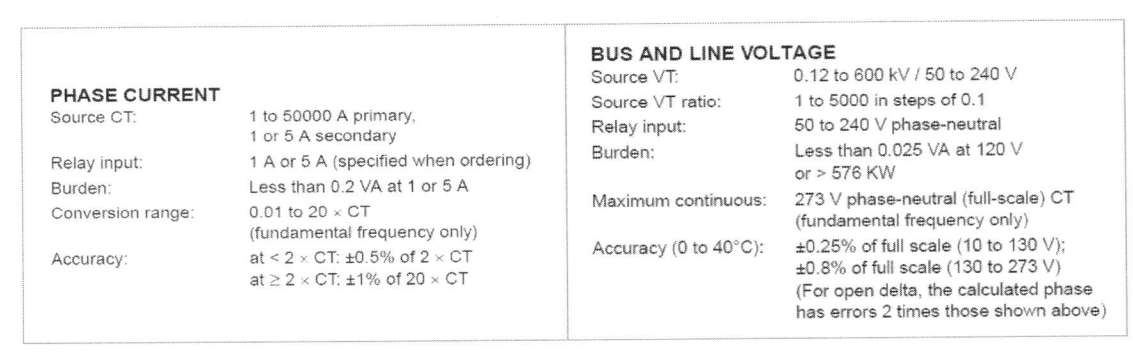

PHASE CURRENT

Source CT:	1 to 50000 A primary, 1 or 5 A secondary
Relay input:	1 A or 5 A (specified when ordering)
Burden:	Less than 0.2 VA at 1 or 5 A
Conversion range:	0.01 to 20 × CT (fundamental frequency only)
Accuracy:	at < 2 × CT: ±0.5% of 2 × CT at ≥ 2 × CT: ±1% of 20 × CT

BUS AND LINE VOLTAGE

Source VT:	0.12 to 600 kV / 50 to 240 V
Source VT ratio:	1 to 5000 in steps of 0.1
Relay input:	50 to 240 V phase-neutral
Burden:	Less than 0.025 VA at 120 V or > 576 KW
Maximum continuous:	273 V phase-neutral (full-scale) CT (fundamental frequency only)
Accuracy (0 to 40°C):	±0.25% of full scale (10 to 130 V); ±0.8% of full scale (130 to 273 V) (For open delta, the calculated phase has errors 2 times those shown above)

Figure 18-13: GE Multilin SR-750 Specifications

We can use these specifications to create a combination test plan. We'll start with the IOC Element (50-Instantaneous Overcurrent) using the following steps:

1. The relay TRIP, R3, and R7 are connected to the test-set status inputs. The test-set is set to ignore any R3 operation because the 50-element does not operate the R3 relay-output.

2. We can set the prefault voltage to the nominal $118.9V_{P-P}$ on all three phases 120° apart in the correct phase-sequence for 2 seconds to prevent the 27 relay from operating.

3. We can set the 1st fault as follows:

    a. Set the fault voltage to any number larger than 108V to prevent the 27 relay from operating. We'll leave it at 118. $9V_{P-P}$ to simplify the test procedure.

b. Set the fault current for an A-B phase test (This is a *Phase* element) with B-phase current 180° from A-phase current.

c. The current magnitude on both phases should be less than the pickup setting minus the relay accuracy.

$$\text{Test Amps} < \text{Pickup} - \text{Accuracy}$$

$$\text{Test Amps} < 45A - (1\% \text{ of } 20)$$

$$\text{Test Amps} < 45A - (0.2)$$

$$\text{Test Amps} < 44.8A$$

$$\text{Test Amps} = 44.7A$$

d. The relay should be allowed time to operate and then automatically transfer to the 2nd fault state. The element time delay is 0.1s and we'll set the maximum duration in the 1st fault state to 0.2s.

4. We can set the 2nd fault as follows:

a. Timer 1 starts when the 2nd fault state starts and stops when the TRIP input operates.

b. Timer 2 starts when the 2nd fault state starts and stops when the R7 input operates.

c. Set the fault voltage to any number larger than 108V to prevent the 27 relay from operating. We'll leave it at 118. $9V_{P-P}$ to simplify the test procedure.

d. Set the fault current for an A-B phase test with B-phase current 180° from A-phase current.

e. The current magnitude on both phases should be more than the pickup setting plus the relay accuracy.

$$\text{Test Amps} > \text{Pickup} + \text{Accuracy}$$

$$\text{Test Amps} > 45A + (0.2)$$

$$\text{Test Amps} > 45.2A$$

$$\text{Test Amps} = 45.3A$$

f. The relay should be allowed time to operate. The element time delay is 0.1s and we'll set the maximum duration to match the 1st fault state which is 0.2s. If the relay does not trip, the test should stop to prevent relay damage.

g. The TRIP or R7 relay-output should stop the test if they operate.

We can create a chart of the test plan as follows using phase angles as the SR-750 would display:

A-B PIOC TEST	PREFAULT	FAULT 1	FAULT 2
AØ Voltage	68.65V@0°	68.65V@0°	68.65V@0°
BØ Voltage	68.65V@120°	68.65V@120°	68.65V@120°
CØ Voltage	68.65V@240°	68.65V@240°	68.65V@240°
AØ Current	0.000A@0°	44.700A@0°	45.300A@0°
BØ Current	0.000A@120°	44.700A@180°	45.300A@180°
CØ Current	0.000A@240°	0.000A@240°	0.000A@240°
Max Duration	2.000s	0.200s	0.200s
Timer 1 Start			Fault 2 Start
Timer 1 Stop			TRIP Operate
Timer 2 Start			Fault 2 Start
Timer 2 Stop			R7 Operate

Create a test plan for the other two phase combinations by rotating the applied current in the two fault states by 120° to ensure the element will operate for any Ø-Ø fault.

B-C PIOC TEST	PREFAULT	FAULT 1	FAULT 2
AØ Voltage	68.65V@0°	68.65V@0°	68.65V@0°
BØ Voltage	68.65V@120°	68.65V@120°	68.65V@120°
CØ Voltage	68.65V@240°	68.65V@240°	68.65V@240°
AØ Current	0.000A@0°	0.000A@0°	0.000A@0°
BØ Current	0.000A@120°	44.700A@120°	45.300A@120°
CØ Current	0.000A@240°	44.700A@300°	45.300A@300°
Max Duration	2.000s	0.200s	0.200s
Timer 1 Start			Fault 2 Start
Timer 1 Stop			TRIP Operate
Timer 2 Start			Fault 2 Start
Timer 2 Stop			R7 Operate

C-A PIOC TEST	PREFAULT	FAULT 1	FAULT 2
AØ Voltage	68.65V@0°	68.65V@0°	68.65V@0°
BØ Voltage	68.65V@120°	68.65V@120°	68.65V@120°
CØ Voltage	68.65V@240°	68.65V@240°	68.65V@240°
AØ Current	0.000A@0°	44.700A@60°	45.300A@60°
BØ Current	0.000A@120°	0.000A@120°	0.000A@120°
CØ Current	0.000A@240°	44.700A @240°	45.300A@240°
Max Duration	2.000s	0.200s	0.200s
Timer 1 Start			Fault 2 Start
Timer 1 Stop			TRIP Operate
Timer 2 Start			Fault 2 Start
Timer 2 Stop			R7 Operate

If we run this test and Timer 1 and Timer 2 are within the relay timing specifications, we can pass the relay element's pickup, timing, and output logic because:

- The pickup is between 44.7A and 45.3A which is only slightly outside the relay's tolerance. There are no adjustments, so we are looking for setting mistakes or gross errors whenever performing tests on a digital relay. The relay pickup is acceptable for service.
- The time delay is set correctly if Timer 1 or Timer 2 has a time which is within the relay tolerance.
- The output logic is working correctly if Timer 1 *and* Timer 2 have nearly identical values.

The same philosophy can be applied to the BUV-1 Element with the following steps using realistic voltage settings as described in *Chapter 15: Line Distance (21) Element Testing*.

1. The relay TRIP, R3, and R7 are connected to the test-set status inputs. The test-set is set to ignore any TRIP operation because the 27-element does not operate the TRIP relay-output.

2. We can set the prefault voltage to the nominal $118.9V_{P-P}$ and prefault current to 1.00A on all three phases 120° apart in the correct phase-sequence for 2 seconds to prevent the 27 relay from operating. The prefault current is set to 1.00A to ensure the 27-element will operate even if its internal logic requires current to flow.

3. We can set the 1st fault as follows:
   a. The fault current equals 1.000A to match the prefault current.
   b. Set the fault voltage for an A-B phase test (This is a *Phase* element). The fault voltage is a combination of the A and B Ø-N voltages added vectorally and should be more than the pickup setting plus the relay accuracy. This is opposite of the 50-element test because we are testing an *under*voltage instead of an *over*current element.

$$\text{Test Volts} > \text{Pickup} + \text{Accuracy}$$

$$\text{Test Volts} > 108V + (2 \times 0.25\% \text{ of full scale})$$

$$\text{Test Volts} > 108V + (2 \times 0.0025 \times 240V)$$

$$\text{Test Volts} > 108V + (1.2)$$

$$\text{Test Volts} > 109.2V$$

$$\text{Test Volts} = 109.3V$$

   c. The relay should be allowed time to operate and then automatically transfer to the 2nd fault state. The element time delay is 60s and we'll set the maximum duration in the 1st fault state to 65s.

4. We can set the 2nd fault as follows:
   a. Timer 1 starts when the 2nd fault state starts and stops when the R3 input operates.
   b. Timer 2 starts when the 2nd fault state starts and stops when the R7 input operates.
   c. The fault current equals 1.000A to match the prefault current.
   d. Set the fault voltage for an A-B phase test. The fault voltage is a combination of the A and B Ø-N voltages added vectorally and should be less than the pickup setting minus the relay accuracy.

$$\text{Test Volts} < \text{Pickup} - \text{Accuracy}$$

$$\text{Test Volts} < 108V - (1.2)$$

$$\text{Test Volts} < 106.8V$$

$$\text{Test Volts} = 106.7V$$

e. The relay should be allowed time to operate and then automatically stop the test if the inputs do not operate. The element time delay is 60s and we'll set the maximum duration to match the 1st fault state which is 65s.

f. The R3 or R7 relay-output should stop the test if they operate.

5. We can create a chart of the test plans as follows using phase angles as the SR-750 would display:

ALL BUV-1 TESTS	PREFAULT	FAULT 1	FAULT 2
Max Duration	2.000s	65.00s	65.00s
Timer 1 Start			Fault 2 Start
Timer 1 Stop			R3 Operate
Timer 2 Start			Fault 2 Start
Timer 2 Stop			R7 Operate
AØ Current	1.000A@0°	1.000A@0°	1.000A@0°
BØ Current	1.000A@120°	1.000A@120°	1.000A@120°
CØ Current	1.000A@240°	1.000A@240°	1.000A@240°
A-B BUV-1 TEST			
AØ Voltage	68.65V@0°	64.54V@2.13°	63.44V@2.76°
BØ Voltage	68.65V@120°	64.54V@117.87°	63.44V@117.24°
CØ Voltage	68.65V@240°	68.65V@240°	68.65V@240°
		A-B = 109.3V@30°	A-B = 106.7V@30°
B-C BUV-1 TEST			
AØ Voltage	68.65V@0°	68.65V@0.00°	68.65V@0.00°
BØ Voltage	68.65V@120°	64.54V@122.13°	63.44V@122.76°
CØ Voltage	68.65V@240°	64.54V@237.87°	63.44V@237.24°
		B-C= 109.3V@90°	B-C = 106.7V@90°
C-A BUV-1 TEST			
AØ Voltage	68.65V@0°	64.54V@357.87°	63.44V@357.24°
BØ Voltage	68.65V@120°	68.65V@120°	68.65V@120°
CØ Voltage	68.65V@240°	64.54@242.13°	63.44V@242.76°
		C-A = 109.3V@210°	C-A = 106.7V@210°

The test plan for the BUV-2 tests are essentially the same as the BUV-1 tests with different voltage magnitudes and a shorter time delays because the BUV-2 time delay is 20s instead of the BUV-1 time delay of 60s. The sum of the Fault 1 and Fault 2 times for the BUV-2 test must be less than the BUV-1 time delay or else the BUV-1 element will operate first. We'll set the fault times to 25s each to allow the relay to operate and the entire test will be 50s so the test will stop before BUV-1 can operate.

Fault 1 Volts > Pickup + Accuracy	Fault 2 Volts > Pickup − Accuracy
Fault 1 Volts > 102V + (1.2)	Fault 2 Volts > 102V − (1.2)
Fault 1 Volts > 103.2V	Fault 2 Volts > 100.0V
Fault 1 Volts = 103.3V	Fault 2 Volts = 100.7V

ALL BUV-1 TESTS	PREFAULT	FAULT 1	FAULT 2
Max Duration	2.000s	25.00s	25.00s
Timer 1 Start			Fault 2 Start
Timer 1 Stop			R3 Operate
Timer 2 Start			Fault 2 Start
Timer 2 Stop			R7 Operate
AØ Current	1.000A@0°	1.000A@0°	1.000A@0°
BØ Current	1.000A@120°	1.000A@120°	1.000A@120°
CØ Current	1.000A@240°	1.000A@240°	1.000A@240°
A-B BUV-1 TEST			
AØ Voltage	68.65V@0°	62.02V@3.61°	60.94V@4.28°
BØ Voltage	68.65V@120°	62.02V@116.39°	60.94V@115.72°
CØ Voltage	68.65V@240°	68.65V@240°	68.65V@240°
		A-B = 103.3V@30°	A-B = 100.7V@30°
B-C BUV-1 TEST			
AØ Voltage	68.65V@0°	68.65V@0.00°	68.65@0.00°
BØ Voltage	68.65V@120°	62.02V@123.61°	60.94V @124.28°
CØ Voltage	68.65V@240°	62.02@236.39°	60.94V @235.72°
		B-C = 103.3V@90°	B-C = 100.7V@90°
C-A BUV-1 TEST			
AØ Voltage	68.65V@0°	62.02V@356.39°	60.94V@355.72°
BØ Voltage	68.65V@120°	68.65V@120°	68.65V@120°
CØ Voltage	68.65V@240°	62.02V@243.61°	60.94V@244.28°
		C-A = 103.3V@210°	C-A= 100.7V@210°

The TOC element (51 or time-overcurrent) can be tested with a similar technique that has been modified to test an inverse curve instead of a fixed time delay. The TOC element would probably never operate if set slightly above the pickup setting, but if our goal for this test is to ensure the TOC element *does not* operate below the pickup setting and *should* operate a specific time when a specific amperage is applied, we can use the identical calculation for Fault 1 and use a multiple of pickup for Fault 2. We can even add enough time in Fault 1 to allow you to see if the relay has picked up in Fault 1 using any of the standard pickup methods to ensure that the pickup is correct.

We also need to make sure that all test magnitudes are less than the IOC element because the instantaneous element would then operate first. That isn't a problem for this test plan because the IOC setting is 9x the pickup, but if the IOC setting was 5xCT we would not be able to perform the standard timing tests at 6x the pickup setting. We could disable the IOC element but that would defeat the purpose of the combination test which tests the relay with in-service settings to prove the relay will operate correctly when required. If we ask ourselves "Why do I need to perform three timing tests at these multiples?" we can imagine several answers:

1. "Because that's the way we've always tested overcurrent relays." If this is your answer, you should think about why we've always performed these tests. E-M relay timing could drift over time due to their nature. Three tests were performed at different points along the curve because you could also change the curve shape when adjusting the relay. Multiple tests were important because it was very possible to get one timing test multiple in tolerance and unknowingly change a third point out-of-tolerance. Microprocessor-based relays use a formula to determine timing and will never drift. One test multiple could be considered acceptable but you should probably perform at least two tests just in case that one timing test unluckily hits a point where different curves cross with the same settings. We can just eliminate the test that exceeds the 50-element.

2. "Because my test sheet requires 3 test points." If this is your answer, then you can change the multiples of your test plan. If the IOC element was set to 5x the TOC pickup, you can perform timing tests at 2x, 4x, and 4.5x pickup. People chose whole numbers in the past because there were no formulas to calculate time delays and it was much easier to pick numbers off a graph when you could follow the line…especially on a log-log graph. You can use your calculator or a spreadsheet to easily determine the time delay at any multiple of pickup.

3. "Because the NERC/FERC plan explicitly states the number of tests and multiples." If this is your answer, you have no choice and must disable the IOC element or assign another output to only operate for a TOC trip for the multiples that exceed the IOC pickup setting. However, many people assume the NERC/FERC plans were mandated

by NERC/FERC, but the NERC/FERC rules only mandate that the organization have a plan and that they implement the plan as specified. Those NERC/FERC plans can be changed at any time by following the appropriate procedures for the relay owner and it would probably make everyone's life easier if they were changed. Always remember that the accepted relay test plan is in place until the date it changes and you will be asked to prove that the old plan was implemented as written until the date of change, even if the plan did not make sense to begin with.

The following calculations and chart prepares us to perform a combination test for the TOC element on all three phases with 2x, 4x, and 6x multiples:

Test Amps < Pickup − Accuracy Test Amps < 5A − (0.2) Test Amps < 4.8A Test Amps = 4.7A	$\text{Test Time} = M \times \left(A + \dfrac{B}{\left(\dfrac{I}{I_{pu}}\right) - C} + \dfrac{D}{\left(\left(\dfrac{I}{I_{pu}}\right) - C\right)^2} + \dfrac{E}{\left(\left(\dfrac{I}{I_{pu}}\right) - C\right)^3} \right)$
$\left(\dfrac{I}{I_{pu}}\right) = \text{Pickup} \times \text{Test Multiple}$ 2x Test Amps = 5A × 2 2x Test Amps = 10A	$\text{Test Time} = 3 \times \left(0.0274 + \dfrac{2.2614}{\left(\dfrac{I}{I_{pu}}\right) - 0.3} + \dfrac{-4.1899}{\left(\left(\dfrac{I}{I_{pu}}\right) - 0.3\right)^2} + \dfrac{9.1272}{\left(\left(\dfrac{I}{I_{pu}}\right) - 0.3\right)^3} \right)$ $\text{Test Time} = 3 \times \left(0.0274 + \dfrac{2.2614}{(2) - 0.3} + \dfrac{-4.1899}{\left((2) - 0.3\right)^2} + \dfrac{9.1272}{\left((2) - 0.3\right)^3} \right)$ $\text{Test Time} = 3 \times \left(0.0274 + \dfrac{2.2614}{1.7} + \dfrac{-4.1899}{(1.7)^2} + \dfrac{9.1272}{(1.7)^3} \right)$ $\text{Test Time} = 3 \times \left(0.0274 + 1.33 + \dfrac{-4.1899}{2.89} + \dfrac{9.1272}{4.913} \right)$ $\text{Test Time} = 3 \times (1.3574 - 1.45 + 1.858)$ $\text{Test Time} = 3 \times (1.7654)$ $\text{Test Time} = 5.296s$
4x Test Amps = 5A × 4 4x Test Amps = 20A	$\text{Test Time} = 3 \times \left(0.0274 + \dfrac{2.2614}{(4) - 0.3} + \dfrac{-4.1899}{\left((4) - 0.3\right)^2} + \dfrac{9.1272}{\left((4) - 0.3\right)^3} \right)$ $\text{Test Time} = 1.538s$
6x Test Amps = 5A × 6 6x Test Amps = 30A	$\text{Test Time} = 3 \times \left(0.0274 + \dfrac{2.2614}{(6) - 0.3} + \dfrac{-4.1899}{\left((6) - 0.3\right)^2} + \dfrac{9.1272}{\left((6) - 0.3\right)^3} \right)$ $\text{Test Time} = 1.0333s$

ALL BUV-1 TESTS	PREFAULT	FAULT 1	FAULT 2
Max Duration	2.000s	60.00s	5.500s
Timer 1 Start			Fault 2 Start
Timer 1 Stop			TRIP Operate
Timer 2 Start			Fault 2 Start
Timer 2 Stop			R7 Operate
AØ Voltage	68.65V@0°	68.65V@0°	68.65V@0°
BØ Voltage	68.65V@120°	68.65V@120°	68.65V@120°
CØ Voltage	68.65V@240°	68.65V@240°	68.65V@240°
A-B 2XTOC TEST			
AØ Current	0.000A@0°	4.700@0°	10.000A@0°
BØ Current	0.000A@120°	4.700A@120°	10.000A@180°
CØ Current	0.000A@240°	0.000A@240°	0.000A@240°
B-C 2XTOC TEST			
AØ Current	0.000A@0°	0.000@0°	0.00A@0°
BØ Current	0.000A@120°	4.700A@120°	10.000A@120°
CØ Current	0.000A@240°	4.700A@240°	10.000A@300°
C-A 4XTOC TEST			
AØ Current	0.000A@0°	4.700@0°	10.000A@0°
BØ Current	0.000A@120°	0.00A@120°	0.00A@120°
CØ Current	0.000A@240°	4.700A@240°	10.000A@240°
A-B 4XTOC TEST			
AØ Current	0.000A@0°	4.700@0°	20.000A@0°
BØ Current	0.000A@120°	4.700A@120°	20.000A@180°
CØ Current	0.000A@240°	0.000A@240°	0.000A@240°
B-C 4XTOC TEST			
AØ Current	0.000A@0°	0.000@0°	0.00A@0°
BØ Current	0.000A@120°	4.700A@120°	20.000A@120°
CØ Current	0.000A@240°	4.700A@240°	20.000A@300°
C-A 2XTOC TEST			
AØ Current	0.000A@0°	4.700@0°	20.000A@0°
BØ Current	0.000A@120°	0.00A@120°	0.00A@120°
CØ Current	0.000A@240°	4.700A@240°	20.000A@240°
A-B 4XTOC TEST			
AØ Current	0.000A@0°	4.700@0°	30.000A@0°
BØ Current	0.000A@120°	4.700A@120°	30.000A@180°
CØ Current	0.000A@240°	0.000A@240°	0.000A@240°
B-C 4XTOC TEST			
AØ Current	0.000A@0°	0.000@0°	0.00A@0°
BØ Current	0.000A@120°	4.700A@120°	30.000A@130°
CØ Current	0.000A@240°	4.700A@240°	30.000A@300°
C-A 2XTOC TEST			
AØ Current	0.000A@0°	4.700@0°	30.000A@0°
BØ Current	0.000A@120°	0.00A@120°	0.00A@120°
CØ Current	0.000A@240°	4.700A@240°	30.000A@240°

All this may seem like a lot of work as described, but remember that most of this work is necessary for the traditional tests. Once the plans have been created, the actual test time will be drastically reduced and the test plans can be re-used in future maintenance plans or modified for similar relays with different pickup and time delays. See the drastic reduction of tests required in the next section as an example.

C) Feeder Protection Test Plan

After reviewing the settings, it's time to decide how we're going to test the relay. Create a test plan with all of the steps required to test the relay. This information doesn't need to be written down and can be incorporated into your test sheet. The following diagrams show three different test plans for our application, one using setting changes, and one without setting changes, and one using the combination test plan. Also, the last two plans can be run in any order because no setting changes are required.

i) Test Plan #1—Setting Changes

1. Record relay nameplate information
2. Review events for abnormalities
3. Set time and date
4. Test input and outputs
5. Test metering
6. Disable 51-element
7. Disable 27-1-element
8. Disable 27-2-element
9. Connect test-set timing input to R7
10. Test 50-element pickup using pickup LED (A-B, B-C, C-A)
11. Test 50-element timing at 1.1 x pickup (A-B, B-C, C-A)
12. Connect test-set timing input to TRIP
13. Test 50-element timing at 1.1 x pickup (A-B, B-C, C-A)
14. Enable 51-element
15. Test 51-element pickup using pickup LED (A-B, B-C, C-A)
16. Test 51-element timing at 2x pickup (A-B, B-C, C-A)
17. Test 51-element timing at 4x pickup (A-B, B-C, C-A)
18. Change test-set timing input to R7
19. Test 51-element timing at 6x pickup (A-B, B-C, C-A)
20. Enable 27-2
21. Test 27-2 pickup using pickup LED (A-B, B-C, C-A)
22. Test 27-2 timing at 0.9x pickup (A-B, B-C, C-A)
23. Connect test-set timing input to R3
24. Test 27-2 timing at 0.9x pickup
25. Enable 27-1-element

26. Test 27-1 pickup using pickup LED (A-B, B-C, C-A)

27. Test 27-1 timing at (27-2 pickup + 1V) (A-B, B-C, C-A)

28. Change test-set timing input to R7

29. Test 27-1 timing at (27-2 pickup + 1V) (A-B, B-C, C-A)

30. Press close pushbutton on relay and monitor R2 trip

31. Review event logs to ensure all styles of trips have been recorded

32. Clear event logs

33. Review settings to ensure no changes have been made

34. Save setting file

35. Save test sheet

ii) Test Plan #2—No Setting Changes

1. Record relay nameplate information

2. Review events for abnormalities

3. Set time and date

4. Test input and outputs

5. Test metering

6. Connect test-set timing input to R7

7. Test 51-element pickup using pickup LED

8. Test 51-element timing at 2x pickup

9. Test 51-element timing at 4x pickup

10. Connect test-set timing input to TRIP

11. Test 51-element timing at 6x pickup

12. Test 50-element pickup using TRIP and timing results

13. Connect test-set timing input to R7

14. Test 50-element timing at 1.1 x pickup

15. Test 27-1 pickup using pickup LED

16. Test 27-1 timing at (27-2 pickup + 1V)

17. Connect test-set timing input to R3

18. Test 27-1 timing at (27-2 pickup + 1V)

19. Test 27-2 pickup using computer status display

20. Test 27-2 timing at 0.9x pickup

21. Change test-set timing input to R7

22. Test 27-2 timing at 0.9x pickup

23. Press close pushbutton on relay and monitor R2 operate

24. Review event logs to ensure all styles of trips have been recorded

25. Clear event logs

26. Save setting file

27. Save test sheet

iii) Test Plan #3—Combination Test

1. Record relay nameplate information
2. Review events for abnormalities
3. Set time and date
4. Test input and outputs
5. Test metering
6. Connect test-set timing inputs to TRIP, R3, and R7
7. Create or load the IOC test described in the previous section and run the test plan.
8. Create or load the BUV-1 test described in the previous section and run the test plan.
9. Create or load the BUV-2 test described in the previous section and run the test plan.
10. Create or load the 1st TOC test described in the previous section and run the test plan.
11. Create or load the 2nd TOC test described in the previous section and run the test plan.
12. Create or load the 3rd TOC test described in the previous section and run the test plan.
13. Press close pushbutton on relay and monitor R2 operate
14. Review event logs to ensure all styles of trips have been recorded
15. Clear event logs
16. Save setting file
17. Save test sheet

D) Generator Protective Relay Test Plan

Generator protective relays have more complex functions than a simple feeder protective relay. Each of these functions protects the generator from a different problem and, if tested in the correct order and set properly, most do not interact with each other. For example, the distance, reverse power, and loss of field protection will only operate if the current flows in the correct direction. The negative sequence element will not pickup during 3-phase balanced tests.

Unless 6 channels are available, the differential protection will have to be disabled until the end. I usually disable the undervoltage protection to remove the nuisance and confusion of undervoltage trips every time the test stops. The two elements most likely to interact is the overvoltage and volts/hertz protection. Disable the volts/hertz protection before testing the overvoltage protection.

Generator protection schemes usually apply different protection elements to different outputs and you may not even need to disable some functions that interact because they might be assigned to different outputs. Here is a typical test plan for generator protective relays:

Generator Protective Relay Test Plan

1. Record relay nameplate information
2. Set time and date
3. Test input and outputs
4. Test metering
5. Disable Differential (87) element
6. Disable Undervoltage (27) element
7. Disable Volts/Hertz (24) element
8. Disable Breaker-fail (50BF) element
9. Test distance element (21) using via 3-phase
10. Test Loss of Field (40) element using via 3-phase
11. Test Reverse Power (32) element using via 3-phase
12. Test Reverse Power (51V) element using single-phase
13. Test Negative Sequence (46) element using single-phase with nominal amps on two phases and zero amps on the third phase.
14. Test Overvoltage (59) element using via 3-phase
15. Enable 24 element
16. Test Volt/Hertz (24) element using via 3-phase
17. Test Frequency (81) element using via 3-phase
18. Test Inadvertent Energization (50/27) element using via 3-phase
19. Enable (50BF) protection
20. Test Breaker-fail (50BF) using single-phase
21. Enable (87) protection
22. Test Differential Protection (87) using 3-phase and singe-phase testing
23. Test Neutral Overvoltage (59N) element
24. Test Neutral 3rd Harmonic Undervoltage (27TN) element
25. Enable 27 protection
26. Test Undervoltage (27) element using via 3-phase
27. Review event logs to ensure all styles of trips have been recorded
28. Clear event logs
29. Review settings to ensure no changes have been made
30. Save setting file
31. Save test sheet

E) Line Distance Protective Relay Test Plan

A typical distance relay test plan follows:

Line Distance Protective Relay Test Plan

1. Record relay nameplate information
2. Set time and date
3. Test input and outputs
4. Test metering
5. Test Zone 1 Distance element (21) using 3-phase
6. Test Zone 2 Distance element using 3-phase
7. Test Zone 3 Distance element using 3-phase
8. Test Zone 1 Distance element (21) using phase-to-phase connections
9. Test Zone 2 Distance element using phase-to-phase connections
10. Test Zone 3 Distance element using phase-to-phase connections
11. Test Zone 1 Distance element (21) using phase-to-neutral connections
12. Test Zone 2 Distance element using phase-to-neutral connections
13. Test Zone 3 Distance element using phase-to-neutral connections
14. Test Phase Time Overcurrent (51) functions using 1-Phase connections unless neutral overcurrent protection is enabled and then use phase-to-phase connections.
15. Test Phase Instantaneous Overcurrent (50) functions using 1-Phase connections unless neutral overcurrent protection is enabled and then use phase-to-phase connections.
16. Test Neutral Time Overcurrent (51) functions using phase-to-neutral connections
17. Test Neutral Instantaneous Overcurrent (50) functions phase-to-neutral connections
18. Test Ground Time Overcurrent (51) functions using 1-Phase ground connections
19. Test Ground Instantaneous Overcurrent (50) functions 1-Phase ground connections
20. Review event logs to ensure all styles of trips have been recorded
21. Clear event logs
22. Review settings to ensure no changes have been made
23. Save setting file
24. Save test sheet

Chapter 19

Testing the Relay

All of the preparation is finally complete, and it's time to start testing. It's been a long time coming, but all of your hard work so far will help make testing a breeze.

1. Perform Relay Self-checks

Some relay manufacturers have self-check commands or display screens that provide a summary of the self-check results. Perform the self-check function, if possible, and record the results. The "STA" command in SEL relays will perform the self-check, and you can cut and paste the results into your setting sheet or test result. Document the relay serial and software version number on your test sheet.

2. Verify Digital Inputs and Outputs

The testing applied in this book includes acceptance testing of the relay and all inputs and outputs should be verified to ensure that they function correctly. This is especially important because relay self-checks cannot detect I/O failures and all I/O should be checked at least once in a relay's lifetime. However, some relay manufacturer's make this very difficult (like the Multilin SR-750 relay in our example) and require many setting changes in order to test the input status. In these cases, it is acceptable to only test the inputs that are being used because major setting changes are required. Test each input and output and record the results on your test sheet. This step also helps reduce connection errors that can be frustrating during later testing when you're trying to figure out why the test is not working.

A) Input Testing

As we discussed in the previous chapter, it is important that you review the manufacturer's literature to determine whether the input uses external power or its own internal power supply. Use the relay's front panel or software to display the input status and apply a jumper or logic input power to each input and ensure the status point changes state. You can skip inputs used in the logic setting as long as they are tested later as part of the logic testing.

B) Output Testing

Most relays have a test function that allows you to operate each relay independently. The "PUL" command in SEL relays will perform this function. GE Multilin relays have a test function to operate relays. Beckwith Electric relays have an output test function in the diagnostic menu from the front screen. Use your test-set input or a meter to monitor each output status and operate each output relay and ensure it changes state. You can skip outputs in-use as their function will be tested during final output testing.

INPUT/OUTPUT CHECKS									
R1	R2	R3	R4	R5	R6	R7			
IN1	IN2	IN3	IN4	IN5	IN6	IN7	IN8	IN9	IN10
IN11	IN12	IN13	IN14	IN5	IN6	IN7	IN8		
COMMENTS:			INPUTS TESTED VIA 32V TERMINAL						
RESULTS ACCEPTABLE:		☑ YES		☐ NO		☐ SEE NOTES			

Figure 19-1: Example Digital Input/Output Test Sheet

3. Verify Current and Voltage Inputs

Like digital inputs discussed previously, the current and voltage inputs cannot be verified by self-check features and are often the most common failures in modern relays. Metering verification is a valuable test that proves the relay is functioning correctly, the wiring is correct between your test-set and the relay, and that you've made the correct connections.

Two single-phase tests will prove that you have connected each phase of your test-set to the correct phases of the relay. Apply single-phase, nominal current and voltage into the relay. Monitor the relay's metering function from the front panel or relay software, and ensure that the voltage and current are measured on the correct phase. Apply current and voltage to another phase, and ensure that the measured current is in the correct phase. If the relay monitors zero-sequence voltage and/or current, record the zero-sequence values on the test sheet. When single-phase voltage current/voltage is applied, the zero-sequence value should match the applied value. Zero sequence components cannot occur in delta connected systems and there will not be zero-sequence measurements for delta connected PTs. You can repeat this test for the third phase, but once two phases have been verified to be connected correctly, the following 3-phase tests can be used for all other measurements.

Apply 3-phase, nominal current and voltage to the relay, record the metering results on your test sheet, and compare them to the CT and PT ratios. If the relay also displays phase angles, record these values and ensure that they are in the correct phase relationship. You might think that correct phase-angle relationships can be used in place of the single-phase tests above, but the relay uses its own reference for phase angles which can mislead you. For example, if all three phases were rolled to the next position (AØ to BØ, BØ to CØ, CØ to AØ) your test-set and the relay would both indicate the correct phase angles for each phase AØ=0°, BØ=-120°, CØ=120° but AØ current/voltage from the test-set would be injected into BØ of the relay. Also, your test-set and the relay could use different references when displaying phase angles that can be confusing as shown in Figure 19-2. For example, the phase relationships displayed by the GE/Multilin SR-750 in our example would be 0°, 120°, 240° LAG and an SEL relay with the same settings and connection would display 0°, -120°, 120°. If the relay monitors positive sequence components, record the current and voltage values on your test sheet. The positive sequence value should match the applied current and voltage and the negative sequence and zero-sequence voltages should be almost zero.

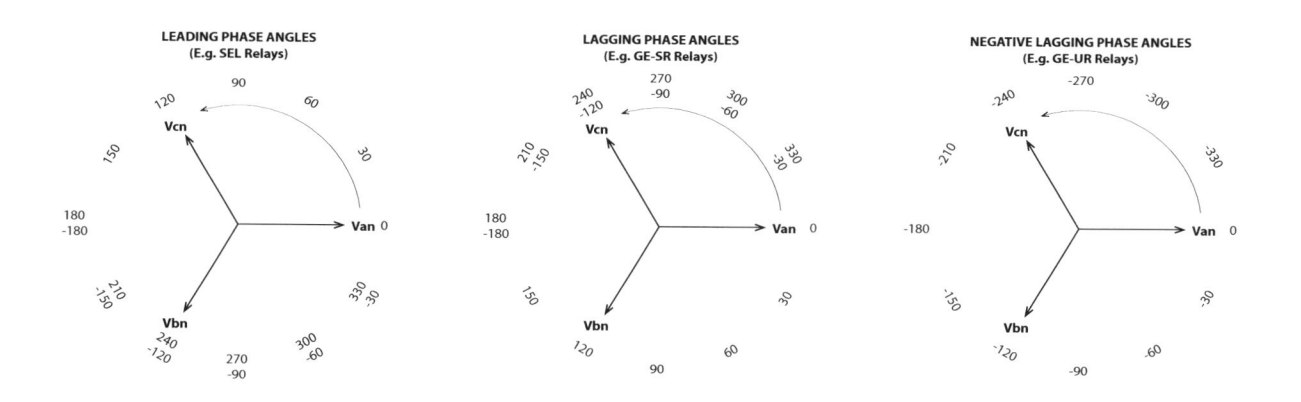

Figure 19-2: Phase Angle Relationships

Watt and VAR measurements can also help determine if the correct connections have been made. When three phase, balanced current and voltage is applied; maximum Watts and almost zero VARs should be measured. Rotate all three currents by 90° and maximum VARs and almost zero Watts should be measured. Any connection problems will skew the Watt and VAR values, and should be corrected.

If the relay monitors negative sequence, reverse any two phases and record the negative sequence values. The negative sequence value should be equal to the applied value, and the positive sequence and zero-sequence values should be nearly zero. Some relays display 3x the negative sequence values. In this case, the negative sequence value will be three times the applied value.

If the relay has a neutral input for voltage and/or current, apply a nominal value and compare metering values to the CT/PT ratios.

A completed test sheet can look like the following:

METERING							
CURRENT (AMPS)							
SEC INJ INPUT	A PH	B PH	C PH	MFG (A)	% ERROR		
5.00	**2007**	**2007**	**2006**	2000	0.35	0.35	0.30
PHASE ANGLE	**31 LAG**	**150 LAG**	**270 LAG**				
SEC INJ INPUT	Pos Seq	Neg Seq	Zero Seq	MFG	% ERROR		
5.00	**2000**	**2001**	**2000**	2000	0.00	0.05	0.00
ROTATION	**ABC**	**ACB**	**A-G**				
VOLTAGE (VOLTS)							
SEC INJ INPUT	A PH	B PH	C PH	MFG (kV)	% ERROR		
120.00	**4.200**	**4.200**	**4.220**	4.200	0.00	0.00	0.48
PHASE ANGLE	**0 LAG**	**120 LAG**	**240 LAG**				
SEC INJ INPUT	Pos Seq (kV)	Neg Seq (kV)	Zero Seq (kV)	MFG (kV)	% ERROR		
120.00	**4.2000**	**4.2100**	**N/A**	4.2000	0.00	0.24	
ROTATION	**ABC**	**ACB**					

3-PHASE METERING							
FREQUENCY (Hz)				**POWER FACTOR**			
SEC INJ INPUT	3 PH (Hz)	MFG (Hz)	% ERROR	SEC INJ INPUT	3 PH	MFG	% ERROR
57.50	**57.50**	57.50	**0.00**	30 Degrees Lag	**0.866**	0.866	**0.00**
60.00	**60.00**	60.00	**0.00**	0 degrees	**1.000**	1.000	**0.00**
62.50	**62.50**	62.50	**0.00**	30 Degrees Lead	**-0.866**	-0.866	**0.00**
POWER (MW)				**VARS (MVAR)**			
SEC INJ INPUT	3 PH (MW)	MFG (MW)	% ERROR	SEC INJ INPUT	3 PH	MFG	% ERROR
1039.23	**14.566**	14.549	**0.12**	1039.23	**14.570**	14.549	**0.14**
-1039.23	**-14.556**	-14.549	**0.05**	-1039.23	**-14.552**	-14.549	**0.02**
1800 @ 90 Deg	**0.25**	0.00	**OK**	1800 @ 0 Deg	**0.360**	0.00	**OK**

COMMENTS:

RESULTS ACCEPTABLE: ☑ YES   ☐ NO   ☐ SEE NOTES

Figure 19-3: Example Metering Test Sheet

4. Element Testing

We've finally reached the part that most people call relay testing. Follow the test plan created in the previous chapter and test all of the elements that create any kind of output from the relay. Try to group similar functions together that will require the minimum number of connection and test-set configuration changes. For example, some relays can have many overcurrent elements enabled. Instead of performing pickup testing and timing testing for each overcurrent element in order, try performing all of the pickup tests first, then perform the timing tests for all elements. You can also speed up your testing and minimize connection changes by testing all elements for one phase and moving on to the next phase.

Instead of following a dogmatic approach to testing, think about the reason for the tests. For example, we perform a time overcurrent test to ensure the settings are correct, we've entered the settings correctly, and the relay will operate correctly on each phase. Following the testing procedure from electromechanical relays, you would test for pickup on all three phases, then time each phase at 2x, 4x, and 6x the pickup value. This is a perfectly acceptable test procedure, but most modern relays use the same timer for all three phases. Testing pickup on all three phases and performing the 2x timing test on AØ, 4x timing test on BØ, and 6x timing test on CØ will prove all of our criteria and remove 6 tests from our test plan. Standard timing test values were used in the past because results were based on graphs, and it was easier and more accurate to determine the expected result from the graph using whole numbers. You can choose any two or three multiples between the pickup value and the 50-element in modern relays to prove the correct curve has been applied.

Many problems during impedance pickup testing occur because modern relays do not believe your test values are actually a fault because the correct prefault polarizing has not occurred. Test impedance element pickup by using 3-phase values to obtain the most accurate pickup results and then perform 3 quick phase-to-phase tests for each phase combination.

Most transformer differential relays internally compensate for Delta-Wye shifts. Differential pickup values can vary depending on your test connection and the internal compensation. Most 3-phase, balanced current pickup tests will match the pickup setting. Perform a timing test for each phase after the 3-phase pickup test to prove each element will create a differential fault.

Document all of your tests on your test sheet. The test sheet for our example would look like the following.

PHASE CURRENT TEST RESULTS							
TIME PICKUP:	1	INST1 PICKUP:	9				
TIME CURVE:	NI	INST1 DELAY:	0,10				
TIME MULTIPLIER:	3,00	INST2 PICKUP:					
TIME RESET:	INST	INST 2 DELAY:					
PHASE CURRENT PICKUP TESTS (in secondary Amps)							
TEST		A PHASE PU (A)	B PHASE PU (A)	C PHASE PU (A)	MFG (A)	% ERROR		
TIME PICKUP		5,01	5,02	4,99	5,00	0,20	0,40	-0,20
INST PICKUP		44,987	44,999	44,975	45,00	-0,03	0,00	-0,06
TIME OVERCURRENT TIMING TESTS (in seconds)							
MULT	AMPS	A PH TRIP (s)	B PH TRIP (s)	C PH TRIP (s)	MFG (s)	% ERROR		
2	10,00	5,315			5,30	0,28		
3	15,00		2,278		2,26		0,80	
4	20,00			1,567	1,54			1,75
INSTANTANEOUS OVERCURRENT TIMING TESTS (in seconds)							
MULT	AMPS	A PH TRIP (s)	B PH TRIP (s)	C PH TRIP (s)	MFG (s)	% ERROR		
1,1	49,50	0,145			0,10	OK		
COMMENTS:								
RESULTS ACCEPTABLE:		✓ YES		☐ NO		☐ SEE NOTES		

Figure 19-4: Example Overcurrent Test Sheet

UNDER VOLTAGE TEST RESULTS					
B-U/V 1 PICKUP:	0,9	B-U/V 2 PICKUP:	0,85		
B -U/V 1 CURVE:	DT	B-U/V 2 CURVE:	DT		
B-U/V 1 DELAY:	60,00	B-U/V 2 DELAY:	20,00		
B-U/V 1 PHASES::	ANY ONE	B-U/V 2 PHASES:	ANY ONE		
B-U/V 1 MIN VS:	0.30	B-U/V 2 MIN V:	0,30		

UNDERVOLTAGE PICK UP TESTS							
TEST	A PHASE PU (V)	B PHASE PU (V)	C PHASE PU (V)	MFG (V)		% ERROR	
B-U/V 1 PICKUP:	107,9	107,95	107,99	108,00	-0,09	-0,05	-0,01
B-U/V 1 MIN V:	35,80			36,00	-0,56		
B-U/V 2 PICKUP:	102,01	101,9	101,95	102,00	0,01	-0,10	-0,05
B-U/V 2 MIN V:		35,9		36,00		-0,28	

UNDERVOLTAGE 1 TIMING TESTS (in seconds)						
MULT	VOLTS		3 PHASE (s)		MFG (s)	% ERROR
0,96	103,68		60,21		60,000	0,35

UNDERVOLTAGE 2 TIMING TESTS (in seconds)						
MULT	VOLTS		3 PHASE (s)		MFG (s)	% ERROR
0,9	91,80		20,13		20,00	0,65

COMMENTS:

RESULTS ACCEPTABLE: ✓ YES ☐ NO ☐ SEE NOTES

Figure 19-5: Example Undervoltage Test Sheet

5. Final Output Tests

Output checks are the most important part of hands-on relay testing. If the outputs are not correctly set, the relay may as well not be installed. More mistakes are made when assigning outputs than any other part of the relay settings. Simple mistakes in output logic schemes can completely disable a relay or cause unintended interaction between devices. These problems are aggravated by the lack of common standards between manufacturers or product lines, relay logic complexity, and poor interpretation of the relay logic. It seems every relay has a different element label, logic scheme, and/or rules for data entry that creates problems that cannot be caught without the final logic test.

Many relays allow output contacts to be easily assigned, and a common trap for relay testers occurs when they use a spare output for all element tests to prevent interference from other elements, and to make sure that they're testing the correct element. They finish testing and remove all test settings thinking the elements have been thoroughly tested. However, there is no guarantee that the relay will actually operate until the output logic is tested.

Think of the relay's output logic as an electrical schematic that must be commissioned with the same attention to detail that you would apply to any other schematic. The first step to output testing is to translate the output logic into a checklist, electrical schematic, or any other representation that you understand. The logic should be broken down into its base components or combinations and you should be able to check each combination off as it is proven.

A) **Example Output Logic**

The output logic in Figure 19-6 is from a typical feeder relay and is shown in the most common logic formats and translated into electrical schematics. This logic was compared to the DC schematic and performs the following functions.

- Output R1 trips a lockout relay that trips the circuit breaker
- Output R2 closes the circuit breaker
- Output R3 trips the circuit breaker directly
- Output R7 provides trip annunciation to the SCADA/DCS/Alarm panel

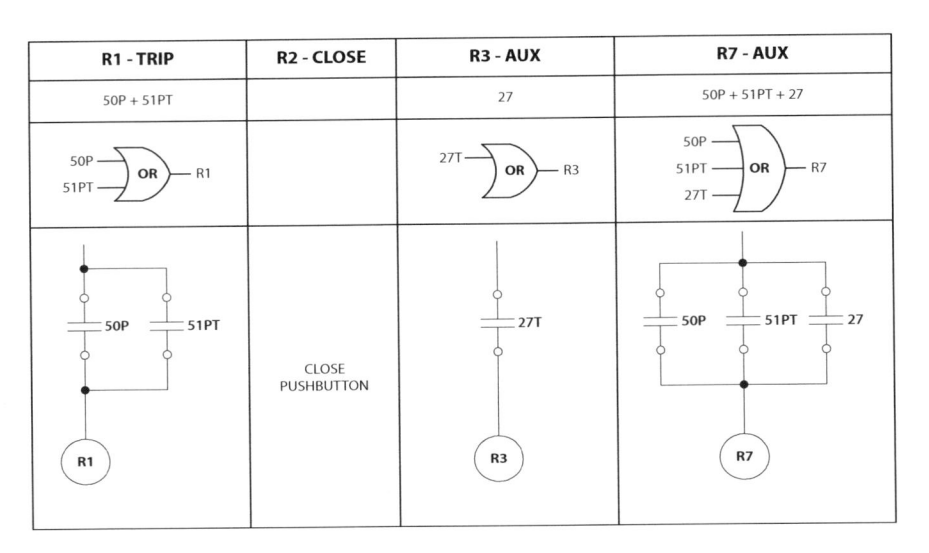

Figure 19-6: Example Output Logic

Testing this logic is relatively straightforward. After all my pickup and timing tests are completed, I perform the following steps:

1. Verify that I'm following the rules discussed above and energize as much of the circuit as possible.

2. Connect my test-set inputs to Output R1 and inject a 50P timing test. Verify:

   - Output R1 operates within the correct time
   - The relay annunciates the correct trip and phase
   - The lockout relay trips (if possible)

3. Reset the lockout, if required, and inject a 51 timing test and verify everything in Step 2.

4. Connect the test-set input across Output R2 and push the close button. Verify the contact operates and closes the circuit breaker, if possible.

5. Connect the test-set inputs to Output R3 and inject a 27 timing test. Verify:

   - Output R3 operates within the correct time
   - The relay annunciates the correct trip and phase
   - The circuit breaker trips (if possible)

6. Repeat Steps 2, 3, and 5 with the test-set inputs connected across R7 and verify DCS/SCADA/Annunciator operation.

I document these tests by creating a spreadsheet with each output contact in a column as shown in Figure 19-7. Each row under the output designations represents an OR gate and I remove the highlight as each step is performed.

FINAL OUTPUT CHECKS								
SOLID STATE TRIP	R1	R2	R3	R5	R6	R7		
NA	51P	CLOSEPB	27-1	NA	NA	27-1		
	50P		27-2			27-2		
						51P		
						50P		
COMMENTS:								
		✓ YES		☐ NO		☐ SEE NOTES		

Figure 19-7: Example Test Sheet for Output Logic

B) Complex Relay Logic

Complex logic schemes require more thought. The breaker-fail protection commonly installed on SEL relays is a great example of a slightly more complex logic scheme. The output logic for our example will be OUT101=SV1T. In order to test this output, we need to find out how SV1T and its related settings are defined. SV1T is derived from the settings in Figure 19-8 from the relay setting application review as discussed in previous chapters:

RELAY SETTINGS	APPLICATION REVIEW
OUT101 = SV1T	IN201 = TRIPS FROM OTHER RELAYS
SV1 = (SV1 + IN201) * (50P1 + 50G1)	
SV1PU = 10	
SV1DO = 10	
50P1P = 0.25	
50G1P = 0.25	

Figure 19-8: Example #2 Settings

We need to break down all of the possible combinations required for OUT101 to operate before we can test it. The first step is to expand the equation until all of the brackets have been removed and we're left with a series of OR functions. SV1 becomes SV1 * 50P1 + SV1 * 50G1 + IN201 * 50P1 + IN201 * 50G1. Figure 19-9 displays the different logic schemes and Figure 19-10 is example test sheet.

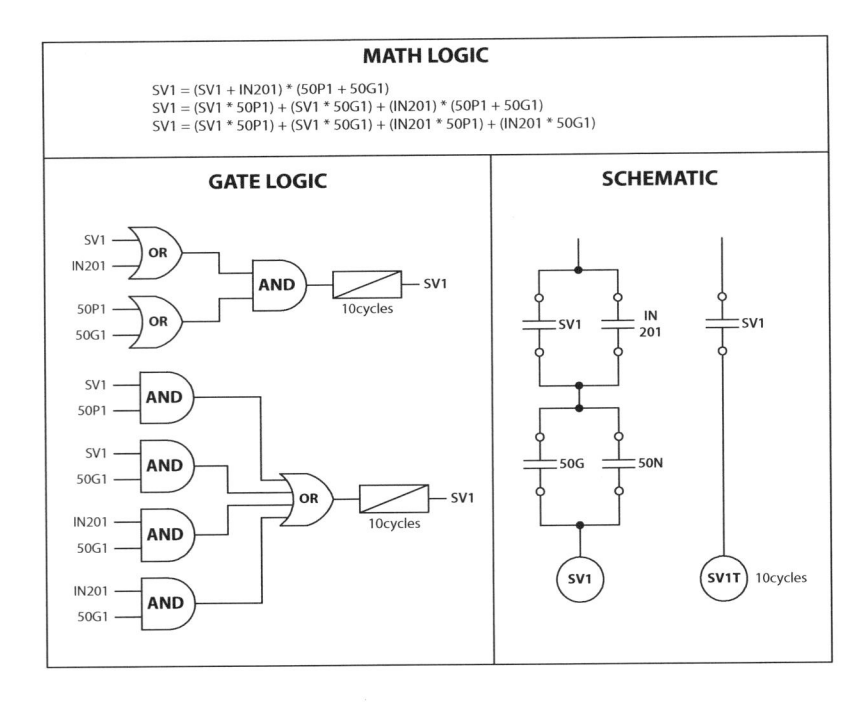

Figure 19-9: Example Breaker-Failure Logic

FINAL OUTPUT CHECKS							
SOLID STATE TRIP	R1	R2	R3	R5	R6	R7	
SV1 * 50P1	NA	NA	NA	NA	NA	NA	
SV1 *50G1							
IN201 * 50P1							
in201 * 50G1							
COMMENTS:							
			✓ YES		☐ NO	☐ SEE NOTES	

Figure 19-10: Example Test Sheet for Breaker-Failure Logic

After the logic has been reduced to its simplest components, output testing is a breeze using the following steps.

1. Apply IN201 and nothing should happen.

2. You could apply 0.3A in any phase to operate the relay, but this will operate both the 50P and 50G elements. First apply 3Ø current higher than 0.25A with IN201 applied. OUT101 should close after 10 cycles have passed. You can now mark "50P1*IN201" as completed.

3. Remove IN201 and if the output remains closed, mark "50P1*SV1" as completed because the SV1 seal-in function has operated correctly.

4. Turn the current off and the output should open.

5. Re-apply current and nothing should happen.

6. Apply 0.20 Amps at 0° in all phases (equals 0.6 Ground Amps). Nothing should happen until you apply IN201. If the output operates 10 cycles after you apply IN201, mark "50G1*IN201" as completed.

7. Remove IN201 and you can mark "50P1*SV1" as completed if the output remains closed.

C) **Output Logic Testing Methods**

There are many ways to prove the output logic settings. The tester can perform all of their other tests and wait until the end before performing a timing test for every element on every output contact as shown in our example. To reduce test-set changes, you could connect your test-set contact sensing to TRIP and perform a 51PT timing test, then move the contact sensing to R7 and repeat the test. Perform a 50P test, then move to TRIP and perform the test again. If your test-set has multiple inputs and/or event reporting, you can connect all relay outputs to the test-set inputs and watch to make sure all the correct relay outputs operate. After proving that the contacts actually operate at least once, you could monitor their status via the software, or open the event records to make sure the correct outputs operate.

Modern test equipment allows you to simulate almost any real-life fault and you should be able to perform a timing test on nearly every element without interference from other elements. If you aren't able to obtain a timing test for an element due to interference from another element, the settings are redundant (except 87-differential elements, of course) and should be noted on your test sheets and brought to the design engineer's attention. Waiting until the end and proving all of the output logic at once is the least confusing method, but can be inefficient as you need to re-configure your test-set between each test. (A relatively minor inconvenience for computer operated testers) This method should always be used if setting changes were performed during testing to ensure the relay will operate when required.

Another method incorporates the element testing into the output testing. This method is more efficient and should only be used if setting changes during testing are non-existent or carefully planned. With this method, you perform each timing test on another output until all of the assigned outputs are used. In our example, the first 51PT timing test (AØ, for example) can be performed using R1 and the next timing test (BØ, for example) is performed on R7. After all of the 51PT timing tests are completed, the first 50-element timing test can be performed on R7, and the next test will be performed on R1. This method is more efficient but requires careful attention to make sure that no element/output combinations are missed.

6. Prepare Relay for Service

You should prepare the relay for service after all of your testing is finished by completing the following steps:

1. Review the event recorder and make sure that the relay has saved records for the tests you have performed. Clear or reset the event recorder to ensure that future users will not be confused by your tests.

2. Review the oscillography or waveform capture list to ensure that some waveforms have been captured during testing. Clear or reset the records to leave a blank slate for future events.

3. Reset all min/max metering data.

4. Some relays record the amperage during faults in an attempt to determine breaker life and should be reset, if it exists, to prevent nuisance alarms during service caused by the high currents during testing.

5. Check and set the date and time to match the local time.

7. Final Report

The final report should document all of your test results, comments, and a final copy of the relay settings to allow the project manager to review the results and final settings. The following items should be included in every test report:

A) Cover Letter

The cover letter should describe the project, provide a brief history about it, and (most importantly) list of all your comments. This letter summarizes all of the test sheets and should be written with non-electrical personnel in mind. Ideally, you should be able to review this document years from now and be able speak with authority about your activities. Any comments should be clearly explained with a brief history of any actions taken regarding the comment and its status at the time of the letter. Organize comments in order of importance and by relay or relay type if the same comment applies to multiple relays. An example comment is "The current transformer ratio on Drawing A and the supplied relay settings did not match. The design engineer was contacted and the correct ratio of 600:5 was applied to the relay settings and confirmed in the field. No further action is required."

This is where you justify your work on the project and why your client's money was well spent.

B) Test Sheet

The test sheet should clearly show all of your test results, the expected result from the manufacturer's manual, and whether the result is acceptable. Review *Chapter 5: Test Sheets and Documentation* for more details.

C) Final Settings

The final, as-left settings should be documented at the end of your test sheet. A digital copy should also be saved, and all relay settings for a project should be made available to the client or design engineer for review and their final documentation. Setting files should be in the relay's native software and in a universal format such as word processor or pdf file to allow the design engineer to make changes, if required, and allow anyone else to review the settings without special software.

Bibliography

Tang, Kenneth, *Dynamic State & Other Advanced Testing Methods for Protection Relays Address Changing Industry Needs*
Manta Test Systems Inc, www.mantatest.com

Tang, Kenneth, *A True Understanding of R-X Diagrams and Impedance Relay Characteristics*
Manta Test Systems Inc, www.mantatest.com

Blackburn, J. Lewis, (October 17, 1997) *Protective Relaying: Principles and Application*
New York. Marcel Dekker, Inc.

Elmore, Walter A., (September 9, 2003) *Protective Relaying: Theory and Applications, Second Edition*
New York. Marcel Dekker, Inc.

Elmore, Walter A., (Editor) (1994) *Protective Relaying Theory and Applications (Red Book)*
ABB

Elmore, Walter A., (September 9, 2003) *Protective Relaying: Theory and Applications, Second Edition*
New York. Marcel Dekker, Inc.

GEC Alstom (Reprint March 1995) *Protective Relays Application Guide (Blue Book), Third Edition*
GEC Alstom T&D

Schweitzer Engineering Laboratories (20011003) *SEL-300G Multifunction Generator Relay Overcurrent Relay Instruction Manual*
Pullman, WA, www.selinc.com

Schweitzer Engineering Laboratories (20010625) *SEL-311C Protection and Automation System Instruction Manual*
Pullman, WA, www.selinc.com

Schweitzer Engineering Laboratories (20010808) *SEL-351A Distribution Protection System, Directional Overcurrent Relay, Reclosing Relay, Fault Locator, Integration Element Standard Instruction Manual*
Pullman, WA, www.selinc.com

Schweitzer Engineering Laboratories (20010606) *SEL-587-0, -1 Current Differential Relay Overcurrent Relay Instruction Manual*
Pullman, WA, www.selinc.com

Schweitzer Engineering Laboratories (20010910) *SEL-387-0, -5, -6 Current Differential Relay Overcurrent Relay Data Recorder Instruction Manual*
Pullman, WA, www.selinc.com

Schweitzer Engineering Laboratories (20011003) *SEL-300G Multifunction Generator Relay Overcurrent Relay Instruction Manual*
Pullman, WA, www.selinc.com

Schweitzer Engineering Laboratories (20010625) *SEL-311C Protection and Automation System Instruction Manual*
Pullman, WA, www.selinc.com

Schweitzer Engineering Laboratories (20010215) *SEL-321-5 Instruction Manual*
Pullman, WA, www.selinc.com

Schweitzer Engineering Laboratories (20010606) *SEL-587-0, -1 Current Differential Relay* Overcurrent Relay Instruction Manual
Pullman, WA, www.selinc.com

GE Power Management (1601-0071-E7) *489 Generator Management Relay Instruction Manual*
Markham, Ontario, Canada, www.geindustrial.com

GE Power Management (1601-0044-AM (GEK-106293B)) *750/760 Feeder Management Relay Instruction Manual*
Markham, Ontario, Canada, www.geindustrial.com

GE Power Management (1601-0070-B1 (GEK-106292)) *745 Transformer Management Relay Instruction Manual*
Markham, Ontario, Canada, www.geindustrial.com

GE Power Management (1601-0110-P2 (GEK-113321A)) *G60 Generator Management Relay: UR Series Instruction Manual*
Markham, Ontario, Canada, www.geindustrial.com

GE Power Management (1601-0089-P2 (GEK-113317A)) *D60 Line Distance Relay: Instruction Manual*
Markham, Ontario, Canada, www.geindustrial.com

GE Power Management (1601-0090-N3 (GEK-113280B)) *T60 Transformer Management Relay: UR Series Instruction Manual*
Markham, Ontario, Canada, www.geindustrial.com

GE Power Management (1601-0044-AM (GEK-106293B)) *750/760 Feeder Management Relay Instruction Manual*
Markham, Ontario, Canada, www.geindustrial.com

GE Power Management (1601-0070-B1 (GEK-106292)) *745 Transformer Management Relay Instruction Manual*
Markham, Ontario, Canada, www.geindustrial.com

GE Power Management (1601-0110-P2 (GEK-113321A)) *G60 Generator Management Relay: UR Series Instruction Manual*
Markham, Ontario, Canada, www.geindustrial.com

GE Power Management, *PK-2 Test Blocks and Plugs*

GE Multilin, (1601-0089-P2) *D60 Line Distance Relay Instruction Manual*
Markham, Ontario, Canada, www.geindustrial.com

Beckwith Electric Co. Inc. *M-3420 Generator Protection Instruction Book*
Largo, FL, www.beckwithelectric.com

Beckwith Electric Co. Inc. *M-3425 Generator Protection Instruction Book*
Largo, FL, www.beckwithelectric.com

Beckwith Electric Co. Inc. *M-3310 Transformer Protection Relay Instruction Book*
Largo, FL, www.beckwithelectric.com

Young, Mike and Closson, James, *Commissioning Numerical Relays*
Basler Electric Company, www.baslerelectric.com

Basler Electric Company (ECNE 10/92) *Generator Protection Using Multifunction Digital Relays*
www.baslerelectric.com

I.E.E.E., (C37.102-1995) *IEEE Guide for AC Generator Protection*

Avo International (Bulletin-1 FMS 7/99) *Type FMS Semiflush-Mounted Test Switches*

Cutler-Hammer Products (Application Data 36-693) *Type CLS High Voltage Power Fuses*
Pittsburg, Pennsylvania

Costello, David and Gregory, Jeff, (AG2000-01) *Application Guide Volume IV Determining the Correct TRCON Setting in the SEL-587 Relay When Applied to Delta-Wye Power Transformers* Pullman, WA, Schweitzer Engineering Laboratories, www.selinc.com

Costello, David and Gregory, Jeff, (AG2000-01) *Application Guide Volume IV Determining the Correct TRCON Setting in the SEL-587 Relay When Applied to Delta-Wye Power Transformers* Pullman, WA, Schweitzer Engineering Laboratories, www.selinc.com

T. Giuliante, ATG Consulting, M. Makki, Softstuf, and Jeff Taffuri, Con Edison; *New Techniques For Dynamic Relay Testing*

Daume, John F., *High Voltage Transmission Protection Practices*
Bonneville Power Administration

Effect of Improper Phase Voltage Quantities When Testing SEL Phase Distance and Overcurrent Elements
Manta Test Systems Inc, www.mantatest.com

Alexander, Ron, *Phasor Diagrams II*
Hands On Relay School
Bonneville Power Administration

Antonova, Galina S. , *PMUs and Synchrophasors*
Hands On Relay School
ABB Inc.

Marx, Stephen E. , P.E., *SYMMETRICAL COMPONENTS 1 & 2*
Hands On Relay School
Bonneville Power Administration, Malin, Oregon

Index

The Relay Testing Handbook
A Series of Nine Indispensable Volumes for Every Relay Tester

Volume 1: Electrical Fundamentals for Relay Testing

Lay the foundation with electrical fundamentals for relay testing.

Volume 2: Relay Testing Fundamentals

Understand the primary philosophies behind relay testing.

Volume 3: Understanding Digital Logic

Apply digital logic to any relay testing scenario.

Volume 4: Creating and Implementing Test Plans

Design testing protocols for modern relay systems.

Volume 5: Testing Voltage Protection (59/27/81)

Learn how to test overvoltage, undervoltage, and frequency.

Volume 6: Testing Overcurrent Protection (50/51/67)

Learn how to test overcurrent, time overcurrent, and directional overcurrent.

Volume 8: Testing Differential Protection (87)

Learn how to test the most common differential protection applications.

Volume 9: Testing Long Distance Protection (21)

Learn how to test the most common distance protection applications.

Volume 7: End-to-End Testing

This stand-alone volume provides the information you need to perform successful end-to-end tests. You will also find an overview of the most common communication-assisted protection schemes in use today.

- -

Order Form (Payment Details on Opposite Page)

Title	Qty	Price	Total
The Relay Testing Handbook (Hard Cover includes volumes 1, 2, 3, 4, 5, 6, 8, 9)		$159.95	$
Volume 1: Electrical Fundamentals for Relay Testing		$23.95	$
Volume 2: Relay Testing Fundamentals		$23.95	$
Volume 3: Understanding Digital Logic		$23.95	$
Volume 4: Creating and Implementing Test Plans		$23.95	$
Volume 5: Testing Voltage Protection (59/27/81)		$23.95	$
Volume 6: Testing Overcurrent Protection (50/51/67)		$23.95	$
Volume 7: End-to-End Testing		$23.95	$
Volume 8: Testing Differential Protection (87)		$33.95	$
Volume 9: Testing Long Distance Protection (21)		$33.95	$
Total including $14.99 Shipping and Handling (Ordering more than 5 copies? Contact info@valenceonline.com for discounts)		$14.99	$

The Relay Testing Handbook
Invest in Your Future as a Relay Tester

Volume 1: Electrical Fundamentals for Relay Testing
Lay the foundation with electrical fundamentals for relay testing.

Volume 2: Relay Testing Fundamentals
Understand the primary philosophies behind relay testing.

Volume 3: Understanding Digital Logic
Apply digital logic to any relay testing scenario.

Volume 4: Creating and Implementing Test Plans
Design testing protocols for modern relay systems.

Volume 5: Testing Voltage Protection (59/27/81)
Learn how to test overvoltage, undervoltage, and frequency.

Volume 6:Testing Overcurrent Protection(50/51/67)
Learn how to test overcurrent, time overcurrent, and directional overcurrent.

Volume 8: Testing Differential Protection (87)
Learn how to test the most common differential protection applications.

Volume 9: Testing Long Distance Protection (21)
Learn how to test the most common distance protection applications.

Volume 7: End-to-End Testing
This stand-alone volume provides the information you need to perform successful end-to-end tests. You will also find an overview of the most common communication-assisted protection schemes in use today.

- -

Payment Details
(Order Form on Opposite Page)

Name:	
Organization:	
Address:	
City :	State:
Zip:	Phone:
Email:	

☐ **My Check or Money Order for** _____ **is enclosed**
   (Make payable to Valence Electrical Training Services, 7450 W. 52nd Ave, PMB#330, Arvada, CO, 80002)
☐ **Please use my email address to send me an online invoice so I can pay with a credit card**

Attentions corporations, universities, colleges, and professional organizations: quantity discounts are available on bulk purchases of this book. Contact us at info@valenceonline.com for details.

Visit www.ValenceOnline.com to order online